Lecture Notes in Computer Science   12159

More information about this series at http://www.springer.com/series/7407

Henning Fernau (Ed.)

Computer Science – Theory and Applications

15th International Computer Science Symposium in Russia, CSR 2020
Yekaterinburg, Russia, June 29 – July 3, 2020
Proceedings

Springer

Editor
Henning Fernau ⓘ
University of Trier
Trier, Germany

ISSN 0302-9743 ISSN 1611-3349 (electronic)
Lecture Notes in Computer Science
ISBN 978-3-030-50025-2 ISBN 978-3-030-50026-9 (eBook)
https://doi.org/10.1007/978-3-030-50026-9

LNCS Sublibrary: SL1 – Theoretical Computer Science and General Issues

This Springer imprint is published by the registered company Springer Nature Switzerland AG
The registered company address is: Gewerbestrasse 11, 6330 Cham, Switzerland

Preface

The 15th International Computer Science Symposium in Russia (CSR 2020) was supposed to be held in Ekaterinburg, Russia. It was organized by the Ural Federal University, Russia. This was the 15th edition of the annual series of meetings, previous editions were held in St. Petersburg (2006), Ekaterinburg (2007), Moscow (2008), Novosibirsk (2009), Kazan (2010), St. Petersburg (2011), Nizhny Novgorod (2012), Ekaterinburg (2013), Moscow (2014), Listvyanka (2015), St. Petersburg (2016), Kazan (2017), Moscow (2018), and Novosibirsk (2019). The symposium covers a broad range of topics in Theoretical Computer Science, ranging from fundamental to application-related.

This year, CSR was planned to be organized as a part of the centennial celebrations of the Ural Federal University. Unfortunately, the corona crisis made it impossible to organize this event in a usual format. At the time of writing this preface, we were investigating the different ways of hosting the conference online.

The distinguished CSR keynote lecture was supposed to be given by Béla Bollobás. The six other CSR invited plenary speakers were, in alphabetical order, Farid Ablaev (Kazan Federal University, Russia), Ulrik Brandes (ETH Zürich, Switzerland), Piotr Faliszewski (AGH University of Science and Technology, Poland), Mateus de Oliveira Oliveira (University of Bergen, Norway), Meirav Zehavi (Ben-Gurion University, Israel), and Binhai Zhu (Montana State University, USA). The Program Committee (PC) included 30 scientists from 14 countries and was chaired by Henning Fernau (University of Trier, Germany). These 30 names included Jérôme Monnot from Dauphine University in Paris, France, who sadly enough passed away in December, 2019.

This volume contains extended abstracts both of the invited lectures and of the accepted papers. We received 49 submissions in total. Each paper was reviewed by at least three PC members on average. As a result, the PC selected 25 papers for presentation on the symposium and publication in these proceedings. The reviewing process was smoothly run using the EasyChair conference system.

The PC also selected two papers to receive awards sponsored by Springer. The awardees are:

- Best Paper Award: Fedor V. Fomin and Vijayaragunathan Ramamoorthi: "On the Parameterized Complexity of the Expected Coverage Problem"
- Best Student Paper Award: Onur Çağırıcı: "On Embeddability of Unit Disk Graphs onto Straight Lines"

We would like to thank all invited speakers for accepting to give a talk at the conference, PC members for giving their time and energy, 19 external reviewers for their expertise, and the publishers Springer and MDPI who provided financial support for this event. We also acknowledge support by the Ural Mathematical Center under agreement No. 075-02-2020-1537/1 with the Ministry of Science and Higher Education of the Russian Federation.

April 2020   Henning Fernau

Organization

Program Committee

Eric Allender	Rutgers University, USA
Sergey Bereg	The University of Texas at Dallas, USA
Davide Bilò	University of Sassari, Italy
Karl Bringmann	Max Planck Institute for Informatics, Germany
Tin-Chih Toly Chen	National Chiao Tung University, Taiwan
Henning Fernau	Universität Trier, Germany
Alexander Grigoriev	Maastricht University, The Netherlands
Yo-Sub Han	Yonsei University, South Korea
Kun He	Huazhong University of Science and Technology, China
Dominik Kempa	University of California, Berkeley, USA
Michael Khachay	Krasovsky Institute of Mathematics and Mechanics, Russia
Margarita Korovina	A. P. Ershov Institute of Informatics Systems, Russia
Dmitry Kosolobov	Ural Federal University, Russia
Andrei Krokhin	University of Durham, UK
Giuseppe Lancia	University of Udine, Italy
Meena Mahajan	The Institute of Mathematical Sciences, HBNI, India
David Manlove	University of Glasgow, UK
Neeldhara Misra	Indian Institute of Science, India
Jerome Monnot	CNRS-LAMSADE, Université Paris-Dauphine, France
Sergio Rajsbaum	UNAM, Mexico
Jörg Rothe	Heinrich-Heine-Universität Düsseldorf, Germany
Markus L. Schmid	Humboldt Universität Berlin, Germany
Vladimir Shenmaier	Sobolev Institute of Mathematics, Russia
Arseny Shur	Ural Federal University, Russia
Ulrike Stege	University of Victoria, Canada
Sergey Verlan	LACL, Université Paris-Est Créteil, France
Mikhail Volkov	Ural Federal University, Russia
Yaokun Wu	Shanghai Jiao Tong University, China
Tomoyuki Yamakami	University of Fukui, Japan
Ryo Yoshinaka	Tohoku University, Japan

Additional Reviewers

Bevern, René Van
de Rezende, Susanna F.
Haghpanah, Mohammadreza
Kerkmann, Anna Maria
Klein, Shmuel Tomi
Kobylkin, Konstantin
Li, Sizhe
Liu, Guodong
Melnikov, Alexander
Mittal, Harshil

Neveling, Marc
Ogorodnikov, Yuri
Puppis, Gabriele
Scheder, Dominik
Sethia, Aditi
Shapira, Dana
Villagra, Marcos
Wang, Xinmao
Weishaupt, Robin

Contents

Quantum Hashing and Fingerprinting for Quantum Cryptography and Computations

Farid Ablayev[1]([⊠])[iD], Marat Ablayev[1,2][iD], and Alexander Vasiliev[1,2][iD]

[1] Kazan Federal University, Kazan 420008, Russian Federation
fablayev@gmail.com, mablayev@gmail.com, vav.kpfu@gmail.com
[2] Kazan Scientific Center of the Russian Academy of Sciences,
Kazan 420029, Russian Federation

Abstract. Fingerprinting and cryptographic hashing have quite different usages in computer science, but have similar properties. Interpretation of their properties is determined by the area of their usage: fingerprinting methods are methods for constructing efficient randomized and quantum algorithms for computational problems, while hashing methods are one of the central cryptographic primitives.

Fingerprinting and hashing methods are being developed from the mid of the previous century, while quantum fingerprinting and quantum hashing have a short history.

In the paper we present computational aspects of quantum fingerprinting, discuss cryptographic properties of quantum hashing. We investigate the pre-image resistance of this function and show that it reveals only $O(1)$ bits of information about the input.

Keywords: Quantum computations · Quantum cryptography · Fingerprinting · Hashing

1 Introduction

Fingerprinting in complexity theory is a procedure that maps a large data item to a much shorter string, its fingerprint, that identifies the original data (with high probability). The key properties of classical fingerprinting methods are: i) they allow to build efficient randomized computational algorithms and ii) the resulting algorithms have bounded error [22].

Rusins Freivalds was one of the first researchers who introduced methods (later called fingerprinting) for constructing efficient randomized algorithms (that are more efficient than any deterministic algorithm) [15,16].

In the quantum case, fingerprinting is a procedure that maps classical data to a quantum state that identifies the original data (with high probability). One of the first applications of the quantum fingerprinting method is due to Ambainis and Freivalds [9]: for a specific language they have constructed a quantum finite

© Springer Nature Switzerland AG 2020
H. Fernau (Ed.): CSR 2020, LNCS 12159, pp. 1–15, 2020.
https://doi.org/10.1007/978-3-030-50026-9_1

automaton with an exponentially smaller size than any classical randomized automaton. An explicit definition of the quantum fingerprinting was introduced by Buhrman et al. in [13] for constructing an efficient quantum communication protocol for equality testing.

Cryptographic hashing is a more formalized procedure and it has a lot of fruitful applications in cryptography. Note that in cryptography, functions satisfying (i) one-way property and (ii) collision resistance property (in different specific meanings) are called hash functions and we propose to do so when we are considering cryptographic aspects of quantum functions with the above properties. So we suggest to call a quantum function that satisfies properties (i) and (ii) (in the quantum setting) a cryptographic quantum hash function or just quantum hash function. Note however, that there is only a thin line between the notions of quantum fingerprinting and quantum hashing. One of the first considerations of a quantum function (that maps classical words into quantum states) as a cryptographic primitive, having both the one-way property and the collision resistance property is due to [18], where the quantum fingerprinting function from [13] was used. Another approach to constructing quantum hash functions from quantum walks was considered in [20,21,28], and it resulted in privacy amplification in quantum key distribution and other useful applications.

In Sect. 3, we consider quantum fingerprinting as a mapping of classical inputs to quantum states that allows to construct efficient quantum algorithms for computing Boolean functions. We consider the quantum fingerprinting function from [13] as well as the quantum fingerprinting technique from [4]. The latter was motivated by the paper [9] and its generalization [10].

Section 4 is based on the results on quantum hashing developed in our research group. We define a notion of quantum hash function which is a quantum one-way function and a quantumly collision resistant function. We show that one-way property and collision resistance property are correlated for a quantum hash function. The more the function is one-way the less it is collision resistant and vice versa. We show that such a correlation can be balanced.

We present an approach for quantum hash function constructions by establishing a connection with small biased sets [23] and quantum hash function constructions: we prove that small sized ϵ-biased sets allow to generate balanced quantum hash functions. Such a connection adds to the long list of small-biased sets' applications.

In particular it was observed in [12,23] that the ϵ-bias property is closely related to the error-correcting properties of linear codes. Note that the quantum fingerprinting function from [13] is based on a binary error-correcting code and so it solves the problem of constructing quantum hash functions for the binary case. For the general case ϵ-bias does not correspond to Hamming distance. Thus, in contrast to the binary case, an arbitrary linear error correcting code cannot be used directly for quantum hash functions.

For the general case of quantum hashing on arbitrary finite abelian groups, we investigate the pre-image resistance of this function. Previously, we have proved the bound on the amount of accessible information about the input using the well-known Holevo theorem [19]. Since no more than $O(s)$ classical bits of information can be extracted from s qubits and the original message contains $n \gg s$ bits, it is impossible to restore the input from the quantum hash. However, using the results of [17] and the properties of ε-biased sets, here we show that the quantum hash function reveals only $O(1)$ bits of information about the input.

2   Preliminaries

Recall that mathematically a qubit is described as a unit vector in the two-dimensional Hilbert complex space $\mathcal{H}^2$. Let $s \geq 1$. Let $\mathcal{H}^d$ be the $d = 2^s$-dimensional Hilbert space, describing the states of s qubits. Another notation for $\mathcal{H}^d$ is $(\mathcal{H}^2)^{\otimes s}$, i.e., $\mathcal{H}^d$ is made up of s copies of a single qubit space $\mathcal{H}^2$

$$(\mathcal{H}^2)^{\otimes s} = \mathcal{H}^2 \otimes \cdots \otimes \mathcal{H}^2 = \mathcal{H}^{2^s}.$$

Conventionally, we use notation $|j\rangle$ for the vector from $\mathcal{H}^d$, which has a 1 on the j-th position and 0 elsewhere. The orthonormal basis $|1\rangle, \ldots, |d\rangle$ is usually referred to as the *standard computational basis*. For an integer $j \in \{0, \ldots, 2^s - 1\}$, let $\sigma_1 \ldots \sigma_s$ be a binary presentation of j. We use the notation $|j\rangle$ to denote the quantum state $|\sigma_1\rangle \cdots |\sigma_s\rangle = |\sigma_1\rangle \otimes \cdots \otimes |\sigma_s\rangle$.

We let $\mathbb{Z}_q$ to be the finite additive group of $\mathbb{Z}/q\mathbb{Z}$, the integers modulo q. Let Σ^k be a set of words of length k over a finite alphabet Σ. Let $\mathbb{X}$ be a finite set. In the paper we let $\mathbb{X} = \Sigma^k$, or $\mathbb{X} = \mathbb{Z}_q$. For $K = |\mathbb{X}|$ and integer $s \geq 1$ we define a $(K; s)$ classical-quantum function (or just quantum function) to be a unitary transformation (determined by an element $w \in \mathbb{X}$) of the initial state $|\psi_0\rangle \in (\mathcal{H}^2)^{\otimes s}$ to a quantum state $|\psi(w)\rangle \in (\mathcal{H}^2)^{\otimes s}$

$$\psi : \{|\psi_0\rangle\} \times \mathbb{X} \to (\mathcal{H}^2)^{\otimes s} \qquad |\psi(w)\rangle = U(w)|\psi_0\rangle,$$

where $U(w)$ is a unitary matrix. We let $|\psi_0\rangle = |0\rangle$ in the paper and use (for short) the following notation (instead of the one above)

$$\psi : \mathbb{X} \to (\mathcal{H}^2)^{\otimes s} \quad \text{or} \quad \psi : w \mapsto |\psi(w)\rangle.$$

3   Quantum Fingerprinting

The ideas of the fingerprinting technique in the quantum setting appeared in [9] for the first time. The authors used a succinct presentation of the classical input by a quantum automaton state, which resulted in an exponential improvement over classical algorithms. Later in [10] the ideas were developed further to give an arbitrarily small probability of error. This was the basis for the general quantum fingerprinting framework proposed in [4].

However, the term "quantum fingerprinting" is mostly used in the scientific literature to address a seminal paper by Buhrman et al. [13], where this notion first appeared explicitly. To distinguish between different versions of the quantum fingerprinting techniques, here we call the fingerprinting function from [13] "binary" (since it uses some binary error-correcting code in its construction), while the fingerprinting from [4] is called "q-ary" as it uses presentation of the input in $\mathbb{Z}_q$.

3.1   Binary Quantum Fingerprinting Function

The quantum fingerprinting function was formally defined in [13], where it was used for quantum equality testing in a quantum communication model. It is based on the notion of a binary error-correcting code.

An (n, k, d) *error-correcting code* is a map $C : \Sigma^k \to \Sigma^n$ such that, for any two distinct words $w, w' \in \Sigma^k$, the Hamming distance $d(C(w), C(w'))$ between code words $C(w)$ and $C(w')$ is at least d. The code is binary if $\Sigma = \{0, 1\}$.

The construction of the quantum fingerprinting function is as follows.

- Let $c > 2$ and $\epsilon < 1$. Let k be a positive integer and $n = ck$. Let $E : \{0, 1\}^k \to \{0, 1\}^n$ be an (n, k, d) binary error-correcting code with Hamming distance $d \geq (1 - \epsilon)n$.
- Define a family of functions $F_E = \{E_1, \ldots, E_n\}$, where $E_i : \{0, 1\}^k \to \mathbb{F}_2$ is defined by the rule: $E_i(w)$ is the i-th bit of the codeword $E(w)$.
- Let $s = \log n + 1$. Define the quantum function $\psi_{F_E} : \{0, 1\}^k \to (\mathcal{H}^2)^{\otimes s}$, determined by a word w as

$$|\psi_{F_E}(w)\rangle = \frac{1}{\sqrt{n}} \sum_{i=1}^{n} |i\rangle |E_i(w)\rangle.$$

Originally, paper [13] used this function to construct a quantum communication protocol that tests equality.

In the same paper it was shown that this result can be improved by choosing an error-correcting code with Hamming distance between any two distinct codewords between $(1 - \epsilon)n/2$ and $(1 + \epsilon)n/2$ for any $\epsilon > 0$ (however, the existence of such codes can only be proved nonconstructively via a probabilistic argument).

But even with such a code, the quantum fingerprinting function above would give

$$|\langle \psi_{F_E}(x) | \psi_{F_E}(y)\rangle| < (1 + \epsilon)/2 \,,$$

which resulted in the following change of construction [13].

Define the classical-quantum function $\psi : \{0, 1\}^k \to (\mathcal{H}^2)^{\otimes s}$, determined by a word w as

$$\psi(w) = \frac{1}{\sqrt{n}} \sum_{i=1}^{n} (-1)^{E_i(w)} |i\rangle.$$

This function gives the following bound for the fingerprints of distinct inputs:

$$|\langle \psi_{F_E}(x) | \psi_{F_E}(y)\rangle| < \epsilon.$$

The further research on this topic mostly used this version of quantum fingerprinting.

3.2 q-ary Quantum Fingerprinting

In this section we show the basic idea of the quantum fingerprinting from [4,6].

Let $\sigma = \sigma_1 \ldots \sigma_n$ be an input string and g be the mapping of $\{0,1\}^n$ onto $\mathbb{Z}_q$ that "encodes" some property of the input we are about to test. We consider g to be the polynomial over $\mathbb{Z}_q$ such that $g(\sigma) = 0 \bmod q$ iff σ has the property encoded by g. For example, if we test the equality of two n-bit binary strings $x_1 \ldots x_n$ and $y_1 \ldots y_n$, we can choose g equal to the following polynomial over $\mathbb{Z}_{2^n}$:

$$\sum_{i=1}^{n} x_i 2^{i-1} - \sum_{i=1}^{n} y_i 2^{i-1}.$$

To test the property encoded by g, we rotate the initial state $|0\rangle$ of a single qubit by an angle of $\theta = \pi g(\sigma)/q$:

$$|0\rangle \rightarrow \cos\theta|0\rangle + \sin\theta|1\rangle.$$

Then this state is measured and the input σ is accepted iff the result of the measurement is $|0\rangle$.

Obviously, this quantum state is $\pm|0\rangle$ iff $g(\sigma) = 0 \bmod q$. In the worst case this algorithm gives the one-sided error of $\cos^2 \pi(q-1)/q$, which can be arbitrarily close to 1.

The above description can be presented as follows using $\log d + 1 = (\log\log q) + 1$ qubits:

$$\underbrace{|0\rangle \otimes \cdots \otimes |0\rangle}_{\log d} \otimes |0\rangle \longrightarrow \frac{1}{\sqrt{d}} \sum_{i=1}^{d} |i\rangle \Big(\cos\theta_i|0\rangle + \sin\theta_i|1\rangle \Big),$$

where $\theta_i = \frac{2\pi s_i g(\sigma)}{q}$ and the set $S = \{s_1, \ldots, s_d\} \subseteq \mathbb{Z}_q$ is chosen in order to guarantee a small probability of error [4,6]. That is, the last qubit is simultaneously rotated in d different subspaces by corresponding angles.

3.3   Quantum Fingerprinting for Computations

The quantum fingerprinting method may be applied for computations in the following manner:

– The initial state of the quantum register is $|0\rangle^{\otimes \log d}|0\rangle$.
– The Hadamard transform creates the equal superposition of the basis states

$$\frac{1}{\sqrt{d}} \sum_{j=1}^{d} |j\rangle|0\rangle.$$

– Based on the input σ, its fingerprint $|\psi(\sigma)\rangle$ is created.

Such a presentation is used in various computational scenarios, depending on the problem we need to solve and depending on computational model we use.

For example, in [13] this technique was used to construct a quantum communication protocol that tests equality in the simultaneous message passing (SMP) model with no shared resources. This protocol requires $O(\log n)$ qubits to compare n-bit binary strings, which is exponentially smaller than any classical deterministic or even randomized protocol in the SMP setting with no shared randomness.

The proposed quantum protocol has a one-sided error of

$$1/2(1 + \langle \psi_{F_E}(\sigma) | \psi_{F_E}(\sigma') \rangle^2),$$

where $|\psi_{F_E}(\sigma)\rangle$ and $|\psi_{F_E}(\sigma')\rangle$ are two different quantum fingerprints. Their inner product

$$|\langle \psi_{F_E}(\sigma) | \psi_{F_E}(\sigma') \rangle|$$

is bounded by ϵ if the Hamming distance of the underlying code is $(1 - \epsilon)n$. For instance, Justesen codes mentioned in the paper give $\epsilon < 9/10 + 1/(15c)$ for any chosen $c > 2$.

In [4–6], we have applied this technique to construct efficient quantum algorithms for a certain class of Boolean functions in the model of read-once quantum branching programs as follows [3].

After creating fingerprint $|\psi_S(\sigma)\rangle$ with

$$|\psi_S(\sigma)\rangle = \frac{1}{\sqrt{d}} \sum_{j=1}^{d} |j\rangle \left(\cos \frac{2\pi s_j g(\sigma)}{q} |0\rangle + \sin \frac{2\pi s_j g(\sigma)}{q} |1\rangle \right),$$

the quantum branching program works as follows.

- The Hadamard transform turns the fingerprint $|\psi_S(\sigma)\rangle$ into the superposition
$$\left(\frac{1}{d} \sum_{l=1}^{d} \cos \frac{2\pi s_l g(\sigma)}{q} \right) |0\rangle^{\otimes \log d} |0\rangle + \dots$$
- The quantum register is measured and the input is accepted iff the result is $|0\rangle^{\otimes \log d} |0\rangle$.

This results into space-efficient quantum algorithms with a small one-sided error for a family of Boolean functions that have linear or quasi-linear polynomial presentations [4–6].

4   Quantum Hashing

In this section we present recent results on quantum hashing developed in our research group.

4.1   One-way δ-Resistance

We present the following definition of a quantum δ-resistant one-way function. Let "information extracting" mechanism **M** be a function $\mathbf{M} : (\mathcal{H}^2)^{\otimes s} \to \mathbb{X}$. Informally speaking, mechanism **M** makes some measurement to state $|\psi\rangle \in (\mathcal{H}^2)^{\otimes s}$ and decodes the result of measurement to $\mathbb{X}$.

Definition 1. *Let X be a random variable distributed over $\mathbb{X}$ like $\{Pr[X = w] : w \in \mathbb{X}\}$. Let $\psi : \mathbb{X} \to (\mathcal{H}^2)^{\otimes s}$ be a quantum function. Let Y be any random variable over $\mathbb{X}$ obtained by some mechanism **M** making measurement to the encoding ψ of X and decoding the result of the measurement to $\mathbb{X}$. Let $\delta > 0$. We call a quantum function ψ a one-way δ-resistant function if*

1. *if it is easy to compute, i.e., a quantum state $|\psi(w)\rangle$ for a particular $w \in \mathbb{X}$ can be determined using a polynomial-time algorithm;*
2. *for any mechanism **M**, the probability $Pr[Y = X]$ that **M** successfully decodes Y is bounded by δ*

$$Pr[Y = X] \leq \delta.$$

For the cryptographic purposes , it is natural to expect (and we do this in the rest of the paper) that the random variable X is uniformly distributed.

A quantum state of $s \geq 1$ qubits can "carry" an infinite amount of information. On the other hand, the fundamental result of quantum informatics known as Holevo's Theorem [19] states that a quantum measurement can only give $O(s)$ bits of information about the state. Here we use the result of [24] motivated by Holevo's Theorem.

Property 1. Let X be a random variable uniformly distributed over $\{0, 1\}^k$. Let $\psi : \{0, 1\}^k \to (\mathcal{H}^2)^{\otimes s}$ be a $(2^k; s)$ quantum function. Let Y be a random variable over $\{0, 1\}^k$ obtained by some mechanism **M** making some measurement of the encoding ψ of X and decoding the result of measurement to $\{0, 1\}^k$. Then the probability of correct decoding is given by

$$Pr[Y = X] \leq \frac{2^s}{2^k}.$$

4.2   Collision ϵ-Resistance

The following definition was presented in [2].

Definition 2. *Let $\epsilon > 0$. We call a quantum function $\psi : \mathbb{X} \to (\mathcal{H}^2)^{\otimes s}$ a collision ϵ-resistant function if for any pair w, w' of different inputs,*

$$|\langle\psi(w)|\psi(w')\rangle| \leq \epsilon.$$

Testing Equality. The crucial procedure for quantum hashing is an equality test for $|\psi(v)\rangle$ and $|\psi(w)\rangle$ that can be used to compare encoded classical messages v and w; see for example [18]. This procedure can be a well-known SWAP-test [13] or something that is adapted for specific hashing functions, like the REVERSE-test; see [7].

4.3   Balanced Quantum (δ, ϵ)-Resistance

The above two definitions and considerations lead to the following formalization of the quantum cryptographic (one-way and collision resistant) function.

Definition 3. *Let $K = |\mathbb{X}|$ and $s \geq 1$. Let $\delta > 0$ and $\epsilon > 0$. We call a function $\psi : \mathbb{X} \to (\mathcal{H}^2)^{\otimes s}$ a quantum (δ, ϵ)-resistant $(K; s)$-hash function (or just quantum (δ, ϵ)-hash function) iff ψ is a one-way δ-resistant and a collision ϵ-resistant function.*

We present below the following two examples to demonstrate how one-way δ-resistance and collision ϵ-resistance are correlated. The first example was presented in [9] in terms of quantum automata.

Example 1. Let us encode numbers v from $\{0, \ldots, 2^k - 1\}$ by a single qubit as follows:

$$\psi : v \mapsto \cos\left(\frac{2\pi v}{2^k}\right)|0\rangle + \sin\left(\frac{2\pi v}{2^k}\right)|1\rangle.$$

Extracting information from $|\psi\rangle$ by measuring $|\psi\rangle$ with respect to the basis $\{|0\rangle, |1\rangle\}$ gives the following result. The function ψ is one-way $\frac{2}{2^k}$-resistant (see Property 1) and collision $\cos\left(\pi/2^{k-1}\right)$-resistant. Thus, the function ψ has a good one-way property, but has a bad collision resistance property for large k.

Example 2. Let $v = \sigma_1 \ldots \sigma_k \in \{0, 1\}^k$. We encode v by k qubits: $\psi : v \mapsto |v\rangle = |\sigma_1\rangle \cdots |\sigma_k\rangle$.

Extracting information from $|\psi\rangle$ by measuring $|\psi\rangle$ with respect to the basis $\{|0 \ldots 0\rangle, \ldots, |1 \ldots 1\rangle\}$ gives the following result. The function ψ is one-way 1-resistant and collision 0-resistant. So, in contrast to Example 1 the encoding ψ from Example 2 is collision free, that is, for different words v and w, the quantum states $|\psi(v)\rangle$ and $|\psi(v)\rangle$ are orthogonal and therefore reliably distinguished; but we lose the one-way property: ψ is easily invertible.

The following result [2] shows that a quantum collision ϵ-resistant $(K; s)$ function needs at least $\log \log K - c(\epsilon)$ qubits.

Property 2. Let $s \geq 1$ and $K = |\mathbb{X}| \geq 4$. Let $\psi : \mathbb{X} \to (\mathcal{H}^2)^{\otimes s}$ be a collision ϵ-resistant quantum hash function. Then

$$s \geq \log \log K - \log \log \left(1 + \sqrt{2/(1 - \epsilon)}\right) - 1.$$

Proof. See [2] for the proof.   □

Properties 1 and 2 provide a basis for building a "balanced" one-way δ-resistance and collision ϵ-resistance properties. That is, roughly speaking, if we need to hash elements w from the domain $\mathbb{X}$ with $|\mathbb{X}| = K$ and if one can build for an $\epsilon > 0$ a collision ϵ-resistant $(K; s)$ hash function ψ with $s \approx \log \log K - c(\epsilon)$ qubits, then the function f is one-way δ-resistant with $\delta \approx (\log K/K)$. Such a function is balanced with respect to Property 2.

To summarize the above considerations we can state the following. A quantum (δ, ϵ)-hash function is a function that satisfies all of the properties that a "classical" hash function should satisfy. Pre-image resistance follows from Property 1. Second pre-image and collision resistance follow, because all inputs are mapped to states that are nearly orthogonal. Therefore, we see that quantum hash functions can satisfy the three properties of a classical cryptographic hash function.

4.4   Quantum (δ, ϵ)-Hash Functions Construction Via Small-Biased Sets

This section is based on the paper [26]. We present here a brief background on ϵ-biased sets as defined in [14] and discuss their connection to quantum hashing. Note that ϵ-biased sets are generally defined for arbitrary finite groups, but here we restrict ourselves to $\mathbb{Z}_q$.

For an $a \in \mathbb{Z}_q$ a character χ_a of $\mathbb{Z}_q$ is a homomorphism $\chi_a : \mathbb{Z}_q \to \mu_q$, where μ_q is the (multiplicative) group of complex q-th roots of unity. That is, $\chi_a(x) = \omega^{ax}$, where $\omega = e^{\frac{2\pi i}{q}}$ is a primitive q-th root of unity. The character $\chi_0 \equiv 1$ is called a trivial character.

Definition 4. *A set $S \subseteq \mathbb{Z}_q$ is called ϵ-biased, if for any nontrivial character $\chi \in \{\chi_a : a \in \mathbb{Z}_q\}$*

$$\frac{1}{|S|} \left| \sum_{x \in S} \chi(x) \right| \le \epsilon.$$

These sets are interesting when $|S| \ll |\mathbb{Z}_q|$ (as $S = \mathbb{Z}_q$ is 0-biased). In their seminal paper Naor and Naor [23] defined these small-biased sets, gave the first explicit constructions of such sets, and demonstrated the power of small-biased sets for several applications.

Remark 1. Note that a set S of $O(\log q/\epsilon^2)$ elements selected uniformly at random from $\mathbb{Z}_q$ is ϵ-biased with positive probability [8].

Many other constructions of small-biased sets followed during the last decades.

Vasiliev [26] showed that ϵ-biased sets generate (δ, ϵ)-resistant hash functions. We present the result of [26] in the following form.

Property 3. Let $S \subseteq \mathbb{Z}_q$ be an ϵ-biased set. Let

$$H_S = \{h_a(x) = ax \pmod q, \quad a \in S, h_a : \mathbb{Z}_q \to \mathbb{Z}_q\}$$

be a set of functions determined by S. Then a quantum function $\psi_S : \mathbb{Z}_q \to (\mathcal{H}^2)^{\otimes \log |S|}$

$$|\psi_S(x)\rangle = \frac{1}{\sqrt{|S|}} \sum_{a \in S} \omega^{h_a(x)} |a\rangle$$

is a (δ, ϵ)-resistant quantum hash function, where $\delta \le |S|/q$.

Proof. The one-way δ-resistance property of ψ_S follows from Property 1: the probability of correct decoding an x from a quantum state $|\psi_S(x)\rangle$ is bounded by $|S|/q$. The efficient computability of such a function follows from the fact that any quantum transformation on s qubits (including the one that creates a quantum hash) can be performed with $O(s^2 4^s)$ elementary quantum gates [25]. Whenever $s = O(\log |S|) = O(\log \log q - \log \epsilon)$, this number of steps is polynomial in $\log q$ (the binary representation of group elements) and $1/\epsilon$.

The collision ϵ-resistance property of ψ_S follows directly from the corresponding property of [26]. Note that

$$|\psi_S(x)\rangle = \frac{1}{\sqrt{|S|}} \sum_{a \in S} \omega^{h_a(x)} |a\rangle = \frac{1}{\sqrt{|S|}} \sum_{a \in S} \chi_x(a) |a\rangle.$$

The remainder of this proof coincides with the proof of the paper [26].   □

Remark 2. It is natural to call the set H_S of functions a *uniform ϵ-biased quantum hash generator* in the context of the definition of quantum hash generator from [1] and the above considerations.

As a corollary of the Property 3 and the above considerations, we can state the following.

Property 4. For a small sized ϵ-biased set $S = \{s_1, \ldots, s_d\} \subset \mathbb{Z}_q$ with $d = O(\log q/\epsilon^2)$, for $\delta \leq O(\frac{\log q}{\epsilon^2 q})$ a quantum hash generator H_S generates the balanced (δ, ϵ)-resistant quantum hash function ψ_S given by

$$|\psi_S(a)\rangle = \frac{1}{\sqrt{d}} \sum_{j=1}^{d} \omega^{a s_j} |j\rangle.$$

5   Quantum Hashing for Finite Abelian Groups

In [26] we have proposed the notion of a quantum hash function, which is defined for arbitrary finite abelian groups.

Let G be a finite abelian group with characters χ_a, indexed by $a \in G$. Let $S \subseteq G$ be an ε-biased set for some $\varepsilon \in (0, 1)$.

Definition 5. *We define a quantum hash function* $\psi_S : G \to (\mathcal{H}^2)^{\otimes \log |S|}$ *as follows:*

$$|\psi_S(a)\rangle = \frac{1}{\sqrt{|S|}} \sum_{j=1}^{|S|} \chi_a(s_j) |j\rangle. \tag{1}$$

We have shown that ψ_S has all the properties of a cryptographic quantum hash function (i.e., it is quantum one-way and collision resistant), which are entirely determined by the ε-biased set $S \subseteq G$.

There are two known special cases of quantum hashing for specific finite abelian groups, which turn out to be the known quantum fingerprinting schemes. In particular, we are interested in hashing binary strings and thus it is natural to consider $G = \mathbb{Z}_2^n$ and $G = \mathbb{Z}_{2^n}$ (or, more generally, any cyclic group $\mathbb{Z}_q$).

Hashing the Elements of the Boolean Cube. For $G = \mathbb{Z}_2^n$, its characters can be written in the form $\chi_a(x) = (-1)^{(a,x)}$, and the corresponding quantum hash function is the following one:

$$|\psi_S(a)\rangle = \frac{1}{\sqrt{|S|}} \sum_{j=1}^{|S|} (-1)^{(a,s_j)} |j\rangle .$$

The resulting hash function is exactly the quantum fingerprinting by Buhrman et al. [13], once we consider an error-correcting code, whose matrix is built from the elements of S. Indeed, as stated in [11] an ε-balanced error-correcting code can be constructed out of an ε-biased set. Thus, the inner product (a, x) in the exponent is equivalent to the corresponding bit of the codeword, and altogether this gives the quantum fingerprinting function, that stores information in the phase of quantum states [27].

Hashing the Elements of the Cyclic Group. For $G = \mathbb{Z}_q$, its characters can be written as $\chi_a(x) = \exp(2\pi iax/q)$, and the corresponding quantum hash function is given by

$$|\psi_S(a)\rangle = \frac{1}{\sqrt{|S|}} \sum_{j=1}^{|S|} \omega^{as_j} |j\rangle .$$

The above quantum hash function is essentially equivalent to the one we have defined earlier in [7], which is in turn based on the quantum fingerprinting function from [4].

6   Pre-image Resistance of Quantum Hashing

In this section we analyze the quantum hash function defined above and prove it has a strong pre-image resistance.

In [17], the authors defined a quantum scheme which is based on quasi-linear codes and maps binary strings to a quantum state. If a scheme uses pure states, accessible information does not exceed $O(1)$ bits. We prove similar properties of a general quantum hash function ψ_S for an arbitrary finite abelian group G and its ε-biased subset $S \subset G$.

For $a \in G$, let $\rho_a = |\psi_S(a)\rangle\langle\psi_S(a)|$ and $\rho'_a = \frac{2^d}{|G|}\rho_a$ be the density operators of a normalized and non-normalized state respectively. Furthermore, for any $|\nu\rangle \in \mathcal{H}^{2^d}$ we define a probability distribution $\mu_\nu(a) = \langle\nu|\rho'_a|\nu\rangle$ that corresponds to a measurement with outcome $|\nu\rangle\langle\nu|$.

The following lemma allows us to estimate the relative entropy between $\mu_\nu(a) = \langle\nu|\rho'_a|\nu\rangle$ and uniform probability distribution over G.

Lemma 1. *Let $|\nu\rangle \in \mathcal{H}^{2^d}$ be a unit vector and $a \in G$ is randomly chosen according to the uniform distribution. Then*

$$\mathbf{E}[\max\{0, \mu_\nu(a) \ln(|G|\mu_\nu(a))\}] < \frac{23}{|G|} . \tag{2}$$

Proof. For all $s \in S$, we define random variables

$$X_s = \chi_a(s)\nu_s \qquad (3)$$

Then $\mu_\nu(a) = \frac{1}{|G|}(\sum_{s \in S} X_s)^2$. $\mathbf{E}[\chi_a(s)] = 0$ and $|\chi_a(s)| \leq 1$ follows from the properties of finite abelian group characters. Then for all $t > 0$,

$$\Pr\left[\mu_\nu(a) \geq \frac{t}{|G|}\right] = \Pr\left[\left|\sum_{s \in S} X_s\right| \geq |\sqrt{t}|\right] \leq 4\exp\left(-\frac{t}{4}\right), \qquad (4)$$

where the last inequality follows from Lemma 2.2 from [17] and from $\|\nu\| = 1$.

Define $g(x) = \max\{0, x\ln(x)\}$ and let $\tilde{\mu}$ be a random variable whose probability distribution is $\Pr[\tilde{\mu} \geq t] = 4\exp(-\frac{t}{4}) = f(t)$ for $t > 8\ln 2$. Then

$$\mathbf{E}[\max\{0, \mu_\nu(a)\ln(|G|\mu_\nu(a))\}] \geq \frac{1}{|G|}\mathbf{E}[g(|G|\mu_\nu(a))] \geq \frac{1}{|G|}\mathbf{E}[g(\tilde{\mu})], \qquad (5)$$

where the first inequality follows from the definition of $g(x)$ and the second one is true by Lemma 2.3 [17].

Therefore,

$$\mathbf{E}[g(\tilde{\mu})] = \int_{8\ln 2}^{\infty} x\ln(x)\left(-\frac{df}{dx}\right) dx = \int_{8\ln 2}^{\infty} \exp\left(\ln(x) + \ln(\ln(x)) - \frac{x}{4}\right) dx < 23 \qquad (6)$$

as required. $\qquad\qquad\square$

Definition 6. *For random variables P and Q having a discrete probability distribution, the Kullback-Leibler divergence is given as follows:*

$$D_{KL}(P \parallel Q) = \sum_i P(i)\ln\frac{P(i)}{Q(i)}. \qquad (7)$$

The following lemma shows that if we use ε-biased sets in our scheme, divergence between $\mu_\nu(a)$ and a random variable x uniformly distributed over G is given by $D_{KL}(\mu_\nu\|x)$ and takes small values.

Lemma 2. *Let $|\nu\rangle \in \mathcal{H}^{2^d}$ be a unit vector. Then*

$$\sum_{a \in G} \mu_\nu(a)\ln(|G|\mu_\nu(a)) < 23. \qquad (8)$$

Proof. We define a random variable

$$\tilde{\mu}(a) = \max\{0, \mu_\nu(a)\ln(|G|\mu_\nu(a))\}. \qquad (9)$$

By Lemma 1, $\mathbf{E}[\tilde{\mu}(a)] < \frac{23}{|G|}$. Therefore,

$$\sum_{a \in G} \mu_\nu(a)\ln(|G|\mu_\nu(a)) < \sum_{a \in G} \tilde{\mu}(a) = |G|\mathbf{E}[\tilde{\mu}(a)] < 23. \qquad (10)$$

$\qquad\qquad\square$

In [17], the accessible information I_{acc} about the input was considered based on the measurement of the quantum state representing this input. It was defined as $I_{acc} = H(J) - H(J|A)$, where A is a random variable describing the choice of input data, J is a random variable that describes the result of measuring the quantum state.

Lemma 3. *Let a be chosen randomly according to uniform distribution over G, then the accessible information I_{acc} of ensemble (ρ_a) does not exceed*

$$\max_{|\nu\rangle} \sum_{a \in G} \mu_\nu(a) \ln\left(|G|\mu_\nu(a)\right) < 23. \tag{11}$$

This lemma rephrases Lemma 3.12 from [17] by using ε-biased sets over finite abelian groups and is given without proof.

Thus, the statements above prove the following theorem.

Theorem 1. *Let $S \subset G$ be an ε-biased set and let ψ_S be a quantum hash function based on S. Then the amount of accessible information about the pre-image of ψ_S is of order $O(1)$.*

Acknowledgments. The research was supported by the government assignment for FRC Kazan Scientific Center of the Russian Academy of Sciences.

References

1. Ablayev, F., Ablayev, M.: On the concept of cryptographic quantum hashing. Laser Phys. Lett. **12**(12), 125204 (2015). http://stacks.iop.org/1612-202X/12/i=12/a=125204
2. Ablayev, F., Ablayev, M.: Quantum hashing via ϵ-universal hashing constructions and classical fingerprinting. Lobachevskii J. Math. **36**(2), 89–96 (2015). https://doi.org/10.1134/S199508021502002X
3. Ablayev, F., Gainutdinova, A., Karpinski, M.: On computational power of quantum branching programs. In: Freivalds, R. (ed.) FCT, vol. 2138, pp. 59–70. Springer, Heidelberg (2001). https://doi.org/10.1007/3-540-44669-9_8, http://arxiv.org/abs/quant-ph/0302022
4. Ablayev, F., Vasiliev, A.: Algorithms for quantum branching programs based on fingerprinting. Electron. Proc. Theoret. Comput. Sci. **9**, 1–11 (2009). https://doi.org/10.4204/EPTCS.9.1, http://arxiv.org/abs/0911.2317
5. Ablayev, F., Vasiliev, A.: Classical and quantum parallelism in the quantum fingerprinting method. In: Malyshkin, V. (ed.) PaCT 2011. LNCS, vol. 6873, pp. 1–12. Springer, Heidelberg (2011). https://doi.org/10.1007/978-3-642-23178-0_1
6. Ablayev, F., Vasiliev, A.: On computational power of quantum read-once branching programs. Electron. Proc. Theoret. Comput. Sci. **52**, 1–12 (2011). https://doi.org/10.4204/EPTCS.52.1
7. Ablayev, F., Vasiliev, A.: Cryptographic quantum hashing. Laser Phys. Lett. **11**(2), 025202 (2014). http://stacks.iop.org/1612-202X/11/i=2/a=025202
8. Alon, N., Roichman, Y.: Random Cayley graphs and expanders. Random Struct. Algorithms **5**(2), 271–284 (1994). https://doi.org/10.1002/rsa.3240050203

9. Ambainis, A., Freivalds, R.: 1-way quantum finite automata: strengths, weaknesses and generalizations. In: Proceeding of the 39th IEEE Conference on Foundation of Computer Science, FOCS 1998, pp. 332–342. IEEE Computer Society, Washington, DC (1998). https://doi.org/10.1109/SFCS.1998.743469, http://arxiv.org/abs/quant-ph/9802062

10. Ambainis, A., Nahimovs, N.: Improved constructions of quantum automata. In: Kawano, Y., Mosca, M. (eds.) Theory of Quantum Computation, Communication, and Cryptography. LNCS, vol. 5106, pp. 47–56. Springer, Heidelberg (2008). https://doi.org/10.1007/978-3-540-89304-2_5, http://arxiv.org/abs/0805.1686

11. Ben-Aroya, A., Ta-Shma, A.: Constructing small-bias sets from algebraic-geometric codes. In: 50th Annual IEEE Symposium on Foundations of Computer Science, FOCS 2009, pp. 191–197, October 2009. https://doi.org/10.1109/FOCS.2009.44

12. Ben-Sasson, E., Sudan, M., Vadhan, S., Wigderson, A.: Randomness-efficient low degree tests and short PCPs via epsilon-biased sets. In: Proceedings of the Thirty-fifth Annual ACM Symposium on Theory of Computing, STOC 2003, pp. 612–621. ACM, New York (2003). https://doi.org/10.1145/780542.780631

13. Buhrman, H., Cleve, R., Watrous, J., de Wolf, R.: Quantum fingerprinting. Phys. Rev. Lett. **87**(16), 167902 (2001). https://doi.org/10.1103/PhysRevLett.87.167902, www.arXiv.org/quant-ph/0102001v1

14. Chen, S., Moore, C., Russell, A.: Small-bias sets for nonabelian groups. In: Raghavendra, P., Raskhodnikova, S., Jansen, K., Rolim, J.D. (eds.) Approximation, Randomization, and Combinatorial Optimization. Algorithms and Techniques. LNCS, vol. 8096, pp. 436–451. Springer, Heidelberg (2013). https://doi.org/10.1007/978-3-642-40328-6_31

15. Freivalds, R.: Probabilistic machines can use less running time. In: IFIP Congress, vol. 839, p. 842 (1977)

16. Freivalds, R.: Fast probabilistic algorithms. In: Becvar, J. (ed.) Mathematical Foundations of Computer Science. LNCS, vol. 74, pp. 57–69. Springer, Heidelberg (1979). https://doi.org/10.1007/3-540-09526-8_5

17. Gavinsky, D., Ito, T.: Quantum fingerprints that keep secrets. Technical report. Cornell University Library arXiv:quant-ph/1010.5342 (2010)

18. Gottesman, D., Chuang, I.: Quantum digital signatures. Technical report. Cornell University Library arXiv:quant-ph/0105032 (2001)

19. Holevo, A.S.: Some estimates of the information transmitted by quantum communication channel (Russian). Probl. Pered. Inform. [Probl. Inf. Transm.] **9**(3), 3–11 (1973)

20. Li, D., Zhang, J., Guo, F.Z., Huang, W., Wen, Q.Y., Chen, H.: Discrete-time interacting quantum walks and quantum hash schemes. Quantum Inf. Process. **12**(3), 1501–1513 (2013). https://doi.org/10.1007/s11128-012-0421-8

21. Li, D., Zhang, J., Ma, X.W., Zhang, W.W., Wen, Q.Y.: Analysis of the two-particle controlled interacting quantum walks. Quantum Inf. Process. **12**(6), 2167–2176 (2013). https://doi.org/10.1007/s11128-012-0516-2

22. Motwani, R., Raghavan, P.: Randomized Algorithms. Cambridge University Press, Cambridge (1995)

23. Naor, J., Naor, M.: Small-bias probability spaces: efficient constructions and applications. In: Proceedings of the Twenty-Second Annual ACM Symposium on Theory of Computing, STOC 1990, pp. 213–223. ACM, New York (1990). https://doi.org/10.1145/100216.100244

24. Nayak, A.: Optimal lower bounds for quantum automata and random access codes. In: 40th Annual Symposium on Foundations of Computer Science, pp. 369–376 (1999). https://doi.org/10.1109/SFFCS.1999.814608

25. Nielsen, M.A., Chuang, I.L.: Quantum Computation and Quantum Information, 1 edn. Cambridge University Press, Cambridge (2000). https://doi.org/10.2277/0521635039
26. Vasiliev, A.: Quantum hashing for finite abelian groups. Lobachevskii J. Math. **37**(6), 753–757 (2016). https://doi.org/10.1134/S1995080216060184
27. de Wolf, R.: Quantum computing and communication complexity. Ph.D. thesis, University of Amsterdam (2001)
28. Yang, Y.G., Xu, P., Yang, R., Zhou, Y.H., Shi, W.M.: Quantum hash function and its application to privacy amplification in quantum key distribution, pseudo-random number generation and image encryption. Sci. Rep. **6**, 19788 (2016). https://doi.org/10.1038/srep19788

Parameterized Analysis of Art Gallery and Terrain Guarding

Akanksha Agrawal and Meirav Zehavi$^{(\boxtimes)}$

Ben-Gurion University of the Negev, Beersheba, Israel
agrawal@post.bgu.ac.il, meiravz@bgu.ac.il

Abstract. The purpose of this invited talk is threefold: provide a brief introduction to both Parameterized Analysis and algorithmic research of visibility problems, and to address a few known results in the intersection. In the first part of the talk, we will discuss basic concepts and definitions in Parameterized Analysis as well as the philosophy behind the field. In the second and third parts of the talk, we will survey some results about the ART GALLERY and TERRAIN GUARDING problems, which have, so far, received only little attention from the viewpoint of Parameterized Analysis. Moreover, we will briefly overview a few of the known positive results on the parameterized complexity of these problems.

Keywords: Parameterized algorithms · Parameterized complexity · Art gallery · Terrain guarding

1 Background on Parameterized Analysis

Design and analysis of algorithms lie at the heart of computer science. Unfortunately, today we know of numerous problems that are NP-hard, which are believed not to admit worst-case efficient (polynomial-time) exact algorithms. However, if we will make a deeper look, we will observe that in many cases the nutshell of hardness lies in either a particular property of the instance, or even just in a small part of it. Parameterized Analysis leads both to deeper understanding of intractability results and to practical solutions for many NP-hard problems. Informally speaking, Parameterized Analysis is a deep mathematical paradigm to answer the following fundamental question:

What makes an NP-hard problem hard?

Specifically, how do different *parameters* (being formal quantifications of structure) of an NP-hard problem relate to its inherent difficulty? Can we exploit these relations algorithmically, and to which extent? Over the past three decades,

Supported by Israel Science Foundation grant no. 1176/18, and United States – Israel Binational Science Foundation grant no. 2018302.

H. Fernau (Ed.): CSR 2020, LNCS 12159, pp. 16–29, 2020.
https://doi.org/10.1007/978-3-030-50026-9_2

Parameterized Analysis has grown to be a mature field of outstandingly broad scope with significant impact from both theoretical and practical perspectives on computation.

Parameterized Algorithms and Complexity. For many NP-hard problems, it possible to find exact (optimal) solutions efficiently. *How can this blatant discrepancy between theory and practice be explained?* An intuitive explanation is that most real-world instances are not worst-case instances, but share structural properties that are (explicitly or implicitly) being utilized in practice. Since the early days of computer science, extensive efforts have been directed at systematic research of tractability results for various problems on specific classes of instances. However, in real-world situations, it is often not possible to define a clear-cut class of instances that we wish to solve; instead of being black and white (belonging to a specific class or not), instances come in various shades of grey (having certain degrees of internal structure). The paradigm of *Parameterized Analysis* (algorithms and complexity) offers the perfect tools to understand the rich and deep theory that underlies this spectrum, and to deal with it algorithmically across a wide and growing number of areas of computer science. This paradigm was introduced in the late 1980s by Downey and Fellows, and it has quickly become a central, ubiquitous and vibrant area of research (see, e.g., the textbooks [23,25,26,35,60,65] dedicate do Parameterized Analysis).

In a nutshell, Parameterized Analysis deals with parameterized problems: a *parameterization* of a problem Π is simply the association of a parameter k with every instance of Π, which captures how "structured" the instance is. A parameter can be any measure that captures "behavior" of inputs or outputs, such as the *treewidth* of a given graph ("how close is the graph to a tree?") or the size of the solution. In particular, parameterization gives rise to the development of algorithms whose performance mainly depends on the value of the parameter—instead of the classical setting, where we often associate tractability with polynomial running times and intractability with superpolynomial ones, *parameterized algorithms naturally "scale" with the amount of structure that implicitly underlies the instance!* Further, understanding the dependency of the complexity of the problem on parameters of it leads to true understanding of its core of hardness.

More concretely, the main objective of Parameterized Analysis is to confine the combinatorial explosion in the running time of an algorithm for Π to the parameter k rather than to let it depend on the entire input size. Formally, a problem is *fixed-parameter tractable (FPT)* with respect to a parameter k if it can be solved by an algorithm, called a *parameterized algorithm*, whose running time is bounded by $f(k) \cdot n^{\mathcal{O}(1)}$ for some computable function f of k, where n is the input size. In particular, f is independent of the input size n. By examination of real datasets, one may choose a parameter k that will often be significantly smaller than the input size n. Over the past thirty years, research in Parameterized Analysis yielded an outstandingly rich and deep theory, with powerful toolkits to classify problems as FPT, W[1]-complete (unlikely to be FPT) or worse (in the so called W-hierarchy of hardness in Parameterized Analysis), and, in particular, for those problems that are FPT, develop fast, or even the fastest possible (with tight conditional lower bounds on their time complexity), parameterized algorithms. Nowadays, as it was phrased by Niedermeier in 2010 [61], "problem parameterization is a pervasive and ubiquitous tool in attacking intractable problems", and "multivariate algorithmics helps to significantly increase the impact of theoretical computer science on practical computing." Indeed, parameterized problems arise in a variety of areas of computer science and mathematics, including Graph Theory, Computational Geometry, Robotics, Computational Social Choice and Voting, Bioinformatics, Artificial Intelligence, Graph Drawing, Psychology and Cognitive Science, and Database Theory.

Kernelization. Preprocessing is an integral part of almost any application, ranging from lossless data compression and navigation systems to microarray data analysis for the classification of cancer types. In fact, in our everyday lives, we often rely on preprocessing, sometimes without even noticing it. The "gold standard" successes in software development for hard problems, such as CPLEX for integer linear programming, depend heavily on sophisticated preprocessing routines. In fact, the idea of preprocessing information to speed up computation can be traced much before the invention of the first computers. The book Mirifci Logarithmorum Canonis Descriptio (A Description of the Admirable Table of Logarithm) authored by Napier (1550–1617), who is credited with the invention of logarithms, was published in 1614. A quote attributed to Laplace states that this table "by shortening the labours, doubled the life of the astronomer." A natural question in this regard is how to measure the quality of data reductions proposed for a specific problem. Yet, for a long time the mathematical analysis of polynomial-time preprocessing algorithms was neglected. One central reason for this anomaly is that unless P = NP, no NP-complete problem Π admits any polynomial-time procedure that, given an instance of Π, is guaranteed to reduce the size of the instance by *even a single bit*! Indeed, if such a procedure existed, then we could have utilized it to solve Π by repeatedly reducing the input size until it becomes a single bit, solvable in polynomial time.

The situation has changed drastically with the advent of Parameterized Analysis. Specifically, the subfield of Parameterized Analysis called *Kernelization* is precisely the area of rigorous research of polynomial-time preprocessing algorithms. Roughly speaking, we say that a problem admits a *kernel* if there exists a polynomial-time algorithm (called a kernelization algorithm) that, given an instance of the problem, translates it into an equivalent instance of the same problem of size $f(k)$ for some computable function f depending only on k. In particular, f is independent of the input size n. It is known that a problem admits a kernel if and only if it is FPT. The central question in the field of Kernelization is which parameterized problems admit kernels of size $f(k)$ where f is *polynomial* in k. Such kernels are called *polynomial kernels*. Here, we note that a problem that is FPT might provably (under plausible assumptions in complexity theory) not admit a polynomial kernel. In case a problem is shown to admit a polynomial kernel, the second question that naturally arises is how small can this kernel be. Due to the profound impact of polynomial-time preprocessing on every aspect of computation, Kernelization has been termed "the lost continent of polynomial time" [33]. Combining tools from Parameterized Analysis and classic (univariate) Computational Complexity, it has become possible to derive both upper and lower bounds on sizes of reduced instances, or so called kernels. As noted by Marx [58], it has become clear that "the existence of polynomial kernels is a mathematically deep and very fruitful research direction." For more information, see the textbook [36] on kernelization.

2   The Art Gallery Problem

Given a *simple* polygon P on n vertices, two points x and y within P are *visible* to each other if the line segment between x and y is contained in P. Accordingly, a set S of points within P is said to *guard* another set Q of points within P if, for every point $q \in Q$, there is some point $s \in S$ such that q and s are visible to each other. The computational problem that arises from this notion is loosely termed the ART GALLERY problem. In its general formulation, the input consists of a simple polygon P, possibly infinite sets G and C of points within P, and a non-negative integer k. The task is to decide whether at most k guards can be placed on points in G so that every point in C is visible to at least one guard. The most well-known cases of ART GALLERY are identified as follows: the X-Y ART GALLERY problem is the ART GALLERY problem where G is the set of all points within P (if X=POINT), all boundary points of P (if X=BOUNDARY), or all vertices of P (if X=VERTEX), and C is defined analogously with respect to Y. The classic variant of ART GALLERY is the POINT-POINT ART GALLERY problem. Moreover, VERTEX-VERTEX ART GALLERY is equivalent to the classic DOMINATING SET problem in the visibility graph of a polygon.

The ART GALLERY problem is a fundamental visibility problem in Discrete and Computational Geometry, which was extensively studied from both combinatorial and algorithmic viewpoints. The problem was first proposed by Victor Klee in 1973, which prompted a flurry of results [62, page 1]. The main combinatorial question posed by Klee was *how many guards are sufficient to see every point of the interior of an n-vertex simple polygon?* Chvátal [19] showed in 1975 that $\lfloor \frac{n}{3} \rfloor$ guards are always sufficient and sometimes necessary for any n-vertex simple polygon (see [34] for a simpler proof by Fisk). After this, many variants of the ART GALLERY problem, based on different definitions of visibility, restricted classes of polygons, different shapes of guards, and mobility of guards, have been defined and analyzed. Several books and extensive surveys were dedicated to ART GALLERY and its variants (see, e.g., [39,62,64,67,68]).

2.1   Known Algorithmic Works

In what follows, we focus only on algorithmic works on X-Y ART GALLERY for X,Y∈{POINT,BOUNDARY,VERTEX}.

Hardness. In 1983, O'Rourke and Supowit [63] proved that POINT-POINT ART GALLERY is NP-hard if the polygon can contain holes. The requirement to allow

holes was lifted shortly afterwards [4]. In 1986, Lee and Lin [56] showed that
VERTEX-POINT ART GALLERY is NP-hard. This result extends to VERTEX-
VERTEX ART GALLERY and VERTEX-BOUNDARY ART GALLERY. Later, numer-
ous other restricted cases were shown to be NP-hard as well. For example, NP-
hardness was established for orthogonal polygons by Katz and Roisman [47] and
Schuchardt and Hecker [66]. We remark that the reductions that show that X-
Y ART GALLERY (for X,Y ∈{POINT, BOUNDARY, VERTEX}) is NP-hard also
imply that these cases cannot be solved in time $2^{o(n)}$ under the Exponential-
Time Hypothesis (ETH).

While it is long known that even very restricted cases of ART GALLERY are
NP-hard, the inclusion of X-Y ART GALLERY, for X,Y ∈{POINT, BOUNDARY},
in NP remained open. (When X=VERTEX, the problem is clearly in NP.) In
2017, Abrahamsen et al. [2] began to reveal the reasons behind this discrepancy for the POINT-POINT ART GALLERY problem: they showed that *exact*
solutions to this problem sometimes require placement of guards on points with
irrational coordinates. Shortly afterwards, they extended this discovery to prove
that POINT-POINT ART GALLERY and BOUNDARY-POINT ART GALLERY are
∃ℝ-complete [3]. Roughly speaking, this result means that *(i)* any system of
polynomial equations over the real numbers can be encoded as an instance of
POINT/BOUNDARY-POINT ART GALLERY, and *(ii)* these problems are not in
the complexity class NP unless NP = ∃ℝ.

Approximation Algorithms. The ART GALLERY problem has been extensively studied from the viewpoint of approximation algorithms [11,12,15,24,28,
40,46,51–53,55] (this list is not comprehensive). Most of these approximation
algorithms are based on the fact that the range space defined by the visibility
regions has bounded VC-dimension for simple polygons [44,45,69], which facilitates the usage of the algorithmic ideas of Clarkson [17,21]. The current state of
the art is as follows. For the BOUNDARY-POINT ART GALLERY problem, King
and Kirkpatrick [52] gave a factor $\mathcal{O}(\log \log \mathsf{OPT})$ approximation algorithm. For
the POINT-POINT ART GALLERY problem, Bonnet and Miltzow [15] gave a
factor $\mathcal{O}(\log \mathsf{OPT})$ approximation algorithm. For X-Y ART GALLERY, where
X,Y∈{POINT,BOUNDARY,VERTEX}, the existence of a constant-factor approximation algorithm is a longstanding open problem [38,40,41]. On the negative
side, all of these variants are known to be APX-hard [29,30]. Yet, restricted
classes of polygons, such as weakly-visible polygons [46], give rise to a PTAS.

Exact Algorithms. For an n-vertex polygon P, one can efficiently find a
set of $\lfloor \frac{n}{3} \rfloor$ vertices that guard all points within P, matching Chvátal's upper
bound [19]. Specifically, Avis and Toussaint [8] presented an $\mathcal{O}(n \log n)$-time
divide-and-conquer algorithm for this task. Later, Kooshesh and Moret [54] gave
a linear-time algorithm based on Fisk's short proof [34]. However, when we seek
an optimal solution, the situation is much more complicated. The first exact
algorithm for POINT-POINT ART GALLERY was published in 2002 in the conference version of a paper by Efrat and Har-Peled [28]. They attribute the result
to Micha Sharir. Before that time, the problem was not even known to be decidable. The algorithm computes a formula in the first order theory of the reals

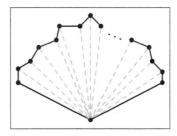

Fig. 1. The solution size $k = 1$, yet the number of reflex vertices r is arbitrarily large.

corresponding to the art gallery instance (with both existential and universal quantifiers), and employs algebraic methods such as the techniques provided by Basu et al. [9], to decide if the formula is true. Given that POINT-POINT ART GALLERY is $\exists\mathbb{R}$-complete [3], it might not be possible to avoid the use of this powerful machinery. However, even for the cases where X=VERTEX, the situation is quite grim; we are not aware of *exact* algorithms that achieve substantially better time complexity bounds than brute-force. Nevertheless, over the years, exact algorithms that perform well in practice were developed. For example, see [16,22,64].

Parameterized Complexity. Two years ago, Bonnet and Miltzow [14] showed that VERTEX-POINT ART GALLERY and POINT-POINT ART GALLERY are W[1]-hard with respect to the *solution size*, k. With straightforward adaptations, their results extend to most of the known variants of the problem, including VERTEX-VERTEX ART GALLERY. Thus, *the classic parameterization by solution size leads to a dead-end*. However, this does not rule out the existence of FPT algorithms for non-trivial structural parametrizations.

2.2   Giannopoulos's Parameterization and Our Contribution

In light of the W[1]-hardness result by Bonnet and Miltzow [14], Giannopoulos [42] proposed to parameterize the ART GALLERY problem by the number r of reflex vertices of the input polygon P. Specifically, Giannopoulos [42] posed the following open problem: *"Guarding simple polygons has been recently shown to be W[1]-hard w.r.t. the number of (vertex or edge) guards. Is the problem FPT w.r.t. the number of reflex vertices of the polygon?"* The motivation behind this proposal is encapsulated by the following well-known proposition, see [62, Sections 2.5-2.6].

Proposition 1 (Folklore). *For any polygon P, the set of reflex vertices of P guards the set of all points within P.*

That is, the minimum number k of guards needed (for any of the cases of ART GALLERY) is upper bounded by the number of reflex vertices r. Clearly, k can be arbitrarily smaller than r (see Fig. 1). Our main result in [5] is that the VERTEX-VERTEX ART GALLERY problem is FPT parameterized by r. This

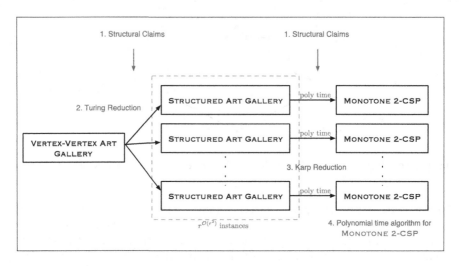

Fig. 2. The four components of our proof.

implies that guarding the vertex set of "almost convex polygons" is easy. In particular, whenever $r^2 \log r = \mathcal{O}(\log n)$, the problem is solvable in polynomial time.

Theorem 1. VERTEX-VERTEX ART GALLERY *is FPT parameterized by* r, *the number of reflex vertices. In particular, it admits an algorithm with running time* $r^{\mathcal{O}(r^2)}n^{\mathcal{O}(1)}$.

Along the way to establish our main result, we prove that a constraint satisfaction problem called MONOTONE 2-CSP is solvable in polynomial time. This result might be of independent interest. Informally, in MONOTONE 2-CSP, we are given k variables and m constraints. Each constraint is of the form $[x \operatorname{sign} f(x')]$ where x and x' are variables, $\operatorname{sign} \in \{\leq, \geq\}$, and f is a *monotone* function. The objective is to assign an integer from $\{0, 1, \ldots, N\}$ to each variable so that all of the constraints will be satisfied. For this problem, we develop a surprisingly simple algorithm based on a reduction to 2-CNF-SAT.

Theorem 2. MONOTONE 2-CSP *is solvable in polynomial time.*

Essentially, the main technical component of our work is an exponential-time reduction that creates an exponential (in r) number of instances of MONOTONE 2-CSP so that the original instance is a Yes-instance if and only if at least one of the instances of MONOTONE 2-CSP is a Yes-instance. Our reduction is done in two stages due to its structural complexity (see Fig. 2). In the first stage of the reduction, we aim to make "guesses" that determine the relations between the "elements" of the problem (that are the "critical" visibility relations in our case) and thereby elucidate and further binarize them (which, in our case, is required to impose order on guards). This part requires exponential time (given that there are exponentially many guesses) and captures the "NP-hardness" of the problem. Then, the second stage of the reduction is to translate each guess into an instance of MONOTONE 2-CSP. This part, while requiring polynomial time, relies on a highly non-trivial problem-specific insight—specifically, here we need to assert that the relations considered earlier can be encoded by constraints that are not only binary, but that the functions they involve are *monotone*.

3   The Terrain Guarding Problem

The study of terrains, also known as x-monotone polygonal chains, has attracted widespread and growing interest over the last two decades, partly due to its resemblance to the ART GALLERY problem. A terrain is a graph $T = (V, E)$ where each vertex $v_i \in V$, $1 \le i \le n$, is associated with a point (x_i, y_i) on the two-dimensional Euclidean plane such that $x_1 \le x_2 \le \ldots \le x_n$, and the edge-set is $E = \{\{v_i, v_{i+1}\} : 1 \le i \le n - 1\}$. The set of points on the terrain includes its vertices, and the points that lie on its edges. In the CONTINUOUS TERRAIN GUARDING problem, the input is a terrain $T = (V, E)$ and a positive integer k, and the task is to decide whether one can place guards on at most k points on a given terrain such that each point on the terrain is seen by at least one guard. Here, we say that a point p sees a point q if no point of the line segment $\overline{pq}$ is strictly below T. This problem is also known as TERRAIN GUARDING in a continuous domain, since we are allowed to place guards on the continuous domain of the given terrain, and we want to cover all points lying on the terrain. Another variant of this problem is the DISCRETE TERRAIN GUARDING problem (or TERRAIN GURADING in a discrete domain), where the input is the same as before, and the objective is to determine whether there is a subset $S \subseteq V$ of size at most k that sees V.

3.1   Known Algorithmic Works

The visibility graphs of terrains exhibit unique properties which render the complexity of the TERRAIN GUARDING problem difficult to elucidate. Some of these properties have already been observed in 1995 by Abello et al. [1], and some of them remain unknown despite recent advances to identify them [32]. Indeed, the TERRAIN GUARDING problem has been extensively studied since 1995, when an NP-hardness proof was claimed but never completed by Chen et al. [18]. Almost 15 years later King and Krohn [50] have finally proved that it is NP-hard. Further, this proof led to the establishment of a $2^{\Omega(n^{1/3})}$ lower bound on the time complexity of any algorithm for this problem under the ETH [13].

Particular attention has been given to the TERRAIN GUARDING problem from the viewpoint of approximation algorithms. In 2005, Ben-Moshe et al. [10] obtained the first constant-factor approximation algorithm for DISCRETE TERRAIN GUARDING. Afterward, the approximation factor was gradually improved in [20,31,49], until a PTAS was proposed by Gibson et al. [43] for DISCRETE TERRAIN GUARDING. Recently, Friedrichs et al. [37] showed that even the CONTINUOUS TERRAIN GUARDING problem admits a PTAS. From the perspective of Parameterized Analysis, besides our work that will be described ahead, we are only aware of the work of Khodakarami et al. [48], who introduced the parameter "the depth of the onion peeling of a terrain" and showed that TERRAIN GUARDING is FPT with respect to this parameter. While by now we have quite satisfactory understanding of the approximability of TERRAIN GUARDING, the parameterized hardness of this problem (with respect to k) is unknown. Indeed, King and Krohn [50] state that "the biggest remaining question regarding the complexity of TERRAIN GUARDING is whether or not it is FPT".

The ORTHOGONAL TERRAIN GUARDING problem is a well-known special case of TERRAIN GUARDING, also known to be NP-hard [13]. In this problem, the terrain is orthogonal: for each vertex v_i, $2 \leq i \leq n-1$, either both $x_{i-1} = x_i$ and $y_i = y_{i+1}$ or both $y_{i-1} = y_i$ and $x_i = x_{i+1}$. In other words, each edge is either a horizontal line segment or a vertical line segment, and each vertex is incident to at most one horizontal edge and at most one vertical edge. The ORTHOGONAL TERRAIN GUARDING problem has already been studied from the perspective of algorithmic theory [27,47,57,59]. Although the PTASes designed in [43] or [37] work for the ORTHOGONAL TERRAIN GUARDING problem as well, there are a few studies on this particular variant of TERRAIN GUARDING, that bring out interesting structural properties specific to this variant. A prime example is the work of Katz and Roisman [47], where they gave a relatively simple 2-approximation algorithm for the problem of guarding all vertices of an orthogonal terrain by vertices. Recently, Lyu and Üngör improved upon this result by developing a linear-time 2-approximation algorithm for ORTHOGONAL TERRAIN GUARDING. The papers [59] and [27] studied restrictions under which ORTHOGONAL TERRAIN GUARDING can be solved in polynomial time.

3.2   Subexponential-Time Parameterized Algorithm for Terrain Guarding and FPT Algorithm for Orthogonal Terrain Guarding

We considered the parameterized complexity of TERRAIN GUARDING in [7]. Although we have not resolved the question of whether or not it is FPT, we achieved two related results that are of independent interest. First, we designed a subexponential-time algorithm for TERRAIN GUARDING in both discrete and continuous domains. For this purpose, we developed an $n^{\mathcal{O}(\sqrt{k})}$-time algorithm for TERRAIN GUARDING in discrete domains. Friedrichs et al. [37] proved that given an instance of TERRAIN GUARDING in a continuous domain, one can construct (in polynomial time) an equivalent instance of TERRAIN GUARDING in a discrete domain. That is, given an instance $(T = (V, E), k)$ of TERRAIN GUARDING in a continuous domain, Friedrichs et al. [37] designed a discretization procedure that outputs an instance $(T' = (V', E'), k)$ of TERRAIN GUARDING in a discrete domain such that $(T = (V, E), k)$ is a yes-instance if and only if $(T' = (V', E'), k)$ is a yes-instance (more precisely, the output refers to an annotated version of the problem). Unfortunately, this reduction blows up the number of vertices of the terrain to $\mathcal{O}(n^3)$, and therefore the existence of a subexponential-time algorithm for TERRAIN GUARDING in discrete domains does not imply that there exists such an algorithm for TERRAIN GUARDING in continuous domains. However, observe that the reduction *does not change* the value of the parameter k. Thus, since we solve TERRAIN GUARDING in discrete domains in time $n^{\mathcal{O}(\sqrt{k})}$ rather than $n^{\mathcal{O}(\sqrt{n})}$, we are able to deduce that TERRAIN GUARDING in continuous domains is solvable in time $n^{\mathcal{O}(\sqrt{k})}$. Observe that, in both discrete and continuous domains, it can be assumed that $k \leq n$: to guard all of the points that lie on a terrain, it is sufficient to place guards only on the vertices of the terrain. Hence, when we solve TERRAIN GUARDING in continuous domains, we assume that $k \leq n$ where n is the number of vertices of the input continuous terrain and *not* of the discrete terrain output by the reduction. The next theorem summarizes our algorithmic contribution.

Theorem 3. TERRAIN GUARDING *in both discrete and continuous domains is solvable in time $n^{\mathcal{O}(\sqrt{k})}$. Thus, it is also solvable in time $n^{\mathcal{O}(\sqrt{n})}$.*

Observe that our result, Theorem 3, demonstrates an interesting dichotomy in the complexities of TERRAIN GUARDING and the ART GALLERY problem: the ART GALLERY problem does not admit an algorithm with running time $2^{o(n)}$ under the ETH, while TERRAIN GUARDING in both discrete and continuous domains is solvable in time $2^{\mathcal{O}(\sqrt{n}\log n)}$. When we measure the running time in terms of both n and k, the ART GALLERY problem does not admit an algorithm with running time $f(k) \cdot n^{o(k/\log k)}$ for any function f [14], while TERRAIN GUARDING in both discrete and continuous domains is solvable in time $n^{\mathcal{O}(\sqrt{k})}$.

Our algorithm is based on a suitable definition of a planar graph. A similar notion of a planar graph, which inspired our work, was previously used in designing a local search based PTAS for the TERRAIN GUARDING problem [43]. In particular, we aim to define a planar graph that has a small domination number and which captures *both* the manner in which a hypothetical solution guards the terrain and some information on the layout of the terrain itself (see Fig. 3). Having this planar graph, we are able to "guess" separators whose exploitation, which involves additional guesses guided by the structure of the graph, essentially results in a divide-and-conquer algorithm. The design of the divide-and-conquer algorithm is non-trivial since given our guesses, it is not possible to divide the problem into two simpler subproblems in the obvious way—that is, we cannot divide the terrain into two disjoint subterrains that can be handled separately. We overcome this difficulty by dividing not the terrain itself, but a set of points of interest on the terrain.

Our second result is the proof that ORTHOGONAL TERRAIN GUARDING of vertices of the orthogonal terrain with vertices is FPT with respect to the parameter k. More precisely, we obtain the following result.

Theorem 4. ORTHOGONAL TERRAIN GUARDING *is solvable in time $k^{\mathcal{O}(k)}n^{\mathcal{O}(1)}$.*

Our algorithm is based on new insights into the relations between the left and right reflex and convex vertices. We integrate these insights in the design of an algorithm that is based on the proof that one can ignore "exposed vertices", which are vertices seen by too many vertices of a specific type, and a non-trivial branching strategy.

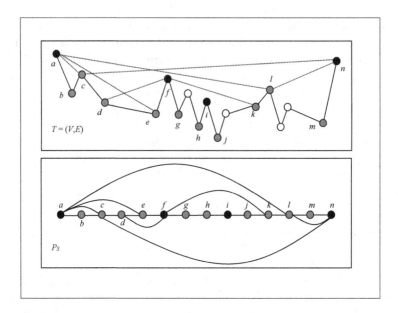

Fig. 3. An instance of TERRAIN GUARDING where only designated vertices must be guarded (in grey), and a hypothetical solution S (in black). Then, a planar graph P_S is defined by taking the edges of the terrain, and an edge between every vertex that must be guarded and the furthest vertices in the solution to the left and right that see it.

We remark that in a recent manuscript [6], we have proved that TERRAIN GUARDING admits a polynomial kernel when parameterized by r.

References

1. Abello, J., Egecioglu, O., Kumar, K.: Visibility graphs of staircase polygons and the weak bruhat order I: from visibility graphs to maximal chains. Discrete Comput. Geom. **14**(3), 331–358 (1995)
2. Abrahamsen, M., Adamaszek, A., Miltzow, T.: Irrational guards are sometimes needed. In: 33rd International Symposium on Computational Geometry (SoCG), pp. 3:1–3:15 (2017)
3. Abrahamsen, M., Adamaszek, A., Miltzow, T.: The art gallery problem is ∃ℝ-complete. In: Proceedings of the 50th Annual ACM SIGACT Symposium on Theory of Computing (STOC), pp. 65–73 (2018)
4. Aggarwal, A.: The art gallery theorem: its variations, applications and algorithmic aspects. Ph.D. thesis, The Johns Hopkins University, Baltimore, Maryland (1986)
5. Agrawal, A., Knudsen, K., Daniel Lokshtanov, S.S., Zehavi, M.: The parameterized complexity of guarding almost convex polygons. In: 36rd International Symposium on Computational Geometry (SoCG) (2020, to appear)
6. Agrawal, A., Kolay, S., Zehavi, M.: Multivariate analysis for guarding terrains. In: Manuscript. p. TBA (2020)
7. Ashok, P., Fomin, F.V., Kolay, S., Saurabh, S., Zehavi, M.: Exact algorithms for terrain guarding. ACM Trans. Algorithms **14**(2), 25:1–25:20 (2018)

8. Avis, D., Toussaint, G.T.: An efficient algorithm for decomposing a polygon into star-shaped polygons. Pattern Recogn. **13**(6), 395–398 (1981)
9. Basu, S., Pollack, R., Roy, M.: On the combinatorial and algebraic complexity of quantifier elimination. J. ACM **43**(6), 1002–1045 (1996)
10. Ben-Moshe, B., Katz, M.J., Mitchell, J.S.B.: A constant-factor approximation algorithm for optimal 1.5d terrain guarding. SICOMP **36**(6), 1631–1647 (2007)
11. Bhattacharya, P., Ghosh, S.K., Pal, S.P.: Constant approximation algorithms for guarding simple polygons using vertex guards. CoRR/arXiv abs/1712.05492 (2017)
12. Bhattacharya, P., Ghosh, S.K., Roy, B.: Approximability of guarding weak visibility polygons. Discrete Appl. Math. **228**, 109–129 (2017)
13. Bonnet, É., Giannopoulos, P.: Orthogonal terrain guarding is NP-complete. J. Comput. Geom. **10**(2), 21–44 (2019)
14. Bonnet, É., Miltzow, T.: Parameterized hardness of art gallery problems. In: Proceedings of the 24th Annual European Symposium on Algorithms (ESA), pp. 19:1–19:17 (2016)
15. Bonnet, É., Miltzow, T.: An approximation algorithm for the art gallery problem. In: Proceedings of the 33rd International Symposium on Computational Geometry (SoCG), pp. 20:1–20:15 (2017)
16. Borrmann, D., et al.: Point guards and point clouds: solving general art gallery problems. In: Symposium on Computational Geometry (SoCG), pp. 347–348 (2013)
17. Brönnimann, H., Goodrich, M.T.: Almost optimal set covers in finite VC-dimension. Discrete Comput. Geom. **14**(4), 463–479 (1995)
18. Chen, D.Z., Estivill-Castro, V., Urrutia, J.: Optimal guarding of polygons and monotone chains. In: CCCG, pp. 133–138 (1995)
19. Chvátal, V.: A combinatorial theorem in plane geometry. J. Comb. Theory Ser. B **18**(1), 39–741 (1975)
20. Clarkson, K.L., Varadarajan, K.R.: Improved approximation algorithms for geometric set cover. Discrete Comput. Geom. **37**(1), 43–58 (2007)
21. Clarkson, K.L.: Algorithms for polytope covering and approximation. In: Dehne, F., Sack, J.-R., Santoro, N., Whitesides, S. (eds.) WADS 1993. LNCS, vol. 709, pp. 246–252. Springer, Heidelberg (1993). https://doi.org/10.1007/3-540-57155-8_252
22. Couto, M.C., de Rezende, P.J., de Souza, C.C.: An exact algorithm for minimizing vertex guards on art galleries. Int. Trans. Oper. Res. **18**(4), 425–448 (2011)
23. Cygan, M., et al.: Parameterized Algorithms. Springer, Heidelberg (2015). https://doi.org/10.1007/978-3-319-21275-3
24. Deshpande, A., Kim, T., Demaine, E.D., Sarma, S.E.: A pseudopolynomial time $O(\log n)$-approximation algorithm for art gallery problems. In: Dehne, F., Sack, J.-R., Zeh, N. (eds.) WADS 2007. LNCS, vol. 4619, pp. 163–174. Springer, Heidelberg (2007). https://doi.org/10.1007/978-3-540-73951-7_15
25. Downey, R.G., Fellows, M.R.: Parameterized Complexity. Springer, Heidelberg (1999). https://doi.org/10.1007/978-1-4612-0515-9
26. Downey, R.G., Fellows, M.R.: Fundamentals of Parameterized Complexity. Texts in Computer Science. Springer, Heidelberg (2013). https://doi.org/10.1007/978-1-4471-5559-1
27. Durocher, S., Li, P.C., Mehrabi, S.: Guarding orthogonal terrains. In: Proceedings of the 27th Canadian Conference on Computational Geometry, CCCG (2015)
28. Efrat, A., Har-Peled, S.: Guarding galleries and terrains. Inf. Process. Lett. **100**(6), 238–245 (2006)
29. Eidenbenz, S., Stamm, C., Widmayer, P.: Inapproximability of some art gallery problems. In: Proceedings of the 10th Canadian Conference on Computational Geometry (CCCG) (1998)

30. Eidenbenz, S., Stamm, C., Widmayer, P.: Inapproximability results for guarding polygons and terrains. Algorithmica **31**(1), 79–113 (2001)
31. Elbassioni, M.K., Krohn, E., Matijevic, D., Mestre, J., Severdija, D.: Improved approximations for guarding 1.5-dimensional terrains. Algorithmica **60**(2), 451–463 (2011)
32. Evans, W., Saeedi, N.: On characterizing terrain visibility graphs. J. Comput. Geom. **6**(1), 108–141 (2015)
33. Fellows, M.R.: The lost continent of polynomial time: preprocessing and kernelization. In: Bodlaender, H.L., Langston, M.A. (eds.) IWPEC 2006. LNCS, vol. 4169, pp. 276–277. Springer, Heidelberg (2006). https://doi.org/10.1007/11847250_25
34. Fisk, S.: A short proof of Chvátal's watchman theorem. J. Comb. Theory Ser. B **24**(3), 374 (1978)
35. Flum, J., Grohe, M.: Parameterized Complexity Theory. Texts in Theoretical Computer Science. An EATCS Series, Springer, Heidelberg (2006). https://doi.org/10.1007/3-540-29953-X
36. Fomin, F., Lokshtanov, D., Saurabh, S., Zehavi, M.: Kernelization: Theory of Parameterized Preprocessing. Cambridge University Press, Cambridge (2018)
37. Friedrichs, S., Hemmer, M., King, J., Schmidt, C.: The continuous 1.5D terrain guarding problem: discretization, optimal solutions, and PTAS. J. Comput. Geom. **7**(1), 256–284 (2016)
38. Ghosh, S.K.: Approximation algorithms for art gallery problems. In: Canadian Information Processing Society Congress, pp. 429–434 (1987)
39. Ghosh, S.K.: Visibility Algorithms in the Plane. Cambridge University Press, Cambridge (2007)
40. Ghosh, S.K.: Approximation algorithms for art gallery problems in polygons. Discrete Appl. Math. **158**(6), 718–722 (2010)
41. Ghosh, S.K., Goswami, P.P.: Unsolved problems in visibility graphs of points, segments, and polygons. ACM Comput. Surv. **46**(2), 22:1–22:29 (2013)
42. Giannopoulos, P.: Open problems: guarding problems. In: Lorentz Workshop on Fixed-Parameter Computational Geometry, Leiden, the Netherlands, p. 12 (2016)
43. Gibson, M., Kanade, G., Krohn, E., Varadarajan, K.: Guarding terrains via local search. J. Comput. Geom. **5**(1), 168–178 (2014)
44. Gilbers, A., Klein, R.: A new upper bound for the VC-dimension of visibility regions. Comput. Geom. **47**(1), 61–74 (2014)
45. Kalai, G., Matoušek, J.: Guarding galleries where every point sees a large area. Israel J. Math. **101**(1), 125–139 (1997)
46. Katz, M.J.: A PTAS for vertex guarding weakly-visible polygons - an extended abstract. CoRR abs/1803.02160 (2018)
47. Katz, M.J., Roisman, G.S.: On guarding the vertices of rectilinear domains. Comput. Geom. **39**(3), 219–228 (2008)
48. Khodakarami, F., Didehvar, F., Mohades, A.: A fixed-parameter algorithm for guarding 1.5D terrains. Theor. Comput. Sci. **595**, 130–142 (2015)
49. King, J.: A 4-approximation algorithm for guarding 1.5-dimensional terrains. In: Correa, J.R., Hevia, A., Kiwi, M. (eds.) LATIN 2006. LNCS, vol. 3887, pp. 629–640. Springer, Heidelberg (2006). https://doi.org/10.1007/11682462_58
50. King, J., Krohn, E.: Terrain guarding is NP-hard. SICOMP **40**(5), 1316–1339 (2011)
51. King, J.: Fast vertex guarding for polygons with and without holes. Comput. Geom. **46**(3), 219–231 (2013)
52. King, J., Kirkpatrick, D.G.: Improved approximation for guarding simple galleries from the perimeter. Discrete Comput. Geom. **46**(2), 252–269 (2011)

53. Kirkpatrick, D.G.: An $O(\log \log \mathrm{OPT})$-approximation algorithm for multi-guarding galleries. Discrete Comput. Geom. **53**(2), 327–343 (2015)
54. Kooshesh, A.A., Moret, B.M.E.: Three-coloring the vertices of a triangulated simple polygon. Pattern Recogn. **25**(4), 443 (1992)
55. Krohn, E., Nilsson, B.J.: Approximate guarding of monotone and rectilinear polygons. Algorithmica **66**(3), 564–594 (2013)
56. Lee, D., Lin, A.: Computational complexity of art gallery problems. IEEE Trans. Inf. Theory **32**(2), 276–282 (1986)
57. Lyu, Y., Üngör, A.: A fast 2-approximation algorithm for guarding orthogonal terrains. In: Proceedings of the 28th Canadian Conference on Computational Geometry, CCCG. pp. 161–167 (2016)
58. Marx, D.: What's next? Future directions in parameterized complexity. In: Bodlaender, H.L., Downey, R., Fomin, F.V., Marx, D. (eds.) The Multivariate Algorithmic Revolution and Beyond. LNCS, vol. 7370, pp. 469–496. Springer, Heidelberg (2012). https://doi.org/10.1007/978-3-642-30891-8_20
59. Mehrabi, S.: Guarding the vertices of an orthogonal terrain using vertex guards. arXiv:1512.08292 (2015)
60. Niedermeier, R.: Invitation to Fixed-Parameter Algorithms, Oxford Lecture Series in Mathematics and its Applications, vol. 31. Oxford University Press, Oxford (2006)
61. Niedermeier, R.: Reflections on multivariate algorithmics and problem parameterization. In: Proceedings of the 27th International Symposium on Theoretical Aspects of Computer Science (STACS), pp. 17–32 (2010)
62. O'Rourke, J.: Art Gallery Theorems and Algorithms, vol. 57. Oxford University Press, Oxford (1987)
63. O'Rourke, J., Supowit, K.J.: Some NP-hard polygon decomposition problems. IEEE Trans. Inf. Theory **29**(2), 181–189 (1983)
64. de Rezende, P.J., de Souza, C.C., Friedrichs, S., Hemmer, M., Kröller, A., Tozoni, D.C.: Engineering art galleries. In: Kliemann, L., Sanders, P. (eds.) Algorithm Engineering. LNCS, vol. 9220, pp. 379–417. Springer, Cham (2016). https://doi.org/10.1007/978-3-319-49487-6_12
65. van Rooij, I., Blokpoel, M., Kwisthout, J., Wareham, T.: Cognition and Intractability: A Guide to Classical and Parameterized Complexity Analysis. Cambridge University Press, Cambridge (2019)
66. Schuchardt, D., Hecker, H.D.: Two NP-hard art-gallery problems for orthopolygons. Math. Logic Q. **41**(2), 261–267 (1995)
67. Shermer, T.C.: Recent results in art galleries (geometry). Proc. IEEE **80**(9), 1384–1399 (1992)
68. Urrutia, J.: Art gallery and illumination problems. Handb. Comput. Geom. **1**(1), 973–1027 (2000)
69. Valtr, P.: Guarding galleries where no point sees a small area. Israel J. Math. **104**(1), 1–16 (1998)

Central Positions in Social Networks

Ulrik Brandes[✉][iD]

Social Networks Lab, ETH Zürich, Zürich, Switzerland
ubrandes@ethz.ch

Abstract. This contribution is an overview of our recent work on the concept of centrality in networks. Instead of proposing new centrality indices, providing faster algorithms, or presenting new rules for when an index can be classified as a centrality, this research shifts the focus to the more elementary question whether a node is in a more central position than another. Viewing networks as data on overlapping dyads, and defining the position of a node as the whole of its relationships to the rest of the network, we obtain a very general procedure for doing centrality analysis; not only on social networks but networks from all kinds of domains. Our framework further suggests a variety of computational challenges.

Keywords: Data science · Social networks · Centrality · Algorithmics

1 Introduction

Today, the study of social networks [7,22,29] is a major application domain of network science [12,26]. It has a long tradition, though, and developed into an established field decades before network science itself [18]. Unsurprisingly, many methods now characteristic of network science have at least one strand of development in the context of social networks.

The strand I will focus on for this contribution is "[o]ne of the primary uses of graph theory in social network analysis," [34] namely the identification of the "most important" actors. Almost exclusively the tool used for this task are *centrality indices* [17,24], i.e., graph invariants mapping each vertex to a (usually non-negative) real number.

Consider, for instance, *closeness centrality*, which for a connected simple undirected graph $G = (V, E)$ and any $i \in V$ may be defined as

$$c_C(i) = \frac{1}{\sum_{t \in V} dist(i, t)},$$

where n is the number of vertices and $dist : V \times V \to \mathbb{N}_0$ denotes the shortest-path distances in the graph. A vertex is thus central to the extent that it is close to the others.

Closeness and maybe a half-dozen other centrality indices are routinely applied to graphs representing networks from diverse application domains, but with little regard for the specific type of relation defining them.

© Springer Nature Switzerland AG 2020
H. Fernau (Ed.): CSR 2020, LNCS 12159, pp. 30–45, 2020.
https://doi.org/10.1007/978-3-030-50026-9_3

Justifications for the use of a particular centrality index grounded in theory are difficult to come by. Indices are selected based on precedence or ascribed high-level features, and rarely is it demonstrated that they actually operationalize the intended concept. To see why such justifications may be called for, consider the following assumptions that are made implicitly when applying closeness centrality:

- *indirect relation:* The focal vertex i is related to other vertices t by shortest-path distance. The assumption that this notion of distance is more appropriate than, say, the expected hitting time of a simple random walk, is rarely stated, let alone tested. Does each additional edge that is needed for reachability reduce the quality of a relationship, and to what degree?
- *homogeneity:* Closeness is a graph invariant because all targets t contribute in the same way. No distinction is made with respect to the actors represented by target vertices, and those involved in shortest paths. Does it not matter that specific vertices are closer or farther away?
- *additivity:* Not only is distance the sole structural criterion determining centrality, but distances are aggregated by summation. This suggests that they are interpreted as being measured on at least an interval scale. Is there a trade-off between the number of targets at a certain distance and the value of said distance?
- *transfer function:* The total distance is inverted to let high centrality scores be associated with short distances. The choice of this non-linear order-reversing transformation has implications for the subsequent use of closeness centrality scores in correlations and other statistics. Is there a decreasing marginal influence of distant vertices, and what is its shape?

In addition to these issues inherent in the definition, it is tempting to treat closeness as a quantitative variable for which statistics involving differences and ratios are meaningful. Pearson correlation and size normalization are but the most straightforward examples of subsequent uses that would benefit from increased justifiability.

The lack of theoretical grounding may ostensibly be a problem of the social sciences. For a computer science audience it is rather convenient to have a given list of indices to focus on. It certainly removes the burden of arguing for the relevance of a computational problem associated with any such index. On the other hand, it also limits the scope of problems that can be identified as potentially interesting. The usual challenges such as efficient algorithms, approximations, bounds, and top-k queries, and the variations obtained by considering restrictions to graph classes, dynamics, or computing environments, are considered largely in association with only a few well-known indices that are based on shortest paths, random walks, and matrix expressions [19].

The aim for this contribution is to promote a more principled view of centrality that nevertheless opens up a wider and more diverse range of computational challenges to address. Moving beyond network analysis as an applied form of graph theory, we position network science within data science. In the next section it is argued that the fundamental distinction between network science and other forms of data science lies in the type of variables considered rather than an underlying theory or application domain. A formal notion of positions in networks is introduced in Sect. 3 and serves as the basis for a novel conceptualization of centrality in Sect. 4. As a result, many new computational challenges arise, some of which are sketched in Sect. 5. We conclude with a summary and some broader implications in Sect. 6.

2 Network Data

In the literature, the notion of a network is often conflated with that of a graph. Sometimes networks are defined as graphs with vertex and/or edge attributes, other times a distinction is made between a network as the empirical phenomenon and the graph as its mathematical representation.

For the purpose of this paper, the distinction between empirical networks and mathematical models is secondary. We do, however, find it important to distinguish formally between networks and graphs, not least because it facilitates a broader view of the problems relevant in network analysis.

2.1 Variables

Network science can be viewed as a particular kind of data science. In this perspective, network data are the values of special types of variables. We consider a *variable* $x : \mathcal{S} \rightarrow \mathcal{X}$ to be a mapping that associates to entities from a (usually finite) *domain* $\mathcal{S}$ with values from an (often numerical) *range* $\mathcal{X}$. Depending on the application area, variables may be referred to as attributes or features, and we often write variables in vector notation $x \in \mathcal{X}^{\mathcal{S}}$, so that $x_i = x(i)$ for $i \in \mathcal{S}$.

The range, in general, is an algebraic structure corresponding to the level of measurement. Common ranges include $\mathcal{X} = \{0, 1\}$ for binary data, a set of labels for categorical data, ranks for ordinal data, and real values $\mathcal{X} = \mathbb{R}$ for quantitative data. While matrices, time intervals, distributions, and other more complicated data are possible, we are going to assume quantitative data for simplicity.

The domain, on the other hand, is usually a set of atomic entities that are comparable with respect to the characteristics expressed in the variables. The data most commonly encountered in empirical research can be organized in $s \times r$-tables, were rows are indexed by the entities $i = 1, \dots, s$ that make up a domain $\mathcal{S}$, and the columns $j = 1, \dots, r$ contain values of r variables $x^{(j)} : \mathcal{S} \rightarrow \mathcal{X}^{(j)}$ defined on the shared domain $\mathcal{S}$, but with potentially different ranges $\mathcal{X}^{(j)}$.

2.2 Network Variables

A special situation occurs when the ranges do not differ because the entries of multiple variables defined on the same domain represent different but comparable properties of the entities. Examples include discrete time series (the same property at different points in time), characteristic vectors (the presence or absence of different features), and co-occurrences (the number of joint observations with different other entities). Note that especially in the last case, rows and columns may even be indexed by the same entities.

The comparability of associations can be made explicit by combining a number of related variables into a single one defined on pairs of entities, or *dyads*, $\mathcal{N} \times \mathcal{A}$. A variable $x : (\mathcal{N} \times \mathcal{A}) \to \mathcal{X}$ thus represents relational data in which entities $i \in \mathcal{N}$ have a number of attributes $j \in \mathcal{A}$ measured on a common scale $\mathcal{X}$.

To allow for the exclusion of ill-defined combinations and unavailable entries, we define *networks* to be variables $x : \mathcal{D} \to \mathcal{X}$ in which dyads $\mathcal{D} \subseteq (\mathcal{N} \times \mathcal{A})$. We refer to $\mathcal{N}$ as the set of *nodes* and $\mathcal{A}$ as the set of *affiliations* but note that they need not be disjoint. The only distinction between networks and the standard variables is therefore the incidence structure on their domain.

As examples, consider two straightfoward kinds of binary network variables common in social network analysis. In affiliation networks, nodes $\mathcal{N}$ represent individuals, affiliations $\mathcal{A}$ represent organizations; because $\mathcal{N} \cap \mathcal{A} = \emptyset$, affiliation networks are *two-mode networks*. In friendship networks, nodes and affiliations $\mathcal{N} = \mathcal{A}$ represent the same set of individuals, and the domain of dyads $\mathcal{D} \subseteq (\mathcal{N} \times \mathcal{A}) \setminus diag(\mathcal{N})$ does not include the diagonal $diag(\mathcal{N}) = \{(i,i) : i \in \mathcal{N}\}$, because friendship of an individual with itself is not defined. Because the same entities are represented both by nodes and affiliations, friendship networks are *one-mode networks*.

2.3 Graph Representations

Techniques for the analysis of networks are often formulated more conveniently in terms of matrices or graphs. A binary network variable $x : \mathcal{D} \to \{0,1\}$ can be represented as a simple undirected graph $G(x)$ in the obvious way if it is symmetric, i.e., $x_{ij} = x_{ji}$ for all $(i,j) \in \mathcal{D}$, and $\mathcal{D} = (\mathcal{N} \times \mathcal{A}) \setminus diag(\mathcal{N})$. Other representations are more involved and may require the introduction of special values to distinguish absent relationships (say, of value zero) from unobserved or impossible ones (when a pair is not in the domain).

For ease of exposition, we generally assume networks of type $x : \mathcal{D} \to \mathbb{R}_{\geq 0}$ to be (almost) totally defined with $\mathcal{D} = (\mathcal{N} \times \mathcal{N}) \setminus diag(\mathcal{N})$ or $\mathcal{D} = (\mathcal{N} \times \mathcal{N})$, so that there is a one-to-one mapping to weighted directed graphs $G(x) = (V, E; w)$, where $V = \mathcal{N}$, $E = \{(i,j) \in \mathcal{D} : x_{ij} > 0\}$, and $w : E \to \mathbb{R}_{>0}$) with $w(i,j) = x_{ij}$.

3 Positions

As outlined in the previous section, the dyad is the unit of observation in network science. The unit of analysis is more commonly the node, at least for social networks and, by definition, for centrality. Since nodes are involved in multiple dyads, network analysis can be viewed as a form of multilevel analysis, in which nodes are aggregations of dyads.

Borgatti, Everett, and Johnson [7] state that "[c]entrality is a property of a nodes' position in a network." While the notion of position is often used rather

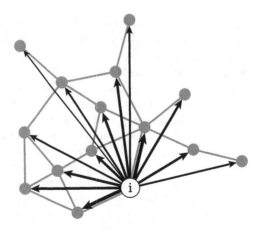

Fig. 1. A network x (gray) and the position (black) of vertex i in the transformed network $dist(x)$ based on shortest-path distances.

metaphorically, a formal definition may serve as a pivotal abstraction for most of network analysis [8].

3.1   Derived Relations

We want to define a node's position so that it characterizes how that node relates to all the other nodes in the network. It would be straightforward to conceive of a network position as the data associated with the dyads containing that node. Analytic concepts such centrality indices, however, generally consider more than just the dyads a node is involved in directly. Fortunately, it turns out that they can be reformulated such that, after explicitly establishing a new, derived, relation, only the dyads involving the focal node are used.

Take, for example, closeness centrality from above. After determining a new network variable $dist(x)$ from shortest-path distances, the closeness of a node i

Table 1. Centrality indices with derived relations they can be based on. While $\sigma(s,t)$ denotes the number of shortest (s,t)-paths, $\sigma(s,t|i)$ is restricted to those that use i as an intermediary, and $\delta(s,i) = \sum_{t \in V} \frac{\sigma(s,t|i)}{\sigma(s,t)}$. Scalar λ is the principal eigenvalue of the graph's adjacency matrix, and $\omega(s,t)$ is the limit, for $k \to \infty$, of the fraction of (s,t)-walks among all walks of length k.

Centrality	Definition	Derived relation
Outdegree	$c_D(i) = \sum_{t \in V} x_{it}$	Identity transform
Closeness	$c_C(i) = \left(\sum_{t \in V} dist(i,t) \right)^{-1}$	Shortest-path distances
Betweenness	$c_B(i) = \sum_{s,t \in V} \frac{\sigma(s,t|i)}{\sigma(s,t)} = \sum_{s \in V} \delta(s,i)$	Dependency of s on i
Eigenvector	$c_E(i) = \lambda^{-1} \sum_{t \in V} c_E(t) = \sum_{t \in V} \omega(i,t)$	Limit share of walks

is a function only of those dyads that are incident to i. This is illustrated in Fig. 1 and does not only hold for closeness but in fact for all commonly used centrality indices. Table 1 gives a few examples of centrality indices and derived dyadic relations they can be expressed in.

We posit that centrality should be considered separately from the derived relations that, as of today, are baked into the calculations of indices. Studying derived relations independently from how their values are aggregated not only suggests many more notions of centrality such as closeness-type centralities based on various properties of random walks [10] but also opens up ways of testing their appropriateness in specific application contexts.

3.2 Network Positions

Definition 1. *Given a network* $x : \mathcal{D} \to \mathcal{X}$ *with* $\mathcal{D} \subseteq (\mathcal{N} \times \mathcal{A})$, *the position of* $i \in \mathcal{N}$ *in* x, *denoted* $pos(i|x)$, *is defined as the restriction of* x *to* $\mathcal{D}(i) = \{(s,t) \in \mathcal{D} : s = i \text{ or } t = i\}$.

In the simplest case, positions correspond to neighborhoods in the graph representation of a network. Since centralities typically depend on network structure more globally, we will be interested more generally in positions $pos(i|y)$ in transformed networks $y = \tau(x)$ that result from a derived relations such as various notions of distance, walk counts, connectivity, dependency, and so on. Other examples include transformations that normalize dyad values or evaluate the prevalence of relationships in substructures such as triads [27]. Positions generalize to multiple relations $pos(i|y_1, \ldots, y_r)$ and may be augmented further with attributes of nodes and affiliations.

Network positions thus integrate the features that characterize how a node relates to the rest of the network. Since the specific set of relations, attributes, and selections may vary, we will use $\textcircled{i}$ to denote the position of a node i when the defining features are clear from context, or irrelevant. Being reminiscent of a conventional node drawing, it symbolizes the fact that we consider a node i in the context of the network it is located in.

The concept of position just presented is novel only in as much as it unifies and extends many previous ones from the literature [35]. For instance, it detaches Burt's notion of position [13] from what he calls individual distances and equivalence of positions, and it is more fine-grained than notions defining positions as sets of nodes that relate to others in similar ways [4,14,16]; with corresponding comparison operators the latter are retained as equivalence classes of individual node positions.

Finally, it is worth pointing out that, being feature vectors in a possibly transformed network, positions also bridge between network analysis and machine learning [21].

4 Centrality

In the words of the late Lin Freeman, "[t]here is certainly no unanimity on exactly what centrality is or its conceptual foundations, and there is very little agreement

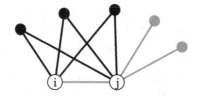

Fig. 2. Neigborhood inclusion: the closed neighborhood of j contains the (open) neighborhood of i, $N(i) \subseteq N[j]$. Gray edges may or may not be present.

on the proper procedure for its measurement." [17] Despite the decades since he wrote this sentence, it still rings true today.

Attempts to bring order to the field range from the conceptual, where the focus is on the relation between indices and the mechanisms they model [3,6], to the axiomatic, where classes of indices are characterized by formal properties they exhibit [1,2,30].

We here follow a different approach to operationalize the notion of centrality. It is based on the observation that indices traditionally considered as some form of centrality have a non-trivial common ground in neighborhood-inclusion [32] and that the generalization from neighborhoods to positions offers more control over properties of centrality concepts than any family of indices can provide.

In the following, we will motivate this approach from first principles.

4.1 Neighborhood Inclusion

Networks are said to exhibit a *core-periphery structure*, if their nodes can be partitioned into a densely connected core and a loosely connected periphery [5].

The ideal form of a core-periphery structure is encountered in a simple undirected graph $G = (V, E)$, if there exists a bipartition $V = C \uplus P$ such that the vertex-induced subgraphs $G[C]$ and $G[P]$ are a clique and an independent set, respectively. In other words, when G is a *split graph*. Note that the bipartition $V = C \uplus P$ discriminates between central and peripheral vertices but can not differentiate vertices in the same group.

Denote by $N(i)$ the neighbors of a vertex $i \in V$, and let $N[i] = N(i) \cup \{i\}$ be its closed neighborhood. Then, vertices $j \in C$ satisfy $N(i) \subseteq N[j]$ for all $i \in P$, i.e., the neighborhood of a central vertex includes all neighbors of a peripheral vertex except for itself, if applicable. This is illustrated in Fig. 2. Since it is the idealized expression of the sentiment that a node should be more central if it is better connected, and therefore commonly implied by axiomatizations of centrality indices, we use the neighborhood-inclusion property to refine the distinction between more or less central vertices beyond those in the core and those in the periphery.

Neighborhood-inclusion yields a preorder (a binary relation that is reflexive and transitive but not necessarily total or antisymmetric) on the vertices of a graph, sometimes referred to as the *vicinal preorder*. We here prefer the term

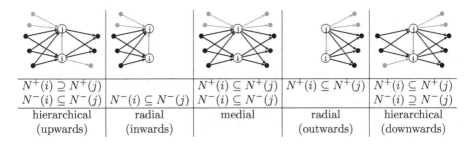

$N^+(i) \supseteq N^+(j)$		$N^+(i) \subseteq N^+(j)$	$N^+(i) \subseteq N^+(j)$	$N^+(i) \subseteq N^+(j)$
$N^-(i) \subseteq N^-(j)$	$N^-(i) \subseteq N^-(j)$	$N^-(i) \subseteq N^-(j)$		$N^-(i) \supseteq N^-(j)$
hierarchical	radial	medial	radial	hierarchical
(upwards)	(inwards)		(outwards)	(downwards)

Fig. 3. Neighborhood-inclusion criteria for directed graphs.

partial ranking over preorder. A total preorder, or simply *ranking*, is also called a *weak order*.

Threshold graphs are a subclass of split graphs also known as *nested split graphs* [25]. One of the many equivalent definitions requires that their neighborhood-inclusion preorder is total, i.e., the vertices can be fully ranked.

In the same way that split graphs represent ideal core-periphery structures, their specialization to threshold graphs thus yields ideal centrality rankings. These rankings are not only consistent with the idea of centrality but indeed preserved by all common centrality indices [32]. The proof is based on the observation that neighborhood inclusion is preserved for a very general class of indirect relations as discussed in the previous section.

Threshold graphs generalize star graphs $K_{1,n-1}$, the only class of graphs for which there has previously been a unanimous interpretation of centrality: "A person located in the center of a star is universally assumed to be structurally more central than any other person in any other position in any other network of similar size." [17]

Proposition 1. *If a network can be represented as a simple undirected graph, every centrality ranking preserves the neighborhood-inclusion preorder.*

Extension of this requirement to networks represented by directed graphs is not canonical and therefore suggests a distinction between different kinds of centrality. Let $G = (V, E)$ be a simple directed graph and denote by $N^-(i)$ and $N^+(i)$ the in- and out-neighborhood of $i \in V$. The question is how to generalize neighborhood inclusion when there are two types of neighborhoods?

Borgatti and Everett [6] propose a conceptual distinction of radial and medial centralities based on how the indices are defined. We use combinations of in- and out-neighborhood inclusion to formalize their idea using only structure, which in turn leads naturally to a third category that we term hierarchical. The labels are assigned based on the following combinations of criteria:

- *radial:* nodes are more connected in a chosen direction, and the other direction is ignored
- *medial:* nodes are more connected with respect to both incoming and outgoing relationships

- *hierarchical:* nodes are more connected in a chosen direction and less constrained in the opposite

The formalizations in terms of neighborhoods are given and illustrated in Fig. 3. Note that, for conceptual reasons beyond the scope of this paper, the conditions are stated in terms of open neighborhoods only. By definition, hierarchical centralities are restrictions of radial centralities, and centralities that are both in- and out-radial are also medial. Medial centralities, however, need not be in- or out-radial.

4.2 Positional Dominance

Any centrality index $c : \mathcal{N} \to \mathbb{R}_{\geq 0}$ implies a ranking of the nodes via $i \leq j \iff c(i) \leq c(j)$ for all $i, j \in \mathcal{N}$. Each of the neighborhood-inclusion conditions from the previous subsection implies a partial ranking of positions, and we propose to consider an index a centrality only if its induced node ranking preserves the partial ranking of their positions.

For a binary network x, let $G(x) = (V, E)$ be its graph representation and let $\textcircled{i}$ denote the positions of all nodes $i \in \mathcal{N}$ in x with respect to the adjacency relation only, i.e., their neighborhoods. If $\leq$ is a partial ranking of positions defined by neighborhood inclusion or one of its extensions to directed graphs from Fig. 3, then the above requirement for $c : \mathcal{N} \to \mathbb{R}_{\geq 0}$ to be a centrality index translates to

$$\textcircled{i} \leq \textcircled{j} \implies c(i) \leq c(j) \quad \text{for all } i, j \in \mathcal{N}.$$

We refer to such partial rankings of positions as *positional dominance* [8,11]. It formalizes the idea that, in some specific sense, the position of one node is no better than that of another node. Instantiations of positional dominance for different positions and notions of "better" are the key instrument to construct interpretable centrality rankings.

Generalization of positional dominance to non-binary networks is immediate when expressing neighborhood inclusion as one would for the characteristic vectors of the vertex sets $N(i)$. The radial centrality condition for outgoing relationships reads

$$\textcircled{i} \leq \textcircled{j} \quad \text{if} \quad x_{it} \leq x_{jt} \quad \text{for all } (i, t) \in \mathcal{D} \text{ with } x_{it} \neq 0$$

where, in particular, the required dyads $(j, t) \in \mathcal{D}$. It does not make a difference what values make up the range $\mathcal{X}$, so long as there is an ordering $\leq$ on $\mathcal{X}$ and an element $0 \in \mathcal{X}$ signifying the absence of a relationship.

It is important to realize that further generalizations are obtained from replacing the observed network x with a transformed network $\tau(x)$. This way, any assumptions about derived relations that may be implicit in a centrality index are made explicit and can be questioned separately from other assumptions.

Positional dominance serves to formulate restrictions on centrality rankings. By allowing for arbitrary ordered value ranges and for moving between different kinds of relations derived from an observed network it provides many degrees of freedom that can be used to adapt the formulation to specific domains and theories.

Positional domiance turns the question which centrality index to use around. Instead of selecting an index, and thus jumping to a conclusion, analysts hypothesize which features make a node central and test consistency of the partial rankings obtained with whatever the centrality was supposed to represent.

4.3 Homogeneity

The main limitation of positional dominance as defined above is the strictness of its requirements. Even with the possibility of defining positions in transformed networks it is still too close to the idealized situation of neighborhood inclusion. In any given network, observed or transformed, few neighborhoods will actually be comparable, so that the resulting partial rankings of positions are sparse. Therefore, they most often contain too little information about relative centrality.

This happens because, to occupy a superior position, a node j has to match every relationship x_{it} of i by an equally strong relationship with that very same third node t.

All standard centrality indices are invariant under automorphisms. Vertices receive higher scores if their relationships as a whole are superior, independent of the particular target vertices. This suggest to generalize positional dominance by allowing to match a relationship to one node with an equally strong relationship to another node.

As an example, we define the dominance relation

$$\textcircled{i} \preccurlyeq \textcircled{j} \quad \text{if} \quad \begin{cases} \text{there exists a permutation } \pi : \mathcal{A} \to \mathcal{A} \text{ such that} \\ x_{it} \le x_{j\pi(t)} \quad \text{for all } (i,t) \in \mathcal{D} \text{ with } x_{it} \ne 0 \end{cases}$$

so that any neighbor of i can be substituted by some neighbor of j. The level of homogeneity introduced can be controlled, for instance, by restricting the permutations that are allowed, or by generalizing them to arbitrary substitution mappings.

Allowing substitutions in the comparison of neighborhoods provides expressive means to generalize positional dominance and take into account domain-specific information and additional data. For instance, substitutions could be allowed only with neighbors that carry similar attributes or rank higher in some property.

4.4 Application

Centrality indices are often used as descriptives, where the centrality of actors is defined by the index, not uncovered. In other cases, centrality indices are

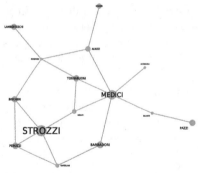

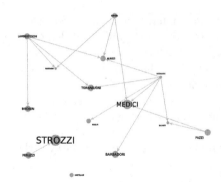

(a) Marriage network among families in 15th century Florence [28]. Node and label size indicate family wealth.

(b) Positional dominance based on adjacency (same as neighborhood inclusion). Edges points from the dominated to the dominating node.

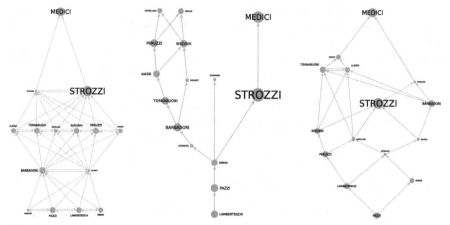

(c) Hasse diagram of positional dominance w.r.t. to adjacency and substitutable neighbors (hence same as degree equivalence). Vertical placement according to degree.

(d) Hasse diagram of positional dominance w.r.t. to adjacency and neighbors substitutable for wealthier ones. Vertical placement only to ensure upward edges.

(e) Hasse diagram of positional dominance w.r.t. to shortest-path distances and homogeneity. Vertical placement according to closeness centrality, which preserves the partial ranking since no edge points downward.

Fig. 4. Different positional dominance relations on the same network. One argument for the rise of the Medici family is that their marriage and business relationships put them in a superior position compared to the richer Strozzi family.

considered as a variable that, for instance, is entered in a regression or is correlated with another variable.

In the absence of theoretical reasons for choosing a specific index, it is difficult to know whether other (existing or even unknown) indices would yield preferable outcomes.

With positional dominance, the approach changes from the ground up. Starting with the most exclusive definition, and thus the sparsest ranking, requirements for dominance are relaxed by explicitly introducing further assumptions until the ranking is sufficiently dense to draw a conclusion, or no further assumptions creating additional dominance relationships are granted.

Figure 4 shows a small example in which a given network is represented by a simple undirected graph with an additional vertex attribute. Positional dominance with respect to adjacency yields few relationships, and by introducing homogeneity (full substitutability) we are overshooting: distinguishing between a marital tie or not (adjacency) and not distinguishing who it is with (homogeneity) leads to a coarse ranked partition. When we distinguish marriage partners by wealth so that neighbors may be substituted only for richer ones, we find, for example, that every family the Strozzi have a marriage relationship with can still be matched by an equally wealthy family married to the Medici. If, instead, we distinguish non-neighbors by their network distance, i.e., positions in a transformed network, the assumption of homogeneity does not lead to large indifference classes, because the relative position of two families where, say, one has more direct ties and the other has more at distance two remains undecided. Additional assumptions such as additivity and the inverse transfer function can resolve the remaining incomparabilities, but only if needed.

The number and combination of assumptions for which there is a theoretical justification can thus be controlled, unlike in a centrality index where adaptation is generally difficult because of side effects. Moreover, it becomes possible to prove statements such as bounds or impossibility results that apply not to one index but more generally to all indices that respect a given dominance relation on positions.

5   Computational Challenges

Replacing centrality indices by notions of positional dominance creates computational challenges in a broad range of topics.

It is worth noting upfront that the range $\mathcal{X}$ of network variables is not restricted to any particular type of values. The data associated with dyads may well consist of life-time intervals (e.g., in phone-call data), time-dependent functions (e.g., in traffic networks), differential equations (e.g., in biological networks), or various kinds of multivariate data (e.g., in economic networks). In many cases, the increase in complexity also necessitates different algorithmic techniques.

The following is just a selection of problem domains, many different and more detailed ones are readily found.

5.1   Transformations

If the network is transformed – which is rather the rule than the exception because relations of interest are often unobservable – the network usually becomes more dense. Transitive derived relations such as reachability, distance, connectivity, and so on are prone to create memory problems. Challenges such as vertex labeling and sketches are commonly studied for relations based on reachability or distance [33]. For many other derived relations, much less seems to be known about oracles, implicit representations, lazy evaluation, and other density mitigation strategies.

Each network transformation represents its own underlying mechanism, yet little is known about the actual differences between the outcomes, especially when determined on restricted graph classes. The dyadic dependency relation of betweenness centrality, for instance, also underlies a form of closeness and thus suggests an alternative generalization to weighted graphs [9].

In which generalizations of threshold graphs do certain families of derived relations yield the same ranking? What are the minimum conditions under which they start to differ?

5.2   Dominance

Determining the partial rankings of positional dominance is a computational challenge in itself. There are straightforward algorithms for positions based on adjacency [11] but can properties of derived relations and different degrees of homogeneity be exploited to obtain faster algorithms? What if only the transitive skeleton is sought?

As a form of sensitivity analysis, it would be interesting to study graph modification problems in which the utility of, say, an edge deletion or addition lies in the number of additional dominance relationships created. How far is a network from an ideal, fully ranked structure? Edit distance to a threshold graph is already intractable [15] but what about other distances such as edge rotation distance?

Our examples in Sect. 3, even under homogeneity assumptions, were focused on element-wise dominance. Alternatives include partial aggregations [23] and stochastic dominance. If the inequalities used to define dominance are violated only just so, approximate notions of dominance with tolerance thresholds may be useful, especially for uncertain data.

5.3   Rankings

Since partial rankings tend to be sparse at least initially, the space of possible completions (weak order extensions) is large. Which assumptions lead to a maximum size reduction? What are appropriate probability models and what is the significance of any particular ranking in that space? For a uniform random model, expected ranks of nodes [31] can be determined by enumeration [20], but are there also more efficient ways?

Problems of ranking similarity and rank aggregation are relevant for the comparison of partial rankings with centrality indices and other variables, as are bounds on the number of inversions when extending a partial ranking.

6   Conclusions

The unifying idea underlying centrality indices is that a node is more central if it has more, better relationships. This is evidenced in verbal descriptions [17] but also in typical axioms [2, 30] and, most importantly, the property of neighborhood inclusion shared by all standard centrality indices [32].

Our novel approach based on positional dominance takes the ideal but rare case of neighborhood inclusion as a starting point to systematically develop more and more general pairwise ranking criteria.

The study of centrality thus shifts from an index-based, one-shot approach to a sequence of smaller, more controllable steps: network transformation, positions, homogeneity, dominance, and, possibly, quantification. Since these steps are conceptually distinct, they allow for modular theorizing and testing.

Positions are a concept of independent interest because they encapsulate the relationships of nodes to the rest of the network, and may therefore enable generalization and restructuring of more forms of analysis such as roles (similar positions) and community structures (close positions).

The consequences for mathematical and computational researchers are two-fold. Firstly, relations between positions suggest the formal study of novel variants of network-analytic concepts without requiring familiarity with potential application domains. Secondly, they involve a more diverse selection of mathematical concepts and may therefore appeal to a broader audience.

References

1. Bloch, F., Jackson, M.O., Tebaldi, P.: Centrality measures in networks. SSRN Electron. J. (2019). https://doi.org/10.2139/ssrn.2749124
2. Boldi, P., Vigna, S.: Axioms for centrality. Internet Math. **10**(3–4), 222–262 (2014). https://doi.org/10.1080/15427951.2013.865686
3. Borgatti, S.P.: Centrality and network flow. Soc. Netw. **27**(1), 55–71 (2005). https://doi.org/10.1016/j.socnet.2004.11.008
4. Borgatti, S.P., Everett, M.G.: Notions of position in social network analysis. Sociol. Methodol. **22**, 1–35 (1992). https://doi.org/10.2307/270991
5. Borgatti, S.P., Everett, M.G.: Models of core/periphery structures. Soc. Netw. **21**(4), 375–395 (1999). https://doi.org/10.1016/S0378-8733(99)00019-2
6. Borgatti, S.P., Everett, M.G.: A graph-theoretic perspective on centrality. Soc. Netw. **28**(4), 466–484 (2006). https://doi.org/10.1016/j.socnet.2005.11.005
7. Borgatti, S.P., Everett, M.G., Johnson, J.C.: Analyzing Social Networks. Sage, Thousand Oaks (2013)
8. Brandes, U.: Network positions. Methodol. Innov. **9**, 1–19 (2016). https://doi.org/10.1177/2059799116630650

9. Brandes, U., Borgatti, S.P., Freeman, L.C.: Maintaining the duality of closeness and betweenness centrality. Soc. Netw. **44**, 153–159 (2016). https://doi.org/10.1016/j.socnet.2015.08.003

10. Brandes, U., Fleischer, D.: Centrality measures based on current flow. In: Diekert, V., Durand, B. (eds.) STACS 2005. LNCS, vol. 3404, pp. 533–544. Springer, Heidelberg (2005). https://doi.org/10.1007/978-3-540-31856-9_44

11. Brandes, U., Heine, M., Müller, J., Ortmann, M.: Positional dominance: concepts and algorithms. In: Gaur, D., Narayanaswamy, N.S. (eds.) CALDAM 2017. LNCS, vol. 10156, pp. 60–71. Springer, Cham (2017). https://doi.org/10.1007/978-3-319-53007-9_6

12. Brandes, U., Robins, G., McCranie, A., Wasserman, S.: What is network science? Netw. Sci. **1**(1), 1–15 (2013). https://doi.org/10.1017/nws.2013.2

13. Burt, R.S.: Positions in networks. Soc. Forces **55**, 93–122 (1976)

14. Doreian, P., Batagelj, V., Ferligoj, A.: Positional analysis of sociometric data. In: Carrington, P.J., Scott, J., Wasserman, S. (eds.) Models and Methods in Social Network Analysis, pp. 77–97. Cambridge University Press (2005)

15. Drange, P.G., Dregi, M.S., Lokshtanov, D., Sullivan, B.D.: On the threshold of intractability. In: Bansal, N., Finocchi, I. (eds.) ESA 2015. LNCS, vol. 9294, pp. 411–423. Springer, Heidelberg (2015). https://doi.org/10.1007/978-3-662-48350-3_35

16. Faust, K.: Comparison of methods for positional analysis: structural and general equivalences. Soc. Netw. **10**(4), 313–341 (1988)

17. Freeman, L.C.: Centrality in social networks: conceptual clarification. Soc. Netw. **1**(3), 215–239 (1979)

18. Freeman, L.C.: The development of social network analysis-with an emphasis on recent events. In: The SAGE Handbook of Social Network Analysis, pp. 26–54 (2011)

19. van der Grinten, A., Angriman, E., Meyerhenke, H.: Scaling up network centrality computations: a brief overview. it - Information Technology (2020). https://doi.org/10.1515/itit-2019-0032

20. Habib, M., Medina, R., Nourine, L., Steiner, G.: Efficient algorithms on distributive lattices. Discrete Appl. Math. **110**(2), 169–187 (2001). https://doi.org/10.1016/S0166-218X(00)00258-4

21. Hamilton, W.L., Ying, R., Leskovec, J.: Representation learning on graphs: methods and applications. Bull. IEEE Comput. Soc. Tech. Committee Data Eng. **40**(3), 52–74 (2017)

22. Hennig, M., Brandes, U., Pfeffer, J., Mergel, I.: Studying Social Networks - A Guide to Empirical Research. Campus, Frankfurt/New York (2012)

23. Kostreva, M.M., Ogryczak, W., Wierzbicki, A.: Equitable aggregations and multiple criteria analysis. Eur. J. Oper. Res. **158**(2), 362–377 (2004). https://doi.org/10.1016/j.ejor.2003.06.010

24. Lü, L., Chen, D., Ren, X.L., Zhang, Q.M., Zhang, Y.C., Zhou, T.: Vital nodes identification in complex networks. Phys. Rep. **650**, 1–6 (2016)

25. Mariani, M.S., Ren, Z.M., Bascompte, J., Tessone, C.J.: Nestedness in complex networks: observation, emergence, and implications. Phys. Rep. **813**, 1–90 (2019). https://doi.org/10.1016/j.physrep.2019.04.001

26. Menczer, F., Fortunato, S., Davis, C.A.: A First Course in Network Science. Cambridge University Press, Cambridge (2020)

27. Ortmann, M., Brandes, U.: Efficient orbit-aware triad and quad census in directed and undirected graphs. Appl. Netw. Sci. **2**(1), 1–17 (2017). https://doi.org/10.1007/s41109-017-0027-2

28. Padgett, J.F., Ansell, C.K.: Robust action and the rise of the Medici, 1400–1434. Am. J. Sociol. **98**(6), 1259–1319 (1993)
29. Robins, G.: Doing Social Network Research. Sage, Thousand Oaks (2015)
30. Sabidussi, G.: The centrality index of a graph. Psychometrika **31**(4), 581–603 (1966). https://doi.org/10.1007/BF02289527
31. Schoch, D.: Centrality without indices: partial rankings and rank probabilities in networks. Soc. Netw. **54**, 50–60 (2018). https://doi.org/10.1016/j.socnet.2017.12.003
32. Schoch, D., Brandes, U.: Re-conceptualizing centrality in social networks. Eur. J. Appl. Math. **27**(6), 971–985 (2016). https://doi.org/10.1017/S0956792516000401
33. Sommer, C.: Shortest-path queries in static networks. ACM Comput. Surv. **46**(4), 45:1–45:31 (2014). https://doi.org/10.1145/2530531
34. Wasserman, S., Faust, K.: Social Network Analysis. Methods and Applications. Cambridge University Press, Cambridge (1994)
35. White, H.C., Boorman, S.A., Breiger, R.L.: Social structure from multiple networks: I. Blockmodels of roles and positions. Am. J. Sociol. **81**(4), 730–780 (1976)

Second-Order Finite Automata

Alexsander Andrade de Melo[1] and Mateus de Oliveira Oliveira[2(✉)]

[1] Federal University of Rio de Janeiro, Rio de Janeiro, Brazil
aamelo@cos.ufrj.br
[2] University of Bergen, Bergen, Norway
mateus.oliveira@uib.no

Abstract. Traditionally, finite automata theory has been used as a framework for the representation of possibly infinite sets of strings. In this work, we introduce the notion of second-order finite automata, a formalism that combines finite automata with ordered decision diagrams, with the aim of representing possibly infinite *sets of sets* of strings. Our main result states that second-order finite automata can be canonized with respect to the second-order language they represent. Using this canonization result, we show that sets of sets of strings represented by second-order finite automata are closed under the usual Boolean operations, such as union, intersection, difference and even under a suitable notion of complementation. Additionally, emptiness of intersection and inclusion are decidable.

We provide two algorithmic applications for second-order automata. First, we show that they can be used to show that several width and size minimization problems for deterministic and nondeterministic ODDs are solvable in fixed-parameter tractable time when parameterized by the width of the input ODD. In particular, our results imply FPT algorithms for corresponding width and size minimization problems for ordered binary decision diagrams with a fixed variable ordering. The previous best algorithms for these problems were exponential on the size of the input ODD even for ODDs of constant width. Second, we show that for each k and w one can count the number of distinct functions accepted by ODDs of width w and length k in time $h_\Sigma(w) \cdot k^{O(1)}$. This improves exponentially on the time necessary to explicitly enumerate all distinct functions, which take time exponential in both the width parameter w and in the length k of the ODDs.

1 Introduction

In its most traditional setting, automata theory has been used as a framework for the representation and manipulation of (possibly infinite) sets of strings.

Alexsander Andrade de Melo acknowledges support from the Brazilian agencies CNPq/GD 140399/2017-8 and CAPES/PDSE 88881.187636/2018-01. Mateus de Oliveira Oliveira acknowledges support from the Bergen Research Foundation and from the Research Council of Norway (Grant Nr. 288761).

H. Fernau (Ed.): CSR 2020, LNCS 12159, pp. 46–63, 2020.
https://doi.org/10.1007/978-3-030-50026-9_4

This framework has been generalized in many ways to allow the representation of sets of more elaborate combinatorial objects, such as trees [5], partial orders [15], graphs [4], pictures [11], etc. Such notions of automata have encountered innumerous applications in fields such as formal verification [3,12], finite model theory [8], concurrency theory [14], parameterized complexity [6,7], etc. Still, these generalized notions of automata share in common the fact that they are designed to represent (possibly infinite) sets of isolated objects.

In this work, we combine traditional finite automata with ordered decision diagrams (ODDs) to introduce a formalism that can be used to represent and manipulate *sets of sets* of strings. We call this combined formalism *second-order automata*. Intuitively, a (Σ, w)-ODD is a sequence $D = B_1 B_2 \ldots B_k$ of (Σ, w)-layers. Each such layer B_i has a set of left-states (a subset of $\{1, \ldots, w\}$), a set of right-states (also a subset of $\{1, \ldots, w\}$), and a set of transitions, labeled with letters in Σ, connecting left states to right states. We require that for each $i \in \{1, \ldots, k-1\}$, the set of right-states of the layer B_i is equal to the set of left states of the layer B_{i+1}. The language of such an ODD D is the set of strings labeling paths from its set of initial states (a subset of the left states of B_1) to its final states (a subset of the right states of B_k). Since the number of distinct (Σ, w)-layers is finite, the set $\mathcal{B}(\Sigma, w)$ of all such layers can itself be regarded as an alphabet. A finite automaton $\mathcal{F}$ over the alphabet $\mathcal{B}(\Sigma, w)$ is said to be a second-order finite automaton if each string $D = B_1 \ldots B_k$ in the language $\mathcal{L}(\mathcal{F})$ of $\mathcal{F}$ is a valid ODD. In this case, the second language of $\mathcal{F}$ is the set of all sets of strings accepted by ODDs in $\mathcal{L}(\mathcal{F})$. We denote the second language of $\mathcal{F}$ by $\mathcal{L}_2(\mathcal{F})$.

Our main result (Theorem 3) states second-order finite automata can be effectively canonized with respect to their second-language. In other words, there is a computable function that sends each second-order finite automaton $\mathcal{F}$ to a second-order automaton $\mathcal{C}_2(\mathcal{F})$ in such a way that they have identical second languages $(\mathcal{L}_2(\mathcal{C}_2(\mathcal{F})) = \mathcal{L}_2(\mathcal{F}))$ and in such a way that any two second-order finite automata $\mathcal{F}$ and $\mathcal{F}'$ with the same second language are sent to the same canonical form $(\mathcal{L}_2(\mathcal{F}) = \mathcal{L}_2(\mathcal{F}') \Rightarrow \mathcal{C}_2(\mathcal{F}) = \mathcal{C}_2(\mathcal{F}'))$. The difficulty in proving this canonization result stems from the fact that many ODDs in the language $\mathcal{L}(\mathcal{F})$ of $\mathcal{F}$ may correspond to *the same* set of strings in the second language $\mathcal{L}_2(\mathcal{F})$. Therefore, obtaining a canonical form for $\mathcal{F}$ with respect to the language $\mathcal{L}_2(\mathcal{F})$ is not equivalent to the problem of obtaining a canonical form for $\mathcal{F}$ with respect to the language $\mathcal{L}(\mathcal{F})$. Indeed, there exist pairs of second-order finite automata $\mathcal{F}$ and $\mathcal{F}'$ with distinct (first) languages $(\mathcal{L}(\mathcal{F}) \neq \mathcal{L}(\mathcal{F}'))$ but with identical second languages $(\mathcal{L}_2(\mathcal{F}) = \mathcal{L}_2(\mathcal{F}))$. We will circumvent this issue by showing that the function $\mathtt{can}[\Sigma, w]$ that sends each (Σ, w)-ODD D to its canonical form $\mathcal{C}(D)$ can be cast as a regularity-preserving transduction. In particular, when applying this transduction to a (Σ, w)-second-order finite automaton $\mathcal{F}$, we obtain a $(\Sigma, 2^w)$-second-order automaton which accepts precisely those canonical forms of ODDs in $\mathcal{L}(\mathcal{F})$. Once this intermediate automaton has been constructed, we can safely use standard determinization and minimization techniques to construct an automaton $\mathcal{C}_2(\mathcal{F})$ which is a canonical form for $\mathcal{F}$ with respect to its second-language.

As a consequence of our canonization result, we have that second-order finite automata have several nice closure properties. In particular, the class of *sets of sets* of strings that can be represented by second-order finite automata is closed under union, intersection, set difference, and even under a suitable notion of complementation. Furthermore, emptiness of intersection and inclusion for the second language of second-order finite automata are decidable.

We provide two algorithmic applications for second-order finite automata. First, we show that several width and size minimization problems for deterministic and non-deterministic ODDs are solvable in fixed parameter tractable time when parameterized by the width of the input ODD. Although ODDs of constant width constitute a simple computational model, they can already be used to represent many interesting functions. It is worth noting that for each width $w \geq 3$, the class of functions that can be represented by ODDs of constant width is at least as difficult to learn in the PAC-learning model as the problem of learning DNFs [9]. Additionally, the study of ODDs of constant width is still very active in the theory of pseudo-random generators [10]. We also note that our FPT algorithm for ODD width minimization parameterized by the width of the input ODD implies that analogous minimization problems can be solved in FPT time on ordered binary decision diagrams (OBDDs) with a *fixed variable ordering* when parameterized by the width of the input OBDD. Such problems have been shown to be NP-hard for OBDDs of unrestricted width [1,2], but no fixed parameter tractable algorithm was known.

As a second application, we show that the problem of counting the number of distinct functions computable by ODDs of a given width w and a given length k can be solved in time $h_\Sigma(w) \cdot k^{O(1)}$, for a suitable function $h_\Sigma(w)$. This improves exponentially on the time necessary to explicitly enumerate all distinct functions computable by such ODDs, since this enumeration process takes time exponential in both the width parameter w and in the length k of the ODDs.

The reminder of this paper is organized as follows. Next, in Sect. 2, we state basic definitions and results concerning finite automata and ordered decision diagrams. Subsequently, in Sect. 3, we formally define the notion of second-order finite automata and state our main result. In Sect. 4, we recall some basic definitions about transductions. In Sect. 5, we show that the process of canonizing ODDs can be cast as regular transductions, and subsequently prove our main result. In Sect. 6, we show how second-order finite automata can be used to provide fixed parameter tractable algorithms for several variants of width and size minimization problems involving ODDs, and show how to count the number of functions computable by ODDs of a given width in FPT time. Finally, in Sect. 7, we conclude the paper with some final considerations.

2 Preliminaries

2.1 Basics

We denote by $\mathbb{N} \doteq \{0, 1, \ldots\}$ the set of natural numbers (including zero), and by $\mathbb{N}_+ \doteq \mathbb{N} \setminus \{0\}$ the set of positive natural numbers. For each $c \in \mathbb{N}_+$, we let $[c] \doteq \{1, 2, \ldots, c\}$ and $[\![c]\!] \doteq \{0, 1, \ldots, c-1\}$. Let X be a finite set. We denote by $\mathcal{P}(X) \doteq \{X' \colon X' \subseteq X\}$ the *power set* of X. A *permutation* of X is a bijection $\pi \colon X \to X$ from X onto itself. We denote by $\mathbb{S}_X$ the set of all permutations of X.

Alphabets and Strings. An *alphabet* is any finite, non-empty set Σ. A *string* over an alphabet Σ is any finite sequence of symbols from Σ. The *empty string*, denoted by λ, is the unique string of length zero. We denote by Σ^* the set of all strings over Σ, including the empty string λ, and by $\Sigma^+ \doteq \Sigma^* \setminus \{\lambda\}$ the set of all non-empty strings over Σ. A *language* over Σ is any subset L of Σ^*. In particular, for each $k \in \mathbb{N}_+$, we let Σ^k be the language of all strings of length k over Σ. We say that an alphabet Σ is ordered if it is endowed with a total order $<_\Sigma \colon \Sigma \times \Sigma$. Unless stated otherwise, we assume that each alphabet considered in this paper is endowed with a fixed ordering. Such an ordering is extended naturally to a lexicographical ordering $\prec_\Sigma \subseteq \Sigma^* \times \Sigma^*$ on the set Σ^*.

Finite Automata. A *finite automaton* (FA) over an alphabet Σ is a tuple $\mathcal{F} = (Q, T, I, F)$, where Q is a finite set of *states*, $T \subseteq Q \times \Sigma \times Q$ is a set of *transitions*, $I \subseteq Q$ is a set of *initial states* and $F \subseteq Q$ is a set of *final states*. The *size* of $\mathcal{F}$ is defined as $|\mathcal{F}| \doteq |Q| + |T| \cdot \log |\Sigma|$. We denote the number of states of $\mathcal{F}$ by $\mathrm{nSt}(\mathcal{F})$, and the number of transitions by $\mathrm{nTr}(\mathcal{F})$.

Let $s \in \Sigma^*$, and $q, q' \in Q$. We say that s reaches q' from q if either $s = \lambda$ and $q = q'$, or if $s = \sigma_1 \ldots \sigma_k$ for some $k \in \mathbb{N}_+$ and there is a sequence

$$\langle (q_0, \sigma_1, q_1), (q_1, \sigma_2, q_2), \ldots, (q_{k-1}, \sigma_k, q_k) \rangle,$$

of transitions such that $q_0 = q$ and $q_k = q'$. We say that $\mathcal{F}$ accepts s if there exist states $q \in I$ and $q' \in F$ such that s reaches q' from q.

The language of $\mathcal{F}$ is defined as the set

$$\mathcal{L}(\mathcal{F}) \doteq \{s \in \Sigma^* \colon s \text{ is accepted by } \mathcal{F}\}$$

of all finite strings over Σ accepted by $\mathcal{F}$. For $\alpha \in \mathbb{N}$, we say that a language $L \subseteq \Sigma^*$ is α-*regular* if there exists a (possibly nondeterministic) finite automaton with at most α states such that $\mathcal{L}(\mathcal{F}) = L$.

We say that $\mathcal{F}$ is *deterministic* if $\mathcal{F}$ contains exactly one initial state, i.e. $|I| = 1$, and for each $q \in Q$ and each $\sigma \in \Sigma$, there exists at most one state $q' \in Q$ such that (q, σ, q') is a transition in T. We say that $\mathcal{F}$ is *complete* if for each $q \in Q$ and each $\sigma \in \Sigma$, there exists at least one state $q' \in Q$ such that (q, σ, q') is a transition in T. For each state $q \in Q$, we let $\mathrm{lex}(q)$ be the lexicographically first string which reaches q from some initial state, according to the order $\prec_\Sigma$. We say that $\mathcal{F}$ is normalized if $Q = \{1, \ldots, r\}$ for some $r \in \mathbb{N}$, and for each $q, q' \in Q$, $q < q'$ if and only if $\mathsf{lex}(q) \prec_\Sigma \mathsf{lex}(q')$. In what follows, we may write $Q(\mathcal{F})$, $T(\mathcal{F})$, $I(\mathcal{F})$ and $F(\mathcal{F})$ to refer to the sets Q, T, I and F, respectively.

The following theorem, stating the existence of canonical forms for finite automata, is one of the most fundamental results in automata theory.

Theorem 1. *Let Σ be an alphabet. For each finite automaton $\mathcal{F}$ over Σ, there is a unique finite automaton $\mathcal{C}(\mathcal{F})$ with minimum number of states with the property that $\mathcal{C}(\mathcal{F})$ is deterministic, complete, normalized, and satisfies $\mathcal{L}(\mathcal{C}(\mathcal{F})) = \mathcal{L}(\mathcal{F})$.*

We note that given a (possibly non-deterministic) finite automaton $\mathcal{F}$ over an alphabet Σ, the canonical form $\mathcal{C}(\mathcal{F})$ of $\mathcal{F}$ can be obtained by the following process. First, one applies Rabin's power-set construction to $\mathcal{F}$ in order to obtain a *deterministic, complete* finite automaton $\mathcal{F}'$ accepting the same language as $\mathcal{F}$. Subsequently, one minimizes $\mathcal{F}'$, using for instance Hopcroft's algorithm [13], in order to obtain a deterministic finite automaton $\mathcal{F}''$ with minimum number of states accepting the same language as $\mathcal{F}$. At this point the automaton $\mathcal{F}''$ is unique *up to renaming of states*. Finally, as a last step, one obtains the canonical form $\mathcal{C}(\mathcal{F})$ by renaming the states of $\mathcal{F}''$ in such a way that the normalization property is satisfied. Note that at this point, the automaton $\mathcal{C}(\mathcal{F})$ is syntactically unique. In particular, for each two finite automata $\mathcal{F}$ and $\mathcal{F}'$, $\mathcal{L}(\mathcal{F}) = \mathcal{L}(\mathcal{F}')$ if and only if $\mathcal{C}(\mathcal{F}) = \mathcal{C}(\mathcal{F}')$.

2.2   Ordered Decision Diagrams

Layers. Let Σ be an alphabet and $w \in \mathbb{N}_+$. A (Σ, w)-*layer* is a tuple $B \doteq (\ell, r, T, I, F, \iota, \phi)$, where $\ell \subseteq \llbracket w \rrbracket$ is a set of *left states*, $r \subseteq \llbracket w \rrbracket$ is a set of *right states*, $T \subseteq \ell \times \Sigma \times r$ is a set of *transitions*, $I \subseteq \ell$ is a set of *initial states*, $F \subseteq r$ is a set of *final states* and $\iota, \phi \in \{0, 1\}$ are Boolean flags satisfying the following two conditions:

1. if $\iota = 0$ then $I = \emptyset$;
2. if $\phi = 0$ then $F = \emptyset$.

In what follows, we may write $\ell(B)$, $r(B)$, $T(B)$, $I(B)$, $F(B)$, $\iota(B)$ and $\phi(B)$ to refer to the sets ℓ, r, T, I and F and to the Boolean flags ι and ϕ, respectively.

We let $\mathcal{B}(\Sigma, w)$ denote the set of all (Σ, w)-layers. Note that, $\mathcal{B}(\Sigma, w)$ is non-empty and has at most $2^{\mathcal{O}(|\Sigma| \cdot w^2)}$ elements. Therefore, $\mathcal{B}(\Sigma, w)$ may be regarded as an alphabet.

Ordered Decision Diagrams. Let $k \in \mathbb{N}_+$. A (Σ, w) *ordered decision diagram* (or simply, (Σ, w)-*ODD*) of *length* k is a string $D \doteq B_1 \cdots B_k \in \mathcal{B}(\Sigma, w)^k$ of length k over the alphabet $\mathcal{B}(\Sigma, w)$ satisfying the following conditions:

1. for each $i \in [k - 1]$, $\ell(B_{i+1}) = r(B_i)$;

2. $\iota(B_1) = 1$ and, for each $i \in \{2, \ldots, k\}$, $\iota(B_i) = 0$;
3. $\phi(B_k) = 1$ and, for each $i \in [k - 1]$, $\phi(B_i) = 0$.

Intuitively, Condition 1 expresses that the set of right states of B_i can be identified with the set of left states of B_{i+1}. Condition 2 guarantees that only the first layer of an ODD is allowed to have initial states. Analogously, Condition 3 guarantees that only the last layer of an ODD is allowed to have final states.

For each $k \in \mathbb{N}_+$, we denote by $\mathcal{B}(\Sigma, w)^{\circ k}$ the set of all (Σ, w)-ODDs of length k. We denote by

$$\mathcal{B}(\Sigma, w)^{\circledast} \doteq \bigcup_{k \in \mathbb{N}_+} \mathcal{B}(\Sigma, w)^{\circ k}$$

the set of all (Σ, w)-ODDs. Given an ODD $D \in \mathcal{B}(\Sigma, w)^{\circledast}$, we let $\text{len}(D)$ denote the length of D.

The *width* of an ODD $D = B_1 \cdots B_k \in \mathcal{B}(\Sigma, w)^{\circ k}$ is defined as

$$\mathrm{w}(D) \doteq \max\{|\ell(B_1)|, \ldots, |\ell(B_k)|, |r(B_k)|\}.$$

We remark that $\mathrm{w}(D) \leq w$.

Length Typed Subsets of Σ^k. Let Σ be an alphabet, and $k \in \mathbb{N}_+$. In this work, it will be convenient assume that subsets of Σ^k are typed with their length. This can be achieved by viewing a subset $L \subseteq \Sigma^k$ as a pair of the form (k, L). We let $\mathcal{P}_k(\Sigma^k) = \{(k, L) : L \subseteq \Sigma^k\}$.

Language Accepted by an ODD. Let $k \in \mathbb{N}_+$, $D = B_1 \cdots B_k$ be an ODD in $\mathcal{B}(\Sigma, w)^{\circ k}$ and $s = \sigma_1 \cdots \sigma_k$ be a string in Σ^k. A *valid sequence* for s in D is a sequence of transitions $\langle (\mathfrak{p}_1, \sigma_1, \mathfrak{q}_1), \ldots, (\mathfrak{p}_k, \sigma_k, \mathfrak{q}_k) \rangle$ such that $\mathfrak{p}_{i+1} = \mathfrak{q}_i$ for each $i \in [k-1]$, and $(\mathfrak{p}_i, \sigma_i, \mathfrak{q}_i) \in T(B_i)$ for each $i \in [k]$. Such a valid sequence is called *accepting* for s if, additionally, $\mathfrak{p}_1$ is an initial state in $I(B_1)$ and $\mathfrak{q}_k$ is a final state in $F(B_k)$. We say that D *accepts* s if there exists an accepting sequence for s in D. The *language* of D, denoted by $\mathcal{L}(D)$, is defined as the (length-typed) set of strings accepted by D, i.e., $\mathcal{L}(D) \doteq (k, \{s \in \Sigma^k : s \text{ is accepted by } D\})$.

Deterministic and Complete ODDs. A (Σ, w)-layer B is called *deterministic* if the following conditions are satisfied:

1. for each $\mathfrak{p} \in \ell(B)$ and each $\sigma \in \Sigma$, there exists at most one right state $\mathfrak{q} \in r(B)$ such that $(\mathfrak{p}, \sigma, \mathfrak{q}) \in T(B)$;
2. if $\iota(B) = 1$, then $I(B) = \ell(B)$ and $|\ell(B)| = 1$.

A (Σ, w)-layer B is called *complete* if, for each $\mathfrak{p} \in \ell(B)$ and each $\sigma \in \Sigma$, there exists at least one right state $\mathfrak{q} \in r(B)$ such that $(\mathfrak{p}, \sigma, \mathfrak{q}) \in T(B)$. We let $\widehat{\mathcal{B}}(\Sigma, w)$ be the subset of $\mathcal{B}(\Sigma, w)$ comprising all deterministic, complete (Σ, w)-layers.

An ODD $D = B_1 \cdots B_k \in \mathcal{B}(\Sigma, w)^{\circ k}$ is called *deterministic* (*complete*, resp.) if, for each $i \in [k]$, B_i is deterministic (complete, resp.). We remark that if D is deterministic, then there exists at most one valid sequence in D for each string in Σ^k. On the other hand, if D is complete, then there exists at least one valid sequence in D for each string in Σ^k. For each $k \in \mathbb{N}_+$, we let $\widehat{\mathcal{B}}(\Sigma, w)^{\circ k}$ be the subset of $\mathcal{B}(\Sigma, w)^{\circ k}$ comprising all deterministic, complete (Σ, w)-ODDs of length k.

Isomorphism of ODDs. Let $D = B_1 \cdots B_k$ and $D' = B'_1 \cdots B'_k$ be two ODDs in $\mathcal{B}(\Sigma, w)^{\circ k}$. An *isomorphism from* D *to* D' is a sequence $\overline{\pi} \doteq \langle \pi_0, \ldots, \pi_k \rangle$ of functions that satisfy the following conditions:

1. $\pi_0 \colon \ell(B_0) \to \ell(B'_0)$ is a bijection from $\ell(B_0)$ to $\ell(B'_0)$;
2. $\pi_0|_{I(B_0)}$ is a bijection from $I(B_0)$ to $I(B'_0)$;
3. for each $i \in [k]$, $\pi_i \colon r(B_i) \to r(B'_i)$ is a bijection from $r(B_i)$ to $r(B'_i)$;
4. $\pi_k|_{F(B_k)}$ is a bijection from $F(B_k)$ to $F(B'_k)$;
5. for each $i \in [k]$, each left state $\mathfrak{p} \in \ell(B_i)$, each symbol $\sigma \in \Sigma$ and each right state $\mathfrak{q} \in r(B_i)$, $(\mathfrak{p}, \sigma, \mathfrak{q}) \in T(B_i)$ if and only if $(\pi_{i-1}(\mathfrak{p}), \sigma, \pi_i(\mathfrak{q})) \in T(B'_i)$.

We note that if $\overline{\pi} = \langle \pi_0, \ldots, \pi_k \rangle$ is an isomorphism from D to D', then $\overline{\pi}^{-1} \doteq \langle \pi_0^{-1}, \ldots, \pi_k^{-1} \rangle$ is an isomorphism from D' to D, where π_i^{-1} denotes the inverse of π_i for each $i \in [\![k+1]\!]$. We say that D and D' are *isomorphic* if there exists an isomorphism $\overline{\pi}$ between D and D'. The following proposition is immediate.

Proposition 1. *If D and D' are isomorphic ODDs, then $\mathcal{L}(D) = \mathcal{L}(D')$.*

Normalized ODDs. Let B be a (Σ, w)-layer. We say that B is *reachable* if, for each $\mathfrak{q} \in r(B)$, there exist $\sigma \in \Sigma$ and $\mathfrak{p} \in \ell(B)$ such that $(\mathfrak{p}\, \sigma, \mathfrak{q})$ is a transition in $T(B)$. If B is reachable, then we let $\chi_B \colon r(B) \to \ell(B) \times \Sigma$ be the map such that, for each $\mathfrak{q} \in r(B)$, $\chi_B(\mathfrak{q}) \doteq \min\{(\mathfrak{p}, \sigma) \colon (\mathfrak{p}, \sigma, \mathfrak{q}) \in T(B)\}$, where the minimum is taken lexicographically, i.e., for each $\mathfrak{p}, \mathfrak{p}' \in \ell(B)$ and each $\sigma, \tau \in \Sigma$, we have that $(\mathfrak{p}, \sigma) < (\mathfrak{p}', \tau)$ if and only if $\mathfrak{p} < \mathfrak{p}'$, or $\mathfrak{p} = \mathfrak{p}'$ and $\sigma <_\Sigma \tau$. (Recall we are assuming that the alphabet Σ is associated with a fixed total order $<_\Sigma \subseteq \Sigma \times \Sigma$.) We say that B is *well-ordered* if it is a reachable, deterministic layer and, for each $\mathfrak{q}, \mathfrak{q}' \in r(B)$, $\mathfrak{q} < \mathfrak{q}'$ if and only if $\chi_B(\mathfrak{q}) < \chi_B(\mathfrak{q}')$. We say that B is *contiguous* if $\ell(B) = [\![w_1]\!]$ and $r(B) = [\![w_2]\!]$ for some $w_1, w_2 \in [\![w]\!]$. (Σ, w)-layer. We say that B is *normalized* if it is both well-ordered and contiguous.

We say that an ODD $D = B_1 \cdots B_k \in \mathcal{B}(\Sigma, w)^{\circ k}$ is *well-ordered/contiguous/normalized* for each $i \in [k]$, B_i is a well-ordered/contiguous/normalized layer. Note that an ODD D is normalized if and only if it is both well-ordered and contiguous. The following theorem is the analog of Theorem 1 in the realm of the theory of ordered decision diagrams.

Theorem 2 (Folklore). *Let D be an ODD in $\mathcal{B}(\Sigma, w)^{\circ k}$. There exists a unique ODD $\mathcal{C}(D) \in \mathcal{B}(\Sigma, 2^w)^{\circ k}$ with minimum number of states with the property that $\mathcal{C}(D)$ is deterministic, complete, normalized and satisfies $\mathcal{L}(\mathcal{C}(D)) = \mathcal{L}(D)$.*

For each ODD $D \in \mathcal{B}(\Sigma, w)^{\circledast}$, we say that the ODD $\mathcal{C}(D)$ specified in Theorem 2 is the *canonical form* of D.

3   Second-Order Finite Automata

In this section, we formally define the main object of study of this work, namely, the notion of second-order finite automata.

Definition 1 (Second-Order Finite Automata). *Let Σ be an alphabet and $w \in \mathbb{N}_+$. A finite automaton $\mathcal{F}$ over the alphabet $\mathcal{B}(\Sigma, w)$ is called a (Σ, w)-second-order finite automaton if*

$$\mathcal{L}(\mathcal{F}) \subseteq \bigcup_{k \in \mathbb{N}_+} \mathcal{B}(\Sigma, w)^{\circ k}.$$

In words, a (Σ, w)-second-order finite automaton $\mathcal{F}$ is a finite automaton over the alphabet $\mathcal{B}(\Sigma, w)$ where each string $D = B_1 \ldots B_k$ in $\mathcal{L}(\mathcal{F})$ is a (Σ, w)-ODD. From now on we may refer to $\mathcal{L}(\mathcal{F})$ as the *first language* of $\mathcal{F}$. Since each string $D \in \mathcal{L}(\mathcal{F})$ is a (Σ, w)-ODD, we can also associated with $\mathcal{F}$ a *second language*, $\mathcal{L}_2(\mathcal{F})$, which is the *set of sets* of strings over Σ accepted by ODDs in $\mathcal{L}(\mathcal{F})$.

$$\mathcal{L}_2(\mathcal{F}) \doteq \{\mathcal{L}(D) \colon D \in \mathcal{L}(\mathcal{F})\}.$$

Note that $\mathcal{L}_2(\mathcal{F})$ is a (possibly infinite) subset of $\bigcup_{k \in \mathbb{N}_+} \mathcal{P}_k(\Sigma^k)$.

The main result of this work (Theorem 3) states that (Σ, w)-second-order automata can be canonized with respect to their *second* languages.

Theorem 3. *Let Σ be an alphabet (endowed with a total ordering $<_\Sigma \subset \Sigma \times \Sigma$), $w \in \mathbb{N}_+$ and $\mathcal{F}$ be a (Σ, w)-second-order finite automaton. One can construct in time $2^{\mathrm{nSt}(\mathcal{F}) \cdot 2^{2^{O(|\Sigma| \cdot w)}}}$ a complete, deterministic, normalized $(\Sigma, 2^w)$-second-order finite automaton $\mathcal{C}_2(\mathcal{F})$ satisfying the following properties.*

1. *$\mathcal{L}(\mathcal{C}_2(\mathcal{F})) = \{\mathcal{C}(D) \colon D \in \mathcal{L}(\mathcal{F})\}$;*
2. *$\mathcal{L}_2(\mathcal{C}_2(\mathcal{F})) = \mathcal{L}_2(\mathcal{F})$;*
3. *For each $w' \in \mathbb{N}_+$ and each (Σ, w')-second-order finite automaton $\mathcal{F}'$, if $\mathcal{L}_2(\mathcal{F}') = \mathcal{L}_2(\mathcal{F})$, then $\mathcal{C}_2(\mathcal{F}') = \mathcal{C}_2(\mathcal{F})$.*

Recall that for each ODD $D \in \mathcal{B}(\Sigma, w)^{\circ k}$, $\mathcal{C}(D) \in \mathcal{B}(\Sigma, 2^w)^{\circ k}$ denotes the unique ODD in with minimum number of states with the property that $\mathcal{C}(D)$ is deterministic, complete, normalized and satisfies $\mathcal{L}(\mathcal{C}(D)) = \mathcal{L}(D)$. Therefore, Property 1 states that the language of the second-order automaton $\mathcal{C}_2(\mathcal{F})$ is formed precisely by the canonical forms of ODDs in $\mathcal{L}(\mathcal{F})$. Property 2 states that $\mathcal{F}$ and $\mathcal{C}_2(\mathcal{F})$ have the same second language. In other words, these two finite automata represent the same *set of sets* of strings over Σ. This property is a direct consequence of Property 1, since each ODD D and its canonical form $\mathcal{C}(D)$ have the same language $\mathcal{L}(D) = \mathcal{L}(\mathcal{C}(D))$. Finally, Condition 3 states that any two second-order automata with identical second languages are mapped to the same canonical form. This is also a direct consequence of Property 1, since any two ODDs D and D' with the same language $\mathcal{L}(D) = \mathcal{L}(D')$ are mapped to the same canonical form $\mathcal{C}(D) = \mathcal{C}(D')$.

It is worth calling attention to the fact that even though $\mathcal{F}$ and $\mathcal{C}_2(\mathcal{F})$ have the same *second* language, i.e., $\mathcal{L}_2(\mathcal{C}_2(\mathcal{F})) = \mathcal{L}_2(\mathcal{F})$, the *first* languages of $\mathcal{F}$ and $\mathcal{C}_2(\mathcal{F})$ may be distinct, i.e., it may be the case that $\mathcal{L}(\mathcal{C}_2(\mathcal{F})) \neq \mathcal{L}(\mathcal{F})$. Additionally, given a subset $\mathcal{S} \subseteq \bigcup_{k \in \mathbb{N}_+} \mathcal{P}_k(\Sigma^k)$, there may exist infinitely many second-order finite automata with distinct first languages, but whose second language is equal to $\mathcal{S}$. Therefore, canonization of a finite automaton $\mathcal{F}$ with respect to its second language $\mathcal{L}_2(\mathcal{F})$ cannot be achieved by simply canonizing $\mathcal{F}$ with respect to its first language $\mathcal{L}(\mathcal{F})$ according to Theorem 1. To circumvent this issue, we will proceed in two steps. First, we will show that for each second-order finite automaton $\mathcal{F}$ one can construct a finite automaton $\mathcal{F}'$ such that $\mathcal{L}(\mathcal{F}') = \{\mathcal{C}(D) : D \in \mathcal{L}(\mathcal{F})\}$. The automaton $\mathcal{C}_2(\mathcal{F})$ can then be defined as $\mathcal{C}(\mathcal{F}')$. The next two sections will be dedicated to the construction of the intermediate finite automaton $\mathcal{F}'$. First, in Sect. 4 we will define a very well-behaved notion of regularity-preserving transduction. Subsequently, in Sect. 5 we will show that the process of constructing a canonical form for an ODD D can be simulated by the application of a sequence of regularity-preserving transductions.

Theorem 3 implies that subsets of $\bigcup_{k \in \mathbb{N}_+} \mathcal{P}_k(\Sigma, w)$ that can be represented by second-order automata are closed under Boolean operations and even under a suitable notion of complementation. Let Σ be an alphabet and $w \in \mathbb{N}_+$. We let $\text{Det}(\Sigma, w) = \{\mathcal{L}(D) : D \in \widehat{\mathcal{B}}(\Sigma, w)^\circledast\}$ be the set of all languages accepted by some deterministic, complete ODD of width w over the alphabet Σ. Given a subset $\mathcal{S} \subseteq \bigcup_{k \in \mathbb{N}_+} \mathcal{P}_k(\Sigma^k)$ we let $\overline{\mathcal{S}}^w \doteq \text{Det}(\Sigma, w) \backslash \mathcal{S}$ be the *width-w complement* of $\mathcal{S}$.

Theorem 4. *Let Σ be an alphabet and $w \in \mathbb{N}$. Let $\mathcal{F}$, $\mathcal{F}_1$ and $\mathcal{F}_2$ be (Σ, w)-second-order finite automata.*

1. *There is a $(\Sigma, 2^w)$-second-order automaton $\mathcal{F}_1 \cap \mathcal{F}_2$ such that*

$$\mathcal{L}_2(\mathcal{F}_1 \cap \mathcal{F}_2) = \mathcal{L}_2(\mathcal{F}_1) \cap \mathcal{L}_2(\mathcal{F}_1).$$

2. *There is a $(\Sigma, 2^w)$-second-order automaton $\mathcal{F}_1 \cup \mathcal{F}_2$ such that*

$$\mathcal{L}_2(\mathcal{F}_1 \cap \mathcal{F}_2) = \mathcal{L}_2(\mathcal{F}_1) \cup \mathcal{L}_2(\mathcal{F}_1).$$

3. *There is a $(\Sigma, 2^w)$-second-order automaton $\mathcal{F}_1 \backslash \mathcal{F}_2$ such that*

$$\mathcal{L}_2(\mathcal{F}_1 \backslash \mathcal{F}_2) = \mathcal{L}_2(\mathcal{F}_1) \backslash \mathcal{L}_2(\mathcal{F}_1).$$

4. *There is a (Σ, w)-second-order automaton $\mathcal{F}(\Sigma, w)$ such that*

$$\mathcal{L}_2(\mathcal{F}(\Sigma, w)) = \text{Det}(\Sigma, w).$$

5. *For each $w' \in \mathbb{N}_+$, there is a $(\Sigma, 2^{\max\{w, w'\}})$-second-order finite automaton $\overline{\mathcal{F}}^{w'}$ such that $\mathcal{L}_2(\overline{\mathcal{F}}^{w'}) = \overline{\mathcal{L}_2(\mathcal{F})}^{w'}$.*

6. *It is decidable whether* $\mathcal{L}_2(\mathcal{F}_1) \cap \mathcal{L}_2(\mathcal{F}_2) = \emptyset$.
7. *It is decidable whether* $\mathcal{L}_2(\mathcal{F}_1) \subseteq \mathcal{L}_2(\mathcal{F}_2)$.

We note that all binary operations above are defined when $\mathcal{F}_1$ is a (Σ, w)-second-order finite automaton and $\mathcal{F}_2$ is a (Σ, w')-second-order finite automaton for distinct w and w'. Just view both finite automata as $(\Sigma, \max\{w, w'\})$-second-order finite automata.

4 Transductions

Tensor Product. Given a set X and a number $\mathfrak{a} \in \mathbb{N}_+$, we let $X^{\times \mathfrak{a}}$ denote the set of all $\mathfrak{a}$-tuples of elements from X. Let Σ be an alphabet and $k \in \mathbb{N}$. If $s_1, \ldots, s_{\mathfrak{a}}$ are strings in Σ^k, where $s_i = \sigma_{i,1} \ldots \sigma_{i,k}$ for each $i \in [\mathfrak{a}]$, then the *tensor product* of $s_1, \ldots, s_{\mathfrak{a}}$ is defined as the string

$$s_1 \otimes \cdots \otimes s_{\mathfrak{a}} \doteq (\sigma_{1,1}, \ldots, \sigma_{\mathfrak{a},1}) \cdots (\sigma_{k,1}, \ldots, \sigma_{k,\mathfrak{a}})$$

of length k over the alphabet $\Sigma^{\times \mathfrak{a}}$. If $L_1, \ldots, L_{\mathfrak{a}}$ are subsets of Σ^*, then the tensor product of $L_1, \ldots, L_{\mathfrak{a}}$ is defined as follows:

$$L_1 \otimes \cdots \otimes L_{\mathfrak{a}} \doteq \{s_1 \otimes \cdots \otimes s_{\mathfrak{a}} : s_i \in L_i \text{ for each } i \in [\mathfrak{a}], |s_i| = |s_j| \text{ for } i \neq j\}.$$

Transductions. Let Σ_1 and Σ_2 be two alphabets. A (Σ_1, Σ_2)-*transduction* is a binary relation $\mathfrak{t} \subseteq \Sigma_1^* \times \Sigma_2^*$. We let

$$\mathsf{Im}(\mathfrak{t}) = \{u \in \Sigma_2^* : \exists s \in \Sigma_1^*, (s, u) \in \mathfrak{t}\}$$

be the image of $\mathfrak{t}$, and

$$\mathsf{Dom}(\mathfrak{t}) = \{s \in \Sigma_1^* : \exists u \in \Sigma_2^*, (s, u) \in \mathfrak{t}\}$$

be the domain of $\mathfrak{t}$. If $\mathfrak{t}$ is a (Σ_1, Σ_2)-transduction and $\mathfrak{t}'$ is a (Σ_2, Σ_3)-transduction, then the *composition* of $\mathfrak{t}$ with $\mathfrak{t}'$ is defined as the (Σ_1, Σ_3)-transduction

$$\mathfrak{t} \circ \mathfrak{t}' \doteq \{(s, v) \in \Sigma_1^* \times \Sigma_3^* : \exists u \in \Sigma_2^*, (s, u) \in \mathfrak{t} \text{ and } (u, v) \in \mathfrak{t}'\}.$$

We say that a (Σ_1, Σ_2)-transduction $\mathfrak{t}$ is *injective* if for each string $s \in \Sigma_1^*$, there exists at most one string $u \in \Sigma_2^*$ such that $(s, u) \in \mathfrak{t}$.

For each language $L \subseteq \Sigma_1^*$ we let $\mathfrak{d}(L) = \{(s, s) : s \in \Sigma_1^*\}$ be the (Σ_1, Σ_1)-transduction derived from L. Given a (Σ_1, Σ_2)-transduction $\mathfrak{t}$, we let

$$\mathfrak{t}(L) \doteq \mathsf{Im}(\mathfrak{d}(L) \circ \mathfrak{t}) = \{u \in \Sigma_2^* : \exists s \in L, (s, u) \in \mathfrak{t}\}$$

be the image of L under $\mathfrak{t}$.

For $\alpha \in \mathbb{N}$, we say that a (Σ_1, Σ_2)-transduction $\mathfrak{t}$ is α-*regular* if the language

$$\mathcal{L}(\mathfrak{t}) = \{s \otimes u : (s, u) \in \mathfrak{t}\} \subseteq (\Sigma_1 \times \Sigma_2)^*$$

is α-regular. We let $\mathcal{F}[\mathfrak{t}]$ be the unique deterministic, complete, normalized finite automaton with minimum number of states accepting $\mathcal{L}(\mathfrak{t})$. Note that if $\mathfrak{t}$ is α-regular, then the automaton $\mathcal{F}[\mathfrak{t}]$ has at most 2^α states.

Proposition 2. *Let Σ_1, Σ_2 and Σ_3 be alphabets, t be an α-regular (Σ_1, Σ_2)-transduction, and t' be a β-regular (Σ_2, Σ_3)-transduction. Let $L \subseteq \Sigma_1^*$ be a γ-regular language.*

1. $t \circ t'$ *is $(\alpha \cdot \beta)$-regular.*
2. $\mathrm{rev}(t)$ *is α-regular.*
3. *The languages $\mathrm{Im}(t)$ and $\mathrm{Dom}(t)$ are α-regular.*
4. *The transduction $\mathfrak{d}(L)$ is γ-regular.*
5. *The language $t(L)$ is $(\alpha \cdot \gamma)$-regular.*

Basic Transductions. Let Σ be an alphabet, and $R \subseteq \Sigma \times \Sigma$ be a binary relation over Σ. We say that a string $s \in \Sigma^*$ is R-*compatible* if $(s_i, s_{i+1}) \in R$ for each $i \in [|s| - 1]$. We let

$$\mathfrak{cp}[R] = \{(s, s) \; : \; s \in \Sigma^+ \text{ is } R\text{-compatible.}\}$$

be the transduction that sends each R-compatible string $s \in \Sigma^+$ to itself. Let Σ_1 and Σ_2 be alphabets and $R \subseteq \Sigma_1 \times \Sigma_2$ be a relation. We let

$$\mathfrak{mm}[R] = \{(s, u) \; : \; s \in \Sigma_1^+, \; u \in \Sigma_2^+, \; (s_i, u_i) \in R \text{ for each } i \in [|s|]\}$$

be the R-*multimap transduction*. If $g : \Sigma_1 \to \Sigma_2$ is a map then we write $\mathfrak{mm}[g]$ to denote the transduction $\mathfrak{mm}[R_g]$ where $R_g = \{(\sigma, g(\sigma)) \; : \; \sigma \in \Sigma_1\}$.

Proposition 3. *Let Σ, Σ_1 and Σ_2 be alphabets, $R \subseteq \Sigma \times \Sigma$ and $R' \subseteq \Sigma_1 \times \Sigma_2$ be relations.*

1. $\mathfrak{cp}[R]$ *is $|\Sigma|$-regular.*
2. $\mathfrak{mm}[R']$ *is 1-regular.*

5   Canonization of ODDs Using Transductions

In this section, we show that for each alphabet Σ and each $w \in \mathbb{N}_+$, the $(\mathcal{B}(\Sigma, w), \widehat{\mathcal{B}}(\Sigma, w))$-transduction $\{(D, \mathcal{C}(D)) \; : \; D \in \mathcal{B}(\Sigma, w)^{\circledast}\}$ is a regular transduction. We will show that this transduction can be cast as the composition of three transductions: a transduction $\mathfrak{det}[\Sigma, w]$ which sends ODDs to deterministic, complete ODDs representing the same language; a transduction $\mathfrak{mer}[\Sigma, w]$ that sends a deterministic, complete ODD to an ODD with minimum number of states representing the same language, and finally, a transduction $\mathfrak{nor}[\Sigma, w]$ that sends a deterministic, complete ODD to its normalized version.

Determinization Transduction. We start by considering, for each $w \in \mathbb{N}_+$, the bijection $\Omega \colon \mathcal{P}([\![w]\!]) \to [\![2^w]\!]$ that sends each subset $X \subseteq [\![w]\!]$ to the natural number $\Omega(X) \doteq \sum_{i \in X} 2^i$. In particular, we note that $\Omega(\emptyset) = 0$ and $\Omega(\{i\}) = 2^i$ for each $i \in X$.

For each alphabet Σ and number $w \in \mathbb{N}_+$, we let $\mathrm{pw}[\Sigma, w] : \mathcal{B}(\Sigma, w) \to \widehat{\mathcal{B}}(\Sigma, 2^w)$ be the map that sends each layer $B \in \mathcal{B}(\Sigma, w)$ to the deterministic, complete layer $\mathrm{pw}(B)$ in $\widehat{\mathcal{B}}(\Sigma, 2^w)$ defined as follows:

$$- \ell(\mathrm{pw}(B)) \doteq \begin{cases} \{\Omega(I(B))\} & \text{if } \iota(B) = 1 \\ \{\Omega(X) \colon X \subseteq \ell(B)\} & \text{otherwise;} \end{cases}$$

$$- r(\mathrm{pw}(B)) \doteq \{\Omega(X) \colon X \subseteq r(B)\};$$

$$- T(\mathrm{pw}(B)) \doteq \begin{cases} \{(\Omega(I(B)), \sigma, \Omega(\mathbf{N}(B, I(B), \{\sigma\})), \sigma \in \Sigma\} & \text{if } \iota(B) = 1 \\ \{(\Omega(X), \sigma, \Omega(\mathbf{N}(B, X, \{\sigma\}))) \colon X \subseteq \ell(B), \sigma \in \Sigma\} & \text{otherwise;} \end{cases}$$

$$- I(\mathrm{pw}(B)) \doteq \begin{cases} \{\Omega(I(B))\} & \text{if } \iota(B) = 1 \\ \emptyset & \text{otherwise;} \end{cases}$$

$$- F(\mathrm{pw}(B)) \doteq \{\Omega(X) \colon X \subseteq r(B), X \cap F(B) \neq \emptyset\};$$

$$- \iota(\mathrm{pw}(B)) \doteq \iota(B);$$

$$- \phi(\mathrm{pw}(B)) \doteq \phi(B).$$

It follows from the fact that Ω is a bijection, that for each subset $X \subseteq \llbracket w \rrbracket$ with $\Omega(X) \in \ell(\mathrm{pw}(B))$, and each symbol $\sigma \in \Sigma$, there exists precisely one right state $\mathfrak{q}^* \in r(\mathrm{pw}(B))$, namely $\Omega(\mathbf{N}(B, X, \{\sigma\}))$, such that $(\Omega(X), \sigma, \mathfrak{q}^*) \in T(\mathrm{pw}(B))$. Furthermore, if $\iota(\mathrm{pw}(B)) = 1$, then $\iota(B) = 1$ and, consequently, $I(\mathrm{pw}(B)) = \ell(\mathrm{pw}(B)) = \{\Omega(I(B))\}$. As a result, we obtain that $\mathrm{pw}(B)$ is indeed a deterministic, complete layer in $\widehat{\mathcal{B}}(\Sigma, 2^w)$.

Now, for each Σ and each $w \in \mathbb{N}_+$, we define the $(\mathcal{B}(\Sigma, w), \widehat{\mathcal{B}}(\Sigma, w))$-transduction $\mathfrak{det}[\Sigma, w] \doteq \mathfrak{mm}[\mathrm{pw}[\Sigma, w]]$. The next lemma states that the transduction $\mathfrak{det}[\Sigma, w]$ sends each ODD D in $\mathcal{B}(\Sigma, w)^{\circledast}$ to a deterministic, complete ODD in $\widehat{\mathcal{B}}(\Sigma, w)^{\circledast}$ representing the same language as D.

Lemma 1 (Determinization Transduction). *For each alphabet Σ and each $w \in \mathbb{N}$, the $(\mathcal{B}(\Sigma, w), \widehat{\mathcal{B}}(\Sigma, 2^w))$-transduction $\mathfrak{det}[\Sigma, w]$ satisfies the following properties.*

1. *$\mathfrak{det}[\Sigma, w]$ is injective.*
2. *$\mathsf{Im}(\mathfrak{det}[\Sigma, w]) = \mathcal{B}(\Sigma, w)^{\circledast}$.*
3. *For each $(D, D') \in \mathfrak{det}[\Sigma, w]$, $\mathcal{L}(D) = \mathcal{L}(D')$.*

Merging Transduction. Let Σ be an alphabet, $w \in \mathbb{N}_+$, B be a complete, deterministic layer in $\widehat{\mathcal{B}}(\Sigma, w)$ and ν be a partition of $r(B)$. Two (not necessarily distinct) left states $\mathfrak{p}, \mathfrak{p}' \in \ell(B)$ are said ν-*equivalent* if for each $\sigma \in \Sigma$ and each $\mathfrak{q} \in r(B)$, $(\mathfrak{p}, \sigma, \mathfrak{q})$ is a transition in $T(B)$ if and only if there exists $\mathfrak{q}' \in r(B)$ such that $\mathfrak{q}$ and $\mathfrak{q}'$ belong to the same cell of ν and $(\mathfrak{p}', \sigma, \mathfrak{q}')$ is a transition in $T(B)$. We note that each left state $\mathfrak{p}$ is trivially ν-equivalent to itself.

A *merging annotation* for B is a pair (μ, ν), where μ is a partition of $\ell(B)$, ν is a partition of $r(B)$, and the following conditions are satisfied.

1. If $\phi(B) = 1$, then $\nu = \{F(B), r(B) \setminus F(B)\}$ whenever $F(B) \neq \emptyset$, and $\nu = \{r(B)\}$ whenever $F(B) = \emptyset$;
2. For each $\mathfrak{p}, \mathfrak{p}' \in \ell(B)$, $\mathfrak{p}$ and $\mathfrak{p}'$ belong to the same cell of μ if and only if $\mathfrak{p}$ and $\mathfrak{p}'$ are ν-equivalent.

Let $D = B_1 \cdots B_k$ be a deterministic, complete ODD in $\widehat{\mathcal{B}}(\Sigma, w)$. A *merging annotation* for D is a sequence

$$\langle (\mu_1, \nu_1) \cdots (\mu_k, \nu_k) \rangle$$

satisfying the following conditions.

1. For each $i \in [k]$, (μ_i, ν_i) is a merging annotation for B_i;
2. For each $i \in [k-1]$, $\nu_i = \mu_{i+1}$.

Lemma 2. *Let Σ be an alphabet, $w \in \mathbb{N}_+$ and $D = B_1 \cdots B_k$ be an ODD in $\widehat{\mathcal{B}}(\Sigma, w)^k$. Then, D admits a unique merging annotation $\langle (\mu_1, \nu_1) \cdots (\mu_k, \nu_k) \rangle$.*

For each alphabet Σ, and each $w \in \mathbb{N}$, we let $M(\Sigma, w)$ be the set of triples (B, μ, ν) such that B is a deterministic, complete layer in $\widehat{\mathcal{B}}(\Sigma, w)$ and (μ, ν) is a merging annotation for B.

For each triple $(B, \mu, \nu) \in M(\Sigma, w)$, we let $\zeta \colon M(\Sigma, w) \to \widehat{\mathcal{B}}(\Sigma, w)$ be the map that sends each triple $(B, \mu, \nu) \in M(\Sigma, w)$ to the layer $\zeta(B, \mu, \nu)$ obtained from B by identifying all states in each cell of μ and in each cell of ν with the smallest state in the respective cell. More formally, for each $(B, \mu, \nu) \in M(\Sigma, w)$, we let $\zeta(B, \mu, \nu)$ be the deterministic, complete layer in $\widehat{\mathcal{B}}(\Sigma, w)$ with left states set $\ell(\zeta(B, \mu, \nu)) \doteq \bigcup_{X \in \mu} \{\min X\}$, right states set $r(\zeta(B, \mu, \nu)) \doteq \bigcup_{X' \in \nu} \{\min X'\}$, initial state set $I(\zeta(B, \mu, \nu)) \doteq I(B)$, final state set $F(\zeta(B, \mu, \nu)) \doteq \ell(\zeta(B, \mu, \nu)) \cap F(B)$, initial flag $\iota(\zeta(B, \mu, \nu)) \doteq \iota(B)$, final flag $\phi(\zeta(B, \mu, \nu)) \doteq \phi(B)$, and transition set

$$T(\zeta(B, \mu, \nu)) \doteq \bigcup_{X \in \mu, X' \in \nu} \{(\min X, \sigma, \min X') : \exists \mathfrak{p} \in X, \exists \mathfrak{q} \in X',$$
$$(\mathfrak{p}, \sigma, \mathfrak{q}) \in T(B), \sigma \in \Sigma \}.$$

For each alphabet Σ, and each $w \in \mathbb{N}_+$, we consider the following binary relations over $\widehat{\mathcal{B}}(\Sigma, w)$.

$$\mathrm{MR}[\Sigma, w] = \{(B, \zeta(B, \mu, \nu)) : (B, \mu, \nu) \in M(\Sigma, w)\}$$

and

$$\mathrm{MC}[\Sigma, w] = \{(\zeta(B, \mu, \nu), \zeta(B', \mu', \nu')) : (B, \mu, \nu), (B', \mu', \nu') \in M(\Sigma, w),$$
$$r(B) = \ell(B'), \nu = \mu'\}$$

Now, for each alphabet Σ and each $w \in \mathbb{N}_+$, we define the transduction

$$\mathfrak{mer}[\Sigma, w] = \mathfrak{mm}[\mathrm{MR}[\Sigma, w]] \circ \mathfrak{cp}[\mathrm{MC}[\Sigma, w]].$$

Lemma 3 (Merging Transduction). *For each alphabet Σ, and each $w \in \mathbb{N}$, the $(\widehat{\mathcal{B}}(\Sigma, w), \widehat{\mathcal{B}}(\Sigma, w))$-transduction $\mathfrak{mer}[\Sigma, w]$ satisfies the following properties.*

1. *$\mathfrak{mer}[\Sigma, w]$ is injective.*
2. *$\mathrm{Im}(\mathfrak{mer}[\Sigma, w]) = \widehat{\mathcal{B}}(\Sigma, w)^{\circledast}$.*
3. *For each $(D, D') \in \mathfrak{mer}[\Sigma, w]$, $\mathcal{L}(D) = \mathcal{L}(D')$, and D' is minimized.*

Normalization Transduction. Let Σ be an alphabet, $w \in \mathbb{N}_+$ and $B \in \mathcal{B}(\Sigma, w)$. Given two bijections $\pi \colon \ell(B) \to [\![|\ell(B)|]\!]$ and $\pi' \colon r(B) \to [\![|r(B)|]\!]$, we let $\langle \pi B \pi' \rangle$ be the layer in $\mathcal{B}(\Sigma, w)$ with left state set $\ell(\langle \pi B \pi' \rangle) \doteq \{\pi(\mathfrak{p}) \colon \mathfrak{p} \in \ell(B)\}$, initial state set $I(\langle \pi B \pi' \rangle) \doteq \{\pi(\mathfrak{p}) \colon \mathfrak{p} \in I(B)\}$, right state set $r(\langle \pi B \pi' \rangle) \doteq \{\pi(\mathfrak{q}) \colon \mathfrak{q} \in r(B)\}$, final state set $F(\langle \pi B \pi' \rangle) \doteq \{\pi'(\mathfrak{q}) \colon \mathfrak{q} \in F(B)\}$, transition set $T(\langle \pi B \pi' \rangle) \doteq \{(\pi(\mathfrak{p}), \sigma, \pi'(\mathfrak{q})) \colon (\mathfrak{p}, \sigma, \mathfrak{q}) \in T(B)\}$, initial flag $\iota(\langle \pi B \pi' \rangle) \doteq \iota(B)$ and final flag $\phi(\langle \pi B \pi' \rangle) \doteq \phi(B)$. We note that if B is deterministic (complete, resp.), then the layer $\langle \pi B \pi' \rangle$ is also deterministic (complete, resp.).

Proposition 4. *Let Σ be an alphabet, $w, k \in \mathbb{N}_+$ and $D = B_1 \cdots B_k$ be a reachable, deterministic ODD in $\mathcal{B}(\Sigma, w)^{\circ k}$. There exists a unique sequence $\overline{\pi} = \langle \pi_0, \pi_1, \ldots, \pi_k \rangle$ such that*

1. *$\pi_0 \colon \ell(B_1) \to [\![|\ell(B_1)|]\!]$ is a bijection from $\ell(B_1)$ to $[\![|\ell(B_1)|]\!]$;*
2. *for each $i \in [k]$, $\pi_i \colon r(B_i) \to [\![|r(B_i)|]\!]$ is a bijection from $r(B_i)$ to $[\![|r(B_i)|]\!]$;*
3. *for each $i \in [k]$, $\langle \pi_{i-1} B_i \pi_i \rangle$ is normalized.*

For each finite set X, we let $\mathbb{S}_X = \{\pi : X \to [\![|X|]\!] \ : \ \pi \text{ is a bijection}\}$ be the set of bijections from X to $[\![|X|]\!]$. For each alphabet Σ and each $w \in \mathbb{N}_+$ we define the following binary relations over $\widehat{\mathcal{B}}(\Sigma, w)$.

$$\mathrm{N}[\Sigma, w] = \{(B, \langle \pi B \pi' \rangle) \ : \ B \in \widehat{\mathcal{B}}(\Sigma, w), \ \pi \in \mathbb{S}_{\ell(B)}, \ \pi' \in \mathbb{S}_{r(B)}\}$$

and

$$\mathrm{C}[\Sigma, w] = \{(\langle \pi B \pi' \rangle, \langle \pi' B' \pi'' \rangle) : B, B' \in \widehat{\mathcal{B}}(\Sigma, w), \ r(B) = \ell(B'),$$
$$\pi \in \mathbb{S}_{\ell(B)}, \ \pi' \in \mathbb{S}_{r(B)}, \ \pi'' \in \mathbb{S}_{r(B')}\}$$

Now, let $\mathfrak{nor}[\Sigma, w] \doteq \mathfrak{mm}[\mathrm{N}[\Sigma, w]] \circ \mathfrak{cp}[\mathrm{C}[\Sigma, w]]$. The next lemma states that $\mathfrak{nor}$ is a transduction that sends reachable, complete, deterministic ODDs to normalized ODDs accepting the same language.

Lemma 4 (Normalization Transduction). *For each alphabet Σ, and each $w \in \mathbb{N}$, the $(\widehat{\mathcal{B}}(\Sigma, w), \widehat{\mathcal{B}}(\Sigma, w))$-transduction $\mathfrak{nor}[\Sigma, w]$ satisfies the following properties.*

1. *$\mathfrak{nor}[\Sigma, w]$ is injective.*
2. *$\mathrm{Im}(\mathfrak{nor}[\Sigma, w]) = \widehat{\mathcal{B}}(\Sigma, w)^{\circledast}$.*
3. *For each $(D, D') \in \mathfrak{nor}[\Sigma, w]$, if D is reachable, then $\mathcal{L}(D) = \mathcal{L}(D')$ and D' is normalized.*

ODD-Canonization Transduction. Now, we combine the three transductions defined in this section in order to define a transduction that can simulate the canonization of ODDs. More precisely, for each alphabet Σ, and each $w \in \mathbb{N}_+$, let $\mathfrak{can}[\Sigma, w]$ be the $(\mathcal{B}(\Sigma, w), \widehat{\mathcal{B}}(\Sigma, 2^w))$-transduction defined as follows

$$\mathfrak{can}[\Sigma, w] = \mathfrak{det}[\Sigma, w] \circ \mathfrak{mer}[\Sigma, 2^w] \circ \mathfrak{nor}[\Sigma, 2^w].$$

Theorem 5. *For each alphabet Σ, and each $w \in \mathbb{N}_+$,*

$$\mathfrak{can}[\Sigma, w] = \{(D, \mathcal{C}(D)) \colon D \in \mathcal{B}(\Sigma, w)^{\circledast}\}.$$

In other words, Theorem 5 states that the transduction $\mathfrak{can}[\Sigma, w]$ sends each ODD $D \in \mathcal{B}(\Sigma, w)^{\circledast}$ to its canonical form $\mathcal{C}(D)$ specified in Theorem 2.

Proof of Theorem 3. Now, we are in a position to prove Theorem 3. First, we note that the transduction $\mathfrak{det}[\Sigma, w]$ is 1-regular, and that the transductions $\mathfrak{mer}[\Sigma, w]$ and $\mathfrak{nor}[\Sigma, w]$ are both $2^{2^{O(|\Sigma| \cdot w)}}$-regular. This implies that $\mathfrak{can}[\Sigma, w]$ is $2^{2^{O(|\Sigma| \cdot w)}}$-regular. We also have that the transduction $\mathfrak{d}(\mathcal{L}(\mathcal{F}))$ is $\mathrm{nSt}(\mathcal{F})$-regular. Therefore, the language

$$\mathfrak{can}[\Sigma, w] \circ \mathfrak{d}(\mathcal{L}(\mathcal{F})) = \{\mathcal{C}(D) \, : \, D \in \mathcal{L}(\mathcal{F})\} \tag{1}$$

is $\mathrm{nSt}(\mathcal{F}) \cdot 2^{2^{O(|\Sigma| \cdot w)}}$-regular. Additionally a (possibly non-deterministic) finite automaton $\mathcal{F}'$ over $\widehat{\mathcal{B}}(\Sigma, 2^w)$ with $\mathrm{nSt}(\mathcal{F}) \cdot 2^{2^{O(|\Sigma| \cdot w)}}$ states accepting the language (1) can be constructed within the same time upper bounds. Finally, we let $\mathcal{C}_2(\mathcal{F})$ be the unique deterministic, complete, normalized, finite automaton over $\widehat{\mathcal{B}}(\Sigma, 2^w)$ accepting the same language as $\mathcal{F}'$. This automaton can be constructed in time $2^{\mathrm{nSt}(\mathcal{F}) \cdot 2^{2^{O(|\Sigma| \cdot w)}}}$. $\qquad\square$

6   Applications

The main technical tool used in the proof of our main result, (Theorem 3) was the fact that for each alphabet Σ, and each width $w \in \mathbb{N}$, the $(\mathcal{B}(\Sigma, w), \widehat{\mathcal{B}}(\Sigma, w))$-transduction $\mathfrak{can}[\Sigma, w] = \{(D, \mathcal{C}(D)) \colon D \in \mathcal{B}(\Sigma, w)\}$ is $f(\Sigma, w)$-regular for $f(\Sigma, w) = 2^{2^{O(|\Sigma| \cdot w)}}$. In this section, we show that this technical result has also other interesting algorithmic applications.

Width-Minimization of Nondeterministic ODDs. It is well known that given a deterministic ODD $D \in \mathcal{B}(\Sigma, w)^{\circ k}$, one can construct in time polynomial both in w and in k the unique ODD $\mathcal{C}(D)$ with minimum number of states with the property that $\mathcal{C}(D)$ is deterministic, complete, normalized, and $\mathcal{L}(D) = \mathcal{L}(\mathcal{C}D)$. Nevertheless, size minimization in the space of *non-deterministic* ODDs and width minimization for both non-deterministic and for deterministic ODDs are computationally hard problems. Next, we show that several width and size minimization problems for ODDs can be solved in FPT time when parameterized by the width of the input ODD.

Let $D = B_1 \ldots B_k$ be an ODD in $\mathcal{B}(\Sigma, w)^{\circ k}$. We let $\mathrm{nSt}(D) = |\ell(B_1)| + \sum_{i \in [k]} |r(B_i)|$ be the number of states in D, $\mathrm{nTr}(D) = |T(B_1)| + \sum_{i \in [k]} |T(B_i)|$ be the number of transitions in D, and $\mathrm{w}(D) = \min\{|\ell(B_1)|, |r(B_1)|, \ldots |r(B_k)|\}$ be the width of D.

Let Σ be an alphabet, $w \in \mathbb{N}_+$, and $\mathcal{S} \subseteq \mathcal{B}(\Sigma, w)$. We let $\mathcal{S}^{\circledast} = \mathcal{S}^+ \cap \mathcal{B}(\Sigma, w)^{\circledast}$ be the set of all (Σ, w)-ODDs whose layers are chosen from the set $\mathcal{S}$.

Theorem 6. *Let Σ be an alphabet, $w, w' \in \mathbb{N}_+$, D be an ODD in $\mathcal{B}(\Sigma, w)^{\circ k}$, and $\mathcal{S} \subseteq \mathcal{B}(\Sigma, w')$. For some computable function $h_\Sigma(w, w')$, one can determine in time $h_\Sigma(w, w') \cdot k$ whether there exists an ODD $D' \in \mathcal{S}^{\circ k}$ such that $\mathcal{L}_1(D') = \mathcal{L}_1(D)$. Additionally, if such an ODD D' exists, the following can be done in time $h_\Sigma(w, w') \cdot k^{O(1)}$.*

1. *One can compute an ODD $D' \in \mathcal{S}^{\circ k}$ of minimum width such that $\mathcal{L}_1(D') = \mathcal{L}_1(D)$.*
2. *One can compute an ODD $D' \in \mathcal{S}^{\circ k}$ with minimum number of states width such that $\mathcal{L}_1(D') = \mathcal{L}_1(D)$.*
3. *One can compute an ODD $D' \in \mathcal{S}^{\circ k}$ with minimum number of transitions such that $\mathcal{L}_1(D') = \mathcal{L}_1(D)$.*

Theorem 6 can be used to solve several types of minimization problems for ODDs in FPT time, when parameterized by the width of the input ODD. For instance, given an ODD $D \in \mathcal{B}(\Sigma, w)^{\circ k}$, if we want to find a (potentially non-deterministic) ODD D' of minimum width with $\mathcal{L}(D') = \mathcal{L}(D)$, then we set $w' = w$ and $\mathcal{S} = \mathcal{B}(\Sigma, w)$. On the other hand, if we want to enforce width minimization among *deterministic* ODDs, we set $w' = 2^w$ and $\mathcal{S}$ as the set of all deterministic $(\Sigma, 2^w)$-layers. Analogously, if we want to enforce minimization over *deterministic, complete* ODDs, then we set $w' = w$ and $\mathcal{S} = \widehat{\mathcal{B}}(\Sigma, w)$.

Counting Functions Computable by ODDs of a Given Width. Each ODD $D \in \mathcal{B}(\Sigma, w)^{\circ k}$ can be regarded as a representation of a function $f_D : \Sigma^k \to \{0, 1\}$. More precisely, for each $s \in \Sigma^k$, $f_D(s) = 1$ if and only if $s \in \mathcal{L}(D)$. The problem of counting functions from Σ^k to $\{0, 1\}$ computable by ODDs of a given width w however is very different from the problem of counting the number of ODDs of width w and length k. This is due to the fact that several ODDs may represent the same function. Nevertheless, our main theorem (Theorem 3) can be used to show that the problem of counting the number of functions of type $\Sigma^k \to \{0, 1\}$ which can be computed by some ODDs of width w, can be solved in time $h_\Sigma(w) \cdot k^{O(1)}$ for some suitable function $h_\Sigma(w)$. More generally, given any subset $\mathcal{S} \subseteq \mathcal{B}(\Sigma, w)$, we can count the number of functions computable by ODDs in $\mathcal{S}^{\circ k}$ in time $h_\Sigma(w)$.

Theorem 7. *Let Σ be an alphabet, $w \in \mathbb{N}$ and $\mathcal{S} \subseteq \mathcal{B}(\Sigma, w)$. For each $k \in \mathbb{N}$, one can count in time $h_\Sigma(w) \cdot k^{O(1)}$ the number of functions from Σ^k to $\{0, 1\}$ that can be computed by some ODD in $\mathcal{S}^{\circ k}$, where $h_\Sigma(w)$ is a suitable computable function depending only on the size of Σ and on w.*

7   Conclusion

In this work, we have introduced the notion of second-order finite automata, a formalism that combines traditional finite automata with ODDs of fixed width in order to represent *set of sets* of strings. Our main result states that second-order finite automata can be canonized with respect to their second languages. In particular, this implies that languages that can be represented by such automata are closed union, intersection, bounded width complementation, and that tests such as emptiness of intersection and inclusion are decidable.

We also provided two algorithmic applications of second order automata. In particular, we have shown that several width and size minimization problems for ODDs can be solved in fixed-parameter tractable time when parameterized by the width of the input ODD. This implies corresponding FPT algorithms for width and size minimization of ordered binary decision diagrams (OBDDs) with a fixed ordering. Finally, we have shown that second-order finite automata can be used to count exactly the number of distinct functions computable by (Σ, w)-ODDs of a given width w and a given length k in time $h_\Sigma(w) \cdot k^{O(1)}$, for some suitable function $h_\Sigma(w)$ while the naive algorithm for this task takes time exponential in both w and in k.

References

1. Bollig, B.: On the width of ordered binary decision diagrams. In: Zhang, Z., Wu, L., Xu, W., Du, D.-Z. (eds.) COCOA 2014. LNCS, vol. 8881, pp. 444–458. Springer, Cham (2014). https://doi.org/10.1007/978-3-319-12691-3_33
2. Bollig, B.: On the minimization of (complete) ordered binary decision diagrams. Theory Comput. Syst. **59**(3), 532–559 (2016)
3. Bouajjani, A., Habermehl, P., Rogalewicz, A., Vojnar, T.: Abstract regular tree model checking. Electron. Notes Theoret. Comput. Sci. **149**(1), 37–48 (2006)
4. Bozapalidis, S., Kalampakas, A.: Graph automata. Theoret. Comput. Sci. **393**(1–3), 147–165 (2008)
5. Courcelle, B.: On recognizable sets and tree automata. In: Algebraic Techniques, pp. 93–126. Elsevier (1989)
6. Courcelle, B.: The monadic second-order logic of graphs. I. Recognizable sets of finite graphs. Inf. Comput. **85**(1), 12–75 (1990)
7. Courcelle, B., Durand, I.: Verifying monadic second order graph properties with tree automata. In: Rhodes, C. (ed.) 3rd European Lisp Symposium, ELS, pp. 7–21. ELSAA (2010)
8. Ebbinghaus, H.D., Flum, J.: Finite automata and logic: a microcosm of finite model theory. In: Finite Model Theory, pp. 107–118. Springer, Heidelberg (1995). https://doi.org/10.1007/978-3-662-03182-7_6
9. Ergün, F., Kumar, R., Rubinfeld, R.: On learning bounded-width branching programs. In: Maass, W. (ed.) Proceedings of the Eighth Annual Conference on Computational Learning Theory, COLT, pp. 361–368. ACM (1995)
10. Forbes, M.A., Kelley, Z.: Pseudorandom generators for read-once branching programs, in any order. In: Thorup, M. (ed.) 2018 IEEE 59th Annual Symposium on Foundations of Computer Science (FOCS), pp. 946–955. IEEE (2018)
11. Giammarresi, D., Restivo, A.: Recognizable picture languages. Int. J. Pattern Recogn. Artif. Intell. **6**(02n03), 241–256 (1992)
12. Godefroid, P.: Using partial orders to improve automatic verification methods. In: Clarke, E.M., Kurshan, R.P. (eds.) CAV 1990. LNCS, vol. 531, pp. 176–185. Springer, Heidelberg (1991). https://doi.org/10.1007/BFb0023731
13. Hopcroft, J.: An $n \log n$ algorithm for minimizing states in a finite automaton. In: Theory of Machines and Computations, pp. 189–196. Elsevier (1971)

14. Priese, L.: Automata and concurrency. Theoret. Comput. Sci. **25**(3), 221–265 (1983)

15. Thomas, W.: Automata theory on trees and partial orders. In: Bidoit, M., Dauchet, M. (eds.) CAAP 1997. LNCS, vol. 1214, pp. 20–38. Springer, Heidelberg (1997). https://doi.org/10.1007/BFb0030586

Isomorphic Distances Among Elections

Piotr Faliszewski[1]([✉]), Piotr Skowron[2], Arkadii Slinko[3], Stanisław Szufa[4], and Nimrod Talmon[5]

[1] AGH University, Kraków, Poland
faliszew@agh.edu.pl
[2] University of Warsaw, Warsaw, Poland
p.skowron@mimuw.edu.pl
[3] University of Auckland, Auckland, New Zealand
a.slinko@auckland.ac.nz
[4] Jagiellonian Univeristy, Kraków, Poland
stanislaw.szufa@doctoral.uj.edu.pl
[5] Ben-Gurion University, Be'er Sheva, Israel
talmonn@bgu.ac.il

Abstract. This paper is an invitation to study the problem of measuring distances between elections, for the case where both the particular names of the candidates and the voters are irrelevant. In other words, we say that two elections are at distance zero (or, that they are isomorphic) if it is possible to make them identical by renaming their candidates and voters, and we are interested in measuring how far are two given elections from being isomorphic. The study of such distances has just begun and in this paper we outline why we believe that it is interesting and what are the natural research directions.

Keywords: Elections · Distances · Isomorphism · Complexity

1 Introduction

An election consists of a set of candidates and a collecton of voters who express preferences regarding these candidates (for example, in the ordinal model the voters rank the candidates from the most to the least appreciated ones). The goal is to choose a candidate or a group of candidates (a committee) in a way that both reflects the preferences of the voters and suits a given application. While elections are most commonly associated with making political decisions (such as choosing countries' presidents or parliaments), they are also used in

Piotr Faliszewski was supported by the funds of the Polish Ministry of Science and Higher Education assigned to AGH University. Piotr Skowron was supported by the Foundation for Polish Science (Homing programme). Arkadii Slinko was supported by Marsden Fund grant 3706352 of The Royal Society of New Zealand. Stanisław Szufa was supported by NCN project 2018/29/N/- ST6/01303. Nimrod Talmon was supported by the Israel Science Foundation (grant No. 630/19).

H. Fernau (Ed.): CSR 2020, LNCS 12159, pp. 64–78, 2020.
https://doi.org/10.1007/978-3-030-50026-9_5

numerous other contexts. For example, we can use the theory of elections to model various competitions (e.g., we can view the participants as the candidates and the judges as the voters), to model business decision-making scenarios (e.g., a company may choose its product portfolio based on an opinion poll), or even to model everyday interactions (e.g., a group of friends may vote on a restaurant for a dinner). Thus it is not surprising that elections are widely studied within such areas as economics, political science, mathematics, operations research, and computer science (mostly within the area of *computational social choice* [6]). In particular, computer scientists seek efficient algorithms for computing election winners (perhaps somewhat surprisingly, for many voting rules the problem of deciding who won an election is NP-hard [4,24,33]), study the complexity of manipulating election results (e.g., via strategic voting [3,9,22], controlling the election structure [5,15,25], or bribing voters [12,18,28]), provide algorithms for verifying election integrity (e.g., via computing candidates' margins of victory [8, 35,41], or through game-theoretic analysis [14,42]), and consider numerous other problems.[1] In particular, researchers also considered various problems related to measuring distances between elections, e.g., in the distance rationalization framework [13,31,32]. In this framework, the focus is on comparing the distances of a given election to a set of consensus elections in order to find the closest one. For this reason the identities of candidates and the identities of voters are kept fixed. On the contrary, we are more interested in comparing the structure of the given elections and, thus, we view the exact names of candidates and voters as irrelevant. We intend this paper to be an invitation and encouragement to study problems that arise due to this view.

1.1   Distances, Election Isomorphism, and Motivation

Suppose we are given two elections over possibly different candidate sets and with possibly different collections of voters, except that in both elections the numbers of candidates and the numbers of voters match (i.e., the elections are of the same size). These elections may represent real-life preferences over some issues, or may have been generated using some statistical model, or may have come from any other source. The particular names of the candidates and identities of voters are irrelevant for us since we would like to compare the internal structure of these elections. For example, if it is possible to rename the candidates and reorder the voters so that the two elections become identical, then we say that they are isomorphic and we insist that the distance between them should be zero. Otherwise, we would like to measure how far they are from being isomorphic (sometimes we will relax this requirement and we will agree that some non-isomorphic elections are also at distance zero).[2] There is a number of reasons why such distances are useful and below we outline a few possible applications:

---

[1] The references here are just examples giving pointers to both classic and new works.

[2] Thus we are effectively interested in pseudodistances over elections. We will, however, typically omit the "pseudo" prefix to maintain simplicity of our writing.

Choosing Elections for Numerical Experiments. There is a growing body of work in computational social choice that is focused on numerical experiments regarding elections. For example, researchers evaluate running times of algorithms [20,39,40], test practically achievable approximation ratios [27,36], or evaluate how frequently a given phenomenon occurs (e.g., how frequently a voting rule is manipulable [16,22,39], or how frequently a certain paradoxical situation occurs [7,29,38]). However, it is not at all clear what data to use for such experiments. One possible approach is to generate elections from various statistical models, but it is useful to know if the generated set of elections is truly representative, or if they all happen to be similar. Thus it is important to be able to measure the distances between the generated elections. Indeed, in our recent work we have taken this distance-based approach to forming a testbed of election instances [37].

Matching Real-Life Elections to Statistical Models. There are two main sources of election data that researchers working on computational social choice use in their experiments: the PrefLib library [30], which collects real-life elections; and various statistical models (referred to as statistical cultures), which provide means of sampling from the space of elections. Both of these sources, however, have some issues specific to them. One problem with real-life election data is that it typically involves just a handful of candidates and, thus, it is insufficient for many experiments. Randomly generated elections can be of any size, but it is not clear which statistical model best reflects the reality for a given application. Thus it would be interesting to match real-life elections with those models that capture their most essential features. Computing distances between various elections is crucial for obtaining such a matching.

Machine Learning Over Preference Data. Many machine-learning algorithms need to be able to evaluate how similar are particular objects to each other. For example, we can imagine a system which learns to recognize elections on which particular algorithms work well (e.g., the system could learn to match elections with the fastest winner-determination algorithms), or a system that recognizes elections with particular features (such as, perhaps, instances of bribery or strategic voting, or types of elections that typically arise in various application domains). In each of these cases a good algorithm for computing distances among elections may be useful.

With the above applications in mind, we are interested in finding distances that are both (1) efficiently computable; and (2) provide meaningful, easy to interpret results. Regarding the former issue, the reader may be already concerned that prior to even considering distances, we need to be able to test if two elections are isomorphic. Yet, even though many problems of this kind—with the GRAPH ISOMORPHISM problem being the most notable example [2]—are not known to have polynomial-time algorithms, this is not the case for ELECTION ISOMORPHISM. Unfortunately, this is where the good news end. So far, all the distances that we have analyzed are either unappealing, computationally intractable [19], or assign distance zero also to elections that are not isomorphic [37] (this latter drawback is less of a problem but, ideally, we would prefer to avoid it). As a consequence, further studies regarding both the algorithmic and normative properties of distances are needed.

1.2 Organization

The paper is organized as follows. First, in Sect. 2, we provide preliminary definitions and the necessary notation to discuss elections. Then, in Sect. 3, we briefly describe the ELECTION ISOMORPHISM problem and discuss several distances among elections, focusing mostly on their computational complexity. In Sect. 4 we outline a few directions for future research, and we conclude in Sect. 5.

2 Preliminaries

We model an election as a pair $E = (C, V)$, where $C = \{c_1, \ldots, c_m\}$ is a set of candidates and $V = (v_1, \ldots, v_n)$ is an (ordered) collection of voters. In the ordinal model, each voter v_i has a preference order $\succ_{v_i}$ in which he or she ranks the candidates from the most to the least preferred one. In the approval model each voter v_i has a set $A_i \subseteq C$ of candidates that he or she finds acceptable. Throughout this paper we focus on the ordinal model, but occasionally we will make references to the approval one.

Let v be a voter with a preference order over candidate set C. We will write v to refer both to the voter v and to his or her preference order (and we will sometimes refer to preference orders as votes); the exact meaning will always be clear from the context. For two candidates a and b, we write $a \succ_v b$ or $v: a \succ b$ to indicate that v prefers a over b (we extend this notation to more candidates in a natural way). For a candidate $c \in C$, we write $\text{pos}_v(c)$ to denote the position of c in $\succ_v$ (so if v ranks c on the first position then $\text{pos}_v(c) = 1$, and if v ranks c on the last position then $\text{pos}_v(c) = |C|$).

Example 1. Consider an election $E = (C, V)$, where $C = \{a, b, c, d\}$ and $V = (v_1, v_2, v_3)$ with voters having the following preference orders:

$$v_1: a \succ b \succ c \succ d, \qquad v_2: a \succ c \succ b \succ d, \qquad v_3: d \succ a \succ c \succ b.$$

Voter v_1 considers candidate a to be the most desirable, then b, then c, and finally d. If the same voters were to cast approval votes, their approval sets might be, e.g., $A_1 = \{a, b, c\}$, $A_2 = \{a\}$, and $A_3 = \{a, d\}$. In this case v_1 approves his or her top three candidates, v_2 approves his or her top candidate only, and v_3 approves his or her top two candidates; we note that the preference orders of these candidates do not define the approval sets uniquely.

For nonnegative integers m and n, we write $\mathcal{E}(m, n)$ to denote the space of all elections with m candidates and n voters.

Bijections and Permutations. Let n be a positive integer. We write $[n]$ to mean the set $\{1,\ldots,n\}$ and S_n to mean the set of all permutations of $[n]$. For an election $E = (C,V)$, with $V = (v_1,\ldots,v_n)$, and permutation $\nu \in S_n$, by $\nu(E)$ we mean the election $(C,(v_{\nu(1)},\ldots,v_{\nu(n)}))$.

Let C and D be two candidate sets of the same cardinality and let $\sigma\colon C \to D$ be a bijection. If v is a vote over the candidates in C, then by $\sigma(v)$ we mean a vote obtained by renaming the candidates according to σ. Formally, for each two candidate $a,b \in C$ it holds that $\sigma(v)\colon \sigma(a) \succ \sigma(b)$ if and only if $v\colon a \succ b$. For an election $E = (C,V)$, where $V = (v_1,\ldots,v_n)$, by $\sigma(E)$ we mean an election $(D,(\sigma(v_1),\ldots,\sigma(v_n)))$. For two sets C and D, $|C| = |D|$, we write $\varPi(C,D)$ to denote the set of all bijections between them.

Distances. Let X be a set. We say that a function $d\colon X \times X \to \mathbb{R}$ is a distance over X if for every $x,y,z \in X$ all of the following hold:

(1) $d(x,y) \geq 0$ (non-negativity),
(2) $d(x,y) = 0$ if and only if $x = y$ (identity of indiscernibles),
(3) $d(x,y) = d(y,x)$ (symmetry),
(4) $d(x,z) \leq d(x,y) + d(y,z)$ (triangle inequality).

We will often consider pseudodistances, i.e., functions that satisfy all the properties of a distance except for the identity of indiscernibles.

Let $x = (x_1,\ldots,x_m)$ and $y = (y_1,\ldots,y_m)$ be two vectors from $\mathbb{R}^m$. The ℓ_1 and ℓ_2 distances between x and y are defined, respectively, as:

$$\ell_1(x,y) = \sum_{i\in[m]} |x_i - y_i| \qquad \text{and} \qquad \ell_2(x,y) = \sqrt{\sum_{i\in[m]} (x_i - y_i)^2}.$$

For the case where x and y are elements of an $(m-1)$-dimensional simplex (i.e., all x_i and y_i are nonnegative and $\sum_{i\in[m]} x_i = \sum_{i\in[m]} y_i = 1$), we are also interested in the earth mover's distance between x and y, denoted $\mathrm{EMD}(x,y)$. $\mathrm{EMD}(x,y)$ is defined as the lowest cost of transforming x into y, where the allowed operations are of the form "move value z from coordinate i to coordinate j at cost $z \cdot |i - j|$" (the costs of the operations sum up). It is well-known that $\mathrm{EMD}(x,y)$ can be computed in polynomial-time using a simple greedy algorithm.

Let C be a set of candidates. The following three distances over the space of preference orders are particularly useful (by u and v we mean two arbitrary preference orders over C):

1. The discrete distance, $d_{\mathrm{disc}}(u,v)$, is 0 if $u = v$ and is 1 otherwise.
2. The swap distance, $d_{\mathrm{swap}}(u,v)$, is the number of swaps of adjacent candidates that one has to perform (one after the other) to transform u into v.
3. The Spearman distance, $d_{\mathrm{Spear}}(u,v)$, is equal to $\sum_{c\in C} |\mathrm{pos}_v(c) - \mathrm{pos}_u(c)|$.

The swap and Spearman distances are closely related and, in particular, for each two preference orders u and v (over the same candidate set) they satisfy the Diaconis-Graham inequality [10]:

$$d_{\mathrm{swap}}(u,v) \leq d_{\mathrm{Spear}}(u,v) \leq 2d_{\mathrm{swap}}(u,v).$$

There are numerous other distances (such as the Cayley distance, a variant of the swap distance where we can swap arbitrary candidates and not only the adjacent ones), but these three will suffice for our discussion.

3   Distances Among Elections

In this section we will consider several types of distances over the space of elections $\mathcal{E}(m, n)$ and we will discuss their computational complexity. As we have argued, we are interested in (pseudo)distances that view isomorphic elections as identical. Formally, we have the following definition (this notion was introduced by Faliszewski et al. [19], but the exact terminology is due to this paper):

Definition 1. *Let* DIS *be a distance over election space* $\mathcal{E}(m, n)$. *We say that* DIS *is symmetric (i.e., neutral and anonymous) if for every two elections* $E_1 = (C, V), E_2 = (D, U) \in \mathcal{E}(m, n)$, *each permutation* $\nu \in S_n$, *and each bijection* $\sigma \colon C \to D$ *it holds that* $\mathrm{DIS}(E_1, E_2) = \mathrm{DIS}(\sigma(E_1), \nu(E_2))$.

In social choice, *neutrality* refers to invariance with respect to renaming candidates, whereas *anonymity* refers to invariance with respect to permuting voters. Szufa et al. [37] referred to symmetric distances as *neutral/anonymous*.

3.1   Election Isomorphism Problem

Before we discuss specific symmetric distances, let us first consider the issue of election isomorphism.

Definition 2 (Faliszewski et al. [19]). *Let* $E_1 = (C, V)$ *and* $E_2 = (D, U)$ *be two elections from* $\mathcal{E}(m, n)$. *We say that* E_1 *and* E_2 *are isomorphic if there is a bijection* $\sigma \colon C \to D$ *and permutation* $\nu \in S_n$ *such that* $\sigma(E_1) = \nu(E_2)$.

Example 2. Consider elections $E_1 = (C, V)$ and $E_2 = (D, U)$ such that $C = \{a, b, c, d\}$, $D = \{x, y, z, w\}$, $V = (v_1, v_2, v_3)$, and $U = (u_1, u_2, u_3)$, with the following preference orders:

$$v_1 \colon a \succ b \succ c \succ d, \qquad u_1 \colon x \succ y \succ z \succ w,$$
$$v_2 \colon b \succ a \succ d \succ c, \qquad u_2 \colon z \succ w \succ y \succ x,$$
$$v_3 \colon c \succ d \succ b \succ a, \qquad u_3 \colon y \succ x \succ w \succ z.$$

E_1 and E_2 are isomorphic, as witnessed by the bijection $\sigma(a) = x$, $\sigma(b) = y$, $\sigma(c) = z$, and $\sigma(d) = w$, and permutation $\nu(1) = 1$, $\nu(2) = 3$, and $\nu(3) = 2$. On the other hand, neither E_1 nor E_2 is isomorphic to the election E from Example 1. To see this, it suffices to note that in elections E_1 and E_2 each voter ranks a different candidate last, whereas this is not the case for E.

In the ELECTION ISOMORPHISM problem we ask if two given elections are isomorphic. While many isomorphism problems are notoriously difficult to classify in terms of their computational complexity (as is the case for GRAPH ISOMORPHISM), ELECTION ISOMORPHISM is easily seen to be polynomial-time solvable [19]: Let $E_1 = (C, V)$ and $E_2 = (D, U)$ be our two input elections, where $C = \{c_1, \ldots c_m\}$, $D = \{d_1, \ldots, d_m\}$, $V = (v_1, \ldots, v_n)$, and $U = (u_1, \ldots, u_n)$. Suppose that E_1 and E_2 are isomorphic and that this is witnessed by bijection $\sigma \colon C \to D$ and permutation $\nu \in S_n$. Then we know that votes $\sigma(v_1)$ and $u_{\nu(1)}$ are identical. Our algorithm simply guesses the value $\nu(1)$ (there are n choices), reads off the bijection σ by comparing the votes v_1 and $u_{\nu(1)}$, and computes the remaining part of ν by finding a matching in the following bipartite graph: On the left side we put votes $v_1, \ldots, v_n$, on the right side we put votes $u_1, \ldots, u_n$, and each two votes v_i and u_j are connected by an edge if $\sigma(v_i) = u_j$. Every perfect matching of such a graph defines a permutation that witnesses that the elections are isomorphic. On the other hand, if we cannot find a perfect matching for any initial value of $\nu(1)$, then the elections are not isomorphic.

Proposition 1 (Faliszewski et al. [19]). ELECTION ISOMORPHISM *is in* P.

This is good news as it means that there are no fundamental obstacles to finding appealing, polynomial-time computable distances, at least for the case of ordinal elections. On the other hand, the situation regarding approval elections is, unfortunately, more challenging. Since in approval elections each voter simply indicates which candidates he or she finds acceptable, the problem of deciding if two approval elections are isomorphic is equivalent to the problem of deciding if two multi-hypergraphs are isomorphic. Since even the complexity status of GRAPH ISOMORPHISM is not resolved [2], we will not discuss approval elections further (however, we mention the works of Arvind et al. [1] and Grohe et al. [23] regarding APPROXIMATE GRAPH ISOMORPHISM and related problems).

3.2   Isomorphic Distances

We say that a symmetric distance is *isomorphism-respecting* (or, *isomorphic*, for short) if it guarantees that two elections are at distance zero if and only if they are isomorphic. One of the most natural ways of defining such distances is by extending distances over votes. In particular, Faliszewski et al. [19] proposed the following approach.

Definition 3 (Faliszewski et al. [19]). *Let d be a distance over preference orders. Let $E_1 = (C, V)$ and $E_2 = (D, U)$ be two elections, where $|C| = |D|$, $V = (v_1, \ldots, v_n)$, and $U = (u_1, \ldots, u_n)$. For a bijection $\sigma \in \Pi(C, D)$, we define:*

$$d^\sigma\text{-ID}(E_1, E_2) = \min_{\nu \in S_n} \sum_{i \in [n]} d(\sigma(v_i), u_{\nu(i)}).$$

We define the d-isomorphic distance between E_1 and E_2 to be $d\text{-ID}(E_1, E_2) = \min_{\sigma \in \Pi(C, D)} d^\sigma\text{-ID}(E_1, E_2)$.

In particular, we have isomorphic distances d_{disc}-ID, d_{swap}-ID, and d_{Spear}-ID.[3] Unfortunately, it seems that all of them suffer from some problems. The first distance, d_{disc}-ID, is polynomial-time computable (by a simple extension of the algorithm for ELECTION ISOMORPHISM), but seems to be too coarse. Indeed, given two elections $E_1, E_2 \in \mathcal{E}(m, n)$, it holds that d_{disc}-ID$(E_1, E_2) \in \{0, \ldots, n-1\}$. This is so, because each pair of matched votes can either contribute distance 0 or 1 and it is always possible to ensure that at least one pair contributes distance 0. While coarseness does not seem to be an issue for d_{swap}-ID and d_{Spear}-ID, these two distances seem to be quite hard to compute.

Theorem 1 (Faliszewski et al. [19]). *Given an integer δ and two elections $E_1, E_2 \in \mathcal{E}(m, n)$, the problems of deciding if d_{swap}-ID$(E_1, E_2) \leq \delta$ and if d_{Spear}-ID$(E_1, E_2) \leq \delta$ are both NP-complete.*

Worse yet, it seems that standard ways of dealing with NP-hard problems are not effective here. In particular, the ability to compute d_{swap}-ID or d_{Spear}-ID approximately implies the ability to solve the GRAPH ISOMORPHISM problem in polynomial time.

Theorem 2 (Faliszewski et al. [19]). *For each $\alpha < 1$, there are neither polynomial-time m^α-approximation nor n^α-approximation algorithms (where m is the number of candidates and n is the number of voters) for d_{swap}-ID and d_{Spear}-ID unless the GRAPH ISOMORPHISM problem is in P.*

Since, in spite of considerable effort, the exact complexity of GRAPH ISOMORPHISM remains elusive, the above theorem indicates that even if there were good approximation algorithms for our distances, finding them would be a major break-through in theoretical computer science. Thus, instead of seeking such algorithms, we rather ask if the conclusion of Thoerem 2 can be strengthened to P = NP (perhaps at the price of a weaker approximation guarantee).[4]

More importantly, instead of seeking approximation algorithms it may be worthwhile to consider parameterized complexity of computing our distances.

Theorem 3 (Faliszewski et al. [19]). *There are FPT algorithms for computing d_{swap}-ID and d_{Spear}-ID for the parameterizations by the number of candidates and by the value of the distance. There is also an FPT algorithm for computing d_{Spear}-ID parameterized by the number of voters, whereas for d_{swap}-ID the corresponding problem is para-NP-hard.[5]*

---

[3] One can verify that, indeed, Definition 3 provides functions d-ID that satisfy the requirements of (pseudo)distances. Further, one can also verify that the Diaconis-Graham inequality also holds for d_{swap}-ID and d_{Spear}-ID.

[4] We thank Gerhard Woeginger for suggesting this idea.

[5] This follows from the fact that being able to compute d_{swap}-ID implies the ability to compute Kemeny rankings, and this problem is NP-hard already for four voters [11]; see also the results of Bartholdi et al. [4] and Hemaspaandra et al. [26].

Unfortunately, the FPT algorithms from the above theorem are quite inefficient and finding faster ones would be desirable. It may also be interesting to consider other parameterizations (e.g., by the largest distance between matched voters).

Faliszewski et al. [19] also proposed formulations of d_{swap}-ID and d_{Spear}-ID as integer linear programs (ILPs). Unfortunately, it turned out that solving these programs is very challenging even for very small elections (for d_{Spear}-ID they computed distances for elections with 6 candidates and 16 voters; for d_{swap}-ID even such tiny elections were too demanding). However, perhaps, the approach of Redko et al. [34] can lead to a practically fast heuristic algorithm.

3.3 Positionwise and Pairwise Distances

As, at least so far, isomorphism-respecting distances are either unappealing or are too difficult computationally, it may be necessary to drop this requirement. To this end, Szufa et al. [37] suggested the *positionwise* distance, which focuses on the positions of the candidates in the votes (and which was inspired by the family of positional scoring rules), and the *pairwise* distance, which focuses on relative orderings of pairs of candidates (and which was inspired by the class of C2 rules in Fishburn's classification [21]).

We discuss the positionwise distance first. Let $E = (C, V)$ be an election. For each candidate $c \in C$ and each possible position in a vote $i \in [|C|]$, we define $\psi_E(c, i)$ to be the fraction of the votes in which c is ranked on the i-th position. We let $\psi_E(c)$ be the vector $(\psi_E(c, 1), \ldots, \psi_E(c, |C|))$.

Definition 4 (Szufa et al. [37]). *Let $E_1 = (C, V)$ and $E_2 = (D, U)$ be two elections from $\mathcal{E}(m, n)$. We define their positionwise distance as* $\mathrm{POS}(E_1, E_2) = \min_{\sigma \in \Pi(C,D)} \sum_{c \in C} \mathrm{EMD}(\Psi_{E_1}(c), \Psi_{E_2}(\sigma(c)))$.

The positionwise distance is, indeed, symmetric, but is not isomorphic.

Example 3. Consider elections $E_1 = (C, V)$ and $E_2 = (D, U)$, where $C = \{a, b, c, d\}$, $D = \{x, y, z, w\}$, $V = (v_1, \ldots, v_4)$, and $U = (u_1, \ldots, u_4)$:

$$
\begin{aligned}
v_1 &: a \succ b \succ c \succ d, & u_1 &: x \succ z \succ w \succ y \\
v_2 &: b \succ c \succ d \succ a, & u_2 &: y \succ w \succ z \succ x \\
v_3 &: c \succ d \succ a \succ b, & u_3 &: z \succ x \succ y \succ w \\
v_4 &: d \succ a \succ b \succ c, & u_4 &: w \succ y \succ x \succ z.
\end{aligned}
$$

These two elections are at positionwise distance zero because in both of them each voter is ranked exactly once at each position. However, they are not isomorphic because in E_2 candidates x and z (as well as y and w) are always ranked next to each other, whereas there is no such pair of candidates in E_1.

On the positive side, it is possible to compute the value $\mathrm{POS}(E_1, E_2)$ in polynomial time as it reduces to finding the lowest cost in the following bipartite graph: The candidates from E_1 are on one side, the candidates from E_2 are on the other side, and for each candidate c from E_1 and each candidate d from E_2 there is a connecting edge with cost $\mathrm{EMD}(\Psi_{E_1}(c), \Psi_{E_2}(d))$.

Proposition 2 (Szufa et al. [37]). *There is a polynomial-time algorithm for computing the positionwise distance between elections.*

Remark 1. Computing the positionwise distance implicitly provides a mapping σ between the candidates from the given elections. Then, for this σ we can compute in polynomial time d^{σ}_{swap}-ID or $d^{\sigma}_{\text{Spear}}$-ID and use them as a proxy for the distance between the given elections. However, doing so does not seem to be a good idea. Indeed, two bijections may provide the same positionwise distance, but different d^{σ}_{swap}-ID or $d^{\sigma}_{\text{Spear}}$-ID distances.

The pairwise distance is defined similarly to the positionwise one, but on top of the notion of a weighted majority relation. For an election $E = (C, V)$, its weighted majority relation is a function M_E such that for each two candidates $a, b \in C$, $M_E(a, b)$ is the fraction of the voters who prefer a over b.

Definition 5 (Szufa et al. [37]). *Let $E_1 = (C, V)$ and $E_2 = (D, U)$ be two elections from $\mathcal{E}(m, n)$. We define their pairwise distance as $\text{PAIR}(E_1, E_2) = \min_{\sigma \in \Pi(C, D)} \sum_{(a,b) \in C \times C} |M_{E_1}(a, b) - M_{E_2}(\sigma(a), \sigma(b))|$.*

The reader can verify that the pairwise distance also is symmetric, but not isomorphic. Further, computing $\text{PAIR}(E_1, E_2)$ is NP-hard (see the work of Szufa et al. [37]; Grohe et al. [23] also provide a very relevant discussion and a set of strong, if negative, results for a very closely related distance over graphs).

Szufa et al. [37] used the positionwise distance to argue about a number of elections generated from several statistical cultures and to provide their visualization. While their results seem appealing, it is not clear how meaningful this distance is (or, how meaningful is the pairwise distance, which they did not use due to its computational complexity). For example, why should we use the EMD distance in the definition of the positionwise distance instead of ℓ_1 or ℓ_2 distances? Intuitively, the EMD distance captures the fact that being ranked on two nearby positions is more similar than being ranked on far-apart positions (indeed, according to EMD, vector $x = (1, 0, 0, 0)$ is closer to vector $y = (0, 1, 0, 0)$ than to vector $z = (0, 0, 0, 1)$, whereas both $\ell_1(x, y) = \ell_1(x, z)$ and $\ell_2(x, y) = \ell_2(x, z)$). However, it would be better to have formal arguments rather than intuitions. Similarly, the pairwise distance is (implicitly) defined based on the ℓ_1 distance, but why not use ℓ_2 or EMD (intuitively, in this case EMD does not seem to be justified, but, again, a formal argument would be better than an intuition).

4   Research Directions

In this section we discuss a few research directions that we believe are important to pursue in order to (1) obtain better understanding of symmetric distances, and (2) to be able to use them for the applications from Sect. 1.

4.1   New Distances

So far, we have considered five different symmetric distances, namely d_{disc}-ID, d_{swap}-ID, d_{Spear}-ID, POS, and PAIR. The first three are natural extensions of distances over preference orders, whereas the other two measure distances between aggregated features of elections. It would be very interesting to provide more approaches to defining symmetric distances, or to exploit these two further. It would be particularly interesting to seek polynomial-time computable distances, but it is not strictly necessary if one can either provide effective approximation[6] or FPT algorithms (or other means of computing them effectively, e.g., using ILP formulations or other such formalisms).

For example, we may try to extend the d_{disc}-ID distance. One interpretation of d_{disc}-ID is that it counts the number of votes that need to be removed from the given elections for them to become isomorphic. Instead, we may allow removing both candidates and voters. While, likely, this would not make the distance much less coarse, perhaps there would be applications where such a distance would be useful. Alternatively, we may consider replacing the discrete distance with some related, but more fine-grained one.

Another idea might be to try to merge the principles behind the position-wise and pairwise distances, while maintaining the polynomial running time. For example, it may be possible to consider relative orderings of pairs of candidates identified by some feature (such as their score under some voting rule) rather than their names.

Finally, in this paper we have considered distances between elections with the same numbers of candidates and voters, but it would be very useful to relax this restriction. In fact, positionwise and pairwise distances can meaningfully compare elections with different numbers of voters, but, dealing with different sizes of candidate sets may be more challenging. We mention that Redko et al. [34] suggest how this can be done for d_{Spear}-ID.

4.2   Effective Algorithms

Currently, we can efficiently compute the d_{disc}-ID and positionwise distances, for which there are polynomial-time algorithms. Our (not so many) attempts to efficiently compute the other distances were not successful. However, as we are only at the beginning of the study of symmetric distances, it is quite possible that there are much better approaches than those considered to date. In cases where there are no polynomial-time algorithms, it is important to seek ones that are superpolynomial in the worst case, but which, nonetheless, usually provide exact solutions quickly. Approximation algorithms are interesting as well, but possibly are more difficult to come by. Indeed, Theorem 2 suggests that seeking approximation algorithms is likely too demanding for some isomorphic distances.

One possible new approach was suggested by Redko et al. [34], who consider a certain new optimal transport problem and show its relations to computing the

---

[6] These approximation algorithms, however, would have to have approximation ratios very close to 1 if the computed distances were to be meaningful.

d_{Spear}-ID distance. Another possibility would be either to seek a different ILP formulations than those used by Faliszewski et al. [19], or to find ways to guide the ILP solvers in computing the solutions based on the current formulations.[7]

4.3   Properties of the Distances

We have briefly argued that d_{disc}-ID may not be a particularly useful symmetric distance because it is too coarse. Similarly, we have also considered the notion of isomorphism-respecting distances. However, in general, we do not yet have a comprehensive theory that would explain which distances are good for which applications and why. One way to start building such a theory is to define a number of desirable properties that the distances may have and test which distances, indeed, have them. We are interested both in quantitative properties (such as the coarseness of d_{disc}-ID) and in normative properties that either hold or not (such as the property of being isomorphism-respecting).

Let DIS be a distance over $\mathcal{E}(m, n)$. It is interesting to consider, e.g., the following quantitative properties:

1. The distance DIS defines an equivalence relation $=_{\text{DIS}}$ so that for $E_1, E_2 \in \mathcal{E}(m, n)$ we have $E_1 =_{\text{DIS}} E_2$ if and only if $\text{DIS}(E_1, E_2) = 0$. The more equivalence classes this relation generates (i.e., the larger is its quotient space), the more appealing DIS seems to be (although this certainly is not the only parameter that one should look at). For isomorphic distances, the number of equivalence classes was provided by Eğecioğlu and Giritligil [17], who considered the problem of generating non-isomorphic elections uniformly at random. For the positionwise and pairwise distances computing this value remains open.

2. We are also interested in the cardinality of the set $\{\text{DIS}(E_1, E_2) \mid E_1, E_2 \in \mathcal{E}(m, n)\}$. Intuitively, the larger it is, the more fine-grained the distance is. Clearly, for d_{swap}-ID and d_{Spear}-ID this value is $O(nm^2)$ as there are n pairs of votes and two matched votes can be at most at distance $O(m^2)$. For d_{disc}-ID it is n.

Other properties may be interesting too. For example, we may ask what happens if given an election E we form a new election E' by swapping two adjacent candidates in a single vote. Is it the case that $D(E, E')$ is the smallest nonzero distance possible between two elections? Or is there some E'' such that $\text{DIS}(E, E'') < \text{DIS}(E, E')$? We believe that building a normative theory of symmetric distances is an important task that deserves attention.

4.4   Evaluating Distances in Practice

In Sect. 1 we highlighted several scenarios in which symmetric distances between elections are useful. Assuming we have enough symmetric distances at our disposal, it is an important task to evaluate their suitability for each of our motivating examples.

---

[7] We thank Tuomas W. Sandholm for suggesting this idea.

In particular, we have recently used the positionwise distance to form a testbed of elections and to visualize it [37]. However, it would be interesting to compare how useful other distances would be for this task. For example, we suspect that d_{disc}-ID would lead to uninteresting results, but it would be good to verify this in practice. It would be even more interesting to find distances that lead to further insights into the nature of elections than those provided by the positionwise distance.

5   Summary

This paper presents the idea of symmetric distances over elections and constitutes an invitation for further researchers to work on this topic. So far, there are only few results and we hope and believe that significant progress is possible. We are interested both in seeking theoretical results (e.g., regarding computational complexity and properties of the distances) and empirical ones (e.g., evaluations of heuristics, applications of the distances to solve various problems).

Acknowledgments. Some of the discussions related to the ideas presented in this paper happened during Dagstuhl Seminar 19381.

References

1. Arvind, V., Köbler, J., Kuhnert, S., Vasudev, Y.: Approximate graph isomorphism. In: Rovan, B., Sassone, V., Widmayer, P. (eds.) MFCS 2012. LNCS, vol. 7464, pp. 100–111. Springer, Heidelberg (2012). https://doi.org/10.1007/978-3-642-32589-2_12
2. Babai, L., Dawar, A., Schweitzer, P., Torán, J.: The graph isomorphism problem (Dagstuhl seminar 15511). Dagstuhl Rep. **5**(12), 1–17 (2015)
3. Bartholdi III, J., Tovey, C., Trick, M.: The computational difficulty of manipulating an election. Soc. Choice Welfare **6**(3), 227–241 (1989)
4. Bartholdi III, J., Tovey, C., Trick, M.: Voting schemes for which it can be difficult to tell who won the election. Soc. Choice Welfare **6**(2), 157–165 (1989)
5. Bartholdi III, J., Tovey, C., Trick, M.: How hard is it to control an election? Math. Comput. Modeling **16**(8/9), 27–40 (1992)
6. Brandt, F., Conitzer, V., Endriss, U., Lang, J., Procaccia, A. (eds.): Handbook of Computational Social Choice. Cambridge University Press, Cambridge (2016)
7. Brandt, F., Geist, C., Strobel, M.: Analyzing the practical relevance of voting paradoxes via Ehrhart theory, computer simulations, and empirical data. In: Proceedings of the 15th International Conference on Autonomous Agents and Multiagent Systems (AAMAS-2016), pp. 385–393 (2016)
8. Bredereck, R., Faliszewski, P., Kaczmarczyk, A., Niedermeier, R., Skowron, P., Talmon, N.: Robustness among multiwinner voting rules. In: Bilò, V., Flammini, M. (eds.) SAGT 2017. LNCS, vol. 10504, pp. 80–92. Springer, Cham (2017). https://doi.org/10.1007/978-3-319-66700-3_7
9. Conitzer, V., Sandholm, T., Lang, J.: When are elections with few candidates hard to manipulate? J. ACM **54**(3), Article 14 (2007)

10. Diaconis, P., Graham, R.: Spearman's footrule as a measure of disarray. J. Roy. Stat. Soc. Ser. B **39**(2), 262–268 (1977)
11. Dwork, C., Kumar, R., Naor, M., Sivakumar, D.: Rank aggregation methods for the web. In: Proceedings of the 10th International World Wide Web Conference (WWW-2001), pp. 613–622. ACM Press, March 2001
12. Elkind, E., Faliszewski, P., Slinko, A.: Swap bribery. In: Mavronicolas, M., Papadopoulou, V.G. (eds.) SAGT 2009. LNCS, vol. 5814, pp. 299–310. Springer, Heidelberg (2009). https://doi.org/10.1007/978-3-642-04645-2_27
13. Elkind, E., Faliszewski, P., Slinko, A.: Distance rationalization of voting rules. Soc. Choice Welfare **45**(2), 345–377 (2015). https://doi.org/10.1007/s00355-015-0892-5
14. Elkind, E., Gan, J., Obraztsova, S., Rabinovich, Z., Voudouris, A.: Protecting elections by recounting ballots. In: Proceedings of the 28th International Joint Conference on Artificial Intelligence (IJCAI-2019), pp. 259–265 (2019)
15. Erdélyi, G., Fellows, M., Rothe, J., Schend, L.: Control complexity in Bucklin and fallback voting: a theoretical analysis. J. Comput. Syst. Sci. **81**(4), 632–660 (2015)
16. Erdélyi, G., Fellows, M., Rothe, J., Schend, L.: Control complexity in Bucklin and fallback voting: an experimental analysis. J. Comput. Syst. Sci. **81**(4), 661–670 (2015)
17. Eğecioğlu, Ö., Giritligil, A.: The impartial, anonymous, and neutral culture model: a probability model for sampling public preference structures. J. Math. Sociol. **37**(4), 203–222 (2013)
18. Faliszewski, P., Hemaspaandra, E., Hemaspaandra, L.: How hard is bribery in elections? J. Artif. Intell. Res. **35**, 485–532 (2009)
19. Faliszewski, P., Skowron, P., Slinko, A., Szufa, S., Talmon, N.: How similar are two elections? In: Proceedings of the 33rd AAAI Conference on Artificial Intelligence (AAAI-2019), pp. 1909–1916 (2019)
20. Faliszewski, P., Slinko, A., Stahl, K., Talmon, N.: Achieving fully proportional representation by clustering voters. J. Heurist. **24**(5), 725–756 (2018). https://doi.org/10.1007/s10732-018-9376-y
21. Fishburn, P.: Condorcet social choice functions. SIAM J. Appl. Math. **33**(3), 469–489 (1977)
22. Goldsmith, J., Lang, J., Mattei, N., Perny, P.: Voting with rank dependent scoring rules. In: Proceedings of the 28th AAAI Conference on Artificial Intelligence (AAAI-2014), pp. 698–704 (2014)
23. Grohe, M., Rattan, G., Woeginger, G.: Graph similarity and approximate isomorphism. In: Proceedings of the 43rd International Symposium on Mathematical Foundations of Computer Science (MFCS-2018), pp. 20:1–20:16 (2018)
24. Hemaspaandra, E., Hemaspaandra, L., Rothe, J.: Exact analysis of Dodgson elections: Lewis Carroll's 1876 voting system is complete for parallel access to NP. J. ACM **44**(6), 806–825 (1997)
25. Hemaspaandra, E., Hemaspaandra, L., Rothe, J.: Anyone but him: the complexity of precluding an alternative. Artif. Intell. **171**(5–6), 255–285 (2007)
26. Hemaspaandra, E., Spakowski, H., Vogel, J.: The complexity of Kemeny elections. Theoret. Comput. Sci. **349**(3), 382–391 (2005)
27. Keller, O., Hassidim, A., Hazon, N.: New approximations for coalitional manipulation in scoring rules. J. Artif. Intell. Res. **64**, 109–145 (2019)
28. Konicki, C., Vassilevska Williams, V.: Bribery in balanced knockout tournaments. In: Proceedings of the 18th International Conference on Autonomous Agents and Multiagent Systems (AAMAS-2019), pp. 2066–2068 (2019)

29. Mattei, N., Forshee, J., Goldsmith, J.: An empirical study of voting rules and manipulation with large datasets. In: Proceedings of the 4th International Workshop on Computational Social Choice (COMSOC-2012) (2012)
30. Mattei, N., Walsh, T.: Preflib: a library for preferences. In: Proceedings of the 3nd International Conference on Algorithmic Decision Theory (ADT-2013), pp. 259–270 (2013)
31. Meskanen, T., Nurmi, H.: Closeness counts in social choice. In: Braham, M., Steffen, F. (eds.) Power, Freedom, and Voting. Springer, Heidelberg (2008). https://doi.org/10.1007/978-3-540-73382-9_15
32. Nitzan, S.: Some measures of closeness to unanimity and their implications. Theory Decis. **13**(2), 129–138 (1981)
33. Procaccia, A., Rosenschein, J., Zohar, A.: On the complexity of achieving proportional representation. Soc. Choice Welfare **30**(3), 353–362 (2008)
34. Redko, I., Vayer, T., Flamary, R., Courty, N.: Co-optimal transport. Technical report arXiv:2002.03731 [stat.ML], February 2020
35. Shiryaev, D., Yu, L., Elkind, E.: On elections with robust winners. In: Proceedings of the 12th International Conference on Autonomous Agents and Multiagent Systems (AAMAS-2013), pp. 415–422 (2013)
36. Skowron, P., Faliszewski, P., Slinko, A.: Achieving fully proportional representation: approximability result. Artif. Intell. **222**, 67–103 (2015)
37. Szufa, S., Faliszewski, P., Skowron, P., Slinko, A., Talmon, N.: Drawing a map of elections in the space of statistical cultures. In: Proceedings of the 19th International Conference on Autonomous Agents and Multiagent Systems (AAMAS-2020) (2020, to appear)
38. Tideman, T., Plassmann, F.: Modeling the outcomes of vote-casting in actual elections. In: Felsenthal, D., Machover, M. (eds.) Electoral Systems: Paradoxes, Assumptions, and Procedures, pp. 217–251. Springer, Heidelberg (2012). https://doi.org/10.1007/978-3-642-20441-8_9
39. Walsh, T.: Where are the hard manipulation problems. J. Artif. Intell. Res. **42**(1), 1–29 (2011)
40. Wang, J., Sikdar, S., Shepherd, T., Zhao, Z., Jiang, C., Xia, L.: Practical algorithms for multi-stage voting rules with parallel universes tiebreaking. In: Proceedings of the 33rd AAAI Conference on Artificial Intelligence (AAAI-2019), pp. 2189–2196 (2019)
41. Xia, L.: Computing the margin of victory for various voting rules. In: Proceedings of the 13th ACM Conference on Electronic Commerce (EC-2012), pp. 982–999. ACM Press, June 2012
42. Yin, Y., Vorobeychik, Y., An, B., Hazon, N.: Optimally protecting elections. In: Proceedings of the 25th International Joint Conference on Artificial Intelligence (IJCAI-2016), pp. 538–545 (2016)

Tandem Duplications, Segmental Duplications and Deletions, and Their Applications

Binhai Zhu$^{(\boxtimes)}$

Gianforte School of Computing, Montana State University,
Bozeman, MT 59717-3880, USA
bhz@montana.edu

Abstract. We review two streams of recent research results in this paper. The first is on converting a sequence A to another sequence B using the minimum number of tandem duplications. This research originates from the *copying systems* in computer science in the early 1980s, and also from biology more than 40 years ago. We review our recent NP-hardness result on this paper, together with several open problems along the line. Segmental duplications and deletions are more discussed recently on cancer research where besides genomes (sequences), the so-called *copy number profile* (a vector where the ith component represents the number of the ith segment appearing in the genome, regardless of their orders) are also used. We again review some of our recent hardness results and preliminary positive results, together with some open problems. This paper is mostly self-contained.

1 Introduction

We review the background and results on tandem duplications and on copy number profiles separately, even though they are related.

1.1 Tandem Duplications

In biology, a tandem duplication is an event which creates two consecutive copies of a segment on a genome during DNA replication. Take a genome as a string, this process converts a string AXB into another string $AXXB$. This process is known to occur either at small scale at the nucleotide level, or at large scale at the genome level [7–9,38,52]. For instance, it is known that the Huntington disease is associated with the duplication of 3 nucleotides CAG [43], whereas at genome level, tandem duplications are known to involve multiple genes during cancer progression [47]. Furthermore, as first pointed by Szostak and Wu in 1980, gene duplication is believed to be the main driving force behind evolution, and the majority of duplications affecting organisms are believed to be of the

© Springer Nature Switzerland AG 2020
H. Fernau (Ed.): CSR 2020, LNCS 12159, pp. 79–102, 2020.
https://doi.org/10.1007/978-3-030-50026-9_6

tandem type [53]. As a result, when the human genomes are sequenced, it is not surprising that around 3% of the human genome are in the form of tandem repeats [33].

In the formal languages community, the natural question arose as early as in 1984 in the context of so-called *copying systems* [19]: given a string S, what is the language that can be obtained starting from S and applying (any number of) tandem duplications, i.e., rules of the form $AXB \rightarrow AXXB$, where X can be any substring of S? (One such operation will generate a *square*, i.e., a substring XX, in the new S.) Combined with results from [4], it was shown that this language is regular if S is on a binary alphabet, but not regular if the alphabet has size three. These results were rediscovered 15 years later in [17,57], coincidentally, motivated by biological applications. In fact, this is related to a celebrated result by Thue in 1906 [54]; namely, to generate a binary sequence T with tandem duplications, it suffices to start with a sequence in the (square-free) ancestor set $\{0, 1, 01, 10, 010, 101\}$. However, if the alphabet size is at least three, then this ancestor set is of infinite size.

Due to these reasons, tandem duplications have received significant attention in the last decades, both in practice and theory. The combinatorial aspects of tandem duplications have been studied extensively by computational biologists [3,24,26,32,55], one question of interest being to reconstruct the evolution of a cluster of tandem repeats by duplications that could have given rise to the observed sequences. In parallel, various formal language communities [17,39,57] have investigated the expressive power of tandem duplications on strings.

In [39], it was shown that the membership, inclusion and regularity testing problems on the language defined by S can all be decided in linear time (still for binary alphabets). In [29,39,40], similar problems are also considered on non-binary alphabets, when the length $|X|$ of duplicated strings is bounded by a constant. More recently, Cho *et al.* [10] introduced a tandem duplication system where the *depth* of a character, i.e., the number of "generations" it took to generate it, is considered. In [28,30], the authors studied the *expressive power* of tandem duplications, a notion based on the subsequences that can be obtained from various types of copying mechanisms.

How to compute the minimum number of duplications required to transform a string S into another string T, which we call the *Tandem Duplication (TD) distance*, was first posed in [39] (pp. 306, Open Problem 3) by Leupold *et al.* and has remained open ever since 2004. In 2017, Alon *et al.* [1] investigated the TD distance problem for binary strings and proved that the maximum TD distance between a square-free string S and a string T of length n is $\Theta(n)$. They also raised the algorithmic question of computing the TD distance between S and T over a larger alphabet again. In 2019, Lafond, Zhu and Zou proved that this problem is in fact NP-hard, settling this 16-year old open problem for an unbounded alphabet [34,35]. We give a brief sketch of this NP-hardness proof in Sect. 3.

On the other hand, we comment that the TD distance is one of the many ways of comparing two genomes represented as strings in computational biology which is beyond this paper. (Other notable examples include reversal [27] and transpositions distances, etc, the latter having recently been shown NP-hard as well in a celebrated paper of Bulteau *et al.* [5]. The TD distance has itself received special attention very recently from a slightly different perspective in cancer evolution involving Copy Number Profiles [48], which we will review next.

1.2 Copy Number Profiles

Copy number variations (CNVs) are gain and loss of DNA contents in a genome. These are caused by duplications and deletions of segments of the corresponding DNA sequence [52]. (On the other hand, CNV detection is beyond this review and interested readers are referred to [42] for a survey and [41] for the most recent development on CNV detection.)

In cancer genomics, a central problem is to investigate the intra-tumor genetic heterogeneity [45, 46, 50], which has a potential to help cancer prognostic [13, 44] and explain drug resistance [14, 15]. In some cancers, for instance, high-grade serous ovarian cancer (HGSOC), the heterogeneity is mainly reflected in genome rearrangement and endoreduplications, which result in aberrant *copy number profiles (CNP)* [6]. This was also found in some other cancer types more recently [12]. (We will define CNP formally in the next section, loosely speaking, it is a vector representing the number of each segment among some specified segments.) In 2014, Schwarz *et al.* proposed a method to infer phylogenetic trees in cancer genomes from unsigned integer copy number profiles [49]. In this method the fundamental problem is to compute the minimum number of segment duplications and deletions to convert some CNP to another CNP. While the latter problem was solved in linear time (with a complex dynamic programming) [51], we noted that sometimes a single duplication or deletion on CNP cannot be implemented with the same number of duplication or deletion operations directly on the corresponding DNA sequence [48].

This interesting phenomenon triggers our algorithmic research on CNV-related research, with only duplication and deletion operations and with a potential application in cancer genomics. Surprisingly, a lot of fundamental questions are unanswered, even taking into account that some related research has been done on sorting genomes with duplication and deletion operations (and usually with some other rearrangement operations like reversals or DCJs). We note that in [20], a more complex distance computation was used as a subroutine to compute an ancestor profile given a set of k profiles. The problem was shown to be NP-hard, though an ILP formulation was given. In fact, Chowdhury *et al.* considered copy number changes at different levels, from single gene, single chromosome to whole genome, to enhance the tumor phylogeny reconstruction [11].

In [48], we initiated the fundamental algorithmic research on converting a genome/sequence G into another sequence H such that the copy number profile of H is equal to a given profile. We showed that this problem, called Minimum Copy Number Generation (MCNG) henceforth, is NP-hard if the objective is to use the minimum number of tandem duplications and deletions, and some greedy method was implemented with decent performance when the sequence length is relatively small. More recently, we prove that the general problem, i.e., when duplications are arbitrary, is also NP-hard; in fact, NP-hard to approximate within a constant factor [36,37]. Moreover, the problem is W[1]-hard when parameterized by the size of the solution. These proofs will involve some variants of the famous Set Cover problem, which will be briefly reviewed in Sect. 4.

At this point, it looks like all these results are negative. One would probably feel pessimistic about the outcome of these research. Wait! We do have some preliminary positive result! In [36,37], the Copy Number Profile Conforming (CNPC for short) was studied. The problem is defined as follows: given two CNP's C_1 and C_2, compute two genomes/sequences S_1 and S_2 such that $cnp(S_1) = cnp(S_2)$ and $d(S_1, S_2)$ is minimized, where $d(-, -)$ could be any genomic distance between two genomes. We made the first progress by showing that if $d(-, -)$ is the breakpoint distance and C_1 and C_2 are both polynomially bounded, then the problem is polynomially solvable. (If only the breakpoint distance is to be computed, then the condition that C_1 and C_2 are polynomially bounded can be withdrawn.)

The paper is organized as follows. In Sect. 2, we give formal definitions for our problems and all other necessary definitions. In Sect. 3, we briefly sketch our NP-hardness proof for the Tandem Duplication Distance problem. In Sect. 4, we review our results for the Minimum Copy Number Generation (MCNG) problem and the Copy Number Profile Conforming (CNPC) problem. We list several open problems in Sect. 5 to conclude the paper.

2 Preliminaries

2.1 Strings and Tandem Duplications

We review the string terminology and notation from [25]. Let $[n]$ denote the set of integers $\{1, 2, \ldots, n\}$. Unless stated otherwise, all the strings in the paper are on an alphabet denoted Σ. If S_1 and S_2 are two strings, we usually denote their concatenation by $S_1 S_2$. For a string S, we write $\Sigma(S)$ for the subset of characters of Σ that have at least one occurrence in S. A string S is called *exemplar* if $|S| = |\Sigma(S)|$, i.e., each character in S occurs exactly once. A *substring* of S is a contiguous sequence of characters within S. A *prefix* (resp. *suffix*) of S is a substring that occurs at the beginning (resp. end) of S, i.e., if $S = S_1 S_2$ for some strings S_1 and S_2, then S_1 is a prefix of S and S_2 a suffix of S. A *subsequence* of S is a string that can be obtained by successively deleting characters from S.

A *tandem duplication* (TD) is an operation on a string S that copies a substring X of S and inserts the copy after the occurrence of X in S. In other words, a TD transforms $S = AXB$ into $AXXB$. Given another string T, we write $S \Rightarrow T$ if there exist strings A, B, X such that $S = AXB$ and $T = AXXB$. More generally, we write $S \Rightarrow_k T$ if there exist $S_1, \ldots, S_{k-1}$ such that $S \Rightarrow S_1 \Rightarrow \ldots \Rightarrow S_{k-1} \Rightarrow T$. We also write $S \Rightarrow_* T$ if there exists some k such that $S \Rightarrow_k T$.

Definition 1. *The TD distance $dist_{TD}(S,T)$ between two strings S and T is the minimum value of k satisfying $S \Rightarrow_k T$. If $S \Rightarrow_* T$ does not hold, then $dist_{TD}(S,T) = \infty$.*

We use the term *distance* here to refer to the number of TD operations from a string S to another string T, but one should be reminded that TD is not formally a metric. For instance, $dist_{TD}$ is not symmetric since duplications can only increase the length of a string.

A *square string* is a string of the form XX, i.e., a concatenation of two identical substrings. Given a string S, a *contraction* is the reverse of a tandem duplication. That is, it takes a square string XX contained in S and deletes one of the two copies of X. We write $T \rightarrowtail S$ if there exist strings A, B, X such that $T = AXXB$ and $S = AXB$. We also define $T \rightarrowtail_k S$ and $T \rightarrowtail_* S$ for contractions analogously as for TDs. Observe that by the symmetry of duplications and contractions, $T \rightarrowtail_k S$ if and only if $S \Rightarrow_k T$ and $T \rightarrowtail_* S$ if and only if $S \Rightarrow_* T$. When there is no possible confusion, we will sometimes write $T \rightarrowtail S$ instead $T \rightarrowtail_* S$.

We have the following problem.

The Tandem Duplication (TD) problem:
Input: Two strings S and T over the same alphabet Σ and an integer k.
Question: Is $dist_{TD}(S,T) \leq k$?

In the Exemplar-TD version of this problem, S is given exemplar. In general, we may call S the *source string* and T the *target string*. We will often use the fact that S and T form a YES instance if and only if T can be transformed into S by a sequence of at most k contractions. See Fig. 1 for a simple example.

	Sequence	Operations
Sequence	$T = \langle a, c, \underline{g, g}, t, a, c, g, t \rangle$	contraction on $\langle g, g \rangle$
	$\langle a, c, g, t, a, c, g, t \rangle$	contraction on $\langle a, c, g, t, a, c, g, t \rangle$
Sequence	$S = \langle a, c, g, t \rangle$	

Fig. 1. An example to transform sequence T to S with two contractions. The corresponding sequence of TDs to convert S to T would duplicate a, c, g, t first, then duplicate the first g.

2.2   Copy Number Profiles, Segmental Duplications and Deletions

A genome G is a string, i.e. a sequence of characters, all of which belong to some alphabet Σ (the characters of G can be interpreted as genes or segments— in the CNP-related context, a genome is composed of segments as duplications and deletions could across the middle of a gene). We use genome and string interchangeably, when the context is clear. We write $G[p]$ to denote the character at position p of G (the first position being 1), and we write $G[i..j]$ for the substring of G from positions i to j, inclusively. For $s \in \Sigma$, we write $G - s$ to denote the subsequence of G obtained by removing all occurrences of s.

When copy number profile is used, we represent an alphabet as an ordered list $\Sigma = (s_1, s_2, \ldots, s_m)$ of distinct characters. Slightly abusing notation, we may write $s \in \Sigma$ if s is a member of this list. We write $n_s(G)$ to denote the number of occurrences of $s \in \Sigma$ in a genome G. A *Copy-Number Profile (or CNP)* on Σ is a vector $\boldsymbol{c} = \langle c_1, \ldots, c_{|\Sigma|} \rangle$ that associates each character s_i of the alphabet with a non-negative integer $c_i \in \mathbb{N}$. We may write $\boldsymbol{c}(s)$ to denote the number associated with $s \in \Sigma$ in $\boldsymbol{c}$. We write $\boldsymbol{c} - s$ to denote the CNP obtained from $\boldsymbol{c}$ by setting $\boldsymbol{c}(s) = 0$.

The *Copy Number Profile (CNP)* of genome G, denoted $cnp(G)$, is the vector of occurrences of all characters of Σ. (Note that in the theory of formal languages, the CNP of a string is also called the *Parikh vector*.) Formally,

$$cnp(G) = \langle n_{s_1}(G), n_{s_2}(G), \ldots, n_{s_m}(G) \rangle.$$

For example, if $\Sigma = (a, b, c)$ and $G = bbcbbcaca$, then $cnp(G) = \langle 2, 4, 3 \rangle$ and $\boldsymbol{c}(b) = 4$.

Deletions and Duplications on Strings

We now describe the two string events of *deletion* and *duplication*. Both are illustrated in Fig. 2.

$$
\begin{array}{cc}
\textit{Sequence} & \textit{Operations} \\
G_1 = abbc \cdot \mathbf{\textit{beab}} \cdot cab & del(5, 8) \\
G_2 = a \cdot \underline{bbcc} \cdot ab & dup(2, 5, 6) \\
G_3 = abbcca \cdot \underline{bbcc} \cdot b &
\end{array}
$$

Fig. 2. Three strings (or toy genomes), G_1, G_2 and G_3. From G_1 to G_2, a deletion is applied to $G_1[5..8]$. From G_2 to G_3, a duplication is applied to $G_2[2..5]$, with the copy inserted after position 6.

Given a genome G, a (segmental) *deletion* on G takes a substring of G and removes it. Deletions are denoted by a pair (i, j) of the positions of the substring to remove. Applying deletion (i, j) to G transforms G into $G[1..i-1]G[j+1..n]$.

A *duplication* on G takes a substring of G, copies it and inserts the copy anywhere in G, except inside the copied substring. A duplication is defined by a triple (i, j, p) where $G[i..j]$ is the string to duplicate and $p \in \{0, 1, \ldots, i-1, j, \ldots n\}$ is the position *after* which we insert (inserting after 0 prepends the copied substring to G). Applying duplication (i, j, p) to G transforms G into $G[1..p]G[i..j]G[p+1..n]$.

An *event* is either a deletion or a duplication. If G is a genome and e is an event, we write $G\langle e \rangle$ to denote the genome obtained by applying e on G. Given a sequence $E = (e_1, \ldots, e_k)$ of events, we define $G\langle E \rangle = G\langle e_1 \rangle \langle e_2 \rangle \ldots \langle e_k \rangle$ as the genome obtained by successively applying the events of E to G. We may also write $G\langle e_1, \ldots, e_k \rangle$ instead of $G\langle (e_1, \ldots, e_k) \rangle$.

The most natural application of the above events is to compare genomes.

Definition 2. *Let G and G' be two strings over alphabet Σ. The* Genome-to-Genome *distance between G and G', denoted $d_{GG}(G, G')$, is the size of the smallest sequence of events E satisfying $G\langle E \rangle = G'$.*

We also define a distance between a genome G and a CNP c, which is the minimum number of events to apply to G to obtain a genome with CNP c.

Definition 3. *Let G be a genome and c be a CNP, both over alphabet Σ. The* Genome-to-CNP *distance between G and c, denoted $d_{GCNP}(G, c)$, is the size of the smallest sequence of events E satisfying $cnp(G\langle E \rangle) = c$.*

The above definition leads to the following problem, which was first studied in [48].

The Minimum Copy Number Generation (MCNG) problem:
Instance: a genome G and a CNP c over alphabet Σ.
Task: compute $d_{GCNP}(G, c)$.

Qingge *et al.* proved that the MCNG problem is NP-hard when all the duplications are restricted to be tandem [48]. They also posed several open questions: Is the problem NP-hard when duplications are arbitrary? Does the problem admit a constant-factor approximation? Is the problem FPT (*fixed-parameter tractable*)? We answered all these questions recently [36,37], and we will review the results in Sect. 4.

Finally, readers are referred to [16,23] for more details regarding the definitions related to (approximation) algorithms and NP-completeness. We refer readers for further details on FPT and W[1]-hardness to the corresponding books [18,22].

3   Results on Tandem Duplications

3.1   Exemplar-TD Is NP-hard

The first difficulty for proving the NP-hardness for Exemplar-TD would be choosing a suitable problem to reduce from. This minimization problem should be of the property that when the optimal cost is obtained, the optimal number of contractions is obtained. This problem is the *Cost-Effective Subgraph* problem, to be defined and discussed next.

The second difficulty in the reduction is that the sequences should be properly designed such that they clearly form a *promise* version for the Exemplar-TD problem; moreover, the contractions look to be ad hoc but they are related to a solution for the Cost-Effective Subgraph problem. (We believe this is one of the reasons why the problem was open for 16 years, as the TDs can in general happen anywhere and it is hard to trace some properties without any proper control.)

The Cost-Effective Subgraph Problem

Let $G = (V, E)$ be a simple graph and c a positive integer. For a subset $X \subseteq V$, let $E(X) = \{(u, v) \in E | u, v \in X\}$ denote the edges inside of X. The *cost* of X is defined as

$$cost(X) = c \cdot (|E(G)| - |E(X)|) + |X| \cdot |E(X)|.$$

The *Cost-Effective Subgraph* problem is to find a subset X of minimum cost. In the decision version of the problem, we are given an integer r and we need to decide if there is a subset X whose cost is at most r. Notice that $X = \emptyset$ or $X = V$ are possible solutions.

The idea is that each edge "outside" of X costs c and each edge "inside" costs $|X|$. Therefore, we pay for each edge not included in X, but if X gets too large, we pay more for edges in X. Hence, we need to find a balance between the size of X and its number of edges. The connection with the TD problem can be roughly described as follows: in our reduction, we will have many substrings which need to be deleted through contractions. We will have to choose an initial set of contractions X and then, each substring will have two ways to be contracted: one way requires c contractions, and the other requires $|X|$.

Clearly, an obvious solution for a *Cost-Effective Subgraph* is to take $X = \emptyset$, which is of cost $c|E(G)|$. Then, another formulation of the problem could be whether there is a subset X of cost at most $c|E(G)| - p$, where p can be seen as a "profit" to maximize. Treating c and p as parameters, we show the NP-hardness and W[1]-hardness in parameters $c + p$ of the *Cost-Effective Subgraph* problem (the parameter r is not studied here). Our reduction to the TD problem does not preserve W[1]-hardness and we only use its NP-hardness, but the W[1]-hardness might be of independent interest.

Theorem 1. *The Cost-Effective Subgraph problem is NP-hard and W[1]-hard for parameter $c + p$.*

Proof. The reduction is from CLIQUE. Let (G, k) be the input for the CLIQUE problem with $G = (V, E)$, let $n = |V|$ and $m = |E|$. Without loss of generality, let k be even (the proof could be easily adapted for odd k). By setting

$$r = cm - p,$$

with $c = \frac{3k}{2}$ and $p = \frac{k}{2}\binom{k}{2}$, we can show that (G, k) is a YES instance for CLIQUE if and only if G contains a set $X \subseteq V$ of cost at most r. We leave out the details for the arguments, which can be found in [34,35].   □

Note that the W[1]-hardness follows as the parameter $c + p = \frac{3k}{2} + \frac{k}{2}\binom{k}{2}$, which is a function of k. On the other hand, r involves both k and m hence is not just a function of k. Therefore, the proof does not imply W[1]-hardness when r is used as the parameter.

The Reduction Ideas

As the reduction is quite involved, we first start with an overview. Let (G, c, r) be a *Cost-Effective Subgraph* instance where c is the parameter and r the optimal cost, and with vertices $V(G) = \{v_1, \ldots, v_n\}$. We construct strings S and T and compute the number of contractions to convert T to S. The source string is $S = x_1 x_2 \ldots x_n$, where each x_i is a distinct character corresponding to vertex v_i. Let S' be constructed by doubling every x_i, i.e., $S' = x_1 x_1 x_2 x_2 \ldots x_n x_n$. Our goal is to put $T = S' E_1 E_2 \ldots E_m$, where each E_i is a substring gadget corresponding to edge $e_i \in E(G)$ that we must remove to go from T to S. Assuming that there is a sequence of contractions that transforms T into S, we make it so that we first want to contract some, but not necessarily all, of the doubled x_i's of S', resulting in another string S''. Let t be the number of x_i's contracted from S' to S''. For instance, we could have $S'' = x_1 x_1 x_2 x_3 x_3 x_4 x_5 x_5$, where only x_2 and x_4 were contracted, and thus $t = 2$. The idea is that these contracted x_i's correspond to the vertices of a cost-effective subgraph. After T is contracted to $S'' E_1 \ldots E_m$, we then force each E_i to use S'' to further contract. For $m = 3$, a contraction sequence that we would like to enforce would take the form

$$\underline{S' E_1 E_2 E_3} \rightarrowtail \underline{S'' E_1} E_2 E_3 \rightarrowtail \underline{S'' E_2} E_3 \rightarrowtail \underline{S'' E_3} \rightarrowtail \underline{S''} \rightarrowtail S,$$

where the underlined substrings are affected by contractions at each step. Moreover, we make it so that when contracting $S'' E_i E_{i+1} \ldots E_m$ into $S'' E_{i+1} \ldots E_m$, we have two options. Suppose that v_j and v_k are the vertices of edge e_i. If, in S'', we had chosen to contract x_j and x_k, we could contract E_i using a sequence of t moves. Otherwise, we must contract E_i using another more costly sequence of c moves. The total cost to eliminate the E_i gadgets will be $c(m - e) + te$, where e is the number of edges that can be contracted using the first choice, i.e., for which both endpoints were chosen in S''.

Intuitively, the main difficulty in the above idea is in forcing an optimal solution to behave as we describe, i.e., enforcing going from S' to S'' first, enforcing the E_i's to use S'', and enforcing the two options to contract E_i with the desired costs. Specifically, we must replace the x_i's by carefully constructed substrings X_i. We must also repeat the sequence of E_i's a certain number p times. We illustrate more technical details next.

The Reduction Details

We first construct an exemplar string $X = x_1 \ldots x_l$ (i.e., a string in which no character occurs twice). We can double its characters and obtain a string $X' =$

$x_1 x_1 \ldots x_l x_l$. The length of X' is only twice that of X and $dist_{TD}(X, X') = l$, i.e., going from X' to X requires l contractions.

Now we show how to construct S and T. First let $d = m + 1$ and $p = m(n+m)^{10}$. The exact values of p is not crucial and it is enough to think of p as "large enough", though it should be noted that p is a multiple of m. The optimal number of contractions from T to S, k, is defined as $k = p/m \cdot d(r + nm) + 4cdn$.

Instead of doubling x_i's as in the intuition paragraph above, we duplicate some characters d times. Moreover, we cannot create a T string that behaves exactly as described above, but we will show that we can append p copies of carefully crafted substring to obtain the desired result. We need d and p to be large enough so that "enough" copies behave as we desire.

For each $i \in [n]$, construct an exemplar string X_i of length d. In this case, enough characters are used so that no two X_i strings contain a common character. Let X_i^d be a string satisfying $dist_{TD}(X_i, X_i^d) = d$.

Then for each $j \in \{0, 1, \ldots, 2p\}$, construct an exemplar string B_j. Again, we ensure that no B_j contains a character from an X_i string, and no two B_j's contain a common character. Except for B_0 and B_1, the B_j strings can consist of a single character. Assume that for B_0 and B_1, we have strings B_0^* and B_1^* such that

$$dist_{TD}(B_0, B_0^*) = dc + 2d - 2,$$
$$dist_{TD}(B_1, B_1^*) = dn + 2d - 1,$$

which can be done using the doubling trick on exemplar strings.

Note that the B_j's are the building blocks of larger strings. For each $q \in [2p]$, define

$$\mathcal{B}_q = B_q B_{q-1} \ldots B_2 B_1 B_0, \qquad \mathcal{B}_q^0 = B_q B_{q-1} \ldots B_2 B_1 B_0^*,$$
$$\mathcal{B}_q^1 = B_q B_{q-1} \ldots B_2 B_1^* B_0, \qquad \mathcal{B}_q^{01} = B_q B_{q-1} \ldots B_2 B_1^* B_0^*.$$

These strings serve as "blockers" and prevent certain contractions from happening. Note that $\mathcal{B}_q^0$ and $\mathcal{B}_q^1$ can be turned into $\mathcal{B}_q$ using $dc + 2d - 2$ contractions and $dn + 2d - 1$ contractions, respectively. Moreover, $\mathcal{B}_q^{01}$ can be turned into $\mathcal{B}_q^0$ using $dn + 2d - 1$ contractions and into $\mathcal{B}_q^1$ using $dc + 2d - 2$ contractions.

Also define the strings

$$\mathcal{X} = X_1 X_2 \ldots X_n, \qquad \mathcal{X}^d = X_1^d X_2^d \ldots X_n^d,$$

and for edge $e_q = v_i v_j$ with $q \in [p]$ whose vertices are v_i and v_j, define

$$\mathcal{X}_{e_q} = X_1^d \ldots X_{i-1}^d X_i X_{i+1}^d \ldots X_{j-1}^d X_j X_{j+1}^d \ldots X_n^d.$$

Therefore in $\mathcal{X}_{e_q}$, all X_k substrings are turned into X_k^d, except X_i and X_j.

Finally, construct a new additional character Δ, which will be used to separate some of the components of our string. We can then define S and T. We have

$$S = \mathcal{B}_{2p} \mathcal{X} \Delta = B_{2p} B_{2p-1} \ldots B_2 B_1 B_0 X_1 X_2 \ldots X_n \Delta.$$

It follows from the definitions of $\mathcal{B}_{2p}, \mathcal{X}$ and Δ that S is exemplar. Now for $i \in [p]$, define

$$E_i := \mathcal{B}_i^{01} \mathcal{X}_{e_i} \Delta \mathcal{B}_{2p}^1 \mathcal{X} \Delta,$$

which we will call the *edge gadget*. Define T as

$$T = \mathcal{B}_{2p}^0 \mathcal{X}^d \Delta \mathcal{B}_{2p}^1 \mathcal{X} \Delta E_1 E_2 \ldots E_p$$
$$= \mathcal{B}_{2p}^0 \mathcal{X}^d \Delta \mathcal{B}_{2p}^1 \mathcal{X} \Delta \left[\mathcal{B}_1^{01} \mathcal{X}_{e_1} \Delta \mathcal{B}_{2p}^1 \mathcal{X} \Delta \right] \left[\mathcal{B}_2^{01} \mathcal{X}_{e_2} \Delta \mathcal{B}_{2p}^1 \mathcal{X} \Delta \right] \ldots \left[\mathcal{B}_p^{01} \mathcal{X}_{e_p} \Delta \mathcal{B}_{2p}^1 \mathcal{X} \Delta \right].$$

Note that the brackets are used as separators for clarity purpose, and they are not the characters in T. We summarize the theorem as follows.

Theorem 2. *The Exemplar-TD problem is NP-complete, even if for the given string S and T, $S \Rightarrow_* T$ is guaranteed to hold.*

Proof. The problem is obviously in NP. Note that $dist_{TD}(S, T) \leq |T|$ since each contraction from T to S removes at least one character. Thus it is easy to verify whether a sequence of contractions serves as a valid certificate in polynomial time.

For hardness, we reduce from the Cost-Effective Subgraph problem, which has been shown NP-hard in Theorem 1. Let (G, c, r) be an instance of Cost-Effective Subgraph, letting $n := |V(G)|$ and $m := |E(G)|$. Here c is the "outsider edge" cost and we ask whether there is a subset $X \subseteq V(G)$ such that $c(m - |E(X)|) + |X||E(X)| \leq r$. We denote $V(G) = \{v_1, \ldots, v_n\}$ and $E(G) = \{e_1, \ldots, e_m\}$. The ordering of vertices and edges is arbitrary but remains fixed for the remainder of the proof. For convenience, we allow the edge indices to loop through 1 to m, and so we put $e_i = e_{i+lm}$ for any integer $l \geq 0$. Thus we may sometimes refer to an edge e_h with an index $h > m$, meaning that e_h is actually the edge $e_{((h-1) \bmod m)+1}$.

From the above construction, S and T can be constructed in polynomial time. It can be shown that G admits a subset of vertices W of cost at most r if and only if T can be contracted to S using at most $k = p/m \cdot d(r + nm) + 4cdn$ contraction operations. Due to space constraints, we leave out the detailed arguments which can be found in [34,35].   □

3.2   Exemplar-TD Is FPT

Recall that the *Exemplar-k-TD* instance is (S, T) with k being the solution size. It is obvious that with a trivial bounded-degree search method, *Exemplar-k-TD* can be solved in $O(n^{2k})$ time, where n is the input size. To obtain an FPT algorithm, we need more properties to avoid such a brute-force kind of search.

Let x and y be two consecutive characters in S (i.e., xy is a substring of S). We say that xy is (S, T)-*stable* if in T, every occurrence of x in T is followed by y and every occurrence of y is preceded by x. That is, the direct successor of every x character is y, and the direct predecessor of every y character is x. An (S, T)-*stable substring* $X = x_1 \ldots x_l$, where $l \geq 2$, is a substring of S such that $x_i x_{i+1}$ is (S,T)-stable for every $i \in [l-1]$. We also define a string with a single character x_i to be a (S,T)-stable substring (provided that x_i appears in S and T). If any substring of S that strictly contains X is not an (S,T)-stable substring, then X is called a *maximal (S,T)-stable substring*. Note that these definitions are independent of S and T, hence so the same definitions apply for (X,Y)-stability, for any strings X and Y.

It can be shown that every maximal (S,T)-stable substring can be replaced by a single character, and that if T can be obtained from S using at most k tandem duplications, then this leaves strings of bounded size. First of all, it can be shown that, roughly speaking, stability is maintained by all tandem duplications when going from S to T. We leave the detailed proofs to [34,35].

Lemma 1. *Suppose that $dist_{TD}(S,T) = k$ and let X be an (S,T)-stable substring. Let $S = S_0, S_1, \ldots, S_k = T$ be any minimum sequence of strings transforming S to T by tandem duplications. Then X is (S,S_i)-stable for every $i \in [k]$.*

The next lemma shows that we can assume that maximal stable substrings never get cut, and thus always get duplicated together. The idea is that any duplication that cuts an X_j can be replaced by an equivalent duplication that does not.

Lemma 2. *Suppose that $dist_{TD}(S,T) = k$, and let $X_1, \ldots, X_l$ be the set of maximal (S,T)-stable substrings. Then there exists a sequence of tandem duplications $D_1, \ldots, D_k$ transforming S into T such that no occurrence of an X_j gets cut by a D_i. In other words, for all $i \in [k]$ and all $j \in [l]$, the tandem duplication D_i does not cut any occurrence of X_j in the string obtained by applying $D_1, \ldots, D_{i-1}$ to S.*

The above lemma implies that we may replace each maximal (S,T)-stable substring X of S and T by a single character, since we may assume that characters of X are always duplicated together (assuming, of course, that S is exemplar). It only remains to show that the resulting strings are small enough. The proof of the following lemma has a very simple intuition. First, S has exactly 1 maximal (S,S)-stable substring. Each time we apply a duplication, we "break" at most 2 stable substrings, which creates 2 new ones. So if we apply k duplications, there are at most $2k+1$ such substrings in the end.

Lemma 3. *If $dist_{TD}(S,T) \leq k$, then there are at most $2k+1$ maximal (S,T)-stable substrings.*

Consequently, we have the following theorem and corollary,

Theorem 3. *An instance (S,T) of Exemplar-k-TD admits a kernel (S',T') in which $|S'| \leq 2k+1$ and $|T'| \leq (2k+1)2^k$.*

The kernelization can be done in polynomial time, as one only needs to identify maximal (S,T)-stable substrings and contract them. Now running the brute-force algorithm on the kernel, we have the following corollary.

Corollary 1. *The exemplar k-tandem duplication problem can be solved in time* $O(((2k+1)2^k)^{2k} + poly(n)) = 2^{O(k^2)} + poly(n)$, *where n is the size of the input.*

There are still many open questions on the TD-distance problems. We will go over them in Sect. 5. We now shift our attention to problems related to Copy Number Profiles.

4   Results on Copy Number Profiles

4.1   Hardness of Approximation for MCNG

We first show that the d_{GCNP} distance is hard to approximate within any constant factor. At a first step, we show this result when only deletions on G are allowed. In fact, this restriction makes the proof significantly simpler. We then extend this result to deletions *and* duplications.

Both proofs are based on a reduction from SET-COVER. Recall that in SET-COVER, we are given a collection of sets $\mathcal{S} = \{S_1, S_2, \ldots, S_n\}$ over universe $U = \{u_1, u_2, \ldots, u_m\} = \bigcup_{S_i \in \mathcal{S}} S_i$, and we need to find a set cover of $\mathcal{S}$ having minimum cardinality (a set cover of $\mathcal{S}$ is a subset $\mathcal{S}^* \subseteq \mathcal{S}$ such that $\bigcup_{S \in \mathcal{S}^*} S = U$). If $\mathcal{S}'$ is a set cover in which no two sets intersect, then $\mathcal{S}'$ is called an *exact cover*.

There is one interesting feature (or constraint) of our reduction g, which transforms a SET-COVER instance $\mathcal{S}$ into a MCNG instance $g(\mathcal{S})$. A set cover $\mathcal{S}^*$ only works on $g(\mathcal{S})$ if $\mathcal{S}^*$ is actually an exact cover, and a solution for $g(\mathcal{S})$ can be turned into a set cover for $\mathcal{S}^*$ that is not necessarily exact. Hence it would be hard to reduce directly from either (the general version of) SET-COVER nor its exact version. We provide a general-purpose lemma for such situations, and our reductions serve as an example of its usefulness.

The proof is based on a result on t-SET-COVER, the special case of SET-COVER in which every given set contains at most t elements. It is known that for any constant $t \geq 3$, the t-SET-COVER problem is hard to approximate within a factor $\ln t - c \ln \ln t$ for some constant c not depending on t [56].

Lemma 4. *Let $\mathcal{B}$ be a minimization problem, and let g be a function that transforms any SET-COVER instance $\mathcal{S}$ into an instance $g(\mathcal{S})$ of $\mathcal{B}$ in polynomial time. Assume that both the following statements hold:*

- *any exact cover $\mathcal{S}^*$ of $\mathcal{S}$ of cardinality at most k can be transformed in polynomial time into a solution of value at most k for $g(\mathcal{S})$;*
- *any solution of value at most k for $g(\mathcal{S})$ can be transformed in polynomial time into a set cover of $\mathcal{S}$ of cardinality at most k.*

Then unless $P = NP$, there is no constant factor approximation algorithm for $\mathcal{B}$.

Proof. Using contradiction, suppose that $\mathcal{B}$ admits a factor b approximation for some constant b. Choose any constant t such that t-SET-COVER is hard to approximate within factor $\ln t - c \ln \ln t$, and such that $b < \ln t - c \ln \ln t$. Note that t might be exponentially larger than b, but is still a constant.

Now, let $\mathcal{S}$ be an instance of t-SET-COVER with universe $U = \{u_1, \ldots, u_m\}$. Consider the intermediate reduction g' that transforms $\mathcal{S}$ into another t-SET-COVER instance $g'(\mathcal{S}) = \{S' \subseteq S : S \in \mathcal{S}\}$. Since t is a constant, $g(\mathcal{S})$ has $O(|\mathcal{S}|)$ sets and this can be carried out in polynomial time.

Now define $\mathcal{S}' = g'(\mathcal{S})$ and consider the instance $B = g(\mathcal{S}') = g(g'(\mathcal{S}))$. By the assumptions of the lemma, a solution for B of value k yields a set cover $\mathcal{S}^*$ for $\mathcal{S}'$. Clearly, $\mathcal{S}^*$ can be transformed into a set cover for instance $\mathcal{S}$: for each $S' \in \mathcal{S}^*$, there exists $S \in \mathcal{S}$ such that $S' \subseteq S$, so we get a set cover for $\mathcal{S}$ by adding this corresponding superset for each $S \in \mathcal{S}^*$. Thus B yields a set cover of $\mathcal{S}$ with at most k sets.

In the other direction, consider a set cover $\mathcal{S}^* = \{S_1, \ldots, S_k\}$ of $\mathcal{S}$ with k sets. This easily translates into an *exact* cover of $\mathcal{S}'$ with k sets by taking the collection

$$\{S_1, S_2 \setminus S_1, S_3 \setminus (S_1 \cup S_2), \ldots, S_k \setminus \bigcup_{i=1}^{k-1} S_i\}\}.$$

By the assumptions of the lemma, this exact cover can then be transformed into a solution of value at most k for instance B.

Therefore, $\mathcal{S}$ has a set cover of cardinality at most k if and only if B has a solution of value at most k. Since there is a correspondence between the solution values of the two problems, a factor b approximation for B would provide a factor $b < \ln t - c \ln \ln t$ approximation for t-SET-COVER, which incurs a contradiction to the result by Trevisan [56].   □

Constructing Genomes and CNPs from SET-COVER Instances

All of our inapproximability results rely on Lemma 4. We need to provide a reduction from SET-COVER to MCNG and prove that both assumptions of the lemma are satisfied.

$$S_1 = \{1, 2, 3\} \quad S_2 = \{1, 2, 4\} \quad S_3 = \{1, 3, 5\}$$

$$G = \langle \beta_{S_1} \rangle \alpha_1 \alpha_2 \alpha_3 \langle \beta_{S_2} \rangle \alpha_1 \alpha_2 \alpha_4 \langle \beta_{S_3} \rangle \alpha_1 \alpha_3 \alpha_5$$

$$c(\alpha_1) = 2 \quad c(\alpha_2) = c(\alpha_3) = 1 \quad c(\alpha_4) = c(\alpha_5) = 0$$

Fig. 3. An example of our construction, with $\mathcal{S} = \{S_1, S_2, S_3\}$ and $U = \{1, 2, 3, 4, 5\}$.

The reduction is in fact the same for deletions-only as well as the case with both deletions and duplications. Given $\mathcal{S}$ and U, we construct a genome G and a CNP c as follows (an example is illustrated in Fig. 3). The alphabet is $\Sigma = \Sigma_\mathcal{S} \cup \Sigma_U$, where $\Sigma_\mathcal{S} := \{\langle \beta_{S_i} \rangle : S_i \in \mathcal{S}\}$ and $\Sigma_U := \{\alpha_{u_i} : u_i \in U\}$. Thus, there is one character for each set of $\mathcal{S}$ and each element of U. Here, each $\langle \beta_{S_i} \rangle$ is a character that will serve as a separator between characters to delete. For a set $S_i \in \mathcal{S}$, define the string $q(S_i)$ as any string that contains each character of $\{\alpha_u : u \in S_i\}$ exactly once. We put

$$G = \langle \beta_{S_1} \rangle q(S_1) \langle \beta_{S_2} \rangle q(S_2) \ldots \langle \beta_{S_n} \rangle q(S_n),$$

i.e., G is the concatenation of the strings $\langle \beta_{S_i} \rangle q(S_i)$. As for the CNP c, put

- $c(\langle \beta_{S_i} \rangle) = 1$ for each $S_i \in \mathcal{S}$;
- $c(\alpha_u) = f(u) - 1$ for each $u \in U$, where $f(u) = |\{S_i \in \mathcal{S} : u \in S_i\}|$ is the number of sets from $\mathcal{S}$ that contain u.

Note that in G, each $\langle \beta_S \rangle$ already has the correct copy-number, whereas each α_u needs exactly one less copy. Our goal is thus to reduce the number of each α_u by 1. This concludes the construction of MCNG instances from SET-COVER instances. We now focus on the hardness of the deletions-only case.

The Deletions-Only Case

Suppose that we are given a set cover instance $\mathcal{S}$ and U, and let G and c be the genome and CNP, respectively, as constructed above.

Lemma 5. *Given an exact cover $\mathcal{S}^*$ for $\mathcal{S}$ of cardinality k, one can obtain a sequence of k deletions transforming G into a genome with CNP c.*

Lemma 6. *Given a sequence of k deletions transforming G into a genome with CNP c, one can obtain a set cover for $\mathcal{S}$ of cardinality at most k.*

Proof. Suppose that the deletion events $E = (e_1, \ldots, e_k)$ transform G into a genome G^* with CNP c. Note that no e_i deletion is allowed to delete a set-character $\langle \beta_{S_i} \rangle \in \Sigma_{\mathcal{S}}$, as there is only one occurrence of $\langle \beta_{S_i} \rangle$ in G and $c(\langle \beta_{S_i} \rangle) = 1$. Thus all deletions remove only α_u characters. In other words, each e_j in E either deletes a substring of G between some $\langle \beta_{S_i} \rangle$ and $\langle \beta_{S_{i+1}} \rangle$ with $1 \leq i < n$, or e_j deletes a substring after $\langle \beta_{S_n} \rangle$. Moreover, exactly one of each α_u occurrences gets deleted from G.

Call $\langle \beta_{S_i} \rangle \in \Sigma_{\mathcal{S}}$ *affected* if there is some event of E that deletes at least one character between $\langle \beta_{S_i} \rangle$ and $\langle \beta_{S_{i+1}} \rangle$ with $1 \leq i < n$, and call $\langle \beta_{S_n} \rangle$ affected if some event of E deletes characters after $\langle \beta_{S_n} \rangle$. Let $\mathcal{S}^* := \{S_i \in \mathcal{S} : \langle \beta_{S_i} \rangle$ is affected$\}$. Then $|\mathcal{S}^*| \leq k$, since each deletion affects at most one $\langle \beta_{S_i} \rangle$ and there are k deletion events. Moreover, $\mathcal{S}^*$ must be a set cover, because each $\alpha_u \in \Sigma_U$ has at least one occurrence that gets deleted and thus at least one set containing u that is included in $\mathcal{S}^*$. This concludes the proof. $\square$

We have shown that both the assumptions required by Lemma 4 are satisfied. Hence the inapproximability follows.

Theorem 4. *Assuming $P \neq NP$, there is no polynomial-time constant factor approximation algorithm for MCNG when only deletions are allowed.*

We comment that the reduction should be adaptable to the duplication-only case, by putting $c(\alpha_u) = f(u) + 1$ for each $u \in U$.

The Case with Both Deletions and Duplications

We now consider both deletions and duplications. The reduction uses the same construction as at the beginning of Sect. 4.1. Thus we assume that we have a SET-COVER instance S over U, and a corresponding instance of MCNG with genome G and CNP c.

In that case, we observe the following: Lemma 5 still holds whether we allow deletion only, or both deletions and duplications. Thus we only need to show that the second assumption of Lemma 4 holds.

On the other hand, this is not as simple as in the deletions-only case. The problem is that some duplications may copy some α_u and $\langle \beta_{S_i} \rangle$ occurrences, and we lose control over what gets deleted, and over what $\langle \beta_{S_i} \rangle$ each α_u corresponds to (in particular, some $\langle \beta_{S_i} \rangle$ might now get deleted, which did not occur in the deletions-only case). Nonetheless, the analogous result can be shown to hold.

Lemma 7. *Given a sequence of k events (deletions and duplications) transforming G into a genome with CNP c, one can obtain a set cover for S of cardinality at most k.*

Due to space constraints, we redirect the reader to [36] for the detailed proof. In a nutshell, given a sequence of events from G to a genome with CNP c, the idea is to find, for each $u \in U$, one occurrence of α_u in G that we have control over. More precisely, even though that occurrence of α_u might spawn duplicates, all its copies (and copies of copies, and so on) will eventually get deleted. The $\langle \beta_{S_i} \rangle$ character preceding this α_u character indicates that S_i can be added to a set cover. The crux of the proof is to show that this α_u character exists for each $u \in U$, and that their corresponding $\langle \beta_{S_i} \rangle$ form a set cover of size at most k.

We summarize our main inapproximability result, which again follows from Lemma 4.

Theorem 5. *Unless $P = NP$, there is no polynomial-time constant factor approximation algorithm for MCNG.*

In the next section, we prove that the MCNG problem, parameterized by the solution size, is W[1]-hard. This answers another open question in [48].

4.2   W[1]-Hardness for MCNG

Since SET-COVER is W[2]-hard, naturally we would like to use the ideas from the above reduction to prove the W[2]-hardness of MCNG. However, the fact that we use t-SET-COVER with constant t in the proof of Lemma 4 is crucial, and t-SET-COVER is in FPT (as the universe would have at most tk elements if such an instance admits a size-k solution). On the other hand, the property that is really needed in the instance of this proof, and in our MCNGreduction, is that we can transform any set cover instance into an exact cover. We capture this intuition in the following, and show that SET-COVER instances that have this property are W[1]-hard to solve.

An instance of SET-COVER-with-EXACT-COVER (SET-COVER-EC for short) is a pair $I = (\mathcal{S}, k)$ where k is an integer and $\mathcal{S}$ is a collection of sets forming a universe U. In this problem, we require that $\mathcal{S}$ satisfies the property that *any* set cover for $\mathcal{S}$ of size at most k is also an exact cover. We are asked whether there exists a set cover for $\mathcal{S}$ of size at most k (in which case this set cover is also an exact cover). Therefore, SET-COVER-EC is a promise problem.

Lemma 8. *The SET-COVER-EC problem is W[1]-hard for parameter k.*

Proof. We show W[1]-hardness using the MULTICOLORED-CLIQUE technique introduced by Fellows *et al.* [21]. In the MULTICOLORED-CLIQUE problem, we are given a graph G, an integer k and a coloring $c : V(G) \to [k]$ such that no two vertices of the same color share an edge. We are asked whether G contains a clique of k vertices (note that such a clique must have a vertex of each color). This problem is W[1]-hard with respect to k.

Given an instance (G, k, c) of MULTICOLORED-CLIQUE, we construct an instance $I = (\mathcal{S}, k')$ of SET-COVER-EC. We put $k' = k + \binom{k}{2}$. For $i \in [k]$, let $V_i = \{v \in V(G) : c(v) = i\}$ and for each pair $i < j \in [k]$, let $E_{ij} = \{uv \in E(G) : u \in V_i, v \in V_j\}$. The universe U of the SET-COVER-EC instance has one element for each color i, one element for each pair $\{i, j\}$ of distinct colors, and two elements for each edge, one for each direction of the edge. That is,

$$U = [k] \cup \binom{[k]}{2} \cup \{(u, v) \in V(G) \times V(G) : uv \in E(G)\}$$

Thus $|U| = k + \binom{k}{2} + 2|E(G)|$. For two colors $i < j \in [k]$, we will denote $U_{ij} = \{(u, v), (v, u) : u \in V_i, v \in V_j, uv \in E_{ij}\}$, i.e. we include in U_{ij} both elements corresponding to each $uv \in E_{ij}$. Now, for each color class $i \in [k]$ and each vertex $u \in V_i$, add to $\mathcal{S}$ the set

$$S_u = \{i\} \cup \{(u, v) : v \in N(u)\},$$

where $N(u)$ is the set of neighbors of u in G. Then for each $i < j \in [k]$, and for each edge $uv \in E_{ij}$, add to $\mathcal{S}$ the set

$$S_{uv} = \{\{i, j\}\} \cup \{(x, y) \in U_{ij} : x \notin \{u, v\}\}.$$

The idea is that S_{uv} can cover every element of U_{ij}, except those ordered pairs whose first element is u or v. Then if we do decide to include S_{uv} in a set cover, it turns out that we will need to include S_u and S_v to cover these missing ordered pairs. See Fig. 4 for an example. For instance if we include S_{u_2, v_3} in a cover, the uncovered (u_2, v_3) and (v_3, u_2) can be covered with S_{u_2} and S_{v_3}. We could show that G has a multicolored clique of size k if and only if $\mathcal{S}$ admits a set cover of size k'. We could also prove that $(\mathcal{S}, k')$ is an instance of SET-COVER-EC, i.e. that any set cover of size at most k' is also an exact cover. We refer the detailed arguments to [36].   □

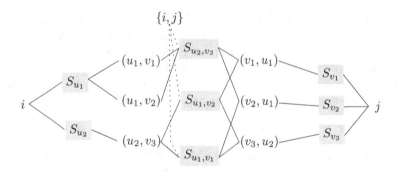

Fig. 4. A graphical example of the constructed sets for the U_{ij} elements of a graph (not shown) with $E_{ij} = \{u_1 v_1, u_1 v_2, u_2 v_3\}$, where the u_l's are in V_i and the v_l's in V_j (sets have a gray background, edges represent containment, the $\{i, j\}$ lines are dotted only for better visualization).

It is now almost immediate that MCNG is W[1]-hard with respect to the natural parameter, namely the number of events to transform a genome G into a genome with a given profile c. We hence have the following theorem.

Theorem 6. *The MCNG problem is W[1]-hard.*

Now that we have finished presenting the negative results on MCNG. An immediate question is whether we could obtain some positive result on a related problem. In the next subsection, we present some positive result for an interesting variation of MCNG.

4.3   The Copy Number Profile Conforming Problem

We define the more general *Copy Number Profile Conforming (CNPC)* problem as follows:

Definition 4. *Given two CNP's $c_1 = \langle u_1, u_2, ..., u_n \rangle$ and $v_2 = \langle v_1, v_2, ..., v_n \rangle$, with $u_i, v_i \geq 0$ and $u_i, v_i \in \mathbb{N}$, the CNPC problem asks to compute two strings S_1 and S_2 with $cnp(S_1) = c_1$ and $cnp(S_2) = c_2$ such that the distance between S_1 and S_2, $d(S_1, S_2)$, is minimized.*

Let $\sum_i u_i = m_1, \sum_i v_i = m_2$, we assume that m_1 and m_2 are bounded by a polynomial of n. (This assumption is needed as the solution of our algorithm could be of size $\max\{m_1, n_2\}$.) We simply say c_1, c_2 are *polynomially bounded*. Note that $d(S_1, S_2)$ is a very general distance measure, i.e., it could be any genome rearrangement distance (like reversal, transposition, and tandem duplication, etc, or their combinations, e.g., tandem duplication + deletion). In this paper, we use the breakpoint distance (and the adjacency number), which is defined as follows. (These definitions are adapted from Angibaud *et al.* [2] and Jiang *et al.* [31], which generalize the corresponding concepts on permutations [58]. See [59] for a recent survey along this line.)

Given two sequences $A = a_1a_2 \cdots a_n$ and $B = b_1b_2 \cdots b_m$ over the sam alphabet, if the multisets $\{a_i, a_{i+1}\} = \{b_j, b_{j+1}\}$ we say that a_ia_{i+1} and b_jb_{j+1} are *matched* to each other. In a maximum matching of 2-substrings in A and B, a matched pair is called an *adjacency*, and an unmatched pair is called a *breakpoint* in A and B respectively. Then, the number of breakpoints in A (resp. B) is denoted as $d_b(A, B)$ (resp. $d_b(B, A)$), and the number of (common) adjacencies between A and B is denoted as $a(A, B)$. For example, if $A = acbdcb, B = abcdabcd$, then $a(A, B) = 3$ and there are 2 and 4 breakpoints in A and B respectively.

Coming back to our problem, we define $d(S_1, S_2) = d_b(S_1, S_2) + d_b(S_2, S_1)$. From the definitions, we have

$$d_b(S_1, S_2) + d_b(S_2, S_1) + 2 \cdot a(S_1, S_2) = (m_1 - 1) + (m_2 - 1),$$

or,

$$d_b(S_1, S_2) + d_b(S_2, S_1) = m_1 + m_2 - 2 \cdot a(S_1, S_2) - 2.$$

Hence, the problem is really to maximize $a(S_1, S_2)$.

Definition 5. *Given n-dimensional vectors $\boldsymbol{u} = \langle u_1, u_2, ..., u_n \rangle$ and $\boldsymbol{w} = \langle w_1, w_2, ..., w_n \rangle$, with $u_i, w_i \geq 0$, and $u_i, w_i \in \mathbb{N}$, we say $\boldsymbol{w}$ is a sub-vector of $\boldsymbol{u}$ if $w_i \leq u_i$ for $i = 1, ..., n$, also denote this relation as $\boldsymbol{w} \leq \boldsymbol{u}$.*

Henceforth, we simply call $\boldsymbol{u}, \boldsymbol{w}$ integer vectors (with the understanding that no item in a vector is negative).

Definition 6. *Given two n-dimensional integer vectors $\boldsymbol{u} = \langle u_1, u_2, ..., u_n \rangle$ and $\boldsymbol{v} = \langle v_1, v_2, ..., v_n \rangle$, with $u_i, v_i \geq 0$, and $u_i, v_i \in \mathbb{N}$, we say $\boldsymbol{w}$ is a common sub-vector of $\boldsymbol{u}$ and $\boldsymbol{v}$ if $\boldsymbol{w}$ is a sub-vector of $\boldsymbol{u}$ and $\boldsymbol{w}$ is also a sub-vector of $\boldsymbol{v}$ (i.e., $\boldsymbol{w} \leq \boldsymbol{u}$ and $\boldsymbol{w} \leq \boldsymbol{v}$). Finally, $\boldsymbol{w}$ is the maximum common sub-vector of $\boldsymbol{u}$ and $\boldsymbol{v}$ if there is no common sub-vector $\boldsymbol{w}' \neq \boldsymbol{w}$ of $\boldsymbol{u}$ and $\boldsymbol{v}$ which satisfies $\boldsymbol{w} \leq \boldsymbol{w}' \leq \boldsymbol{u}$ or $\boldsymbol{w} \leq \boldsymbol{w}' \leq \boldsymbol{v}$.*

An example is illustrated as follows. Take $\boldsymbol{u} = \langle 3, 2, 1, 0, 5 \rangle$, $\boldsymbol{v} = \langle 2, 1, 3, 1, 4 \rangle$, $\boldsymbol{w}' = \langle 2, 1, 0, 0, 3 \rangle$ and $\boldsymbol{w} = \langle 2, 1, 1, 0, 4 \rangle$. Both $\boldsymbol{w}$ and $\boldsymbol{w}'$ are common sub-vectors for $\boldsymbol{u}$ and $\boldsymbol{v}$, $\boldsymbol{w}'$ is not the maximum common sub-vector of $\boldsymbol{u}$ and $\boldsymbol{v}$ (since $\boldsymbol{w}' \leq \boldsymbol{w}$) while $\boldsymbol{w}$ is.

Given a CNP $\boldsymbol{u} = \langle u_1, u_2, ..., u_n \rangle$ and alphabet $\Sigma = (x_1, x_2, ..., x_n)$, for $i \in \{1, 2\}$, we use $S(\boldsymbol{u})$ to denote the multiset of letters (genes) corresponding to $\boldsymbol{u}$; more precisely, u_i denotes the number of x_i's in $S(\boldsymbol{u})$. Similarly, given a multiset of letters Z, we use $s(Z)$ to denote a string where all the letters in Z appear exactly once (counting multiplicity; i.e., $|Z| = |s(Z)|$). $s(Z)$ is similarly defined when Z is a CNP. We present Algorithm 1 as follows.

1. Compute the maximum common sub-vector $\boldsymbol{v}$ of $\boldsymbol{c}_1$ and $\boldsymbol{c}_2$.
2. Given the gene alphabet Σ, compute $S(\boldsymbol{v})$, $S(\boldsymbol{c}_1)$ and $S(\boldsymbol{c}_2)$. Let $X = S(\boldsymbol{c}_1) - S(\boldsymbol{v})$ and $Y = S(\boldsymbol{v}_2) - S(\boldsymbol{v})$.
3. If $S(\boldsymbol{v}) = \emptyset$, then return two arbitrary strings $s(\boldsymbol{c}_1)$ and $s(\boldsymbol{c}_2)$ as S_1 and S_2, exit; otherwise, continue.

4. Find $\{x, y\}$, $x, y \in \Sigma$ and $x \neq y$, such that $x \in S(v)$ and $y \in S(v)$, and exactly one of x, y is in X (say $x \in X$), and the other is in Y (say $y \in Y$). If such an $\{x, y\}$ cannot be found then return two strings S_1 and S_2 by concatenating letters in X and Y arbitrarily at the ends of $s(v)$ respectively, exit; otherwise, continue.
5. Compute an arbitrary sequence $s(v)$ with the constraint that the first letter is x and the last letter is y. Then obtain $s_1 = s(v) \circ x$ and $s_2 = y \circ s(v)$ ($\circ$ is the concatenation operator).
6. Finally, insert all the elements in $X - \{x\}$ arbitrarily at the two ends of s_1 to obtain S_1, and insert all the elements in $Y - \{y\}$ arbitrarily at the two ends of s_2 to obtain S_2.
7. Return S_1 and S_2.

Let $\Sigma = \{a, b, c, d, e\}$. Also let $c_1 = \langle 2, 2, 2, 4, 1 \rangle$ and $c_2 = \langle 4, 4, 1, 1, 1 \rangle$. We walk through the algorithm using this input as follows.

1. The maximum common sub-vector v of c_1 and c_2 is $v = \langle 2, 2, 1, 1, 1 \rangle$.
2. Compute $S(v) = \{a, a, b, b, c, d, e\}$, $S(c_1) = \{a, a, b, b, c, c, d, d, d, d, e\}$ and $S(c_2) = \{a, a, a, a, b, b, b, b, c, d, e\}$. Compute $X = \{c, d, d, d\}$ and $Y = \{a, a, b, b\}$.
3. Identify d and a such that $d \in S(v)$ and $a \in S(v)$, and $d \in X$ while $a \in Y$.
4. Compute $s(v) = dabbcea$, $s_1 = dabbcea \cdot d$ and $s_2 = a \cdot dabbcea$.
5. Insert elements in $X - \{d\} = \{c, d, d\}$ arbitrarily at the right end of s_1 to obtain S_1, and insert all the elements in $Y - \{a\} = \{a, b, b\}$ at the right end of s_2 to obtain S_2.
6. Return $S_1 = dabbcea \cdot d \cdot cdd$ and $S_2 = a \cdot dabbcea \cdot abb$.

Theorem 7. *Let c_1, c_2 be polynomially bounded. The number of common adjacencies generated by Algorithm 1 is optimal with a value either n^* or $n^* - 1$, where $n^* = \sum_{i=1}^{n} v_i$ with the maximum common sub-vector of c_1 and c_2 being $v = \langle v_1, v_2, ..., v_n \rangle$.*

Note that if we only want the breakpoint distance between S_1 and S_2, then the polynomial boundness condition of c_1 and c_2 can be withdrawn as we can decide whether $\{x, y\}$ exists by searching directly in the CNPs (vectors).

5 Concluding Remarks and Open Problems

We conclude this paper with a list of open problems.

1. Is the k-TD-distance problem FPT? Note that in Exemplar-k-TD we have some useful property since the source string S is exemplar.
2. Let $|\Sigma| = O(1)$, is the TD-distance problem still NP-hard? It would be especially interesting to know the answer when $|\Sigma| = 2$. In our construction, $|\Sigma|$ is a polynomial function of the input size. A natural way is to encode each character into a sequence on a bounded alphabet (e.g., 2), but how to do it so that the contractions still behave in our way is not trivial.

3. If $|\Sigma| \geq 3$, is the problem of deciding whether sequence S can be converted to T using tandem duplications (regardless of the number) NP-hard?
4. For the Copy Number Profile Conforming problem, if the distance $d(S_1, S_2)$ is any genomic distance except the breakpoint distance, is the problem still polynomially solvable?
5. While untouched in this paper, what is the complexity of the following problem: given genomes G_1, G_2 and an integer k, is $d_{GG}(G_1, G_2) \leq k$? Note that in this case, if we do not care about the minimum number of (duplication and deletion) events, then it is always possible to convert G_1 to G_2 as long as they are over the same alphabet.

Acknowledgments. I would like to thank my collaborators for these research: Manuel Lafond, Letu Qingge and Peng Zou. I also thank Prof. Henning Fernau and the organizers of CSR'2020 to give me the chance to survey these research.

References

1. Alon, N., Bruck, J., Hassanzadeh, F.F., Jain, S.: Duplication distance to the root for binary sequences. IEEE Trans. Inf. Theory **63**(12), 7793–7803 (2017)
2. Angibaud, S., Fertin, G., Rusu, I., Thevenin, A., Vialette, S.: On the approximability of comparing genomes with duplicates. J. Graph Algorithms Appl. **13**(1), 19–53 (2009)
3. Benson, G., Dong, L.: Reconstructing the duplication history of a tandem repeat. In: Proceedings of ISMB 1999, pp. 44–53 (1999)
4. Bovet, D.P., Varricchio, S.: On the regularity of languages on a binary alphabet generated by copying systems. Inf. Process. Lett. **44**(3), 119–123 (1992)
5. Bulteau, L., Fertin, G., Rusu, I.: Sorting by transposition is difficult. SIAM J. Discrete Math. **26**(3), 1148–1180 (2012)
6. The Cancer Genome Atlas Research Network: Integrated genomic analyses of ovarian carcinoma. Nature **474**, 609–615 (2011)
7. Charlesworth, B., Sniegowski, P., Stephan, W.: The evolutionary dynamics of repetitive DNA in eukaryotes. Nature **371**(6494), 215–220 (1994)
8. Chaudhuri, K., Chen, K., Mihaescu, R., Rao, S.: On the tandem duplication-random loss model of genome rearrangement. In: Proceedings of the 17th Annual ACM-SIAM Symposium on Discrete Algorithms (SODA 2006), pp. 564–570 (2006)
9. Chen, Z., Wang, L., Wang, Z.: Approximation algorithms for reconstructing the duplication history of tandem repeats. Algorithmica **54**(4), 501–529 (2009)
10. Cho, D.-J., Han, Y.-S., Kim, H.: Bound-decreasing duplication system. Theoret. Comput. Sci. **793**, 152–168 (2019)
11. Chowdhury, S., Shackney, S., Heselmeyer-Haddad, K., Ried, T., Schaeffer, A., Schwartz, R.: Algorithms to model single gene, single chromosome, and whole genome copy number changes jointly in tumor phylogenetics. Plos Comput. Biol. **10**(7), e1003740 (2014)
12. Ciriello, G., Killer, M., Aksoy, B., Senbabaoglu, Y., Schultz, N., Sanders, C.: Emerging landscape of oncogenic signatures across human cancers. Nat. Genet. **45**, 1127–1133 (2013)
13. Cooke, S., et al.: Intra-tumour genetic heterogeneity and poor chemoradiotherapy response in cervical cancer. Br. J. Cancer **104**(2), 361–368 (2011)

14. Cooke, S., Brenton, J.: Evolution of platinum resistance in high-grade serous ovarian cancer. Lancet Oncol. **12**(12), 1169–1174 (2011)
15. Cowin, P., et al.: LRP1B deletion in high-grade serous ovarian cancers is associated with acquired chemotherapy resistance to liposomal doxorubicin. Cancer Res. **72**(16), 4060–4073 (2012)
16. Cormen, T., Leiserson, C., Rivest, R., Stein, C.: Introduction to Algorithms, Second edn. MIT Press, Cambridge (2001)
17. Dassow, J., Mitrana, V., Paun, G.: On the regularity of the duplication closure. Bull. EATCS **69**, 133–136 (1999)
18. Downey, R., Fellows, M.: Parameterized Complexity. Springer, Heidelberg (1999). https://doi.org/10.1007/978-1-4612-0515-9
19. Ehrenfeucht, A., Rozenberg, G.: On regularity of languages generated by copying systems. Discrete Appl. Math. **8**(3), 313–317 (1984)
20. El-Kebir, M., et al.: Copy-number evolution problems: complexity and algorithms. In: Frith, M., Storm Pedersen, C.N. (eds.) WABI 2016. LNCS, vol. 9838, pp. 137–149. Springer, Cham (2016). https://doi.org/10.1007/978-3-319-43681-4_11
21. Fellows, M., Hermelin, D., Rosamond, F., Vialette, S.: On the parameterized complexity of multiple-interval graph problems. Theoret. Comput. Sci. **410**(1), 53–61 (2009)
22. Flum, J., Grohe, M.: Parameterized Complexity Theory. Springer, Heidelberg (2006). https://doi.org/10.1007/3-540-29953-X
23. Garey, M.R., Johnson, D.S.: Computers and Intractability: A Guide to the Theory of NP-Completeness. Freeman W. H., New York (1979)
24. Gascuel, O., Hendy, M.D., Jean-Marie, A., McLachlan, R.: The combinatorics of tandem duplication trees. Syst. Biol. **52**(1), 110–118 (2003)
25. Gusfield, D.: *Algorithms on Strings, Trees and Sequences: Computer Science and Computational Biology*. Cambridge University Press, Cambridge (1997)
26. Gusfield, D., Stoye, J.: Linear time algorithms for finding and representing all the tandem repeats in a string. J. Comput. Syst. Sci. **69**(4), 525–546 (2004)
27. Hannenhalli, S., Pevzner, P.: Transforming men into mice (polynomial algorithm for genomic distance problem). In: Proceedings of FOCS 1995, pp. 581–592 (1995)
28. Hassanzadeh, F., Schwartz, M., Bruck, J.: The capacity of string-duplication systems. IEEE Trans. Inf. Theory **62**(2), 811–824 (2016)
29. Ito, M., Leupold, P., Shikishima-Tsuji, K.: Closure of language classes under bounded duplication. In: Ibarra, O.H., Dang, Z. (eds.) DLT 2006. LNCS, vol. 4036, pp. 238–247. Springer, Heidelberg (2006). https://doi.org/10.1007/11779148_22
30. Jain, S., Hassanzadeh, F., Bruck, J.: Capacity and expressiveness of genomic tandem duplication. IEEE Trans. Inf. Theory **63**(10), 6129–6138 (2017)
31. Jiang, H., Zheng, C., Sankoff, D., Zhu, B.: Scaffold filling under the breakpoint and related distances. IEEE/ACM Trans. Comput. Biol. Bioinform. **9**(4), 1220–1229 (2012)
32. Landau, G., Schmidt, J., Sokol, D.: An algorithm for approximate tandem repeats. J. Comput. Biol. **8**(1), 1–18 (2001)
33. Lander, E.S., et al.: Initial sequencing and analysis of the human genome. Nature **409**(6822), 860–921 (2001)
34. Lafond, M., Zhu, B., Zou, P.: The tandem duplication distance is NP-hard. CoRR abs/1906.05266, June 2019
35. Lafond, M., Zhu, B., Zou, P.: The tandem duplication distance is NP-hard. In: Proceedings of STACS 2020. LiPIcs, vol. 154, pp. 15:1–15:15 (2020)
36. Lafond, M., Zhu, B., Zou, P.: Genomic problems involving copy number profiles: complexity and algorithms. CoRR abs/2002.04778, February 2020

37. Lafond, M., Zhu, B., Zou, P.: Genomic problems involving copy number profiles: complexity and algorithms. In: Proceedings of CPM 2020. LiPIcs, vol. 161, pp. 22:1–22:25 (2020)
38. Letunic, I., Copley, R., Bork, P.: Common exon duplication in animals and its role in alternative splicing. Hum. Mol. Genet. **11**(13), 1561–1567 (2002)
39. Leupold, P., Mitrana, V., Sempere, J.M.: Formal languages arising from gene repeated duplication. In: Jonoska, N., Paun, G., Rozenberg, G. (eds.) Aspects of Molecular Computing. LNCS, vol. 2950, pp. 297–308. Springer, Heidelberg (2003). https://doi.org/10.1007/978-3-540-24635-0_22
40. Leupold, P., Carlos, M.V., Mitrana, V.: Uniformly bounded duplication languages. Discrete Appl. Math. **146**(3), 301–310 (2005)
41. Li, S., Dou, X., Ge, R., Qian, M., Wan, L.: A remark on copy number variation detection. Plos One **13**(4), e0196226 (2018)
42. Li, W., Olivier, M.: Current analysis platforms and methods for detecting copy number variation. Physiol. Genomics **45**(1), 1–16 (2013)
43. Macdonald, M., et al.: A novel gene containing a trinucleotide repeat that is expanded and unstable on Huntington's disease. Cell **72**(6), 971–983 (1993)
44. Maley, C., et al.: Genetic clonal diversity predicts progression to esophageal adenocarcinoma. Nat. Genet. **38**(4), 468–473 (2006)
45. Marusyk, A., Almendro, V., Polyak, K.: Intra-tumour heterogeneity: a looking glass for cancer. Nat. Rev. **13**, 323–334 (2012)
46. Navin, N., et al.: Inferring tumor progression from genomic heterogeneity. Genome Res. **20**, 68–80 (2010)
47. Oesper, L., Ritz, A., Aerni, S., Drebin, R., Raphael, B.: Reconstructing cancer genomes from paired-end sequencing data. BMC Bioinform. **13**(Suppl 6), S10 (2012)
48. Qingge, L., He, X., Liu, Z., Zhu, B.: On the minimum copy number generation problem in cancer genomics. In: Proceedings of ACM BCB 2018, pp. 260–269. ACM (2018)
49. Schwarz, R., Trinh, A., Sipos, B., Brenton, J., Goldman, N., Markowetz, F.: Phylogenetic quantification of intra-tumour heterogeneity. Plos Comput. Biol. **10**(4), e1003535 (2014)
50. Shah, S., et al.: Mutational evolution in a lobular breast tumor profiled at single nucleotide resolution. Nature **461**(7265), 809–813 (2009)
51. Shamir, R., Zehavi, M., Zeira, R.: A linear-time algorithm for the copy number transformation problem. In: Proceedings of CPM 2016. LiPIcs, vol. 54, pp. 16:1–16:13 (2016)
52. Sharp, A., et al.: Segmental duplications and copy-number variation in the human genome. Am. J. Hum. Genet. **77**(1), 78–88 (2005)
53. Szostak, J.W., Wu, R.: Unequal crossing over in the ribosomal DNA of Saccharomyces cerevisiae. Nature **284**(5755), 426–430 (1980)
54. Thue, A.: Über unendliche Zeichenreihen (Mathematisk-Naturvidenskabelig Klasse). Videnskabsselskabet, Freetown Christiania, Denmark (1906)
55. Tremblay-Savard, O., Bertrand, D., El-Mabrouk, N.: Evolution of orthologous tandemly arrayed gene clusters. BMC Bioinform. **12**(S-9), S2 (2011)
56. Trevisan, L.: Non-approximability results for optimization problems on bounded degree instances. In: Proceedings of 33rd ACM Symposium on Theory of Computing (STOC 2001), pp. 453–461. ACM (2001)
57. Wang, M.W.: On the irregularity of the duplication closure. Bull. EATCS **70**, 162–163 (2000)

58. Watterson, G.A., Ewens, W.J., Hall, T.E., Morgan, A.: The chromosome inversion problem. J. Theoret. Biol. **99**(1), 1–7 (1982)
59. Zhu, B.: A retrospective on genomic preprocessing for comparative genomics. In: Chauve, C., El-Mabrouk, N., Tannier, E. (eds.) Models and Algorithms for Genome Evolution, vol. 19, pp. 183–206. Springer, Heidelberg (2013). https://doi.org/10.1007/978-1-4471-5298-9_9

Faster 2-Disjoint-Shortest-Paths Algorithm

Maxim Akhmedov[1,2](✉)(iD)

[1] Department of Mathematical Logic and Algorithms, Moscow State University,
Moscow, Russia
akhmedov@lpcs.math.msu.su
[2] Yandex LLC, Moscow, Russia
max42@yandex-team.ru

Abstract. Consider the following kDSP problem: given a graph G and k pairs of terminal vertices $(s_1, t_1), (s_2, t_2), \ldots, (s_k, t_k)$, check if there exists a k-tuple of pairwise disjoint shortest s_i–t_i paths between these k pairs of terminal vertices. Algorithmically, the case of two vertex-disjoint paths turns out to be the most interesting one. For this setting, Eilam-Tzoreff established an algorithm running in $\mathcal{O}(|V|^8)$ time, which uses dynamic programming (DP) and applies to both directed and undirected graphs and arbitrary positive edge weights (lengths). In this paper, we examine the DP relations arising in this problem and reduce the time complexity to $\mathcal{O}(|V|^6)$ for the unit-length case and to $\mathcal{O}(|V|^7)$ for the case of general weights.

Keywords: Dynamic programming · Shortest paths · Graph theory · Linear algebra

1 Introduction

We consider the following three settings of combinatorial problems. In all problems, given a graph $G = (V, E)$ and k pairs of vertices (s_i, t_i) $(s_i, t_i \in V, 1 \leq i \leq k)$, we should find out if there exists a k-tuple of paths P_i such that certain conditions hold. Possible conditions are listed below.

kDP problem:

1. path P_i goes from s_i to t_i;
2. all paths are pairwise disjoint.

kDSP problem:

1. path P_i goes from s_i to t_i;
2. all paths are pairwise disjoint;
3. each P_i is one of the shortest paths from s_i to t_i.

© Springer Nature Switzerland AG 2020
H. Fernau (Ed.): CSR 2020, LNCS 12159, pp. 103–116, 2020.
https://doi.org/10.1007/978-3-030-50026-9_7

min-sum kDSP problem:

1. path P_i goes from s_i to t_i;
2. all paths are pairwise disjoint;
3. the total length of all P_i is smallest possible.

This description still leaves some degrees of freedom: graph G above may either be directed or undirected; the disjointness property may require paths to be either node-disjoint or edge-disjoint; finally the graph may be either weighted (with positive integer lengths assigned to edges) or unweighted (having only unit-length edges). Clearly, there are 8 problem versions for kDSP and *min-sum* kDSP and 4 problem versions for kDP (as the presence of weights does not matter). For example, one may consider the undirected weighted vertex-disjoint version of kDSP.

The three problem settings are closely related to each other. One may verify that an algorithm solving *min-sum* kDSP may be reduced to be an algorithm for each of kDP and kDSP. The first reduction is trivial, while the second one can be done by checking if the total length of paths in the algorithm output is equal to the sum of shortest path lengths over all pairs (s_i, t_i). Thus kDP and kDSP are easier than *min-sum* kDSP. There is no direct reduction between kDP and kDSP, but questions arising in these problems tend to be connected.

The kDP problem is well-studied. The directed kDP problem is NP-complete for $k \geq 2$ due to Fortune, Hopcroft and Wyllie [5]. The undirected kDP is known to have a polynomial-time algorithm for any fixed k [10]. For $k = 2$, there exists an algorithm by Gustedt [7] with $\mathcal{O}(|E| \log |V|)$ running time. If k is a part of problem input, undirected kDP problem is NP-complete even for planar graphs due to Lynch [9].

The *min-sum* kDSP problem has been recently studied by different authors; there is a 2014 result of Björklund and Husfeldt [1] providing a Monte Carlo algorithm for the unweighted case running in time $\mathcal{O}(|V|^{11})$, using an algebraic approach with permanents over quotient rings. There is another approach by Hirai and Namba [8] for the same problem based on hafnians modulo 2^k and a classic reduction from T-paths to matchings due to Gallai [6] yielding another polynomial bound for *min-sum* $kDSP$ for fixed k. For the planar case, there is a result by Datta et al. [3] yielding an $\mathcal{O}(n^\omega)$ randomized sequential algorithm with the restriction that all terminals lie on a single face of a pair of faces. There is also a result for cubic planar graphs due to Björklund and Husfeldt [2], providing a deterministic algorithm with sequential time complexity of $\mathcal{O}(|V|^{\omega/2+2}L^2)$, where edge weights are bounded by L.

As well as kDP, kDSP immediately becomes NP-complete if k is a part of the input. This is even true in the planar unit-length case, irrespectively of whether the graph is directed or undirected, or of whether we choose vertex-disjoint paths or edge-disjoint paths [4].

In this article we will concentrate on the 2DSP problem. For the weighted undirected vertex-disjoint case, Eilam-Tzoreff provided a polynomial-time algorithm based on a dynamic programming approach. He also provided a linear-time reduction from the edge-disjoint case to the vertex-disjoint case. The algorithm of Eilam-Tzoreff has a running time of $\mathcal{O}(|V|^8)$. Suchm a running time bound motivates a natural question — is it possible to solve the problem faster?

We obtain an algorithm with running time of $\mathcal{O}(|V|^6)$ for the unit-length case of 2DSP and an algorithm with running time of $\mathcal{O}(|V|^7)$ for the weighted case of 2DSP (in both cases we consider the vertex-disjoint undirected formulation). Our algorithms may be viewed as modifications of the Eilam-Tzoreff algorithm with two improvements. The first one is somewhat standard to dynamic programming and consists of choosing the appropriate computation order enabling us to factor the problem into two independent subproblems with better running time.

The second improvement is novel and works as follows: we interpret the computationally hardest subroutine of the algorithm as taking the value $x^T \beta y$ of a certain bilinear form β at some pair of vectors x, y, and then analyze the triples β, x, y arising during the computation. It turns out that by pre-evaluating the partial products $x^T \beta$, we may reduce the running time complexity even further.

As a proof of concept, the obtained algorithms were implemented in C++. With their use, the correctness of the algorithm was checked for all possible unweighted graphs of small size ($|V| \leq 8$) and for a significant number of connected graphs of larger size ($|V| = 10, 20, 30$).

The source code of our implementations and the LaTeX sources of this article (both in English and in Russian) are available at https://github.com/zlobober/thesis.

2   Definitions and the DP Formululation

Let us introduce some definitions and notation that we employ. From now on, let $G = (V, E)$ be an undirected loopless graph, $w : E \to \mathbb{R}_{>0}$ be an edge weight function.

The *length* of a path formed by edges $e_1, e_2, \ldots, e_k$ is $w(e_1) + w(e_2) + \ldots + w(e_k)$.

Let $x, y \in V$ be two vertices belonging to the same connected component of G. Define $l(x, y)$ to be the length of the shortest path between x and y. For x and y belonging to distinct connected components, define $l(x, y) := +\infty$.

In particular, for any $x \in V$ it is true that $l(x, x) = 0$.

Fix $x, y \in V$. Define $L(x, y) \subseteq V$ to be the set of all $v \in V$ belonging to at least one shortest path between x and y.

Clearly, $v \in L(x, y)$ iff $l(x, v) + l(v, y) = l(x, y)$. Also for x and y from distinct connected components, $L(x, y) = \varnothing$. Finally, $L(v, v) = \{v\}$ for all $v \in V$.

We also introduce a few special definitions that will significantly simplify the DP formulae.

Fix $x, y \in V$. Let $F(x, y) = \{v \in L(x, y) \mid (x, v) \in E\}$, i.e., $F(x, y)$ is formed by all of the vertices that are successors of x in at least one shortest path from x to y.

For the sake of convenience, we also introduce a 3-argument variant of F:

Fix $x, y, z \in V$. Let $F(x, y, z) = F(x, y) \cap F(x, z)$, i.e., $F(x, y, z)$ is formed by all of the vertices that are successors of x in at least one shortest path from x to y and at least one shortest path from x to z.

Fix $s_1, t_1, s_2, t_2 \in V$. Define $2DSP(s_1, t_1, s_2, t_2)$ to be equal to 1 if there exist two vertex-disjoint shortest paths P_1 and P_2 from s_1 to t_1 and P_2 from s_2 to t_2, respectively; and let it be 0, otherwise.

The quadruple $(s_1, t_1, s_2, t_2) \in V^4$ is called *rigid* iff $s_1, t_1 \in L(s_2, t_2)$ and $s_2, t_2 \in L(s_1, t_1)$.

Example. Consider the graph G from Fig. 1.

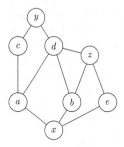
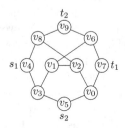

Fig. 1. Graph G **Fig. 2.** Graph H

For this graph it is true that $L(x, y) = \{x, a, b, c, d, y\}$, $L(x, z) = \{x, b, e, z\}$, $F(x, y) = \{a, b\}$, $F(x, z) = \{b, e\}$, $F(x, y, z) = \{b\}$. The quadruple (x, y, b, c) is rigid as $b, c \in L(x, y)$ and $x, y \in L(b, c)$. $2DSP(x, y, b, c) = 0$, as a and d form a cut between x and y. At the same time, $2DSP(x, z, b, c) = 1$, as there are paths (x, e, z) and (b, d, a, c) which are vertex-disjoint.

We will also introduce the following notation simplifying formulae involving predicates. We allow some of the arguments of $2DSP$ to be vertex sets instead of single vertices. In this case, we consider the resulting expression to be the logical disjunction of $2DSP$ over all quadruples, where each component belongs to the corresponding argument set. For example,

$$2DSP(F(s_1, t_1), t_1, s_2, t_2) = \bigvee_{x \in F(s_1, t_1)} 2DSP(x, t_1, s_2, t_2). \tag{1}$$

We rely on the following structural result from [4]:

Theorem 1. *For any $s_1, t_1, s_2, t_2 \in V$ one of the following cases applies:*

1. *If $s_1 = t_1$, $s_2 = t_2$, then $2DSP(s_1, t_1, s_2, t_2)$ is 1 iff $s_1 \neq s_2$.*
2. *Otherwise, assume (s_1, t_1, s_2, t_2) is not rigid; consider a vertex of a quadruple for which the rigidness condition does not hold. Without loss of generality, suppose that s_1 is such a vertex, i.e., $s_1 \notin L(s_2, t_2)$ and $s_1 \neq t_1$. Then $2DSP(s_1, t_1, s_2, t_2) = 2DSP(F(s_1, t_1), s_2, t_2)$.*
3. *Otherwise, define $C = L(s_1, s_2) \cup L(s_2, t_1) \cup L(t_1, t_2) \cup L(t_2, s_1)$. Then $2DSP(s_1, t_1, s_2, t_2) = Q_2(s_1, t_1, s_2, t_2) \vee Q_4(s_1, t_1, s_2, t_2)$ where Q_2 and Q_4 are defined as follows:*

$$Q_2(s_1, t_1, s_2, t_2) =$$

$$
\begin{array}{lllll}
2DSP(& s_1, & F(t_1, s_1, s_2), & s_2, & F(t_2, s_2, s_1)) \ \vee \\
2DSP(& F(s_1, t_1, s_2), & t_1, & s_2, & F(t_2, s_2, t_1)) \ \vee \\
2DSP(& s_1, & F(t_1, s_1, t_2), & F(s_2, t_2, s_1), & t_2) \ \vee \\
2DSP(& F(s_1, t_1, t_2), & t_1, & F(s_2, t_2, t_1), & t_2) \ \vee \\
\\
2DSP(& F(s_1, t_1) \setminus C, & t_1, & F(s_2, t_2) \setminus C, & t_2) \ \vee \\
2DSP(& F(s_1, t_1) \setminus C, & t_1, & s_2, & F(t_2, s_2) \setminus C) \ \vee \\
2DSP(& s_1, & F(t_1, s_1) \setminus C, & F(s_2, t_2) \setminus C, & t_2) \ \vee \\
2DSP(& s_1, & F(t_1, s_1) \setminus C, & s_2, & F(t_2, s_2) \setminus C) \ \vee \\
\\
2DSP(& s_1, & F(t_1, s_1, s_2), & s_2, & F(t_2, s_2) \setminus C) \ \vee \\
2DSP(& s_1, & F(t_1, s_1, t_2), & F(s_2, t_2) \setminus C, & t_2) \ \vee \\
2DSP(& F(t_1, s_1, s_2), & t_1, & s_2, & F(t_2, s_2) \setminus C) \ \vee \\
2DSP(& F(t_1, s_1, t_2), & t_1, & F(s_2, t_2) \setminus C, & t_2) \ \vee \\
2DSP(& s_1, & F(t_1, s_1) \setminus C, & s_2, & F(t_2, s_2, s_1)) \ \vee \\
2DSP(& F(s_1, t_1) \setminus C, & t_1, & s_2, & F(t_2, s_2, t_1)) \ \vee \\
2DSP(& s_1, & F(t_1, s_1) \setminus C, & F(s_2, t_2, s_1), & t_2) \ \vee \\
2DSP(& F(s_1, t_1) \setminus C, & t_1, & F(s_2, t_2, t_1), & t_2) & (2)
\end{array}
$$

$$Q_4(s_1, t_1, s_2, t_2) = \bigvee_{\substack{x \in F(s_1, s_2) \\ y \in F(t_1, t_2) \\ u \in F(s_2, t_1) \\ v \in F(t_2, s_1) \\ l(s_1, x) + l(x, y) + l(y, t_1) = l(s_1, t_1) \\ l(s_1, u) + l(u, v) + l(v, t_2) = l(s_2, t_2)}} 2DSP(x, y, u, v) \quad \vee \quad \bigvee_{\substack{x \in F(s_1, t_2) \\ y \in F(t_1, s_2) \\ u \in F(s_2, s_1) \\ v \in F(t_2, t_1) \\ l(s_1, x) + l(x, y) + l(y, t_1) = l(s_1, t_1) \\ l(s_2, u) + l(u, v) + l(v, t_2) = l(s_2, t_2)}} 2DSP(x, y, u, v) \quad (3)$$

We will formulate (without proof) two important results following from the original article here.

Proposition 1. *For a rigid quadruple* (s_1, t_1, s_2, t_2)*, it is true that* $l(s_1, t_1) = l(s_2, t_2)$*.*

Theorem 2. *Algorithm 4 calculates* $2DSP(s_1, t_1, s_2, t_2)$ *for all quadruples* (s_1, t_1, s_2, t_2) *in* $\mathcal{O}(|V|^8)$ *time using* $\Theta(|V|^4)$ *memory.*

3   Reducing the Running Time to $\mathcal{O}(|V|^7)$

In the following two sections we assume edges to have unit length, i.e., the graph to be unweighted.

We start by optimizing the DP transitions given by (3) as all the remaining transitions require $\mathcal{O}(|V|^6)$ time. Denote the first and the second expression in (3) by $A_4(s_1, t_1, s_2, t_2)$ and $B_4(s_1, t_1, s_2, t_2)$:

$$A(s_1, t_1, s_2, t_2) = \bigvee_{\substack{x \in F(s_1, s_2) \\ y \in F(t_1, t_2) \\ u \in F(s_2, t_1) \\ v \in F(t_2, s_1) \\ l(x,y)+2=l(s_1,t_1) \\ l(u,v)+2=l(s_2,t_2)}} 2DSP(x, y, u, v) \qquad B(s_1, t_1, s_2, t_2) = \bigvee_{\substack{x \in F(s_1, t_2) \\ y \in F(t_1, s_2) \\ u \in F(s_2, s_1) \\ v \in F(t_2, t_1) \\ l(x,y)+2=l(s_1,t_1) \\ l(u,v)+2=l(s_2,t_2)}} 2DSP(x, y, u, v)$$

$$(4) \qquad\qquad\qquad\qquad\qquad\qquad (5)$$

Proposition 2. *The following equation holds:*

$$B(s_1, t_1, s_2, t_2) = A(s_2, t_2, s_1, t_1). \qquad (6)$$

Proof. The equation immediately follows from the existence of a bijection between quadruples (x, y, u, v) s.t. $x \in F(s_2, s_1)$, $y \in F(t_2, t_1)$, $u \in F(s_1, t_2)$, $v \in F(t_1, s_2)$, $l(x, y) + 2 = l(s_2, t_2)$, $l(u, v) + 2 = l(s_1, t_1)$ (that form the disjunction domain in the expression for $A(s_2, t_2, s_1, t_1)$) and quadruples (x', y', u', v') s.t. $x' \in F(s_1, t_2)$, $y' \in F(t_1, s_2)$, $u' \in F(s_2, s_1)$, $v' \in F(t_2, t_1)$, $l(x', y') + 2 = l(s_1, t_1)$, $l(u', v') + 2 = l(s_2, t_2)$. The sought bijection is defined by $x' = u$, $y' = v$, $u' = x$, $v' = y$. $\qquad\square$

Thus, $Q_4(s_1, t_1, s_2, t_2)$ may be expressed using just predicate A:

$$Q_4(s_1, t_1, s_2, t_2) = A(s_1, t_1, s_2, t_2) \vee A(s_2, t_2, s_1, t_1) \qquad (7)$$

We rewrite the conditions on (x, y, u, v) in (4) in the following equivalent form:

Proposition 3.

$$\begin{cases} x \in F(s_1, s_2) \\ y \in F(t_1, t_2) \\ u \in F(s_2, t_1) \\ v \in F(t_2, s_1) \\ l(x, y) + 2 = l(s_1, t_1) \\ l(u, v) + 2 = l(s_2, t_2) \end{cases} \Longleftrightarrow \begin{cases} x \in F(s_1, t_1, s_2) \\ y \in F(t_1, s_1, t_2) \\ u \in F(s_2, t_2, t_1) \\ v \in F(t_2, s_2, s_1) \\ l(x, y) + 2 = l(s_1, t_1) \\ l(u, v) + 2 = l(s_2, t_2) \end{cases} \Longleftrightarrow \begin{cases} x \in F(s_1, t_1, s_2) \\ y \in F(t_1, x, t_2) \\ u \in F(s_2, t_2, t_1) \\ v \in F(t_2, u, s_1) \end{cases}$$

$$(8)$$

Proof. First, note that $l(x, y) + 2 = l(s_1, t_1)$ and $l(u, v) + 2 = l(s_2, t_2)$ immediately implies that $x, y \in L(s_1, t_1)$ and $u, v \in L(s_2, t_2)$, so we may replace $F(s_1, s_2)$ with $F(s_1, t_1, s_2)$ as a domain for x, and perform the similar action for y, u and v. This proves the first equivalence.

The second equivalence may be proven by exploiting the fact that $l(x, y) + 2 = l(s_1, t_1)$ (which means that x and y belong to the same shortest path between s_1 and t_1) is equivalent to $x \in F(s_1, t_1) \wedge y \in F(t_1, x)$, and similarly for s_2, t_2, u and v. $\qquad\square$

Now we can rewrite (4) as follows:

$$A(s_1, t_1, s_2, t_2) = \bigvee_{\substack{x \in F(s_1,t_1,s_2) \\ u \in F(s_2,t_2,t_1)}} \bigvee_{\substack{y \in F(t_1,x,t_2) \\ v \in F(t_2,u,s_1)}} 2DSP(x, y, u, v)$$

Note the important property of the equation above: the expression inside the first disjunction does not depend on s_2. Let us denote it by $ExYV(x, u, s_1, t_1, t_2)$ (an acronym for "there exist y and v"):

$$A(s_1, t_1, s_2, t_2) = \bigvee_{\substack{x \in F(s_1,t_1,s_2) \\ u \in F(s_2,t_2,t_1)}} ExYV(x, u, s_1, t_1, t_2) = ExYV(F(s_1, t_1, s_2), F(s_2, t_2, t_1), s_1, t_1, t_2)$$

$$ExYV(x, u, s_1, t_1, t_2) = \bigvee_{\substack{y \in F(t_1,x,t_2) \\ v \in F(t_2,u,s_1)}} 2DSP(x, y, u, v) = 2DSP(x, F(t_1, x, t_2), u, F(t_2, u, s_1)) \qquad (9)$$

The formulae above suggest the following optimization: $A(s_1, t_1, s_2, t_2)$ only depends on $\mathcal{O}(|V|^2)$ values of $ExYV(x, u, s_1, t_1, t_2)$. So, if we have all the necessary values of $ExYV(x, u, s_1, t_1, t_2)$ ready by the moment we calculate the value of $A(s_1, t_1, s_2, t_2)$, then the total running time needed for calculating all $A(s_1, t_1, s_2, t_2)$ becomes $\mathcal{O}(|V|^6)$.

Finally, note that by (9) we may calculate each value of $ExYV(x, u, s_1, t_1, t_2)$ in $\mathcal{O}(|V|^2)$ time. This means that $ExYV(x, u, s_1, t_1, t_2)$ is a suitable auxiliary predicate that, on the one hand, may be calculated efficiently (in $\mathcal{O}(|V|^7)$ time), and on the other hand, is reused multiple times while calculating values of predicate $A(s_1, t_1, s_2, t_2)$, providing the sought running time optimization.

The only remaining step is to verify that we can organize the calculations in an appropriate order; i.e., by the moment we use each of the values in the right-hand sides of (7) and (9), it is already known. To show this, we provide a pseudocode of Algorithm 1 implementing this approach.

Due to the restriction on paper size, the proof of the following theorem (which is quite technical and straightforward) is omitted.

So, the final result of this section is the following

Theorem 3. *Algorithm 1 calculates the predicate $2DSP(s_1, t_1, s_2, t_2)$ for all quadruples (s_1, t_1, s_2, t_2) in time $\mathcal{O}(|V|^7)$ using $\Theta(|V|^5)$ memory.*

4   Reducing Running Time to $\mathcal{O}(|V|^6)$

It is easy to see that the most inefficient part of Algorithm 1 is the procedure CALCULATEALLEXYVVALUES whose total running time is $\mathcal{O}(|V|^7)$. In this section we optimize the running time of calculating $ExYV$ for all arguments to $\mathcal{O}(|V|^6)$.

Let us rephrase (9) by rewriting it in terms of bilinear forms. In this section we fix some basis and identify bilinear forms with their matrices; also vectors are assumed to be written as columns.

Algorithm 1. Calculation of $2DSP(s_1, t_1, s_2, t_2)$ in $\mathcal{O}(|V|^7)$

1: **procedure** CALCULATEALL2DSPVALUES(V, E)
2:    $l \leftarrow$ matrix of pairwise distances in G
         ▷ Group pairs of vertices according to the distance between them.
3:    $P_i \leftarrow$ empty list for all $i = 0, \ldots, |V| - 1$
4:    **for** $s, t \in V$ **do**
5:        Append (s, t) to $P_{l(s,t)}$
         ▷ Calculate values of the predicate in lexicographical order of pairs
         ▷ $(d_{\min}, d_{\max})$, where $d_{\min}$ is for the smallest of the distances
         ▷ between terminal pairs and $d_{\max}$ is for the largest of them.
6:    **for** $d_{\min} \leftarrow 0, \ldots, |V| - 1$ **do**
7:        CALCULATEALLEXYVVALUES($d_{\min}$)
8:        **for** $d_{\max} \leftarrow d_{\min}, \ldots, |V| - 1$ **do**
9:            **for** $(s_1, t_1) \in P_{d_{\min}}$ **do**
10:               **for** $(s_2, t_2) \in P_{d_{\max}}$ **do**
11:                   $2DSP(s_1, t_1, s_2, t_2) \leftarrow$ CALCULATESINGLE2DSPVALUE(s_1, t_1, s_2, t_2)
                      ▷ Finally use the symmetric nature of our predicate.
12:                   $2DSP(s_2, t_2, s_1, t_1) \leftarrow 2DSP(s_1, t_1, s_2, t_2)$

13: **procedure** CALCULATEALLEXYVVALUES(d)
14:    **for** $(s_1, t_1 \in P_d)$ **do**
15:        **for** $x \in F(s_1, t_1)$ **do**
16:            **for** $u, t_2 \in V$ **do**
17:                $ExYV(x, u, s_1, t_1, t_2) \leftarrow 2DSP(x, F(t_1, x, t_2), u, F(t_2, u, s_1))$

18: **procedure** CALCULATESINGLE2DSPVALUE(s_1, t_1, s_2, t_2)
19:    **if** $s_1 = t_1 \wedge s_2 = t_2$ **then**
20:        **return** $s_1 = s_2$
21:    **else if** (s_1, t_1, s_2, t_2) is not rigid **then**
22:        Calculate and return the value according to case (2) of Th 1
23:    **else**
24:        Calculate $Q_2(s_1, t_1, s_2, t_2)$ using (2)
25:        $Q_4(s_1, t_1, s_2, t_2) = ExYV(F(s_1, t_1, s_2), F(s_2, t_2, t_1), s_1, t_1, t_2) \vee$
                $ExYV(F(s_2, t_2, s_1), F(s_1, t_1, t_2), s_2, t_2, t_1)$
26:        **return** $Q_2(s_1, t_1, s_2, t_2) \vee Q_4(s_1, t_1, s_2, t_2)$

Definition 1. *Let $V = \{v_0, v_1, \ldots, v_{n-1}\}$ be a certain fixed numbering of vertices and $S \subseteq V$. Then the characteristic vector $\chi[S]$ of set S is the column vector with 1 at the i-th place if $v_i \in S$ and 0 otherwise.*

We claim (9) amounts in computing the value of a certain bilinear form at vectors $\chi[F(t_1, x, t_2)]$ and $\chi[F(t_2, u, s_1)]$ and checking if the result is zero. A large caveat here is not to rely on any of yet-to-be-calculated values while constructing the matrix of this bilinear form.

Suppose the values of $2DSP(x, y, u, v)$ for all (x, y, u, v) s.t. $\min\{l(x, y), l(u, v)\} < d$ are already calculated. Define the matrix $2DSP(d, x, u) = (2DSP(d, x, u)_{i,j})_{i,j=0,\ldots,n-1}$ as follows:

$$2DSP(d, x, u)_{i,j} = \begin{cases} 2DSP(x, v_i, u, v_j) & \text{if } l(x, v_i) < d \text{ and } l(u, v_j) < d \\ 0 & \text{otherwise} \end{cases} \quad (10)$$

Now we can rewrite (9).

Proposition 4. *Suppose the values of $2DSP(x, y, u, v)$ for all (x, y, u, v) s.t. $\min\{l(x, y), l(u, v)\} < d$ are already calculated. Suppose that $l(x, t_1) = l(u, t_2) = d - 1$. Then the following equation holds:*

$$ExYV(x, u, s_1, t_1, t_2) = \begin{cases} 1 & \text{if } \chi[F(t_1, x, t_2)]^T \; 2DSP(d, x, u) \; \chi[F(t_2, u, s_1)] > 0 \\ 0 & \text{otherwise} \end{cases}$$

$$(11)$$

Proof. Suppose there exist y and v producing a positive value of (9), namely $y = v_i$ and $v = v_j$. Note that $\chi[F(t_1, x, t_2)]_i = \chi[F(t_2, u, s_1)]_j = 1$ as $y \in F(t_1, x, t_2)$ and $v \in F(t_2, u, s_1)$ due to variable domain in (9). Also note that $2DSP(d, x, u)_{i,j} = 2DSP(x, y, u, v) = 1$ (as $l(x, y) = l(u, v) = d - 2 < d$). Hence, the value of the bilinear form at our pair of characteristic vectors is at least 1.

By using exactly the same argument in the reverse direction we show that the right-hand side being at least 1 implies that there exist the desired y and v, finishing the proof. □

We have not achieved any significant speedup yet as we have only rewritten the same formulae in a different way. The key idea is to carefully estimate the number of pairs of bilinear forms and its right-hand vector argument.

Proposition 5. *There are $\mathcal{O}(|V|^3)$ bilinear forms that we consider during our algorithm.*

Proof. The claim immediately follows from the fact that each form corresponds to a triple of (d, x, u), each of whose component takes $|V|$ possible values. □

Proposition 6. *There are $\mathcal{O}(|V|^4)$ pairs of bilinear forms and its right-hand vector arguments that we consider during our algorithm.*

Proof. A pair of the above type is $(2DSP(d, x, u), \chi[F(t_2, u, s_1)])$ with the extra condition of $l(u, t_2) = d - 1$. So, the desired number of pairs is bounded by the number of quadruples (x, u, t_2, s_1), which is $|V|^4$. □

Define the value of $\psi(x, u, t_2, s_1)$ as follows:

$$\psi(x, u, t_2, s_1) = 2DSP(l(u, t_2) + 1, x, u) \; \chi[F(t_2, u, s_1)] \qquad (12)$$

The formula above provides an upper time bound of $\mathcal{O}(|V|^6)$ for calculating all $\psi(x, u, t_2, s_1)$, as each single value may be calculated in $\mathcal{O}(|V|^2)$ by multiplying a matrix by a vector in a straightforward manner.

Formula (11) takes the following form:

$$ExYV(x, u, s_1, t_1, t_2) = \chi[F(t_1, x, t_2)]^T \psi(x, u, t_2, s_1) \qquad (13)$$

Using the new formula, we may calculate $ExYV(x, u, s_1, t_1, t_2)$ in time $\mathcal{O}(|V|)$, so the total time of calculating all $ExYV(x, u, s_1, t_1, t_2)$ becomes $\mathcal{O}(|V|^6)$.

As in the previous section, we provide the pseudocode for our approach (Algorithm 2) and analyze its running time and space complexity.

Algorithm 2. Calculation of all $ExYV(x, u, s_1, t_1, t_2)$ in $\mathcal{O}(|V|^6)$

1: **procedure** CALCULATEALLEXYVVALUES(d)
2:   **for** $(u, t_2) \in P_{d-1}$ **do**
3:     **for** $x, s_1 \in V$ **do**
4:       $\psi(x, u, t_2, s_1) \leftarrow 2DSP(l(u, t_2) + 1, x, u) \, \chi[F(t_2, u, s_1)]$
5:   **for** $(s_1, t_1 \in P_d)$ **do**
6:     **for** $x \in F(s_1, t_1)$ **do**
7:       **for** $u, t_2 \in V$ **do**
8:         $ExYV(x, u, s_1, t_1, t_2) \leftarrow \chi[F(t_1, x, t_2)]^T \psi(x, u, t_2, s_1)$

Proposition 7. *The total running time of* CALCULATEALLEXYVVALUES *for all* $d = 0, \ldots, |V| - 1$ *is* $\mathcal{O}(|V|^6)$.

Proposition 8. *The total memory usage of Algorithm 2 is* $\mathcal{O}(|V|^5)$.

Hence, the final result of this section is the following theorem:

Theorem 4. *Algorithm 2 calculates predicate* $2DSP(s_1, t_1, s_2, t_2)$ *for all quadruples* (s_1, t_1, s_2, t_2) *in time* $O(|V|^6)$ *using* $\Theta(|V|^5)$ *memory.*

Let us also note an interesting feature of the obtained algorithm. In line 4 we repeatedly multiply the same matrix by a large number of column vectors. For each matrix, if we group $\mathcal{O}(|V|)$ possible vectors into a separate matrix, the desired procedure becomes expressible in terms of matrix multiplication, which may be done more efficiently in time $\mathcal{O}(|V|^\omega)$ with $\omega < 2.3727$ [11]. But even if we use this observation and reduce the running time of this part of the algorithm to $\mathcal{O}(|V|^{3+\omega})$, computing the values of Q_2 (which takes $\mathcal{O}(|V|^6)$ time in total) becomes the bottleneck.

5 Experimental Evaluation

First of all, we would like to make a disclaimer. We do not claim that our algorithm runs significantly faster on any graph of reasonable size (tens of vertices) as on small inputs the running time depends more on hidden constant, rather than on the power of complexity polynomial. Still, we believe that experimental evaluation is an important part of presenting the algorithm since it allows to test theoretical results on lots of small cases and sometimes shows mistakes that may be hard to found by checking the formal proof.

We implemented four algorithms:

- Original Algorithm 4 working in $\mathcal{O}(|V|^8)$;
- Algorithm 1 working in $\mathcal{O}(|V|^7)$;
- Algorithm 2 working in $\mathcal{O}(|V|^6)$;
- Brute-force Algorithm 3 working in exponential running time $\mathcal{O}(2^{|V|} \cdot |V|)$, not relying on Theorem 1.

Algorithm 3. Calculation of all $2DSP(s_1, t_1, s_2, t_2)$ in $O(2^{|V|} \cdot |V|)$

1: **procedure** CALCULATEALL2DSPVALUES(V, E)
2:     $l \leftarrow$ matrix of pairwise distances in G;
3:     **for** $s_1, t_1, s_2, t_2, \in V$ **do**
4:         $2DSP(s_1, t_1, s_2, t_2) =$ CALCULATESINGLE2DSPVALUE(s_1, t_1, s_2, t_2);

5: **procedure** CALCULATESINGLE2DSPVALUE(s_1, t_1, s_2, t_2)
6:     $2DSP(s_1, t_1, s_2, t_2) \leftarrow 0$;
7:     **for** shortest path P between s_1 and t_1 **do**
8:         **if** distance between s_2 and t_2 in $G \setminus P = l(s_2, t_2)$ **then**
9:             $2DSP(s_1, t_1, s_2, t_2) \leftarrow 1$;

All mentioned algorithms were implemented as the routines in a single program, allowing their simulateneous evaluation on the same graph G (either provided or randomly generated from some probability distribution). The program was run on all enumerated connected graphs consisting of no more than 8 vertices that helped find an enormous number of mistakes in the implementation of formula (2). The number of enumerated connected graphs on 9 vertices exceeds $6 \cdot 10^{10}$, so the experimental evaluation of the algorithms on all 9 vertex graphs with only one execution thread is not practically possible.

As a next step, we ran 10^4 instances of a program at the computational cluster of the company Yandex, each of which evaluated 10^5 random 10-vertex graphs and assered that all implemented algorithms produce the same result. Such stress test discovered a few more mistakes in the implementation of the original Algorithm 4 and its optimized versions. One interesting detail was that all the mistakes were located exactly in the hardest case of rigid quadruple (s_1, t_1, s_2, t_2) (formula (3)). Hence, we can conclude that the minimum size of the graph that triggers the hardest case of the approach of Eilam-Tzoreff is either 9 or 10. An example of a "complex" graph H consisting of 10 vertices, for which the transitions defined by the formula (3) are important, is provided in Fig. 2.

After fixing all mistakes, we evaluated the algorithms using 10^4 program instances processing 1000 random graphs with 20 vertices (without using the Algorithm 3) and then processing 100 graphs with 30 vertices (without using Algorithms 3, 2). The final evaluation did not show any discrepancy between the implemented algorithm results.

6   Extension to the Weighted Case

Algorithm 1 can be easily generalized to the case of a graph $G = (V, E)$ endowed with positive edge weigths $w : E \to \mathbb{R}_+$ by exploiting the fact that the algorithm only compares the distances between the pairs of vertices and performs arithmetic operations on them (in contrast to Algorithm 2 where distances become the parameter of a bilinear form).

We provide the following theorem without proof:

Theorem 5. *Given a graph $G = (V, E)$ with positive edge weights w, There exists an algorithm which computes the values of predicate $2DSP(s_1, t_1, s_2, t_2)$ for all quadruples (s_1, t_1, s_2, t_2) in $\mathcal{O}(|V|^7)$ time and $\mathcal{O}(|V|^5)$ space.*

7   Conclusion and Further Work

We originated from the algorithm of Eilam-Tzoreff and significantly improved its running time to $\mathcal{O}(|V|^6)$ for unweighted graphs and to $\mathcal{O}(|V|^7)$ for weighted graphs.

We have also performed an extensive experimental evaluation of the original algorithm as well as its optimized versions. Taking into account the fact that the main Theorem 1 (that forms the basis for all of the algorithms considered here) is rather complex both in sense of its statement and the proof (cf. [4]), its experimental validation is of certain importance.

Let us also note a wide spectrum of possible further work in this area. First of all, one may improve the running time for unweighted or weighted undirected $2DSP$ even further, for example, by extending the linear algebra ideas presented in this work (which seems to be a perspective direction, considering the fact that what seems to be hardest part of the algorithm may indeed be optimized to the running time of $\mathcal{O}(|V|^{3+\omega})$). Next, there are open questions regarding the existence of a polynomial algorithm for the directed case of $2DSP$ (weighted or unit-length) and for the undirected version of kDSP for $k \geq 3$ (in any formulation).

As a final remark, note that all the given complexity upper bounds are expressed in terms of number of vertices. It is possible that, by performing a more rigorous analysis, one may obtain a better upper bound in terms of the number of vertices as well as the number of edges in the graph, which will yield tighter bounds for sparse graphs.

8   Algorithm Pseudocodes

This section contains pseudocode of the original algorithm by Eilam-Tzoref [4].

Algorithm 4. Calculation of all values of $2DSP(s_1, t_1, s_2, t_2)$ in $\mathcal{O}(|V|^8)$

1: **procedure** CALCULATEALL2DSPVALUES(V, E)
2:     $l \leftarrow$ matrix of pairwise distances in G
        $\triangleright$ Group pairs of vertices according to the distance between them.
3:     $P_i \leftarrow$ empty list for all $i = 0, \ldots, |V| - 1$
4:     **for** $s, t \in V$ **do**
5:         Append (s, t) to $P_{l(s,t)}$
6:     **for** $d_{\min} \leftarrow 0, \ldots, |V| - 1$ **do**
7:         **for** $d_{\max} \leftarrow d_{\min}, \ldots, |V| - 1$ **do**
8:             **for** $(s_1, t_1) \in P_{d_{\min}}$ **do**
9:                 **for** $(s_2, t_2) \in P_{d_{\max}}$ **do**
10:                     $2DSP(s_1, t_1, s_2, t_2) \leftarrow$ CALCULATESINGLE2DSPVALUE(s_1, t_1, s_2, t_2)
                        $\triangleright$ Finally use the symmetric nature of our predicate.
11:                     $2DSP(s_2, t_2, s_1, t_1) \leftarrow 2DSP(s_1, t_1, s_2, t_2)$
12: **procedure** CALCULATESINGLE2DSPVALUE(s_1, t_1, s_2, t_2)
13:     **if** $s_1 = t_1 \wedge s_2 = t_2$ **then**
14:         **return** $s_1 = s_2$
15:     **else if** (s_1, t_1, s_2, t_2) is not rigid **then**
16:         Calculate and return the value according to case (2) of Th 1
17:     **else**
18:         Calculate $Q_2(s_1, t_1, s_2, t_2)$ using formula (2)
19:         Calculate $Q_4(s_1, t_1, s_2, t_2)$ using formula (3)
20:         **return** $Q_2(s_1, t_1, s_2, t_2) \vee Q_4(s_1, t_1, s_2, t_2)$

Acknowledgements. I would like to express gratitude to Maxim Babenko for pointing me to the original article [4] and suggesting to investigate the possibility of optimization, for a lot of helpful discussions and for helping me with the preparation of the paper text.

References

1. Björklund, A., Husfeldt, T.: Shortest two disjoint paths in polynomial time. In: Esparza, J., Fraigniaud, P., Husfeldt, T., Koutsoupias, E. (eds.) ICALP 2014. LNCS, vol. 8572, pp. 211–222. Springer, Heidelberg (2014). https://doi.org/10.1007/978-3-662-43948-7_18

2. Björklund, A., Husfeldt, T.: Counting shortest two disjoint paths in cubic planar graphs with an NC algorithm. ArXiv abs/arXiv:1806.07586 (2018)

3. Datta, S., Iyer, S., Kulkarni, R., Mukherjee, A.: Shortest k-disjoint paths via determinants. In: Ganguly, S., Pandya, P. (eds.) 38th IARCS Annual Conference on Foundations of Software Technology and Theoretical Computer Science (FSTTCS 2018). Leibniz International Proceedings in Informatics (LIPIcs), vol. 122, pp. 19:1–19:21. Schloss Dagstuhl-Leibniz-Zentrum fuer Informatik, Dagstuhl (2018). https://doi.org/10.4230/LIPIcs.FSTTCS.2018.19. http://drops.dagstuhl.de/opus/volltexte/2018/9918

4. Eilam-Tzoreff, T.: The disjoint shortest paths problem. Discrete Appl. Math. **85**(2), 113–138 (1998). https://doi.org/10.1016/S0166-218X(97)00121-2. http://www.sciencedirect.com/science/article/pii/S0166218X97001212

5. Fortune, S., Hopcroft, J., Wyllie, J.: The directed subgraph homeomorphismproblem. Theor. Comput. Sci. **10**(2), 111–121 (1980). https://doi.org/10.1016/0304-3975(80)90009-2. http://www.sciencedirect.com/science/article/pii/030439758090 0092

6. Gallai, T.: Maximum-minimum Sätze und verallgemeinerte Faktoren vonGraphen. Acta Math. Acad. Scientiarum Hung. **12**(1), 131–173 (1964). https://doi.org/10. 1007/BF02066678

7. Gustedt, J.: The general two-path problem in time $\mathcal{O}(m \log n)$. Technical report 394, TU Berlin (1994)

8. Hirai, H., Namba, H.: Shortest $(a + b)$-path packing via Hafnian. Algorithmica **80**(8), 2478–2491 (2018). https://doi.org/10.1007/s00453-017-0334-0

9. Lynch, J.F.: The equivalence of theorem proving and the interconnection problem. SIGDA Newsl. **5**(3), 31–36 (1975). https://doi.org/10.1145/1061425.1061430

10. Robertson, N., Seymour, P.D.: Disjoint paths–a survey. SIAM J. Algebraic Discrete Methods **6**(2), 300–305 (1985). https://doi.org/10.1137/0606030

11. Williams, V.V.: Multiplying matrices faster than Coppersmith-Winograd. In: Proceedings of the Forty-Fourth Annual ACM Symposium on Theory of Computing, STOC 2012, pp. 887–898. ACM, New York (2012). https://doi.org/10.1145/2213977.2214056

An Improvement to Chvátal
and Thomassen's Upper Bound
for Oriented Diameter

Jasine Babu, Deepu Benson$^{(\boxtimes)}$, Deepak Rajendraprasad, and Sai Nishant Vaka

Indian Institute of Technology Palakkad, Palakkad, India
{jasine,deepak}@iitpkd.ac.in, bensondeepu@gmail.com

Abstract. An orientation of an undirected graph G is an assignment of exactly one direction to each edge of G. The oriented diameter of a graph G is the smallest diameter among all the orientations of G. The maximum oriented diameter of a family of graphs $\mathscr{F}$ is the maximum oriented diameter among all the graphs in $\mathscr{F}$. Chvátal and Thomassen [JCTB, 1978] gave a lower bound of $\frac{1}{2}d^2 + d$ and an upper bound of $2d^2 + 2d$ for the maximum oriented diameter of the family of 2-edge connected graphs of diameter d. We improve this upper bound to $1.373d^2 + 6.971d - 1$, which outperforms the former upper bound for all values of d greater than or equal to 8. For the family of 2-edge connected graphs of diameter 3, Kwok, Liu and West [JCTB, 2010] obtained improved lower and upper bounds of 9 and 11 respectively. For the family of 2-edge connected graphs of diameter 4, the bounds provided by Chvátal and Thomassen are 12 and 40 and no better bounds were known. By extending the method we used for diameter d graphs, along with an asymmetric extension of a technique used by Chvátal and Thomassen, we have improved this upper bound to 21.

Keywords: Oriented diameter · Strong orientation · One-way traffic problem

1 Introduction

An *orientation* of an undirected graph G is an assignment of exactly one direction to each of the edges of G. A given undirected graph can be oriented in many different ways (2^m, to be precise, where m is the number of edges). One of the earliest studies regarding graph orientations were carried out by H.E. Robbins in 1939. He was trying to answer a question posed by Stanislaw Ulam. *"When may the arcs of a graph be so oriented that one may pass from any vertex to any other, traversing arcs in the positive sense only?"*. This led to the seminal work [1] of Robbins in which he proved the following theorem, *"A graph is orientable if and only if it remains connected after the removal of any arc"*.

A directed graph $\vec{G}$ is called *strongly connected* if it is possible to reach any vertex starting from any other vertex using a directed path. An undirected

© Springer Nature Switzerland AG 2020
H. Fernau (Ed.): CSR 2020, LNCS 12159, pp. 117–129, 2020.
https://doi.org/10.1007/978-3-030-50026-9_8

graph G is called *strongly orientable* if it has a strongly connected orientation. A *bridge* in a connected graph is an edge whose removal will disconnect the graph. A *2-edge connected* graph is a connected graph which does not contain any bridges. The theorem of Robbins stated earlier says that it is possible for a graph G to be strongly oriented if and only if G is 2-edge connected. Though Robbins stated the necessary and sufficient conditions for a graph to have a strong orientation, no comparison between the diameter of a graph and the diameter of an orientation of this graph was given in this study. This was taken up by Chvátal and Thomassen in 1978 [2].

In order to discuss these quantitative results, we introduce some notation. Let G be an undirected graph. The *distance* between two vertices u and v of G, $d_G(u, v)$ is the number of edges in a shortest path between u and v. For any two subsets A, B of $V(G)$, let $d_G(A, B) = \min\{d_G(u, v) : u \in A, v \in B\}$. The *eccentricity* of a vertex v of G is the maximum distance between v and any other vertex u of G. The *diameter* of G is the maximum of the eccentricities of its vertices. The *radius* of G is the minimum of the eccentricities of its vertices. Let $\vec{G}$ be a directed graph and $u, v \in V(\vec{G})$. Then the *distance* from a vertex u to v, $d_{\vec{G}}(u, v)$, is defined as the length of a shortest directed path from u to v. For any two subsets A, B of $V(\vec{G})$, let $d_{\vec{G}}(A, B) = \min\{d_{\vec{G}}(u, v) : u \in A, v \in B\}$. The *out-eccentricity* of a vertex v of $\vec{G}$ is the greatest distance from v to a vertex $u \in V(\vec{G})$. The *in-eccentricity* of a vertex v of $\vec{G}$ is the greatest distance from a vertex $u \in V(\vec{G})$ to v. The *eccentricity* of a vertex v of $\vec{G}$ is the maximum of its out-eccentricity and in-eccentricity. The *diameter* of $\vec{G}$, denoted by $d(\vec{G})$, is the maximum of the eccentricities of its vertices. The *radius* of $\vec{G}$ is the minimum of the eccentricities of its vertices. The *oriented diameter* of an undirected graph G, denoted by $\vec{d}(G)$, is the smallest diameter among all strong orientations of G. That is, $\vec{d}(G) := \min\{d(\vec{G}) : \vec{G} \text{ is an orientation of G}\}$. The *oriented radius* of an undirected graph G is the smallest radius among all strong orientations of G. The maximum oriented diameter of the family $\mathscr{F}$ of graphs is the maximum oriented diameter among all the graphs in $\mathscr{F}$. Let $f(d)$ denote the maximum oriented diameter of the family of 2-edge connected diameter d graphs. That is, $f(d) := \max\{\vec{d}(G) : G \in \mathscr{F}\}$, where $\mathscr{F}$ is the family of 2-edge connected graphs with diameter d.

The following theorem by Chvátal and Thomassen [2] gives an upper bound for the oriented radius of a graph.

Theorem 1 [2]. *Every 2-edge connected graph of radius r admits a strong orientation of radius at most $r^2 + r$.*

The above bound was also shown to be tight. In the same paper, they also proved that the problem of deciding whether an undirected graph admits an orientation of diameter 2 is NP-hard. Motivated by the work of Chvátal and Thomassen [2], Chung, Garey and Tarjan [3] proposed a linear-time algorithm to check whether a mixed multigraph has a strong orientation or not. They have also proposed a polynomial time algorithm which provides a strong orientation (if it exists) for a mixed multigraph with oriented radius at most $4r^2 + 4r$. Studies have also been carried out regarding the oriented diameter of specific subclasses of graphs like AT-free graphs, interval graphs, chordal graphs and planar graphs [4–6]. Bounds on oriented diameter in terms of minimum degree is also available in the literature [7,8].

The following bounds for $f(d)$ were given by Chvátal and Thomassen [2].

Theorem 2 [2]. $\frac{1}{2}d^2 + d \leq f(d) \leq 2d^2 + 2d$.

Chvátal and Thomassen [2] has also proved that $f(2) = 6$. By Theorem 2, $8 \leq f(3) \leq 24$. In 2010, Kwok, Liu and West [9] improved these bounds to $9 \leq f(3) \leq 11$. To prove the upper bound of 11, Kwok, Liu and West partitioned the vertices of G into a number of sets based on the distances from the endpoints of an edge which is not part of any 3-cycle. Our study on the oriented diameter of 2-edge connected graphs with diameter d uses this idea of partitioning the vertex set into a number of sets based on their distances from a specific edge.

Our Results

In this paper we establish two improved upper bounds. Firstly in Sect. 2, we show that $f(d) \leq 1.373d^2 + 6.971d - 1$ (Theorem 4). This is the first general improvement to Chvátal and Thomassen's upper bound $f(d) \leq 2d^2 + 2d$ from 1978. For all $d \geq 8$, our upper bound outperforms that of Chvátal and Thomassen. Their lower bound $f(d) \geq \frac{1}{2}d^2 + d$ still remains unimproved. We do not believe that our upper bound is tight. Secondly in Sect. 3, for the case of $d = 4$, we further sharpen our analysis and show that $f(4) \leq 21$ (Theorem 5). This is a considerable improvement from 40, which follows from Chvátal and Thomassen's general upper bound. Here too, our upper bound is not yet close to the lower bound of 12 given by Chvátal and Thomassen and we believe that there is room for improvement in the upper bound.

2   Oriented Diameter of Diameter d Graphs

A subset D of the vertex set of G is called a *k-step dominating set* of G if every vertex not in D is at a distance of at most k from at least one vertex of D. An oriented subgraph $\vec{H}$ of G is called a k-step dominating oriented subgraph if $V(\vec{H})$ is a k-step dominating set of $V(G)$. To obtain upper bounds for the oriented diameter of a graph G with n vertices and minimum degree $\delta \geq 2$, Bau and Dankelmann [7] and Surmacs [8] first constructed a 2-step dominating oriented subgraph $\vec{H}$ of G. They used this together with the idea in the proof of Theorem 1 on $\vec{H}$ to obtain the upper bounds of $\frac{11n}{\delta+1} + 9$ and $\frac{7n}{\delta+1}$, respectively, for the oriented diameter of graphs with minimum degree $\delta \geq 2$. We are using the algorithm ORIENTEDCORE described below to produce a 2-edge connected oriented subgraph $\vec{H}$ of G with some distance guarantees between the vertices in $\vec{H}$ (Lemma 1) and some domination properties (Lemma 2).

2.1 Algorithm ORIENTEDCORE

Input: A 2-edge connected graph G and a specified edge pq in G.

Output: A 2-edge connected oriented subgraph $\vec{H}$ of G.

Terminology: Let d be the diameter of G, let k be the length of a smallest cycle containing pq in G and let $h = \lfloor k/2 \rfloor$. Notice that $k \leq 2d + 1$ and $h \leq d$. Define $S_{i,j} = \{v \in V(G) : d_G(v, p) = i, d_G(v, q) = j\}$. Since $S_{i,j}$ is non-empty only if $0 \leq i, j \leq d$ and $|i - j| \leq 1$, we implicitly assume these restrictions on the subscripts of $S_{i,j}$ wherever we use it. For a vertex $v \in S_{i,j}$, its *level* $L(v)$ is $(j - i)$ and its *width* $W(v)$ is $\max(i, j)$. We will always refer to an edge $\{u, v\}$ between two different $S_{i,j}$'s as uv when either $L(u) > L(v)$ or $L(u) = L(v)$ and $W(u) < W(v)$ (downward or rightward in Fig. 1). Moreover the edge uv is called *vertical* in the first case and *horizontal* in the second.

Observations based on the first edge of shortest paths from a vertex v to p or v to q: Every vertex $v \in S_{i,i+1}$, $1 \leq i \leq d - 1$, is incident to a horizontal edge uv with $u \in S_{i-1,i}$. Every vertex $v \in S_{i+1,i}$, $1 \leq i \leq d-1$, is incident to a horizontal edge uv with $u \in S_{i,i-1}$. Every vertex $v \in S_{i,i}$, $1 \leq i \leq d$, is incident either to a horizontal edge uv with $u \in S_{i-1,i-1}$ or two vertical edges uv and vx with $u \in S_{i-1,i}$ and $x \in S_{i,i-1}$. Consequently for any v in Level 1, all the shortest p–v path consists of Level 1 horizontal edges only and for any vertex v in Level -1, all the shortest v–q path consists of Level -1 horizontal edges alone. For any vertex v in Level 0, all the shortest p–v path consists of horizontal edges in levels 1 and 0 and exactly one vertical edge; while all the shortest v–q path consists of horizontal edges in levels 0 and -1 and exactly one vertical edge.

Stage 1. Initialise $\vec{H}$ to be empty. For each vertical edge uv with $L(u) = 1$ and $L(v) \in \{0, -1\}$, and for each shortest p–u path P_u and shortest v–q path P_v, do the following: Let P be the p–q path formed by joining P_u, the edge uv and P_v. Orient the path P as a directed path $\vec{P}$ from p to q and add it to $\vec{H}$. Notice that even though two such paths can share edges, there is no conflict in the above orientation since, in Stage 1, every vertical edge is oriented downward, every horizontal edge in Level 1 is oriented rightward and every horizontal edge in levels 0 and -1 is oriented leftward.

Stage 2. For each vertical edge uv with $L(u) = 0$ and $L(v) = -1$ not already oriented in Stage 1, and for each shortest p–u path P_u and shortest v–q path P_v do the following: Let x be the last vertex in P_u (nearest to u) that is already in $V(\vec{H})$ and let P'_u be the subpath of P_u from x to u. Similarly let y be the first vertex in P_v (nearest to v) that is already in $V(\vec{H})$ and let P'_v be the subpath of P_v from v to y. Let P be the x–y path formed by joining P'_u, the edge uv and P'_v. Orient the path P as a directed path $\vec{P}$ from x to y and add it to $\vec{H}$. Notice that P does not share any edge with a path added to $\vec{H}$ in Stage 1, but it can share edges with paths added in earlier steps of Stage 2. However there is no conflict in the orientation since, in Stage 2, every vertical edge is oriented downward, every horizontal edge in Level 0 is oriented rightward, every horizontal edge in Level -1 is oriented leftward, and no horizontal edges in Level 1 is added.

Stage 3. Finally orient the edge pq from q to p and add it to $\vec{H}$. This completes the construction of $\vec{H}$, the output of the algorithm.

Distances in $\vec{H}$. First we analyse the (directed) distance from p and to q of vertices added to $\vec{H}$ in Stage 1. The following bounds on distances in $\vec{H}$ follow from the construction of each path P in Stage 1. Let w be any vertex that is added to $\vec{H}$ in Stage 1. Then

$$
d_{\vec{H}}(p, w) \leq \begin{cases} i, & w \in S_{i,i+1}, \\ h, & w \in S_{h,h}, \\ 2d - i, & w \in S_{i,i}, \ i > h, \ \text{and} \\ 2d - i, & w \in S_{i+1,i}. \end{cases} \tag{1}
$$

$$
d_{\vec{H}}(w, q) \leq \begin{cases} 2d - i, & w \in S_{i,i+1}, \\ h, & w \in S_{h,h}, \\ 2d - i, & w \in S_{i,i}, \ i > h, \ \text{and} \\ i, & w \in S_{i+1,i}. \end{cases} \tag{2}
$$

It is easy to verify the above equations using the facts that w is part of a directed p–q path of length at most $2d$ (at most $2h$ if $w \in S_{h,h}$) in $\vec{H}$.

No new vertices from Level 1 or $S_{h,h}$ are added to $\vec{H}$ in Stage 2. Still the distance bounds for vertices added in Stage 2 are slightly more complicated since a path P added in this stage will start from a vertex x in Level 0 and end in a vertex y in Level -1, which are added to $\vec{H}$ in Stage 1. But we can complete the analysis since we already know that $d_{\vec{H}}(p, x) \leq 2d - h - 1$ and $d_{\vec{H}}(y, q) \leq i$ where i is such that $y \in S_{i+1,i}$ from the analysis of Stage 1. Let w be any vertex that is added to $\vec{H}$ in Stage 2. Then

$$
d_{\vec{H}}(p, w) \leq \begin{cases} (2d - h - 1) + (i - h - 1) \\ \quad = 2d - 2h - 2 + i, & w \in S_{i,i}, \ i > h, \ \text{and} \\ (2d - h - 1) + (d - h - 1) + (d - i) \\ \quad = 4d - 2h - 2 - i, & w \in S_{i+1,i}. \end{cases} \tag{3}
$$

The distance from w to q in $\vec{H}$ is not affected even though we trim the path P_v at y since y already has a directed shortest path to q from Stage 1. Hence

$$
d_{\vec{H}}(w, q) \leq \begin{cases} 2d - i, & w \in S_{i,i}, \ i > h, \ \text{and} \\ i, & w \in S_{i+1,i}. \end{cases} \tag{4}
$$

The first part of the next lemma follows from taking the worst case among (1) and (3). Notice that $\forall i > h$, $(2h + 2 - i \leq i)$ and $(4d - 2h - 2 \geq 2d)$ when $h < d$. New vertices are added to $\vec{H}$ in Stage 2 only if $h < d$. The second part follows from (2) and (4). The subsequent two claims are easy observations.

Lemma 1. *Let G be a 2-edge connected graph, pq be any edge of G and let $\vec{H}$ be the oriented subgraph of G returned by the algorithm* ORIENTEDCORE. *Then for every vertex $w \in V(\vec{H})$ we have*

$$d_{\vec{H}}(p,w) \leq \begin{cases} i, & w \in S_{i,i+1}, \\ h, & w \in S_{h,h}, \\ 2d - 2h - 2 + i, & w \in S_{i,i}, \ i > h, \ and \\ 4d - 2h - 2 - i, & w \in S_{i+1,i}. \end{cases} \tag{5}$$

$$d_{\vec{H}}(w,q) \leq \begin{cases} 2d - i, & w \in S_{i,i+1}, \\ h, & w \in S_{h,h}, \\ 2d - i, & w \in S_{i,i}, \ i > h, \ and \\ i, & w \in S_{i+1,i}. \end{cases} \tag{6}$$

Moreover, $d_{\vec{H}}(q,p) = 1$ and $d_{\vec{H}}(p,q) \leq k - 1$.

We can see that if $S_{h,h}$ is non-empty, then all the vertices in $S_{h,h}$ are captured into $\vec{H}$.

Notice that when $k \geq 4$, $S_{1,2}$ and $S_{2,1}$ are non empty. Thus the bound on the diameter of $\vec{H}$ follows by the triangle inequality $d_{\vec{H}}(x,y) \leq d_{\vec{H}}(x,q) + d_{\vec{H}}(q,p) + d_{\vec{H}}(p,y)$ and the fact that $\forall k \geq 4$ the worst bounds for $d_{\vec{H}}(x,q)$ and $d_{\vec{H}}(p,y)$ from Lemma 1 are when $x \in S_{1,2}$ and $y \in S_{2,1}$. Hence the following corollary.

Corollary 1. *Let G be a 2-edge connected graph, pq be any edge of G and let $\vec{H}$ be the oriented subgraph of G returned by the algorithm* ORIENTEDCORE. *If the length of the smallest cycle containing pq is greater than or equal to 4, then the diameter of $\vec{H}$ is at most $6d - 2h - 3$.*

Domination by $\vec{H}$. Let us call the vertices in $V(\vec{H})$ as *captured* and those in $V(G) \backslash V(\vec{H})$ as *uncaptured*. For each $i \in \{1, 0, -1\}$ let L_i^c and L_i^u denote the captured and uncaptured vertices in level i, respectively. Since L_i^c contains every level i vertex incident with a vertical edge, L_i^c separates L_i^u from rest of G. Let d_i denote the maximum distance between a vertex in L_i^u and the set L_i^c. If $u_i \in L_i^u$ and $u_j \in L_j^u$ such that $d_G(u_i, L_i^c) = d_i$ and $d_G(u_j, L_j^c) = d_j$ for distinct $i, j \in \{1, 0, -1\}$, the distance $d_G(u_i, u_j)$ is bounded above by d, the diameter of G, and bounded below by $d_i + 1 + d_j$. Hence $d_i + d_j \leq d - 1$ for every distinct $i, j \in \{1, 0, -1\}$.

For any vertex $u \in L_0^u$, the last Level 0 vertex in a shortest (undirected) u–q path is in L_0^c. Hence if Level 0 is non-empty then $d_0 \leq (d - h)$. In order to bound d_1 and d_{-1}, we take a close look at a shortest cycle C containing the edge pq. Let $C = (v_1, \ldots, v_k, v_1)$ with $v_1 = q$ and $v_k = p$. Each v_i is in $S_{i,i-1}$ when $2i < k+1$, $S_{i-1,i-1}$ if $2i = k+1$ and $S_{k-i,k-i+1}$ when $2i > k+1$. Let $t = \lceil k/4 \rceil$. The Level -1 vertex v_t is special since it is at a distance t from Level 1 and thus L_1^c. If u_1 is a vertex in L_1^u such that $d_G(u_1, L_1^c) = d_1$, the distance $d_G(u_1, v_t)$ is bounded above by d and below by $d_1 + t$. Hence $d_1 \leq d - t$. Similarly we can see that $d_{-1} \leq (d - t)$.

Putting all these distance bounds on domination together, we get the next lemma.

Lemma 2. *Let G be a 2-edge connected graph, pq be any edge of G and let $\vec{H}$ be the oriented subgraph of G returned by the algorithm ORIENTEDCORE. For each $i \in \{1, 0, -1\}$, let d_i denote the maximum distance of a level i vertex not in $V(\vec{H})$ to the set of level i vertices in $V(\vec{H})$. Then $d_0 \leq d - \lfloor k/2 \rfloor$, $d_1, d_{-1} \leq d - \lceil k/4 \rceil$ and for any distinct $i, j \in \{1, 0, -1\}$, $d_i + d_j \leq d - 1$.*

2.2 The Upper Bound

Consider a 2-edge connected graph G with diameter d. Let $\eta(G)$ denote the smallest integer such that every edge of a graph G belongs to a cycle of length at most $\eta(G)$. Huang, Li, Li and Sun [10] proved the following theorem.

Theorem 3 [10]. *$\vec{d}(G) \leq 2r(\eta - 1)$ where r is the radius of G and $\eta = \eta(G)$.*

We know that $r \leq d$ and hence we have $\vec{d}(G) \leq 2d(\eta - 1)$ as our first bound. Let pq be an edge in G such that the length of a smallest cycle containing pq is η. If $\eta \leq 3$, then $\vec{d}(G) \leq 4d$ which is smaller than the bound claimed in Theorem 4. So we assume $\eta \geq 4$. By Corollary 1, G has an oriented subgraph $\vec{H}$ with diameter at most $6d - 2\lfloor \frac{\eta}{2} \rfloor - 3$. Moreover by Lemma 2, $\vec{H}$ is a $(d - \lceil \frac{\eta}{4} \rceil)$-step dominating subgraph of G. Let G_0 be a graph obtained by contracting the vertices in $V(\vec{H})$ into a single vertex v_H. We can see that G_0 has radius at most $(d - \lceil \frac{\eta}{4} \rceil)$. Thus by Theorem 1, G_0 has a strong orientation $\vec{G_0}$ with radius at most $(d - \lceil \frac{\eta}{4} \rceil)^2 + (d - \lceil \frac{\eta}{4} \rceil)$. Since $d \leq 2r$, we have $d(\vec{G_0}) \leq 2(d - \lceil \frac{\eta}{4} \rceil)^2 + 2(d - \lceil \frac{\eta}{4} \rceil)$. Notice that $\vec{G_0}$ and $\vec{H}$ do not have any common edges. Hence G has an orientation with diameter at most $2(d - \lceil \frac{\eta}{4} \rceil)^2 + 2(d - \lceil \frac{\eta}{4} \rceil) + (6d - 2\lfloor \frac{\eta}{2} \rfloor - 3)$ by combining the orientations in $\vec{H}$ and $\vec{G_0}$. Let $\eta = 4\alpha d$. Hence we get $\vec{d}(G) \leq \min\{f_1(\alpha, d), f_2(\alpha, d)\}$ where $f_1(\alpha, d) = 8\alpha d^2 - 2d$ and $f_2(\alpha, d) = 2(1 - \alpha)^2 d^2 + 8d - 6\alpha d - 1$. Notice that $0 < \frac{3}{4d} \leq \alpha < 1$ and for $\alpha = 3 - 2\sqrt{2}$, $8\alpha d^2 = 2(1 - \alpha)^2 d^2$. Further notice that, for each fixed value of d, f_1 is an increasing function of α and f_2 is a decreasing function of α in the interval $(0, 1)$. Hence, if $\alpha \in [3 - 2\sqrt{2}, 1)$, $\vec{d}(G) \leq f_2(\alpha, d) \leq f_2(3 - 2\sqrt{2}, d) \leq 1.373d^2 + 6.971d - 1$ and if $\alpha \in (0, 3 - 2\sqrt{2}]$, $\vec{d}(G) \leq f_1(\alpha, d) \leq f_1(3 - 2\sqrt{2}, d) \leq 1.373d^2 - 2d$. Since $\alpha \in (0, 1)$, we obtain the following theorem.

Theorem 4. *$f(d) \leq 1.373d^2 + 6.971d - 1$.*

For any $d \geq 8$, the above upper bound is an improvement over the upper bound of $2d^2 + 2d$ provided by Chvátal and Thomassen.

3   Oriented Diameter of Diameter 4 Graphs

Throughout this section, we consider G to be an arbitrary 2-edge connected diameter 4 graph. We will show that the oriented diameter of G is at most 21 and hence $f(4) \leq 21$. The following lemma by Chvátal and Thomassen [2] is used when $\eta(G) \leq 4$.

Lemma 3 [2]. *Let Γ be a 2-edge connected graph. If every edge of Γ lies in a cycle of length at most k, then it has an orientation $\vec{\Gamma}$ such that*

$$d_{\vec{\Gamma}}(u,v) \leq ((k-2)2^{\lfloor (k-1)/2 \rfloor} + 1)d_\Gamma(u,v) \quad \forall u,v \in V(\vec{\Gamma})$$

Hence if all edges of the graph G lie in a 3-cycle or a 4-cycle, the oriented diameter of G will be at most 20. Hence we can assume the existence of an edge pq which is not part of any 3-cycle or 4-cycle as long as we are trying to prove an upper bound of 20 or more for $f(4)$. We apply algorithm ORIENTDCORE on G with the edge pq to obtain an oriented subgraph $\vec{H}_1$ of G. Figure 1 shows a coarse representation of $\vec{H}_1$.

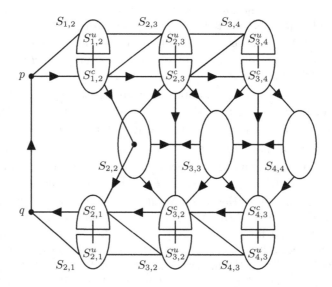

Fig. 1. A coarse representation of $\vec{H}_1$ which shows the orientation of edges between various subsets of $V(G)$. A single arrow from one part to another indicates that all the edges between these parts are oriented from the former to latter. A double arrow between two parts indicates that the edges between the two parts are oriented in either direction or unoriented. An unoriented edge between two parts indicate that no edge between these two parts are oriented.

3.1 Oriented Diameter and 2-Step Domination Property of $\vec{H}_1$

Let $\vec{H}_1$ be the oriented subgraph of G returned by the algorithm ORIENTED-CORE. Since the smallest cycle containing pq is of length greater than or equal to 5, by Corollary 1, we can see that the diameter of $\vec{H}_1$ is at most 17. Moreover, from Eqs. 5 and 6 of Lemma 1, we get the upper bounds on the distances of $\vec{H}_1$ in Table 1. Hence, the following corollary.

Corollary 2. $d(\vec{H}_1) \leq 17$. Moreover $\forall w \in V(\vec{H}_1)$, $d_{\vec{H}_1}(p,w)$ and $d_{\vec{H}_1}(w,q)$ obey the bounds in Table 1.

Remark 1. If $k > 5$ ($h > 2$), then $S_{2,2}$ is empty. Moreover if $S_{2,2}$ is non-empty, then all the vertices in $S_{2,2}$ are captured into $\vec{H}_1$.

Table 1. Upper bounds on the distances of $\vec{H}_1$

For w in	S_{12}^c	S_{23}^c	S_{34}^c	S_{22}	S_{33}^c	S_{44}^c	S_{21}^c	S_{32}^c	S_{43}^c
$d_{\vec{H}_1}(p,w) \leq$	1	2	3	2	5	6	9	8	7
$d_{\vec{H}_1}(w,q) \leq$	7	6	5	2	5	4	1	2	3

Furthermore, applying Lemma 2 on $\vec{H}_1$ shows that $\vec{H}_1$ is a 2-step dominating subgraph of G. Let G_0 be a graph obtained by contracting the vertices in $V(\vec{H}_1)$ into a single vertex v_H. We can see that G_0 has radius at most 2. Thus by Theorem 1, G_0 has a strong orientation $\vec{G}_0$ with radius at most 6. Since $d \leq 2r$, we have $d(\vec{G}_0) \leq 12$. Since $\vec{G}_0$ and $\vec{H}_1$ do not have any common edges we can see that G has an orientation with diameter at most 29 by combining the orientations in $\vec{H}_1$ and $\vec{G}_0$. But we further improve this bound to 21 by constructing a 1-step dominating oriented subgraph $\vec{H}_2$ of G. We propose the following asymmetric variant of a technique by Chvátal and Thomassen [2] for the construction and analysis of $\vec{H}_2$.

3.2 Asymmetric Chvátal-Thomassen Lemma

For any subset A of $V(G)$, let $N(A)$ denote the set of all vertices with an edge incident on some vertex in A. Let H be a subgraph of G. An *ear* of H in G is a sequence of edges $uv_1, v_1v_2, \dots, v_{k-1}v_k, v_k v$ such that $u, v \in V(H)$, $k \geq 1$ and none of the vertices $v_1, \dots, v_k$ and none of the edges in this sequence are in H. In particular we allow $u = v$.

Lemma 4 (Asymmetric Chvátal-Thomassen Lemma). *Let G be an undirected graph and let $A \subseteq B \subseteq V(G)$ such that*

(i) *B is a k-step dominating set in G,*
(ii) *G/B is 2-edge connected, and*

(iii) $N(A) \cup B$ *is a $(k-1)$-step dominating set of G.*

Then there exists an oriented subgraph $\vec{H}$ of $G \backslash G[B]$ such that

(i) $N(A) \backslash B \subseteq V(\vec{H})$ *and hence $V(\vec{H}) \cup B$ is a $(k-1)$-step dominating set of G, and*

(ii) $\forall v \in V(\vec{H})$, *we have $d_{\vec{H}}(A, v) \leq 2k$ and either $d_{\vec{H}}(v, A) \leq 2k$ or $d_{\vec{H}}(v, B \backslash A) \leq 2k - 1$.*

Proof. We construct a sequence $\vec{H}_0, \vec{H}_1, \ldots$ of oriented subgraphs of $G \backslash G[B]$ as follows. We start with $\vec{H}_0 = \emptyset$ and add an oriented ear $\vec{Q}_i$ that starts in A and ends in B in each step. Let $i \geq 0$. If $N(A) \backslash B \subseteq V(\vec{H}_i)$, then we stop the construction and set $\vec{H} = \vec{H}_i$. Since $N(A) \cup B$ is a $(k-1)$-step dominating set of G, the first conclusion of the lemma is satisfied when the construction ends with $N(A) \backslash B \subseteq V(\vec{H})$. If $N(A) \backslash B \not\subseteq V(\vec{H}_i)$, then let $v \in (N(A) \backslash B) \backslash V(\vec{H}_i)$ and let u be a neighbour of v in A. Since G/B is 2-edge connected, there exists a path in $G' = (G/B) \backslash \{uv\}$ from v to B. Let P_i be a shortest v–B path in G'. If multiple such shortest paths are available, we will choose one that ends in A if available. We further ensure that once P_i hits a vertex in an oriented ear $\vec{Q}_j$ that was added in a previous step, P_i continues further to B along the shorter arm of Q_j. It can be verified that P_i is still a shortest v–B path in G'. The ear Q_i is the union of the edge uv and the path P_i. If P_i hits B without hitting any previous ear, then we orient Q_i as a directed path $\vec{Q}_i$ from u to B. If $Q_i \cap Q_j \neq \emptyset$, then we orient Q_i as a directed path $\vec{Q}_i$ by extending the orientation of $Q_i \cap Q_j$. Notice that, in both these cases, the source vertex of $\vec{Q}_i$ is in A. We add $\vec{Q}_i$ to $\vec{H}_i$ to obtain $\vec{H}_{i+1}$.

Let $Q_i = (v_0, \ldots, v_q)$ with $v_0 \in A$ and $v_q \in B$ be the ear added in the i-th stage above. Since $(v_1, \ldots, v_q)$ is a shortest v_1–B path in $G' = (G/B) \backslash \{v_0 v_1\}$ and since B is a k-step dominating set, $q \leq 2k + 1$. Moreover, if $v_q \in B \backslash A$, then by our choice of path, vertices v_2 to v_{q-1} are not in $N(A) \cup B$. Hence $q \leq 2k$, since $N(A) \cup B$ is a $(k-1)$-step dominating set. These bounds on the length of Q_i along with the observation that the source vertex of $\vec{Q}_i$ is in A, verifies the second conclusion of the lemma. □

Remark 2. If we flip the orientation of $\vec{H}$ we get the bounds $d_{\vec{H}}(v, A) \leq 2k$ and either $d_{\vec{H}}(A, v) \leq 2k$ or $d_{\vec{H}}(B \backslash A, v) \leq 2k - 1$, $\forall v \in V(\vec{H})$ in place of Conclusion (ii) of Lemma 4.

Setting $A = B$ in Lemma 4 gives the key idea which is recursively employed by Chvátal and Thomassen to prove Theorem 1 [2].

3.3   A 1-Step Dominating Oriented Subgraph $\vec{H}_2$ of G

Let $\vec{H}_1$ be the oriented subgraph of G returned by the algorithm ORIENTED-CORE. We will add further oriented ears to $\vec{H}_1$ to obtain a 1-step dominating oriented subgraph $\vec{H}_2$ of G. We have already seen that $\vec{H}_1$ is a 2-step dominating oriented subgraph of G. By Lemma 2, we also have $d_i + d_j \leq 3$ for any distinct $i, j \in \{1, 0, -1\}$.

Now consider the first case where we have vertices in Level 0 which are at a distance 2 from $S_{2,2}$. Notice that in this case, $d_0 = 2$ and hence $d_1, d_{-1} \leq 1$. Let $B = L_0^c$, $A = S_{2,2}$ and $G_0 = G[L_0]$. By Remark 1, $A \subseteq B$. Notice that $B = L_0^c$ is a cut-set separating L_0^u from the rest of G and hence the graph G_0/B is 2-edge connected. Since $S_{3,3}^u \subseteq N(S_{2,2})$, we can see that $N(A) \cup B = N(S_{2,2}) \cup L_0^c$ is a 1-step dominating subgraph of G_0. Therefore we can apply Lemma 4 on G_0. Every edge of the oriented subgraph of $G_0 \backslash G_0[B]$ obtained by applying Lemma 4 is reversed to obtain the subgraph $\vec{H}_2^0$. Now consider the vertices captured into $\vec{H}_2^0$. From Lemma 4 and Remark 2, we get the following bounds $d_{\vec{H}_2^0}(v, A) \leq 4$ and either $d_{\vec{H}_2^0}(A, v) \leq 4$ or $d_{\vec{H}_2^0}(B \backslash A, v) \leq 3$, $\forall v \in V(\vec{H}_2^0)$. Here $B \backslash A = S_{3,3}^c \cup S_{4,4}^c$ and from Table 1, we have the bounds $d_{\vec{H}_1}(p, x) \leq 5$, $\forall x \in S_{3,3}^c$, $d_{\vec{H}_1}(p, y) \leq 6$, $\forall y \in S_{4,4}^c$ and $d_{\vec{H}_1}(p, z) = 2$, $\forall z \in S_{2,2}$. Hence $d_{\vec{H}_1 \cup \vec{H}_2^0}(p, v) \leq 9$, $\forall v \in V(\vec{H}_2^0)$. Since $d_{\vec{H}_2^0}(v, A) \leq 4$ and $d_{\vec{H}_1}(x, q) = 2$, $\forall x \in A$, we also have $d_{\vec{H}_1 \cup \vec{H}_2^0}(v, q) \leq 6$, $\forall v \in V(\vec{H}_2^0)$. Let $\vec{H}_2 = \vec{H}_1 \cup \vec{H}_2^0$. By the above discussion, in combination with the distances in Table 1 and Corollary 2, we get $d(\vec{H}_2) \leq 17$.

Now consider the second case where $d_1 = 2$ or $d_{-1} = 2$. Since $d_1 + d_{-1} \leq 3$, uncaptured vertices at a distance 2 from $\vec{H}_1$ can exist either in Level 1 or in Level -1 but not both. By flipping the role of the vertices p and q in Algorithm ORIENTEDCORE if necessary, without loss of generality, we can assume vertices which are at a distance 2 from $\vec{H}_1$ exists only in Level -1 and not in Level 1. Recall that $L_{-1}^c = \{q\} \cup S_{2,1}^c \cup S_{3,2}^c \cup S_{4,3}^c$. Let $G_{-1} = G[L_{-1}]$ and $B = L_{-1}^c$. Further let r be any vertex in $S_{1,2}^c$ and $A = \{v \in B : d_G(r, v) = 2\}$. Since $q \in A$, A is never empty. Note that $A \subseteq B \subseteq V(G_{-1})$. Also G_{-1}/B is 2-edge connected since $B = L_{-1}^c$ is a cut-set which separates L_{-1}^u from the rest of G. Now consider a vertex z in Level -1 which is exactly at a distance 2 from B. Since the graph G is of diameter 4, there exists a 4-length path P from z to r. Since B separates L_{-1}^u from r, P intersects B, say at a vertex b. Further, we have $d_G(b, r) = 2$ and thus $b \in A$. Hence z has a 2-length path to a vertex $b \in A$. Thus $N(A) \cup B$ is a 1-step dominating subgraph of G_{-1}. Hence we can apply Lemma 4 on G_{-1} to obtain $\vec{H}_2^{-1}$, an oriented subgraph of $G_{-1} \backslash G_{-1}[B]$. Now consider the vertices captured into $\vec{H}_2^{-1}$. From Lemma 4, we get the following bounds $\forall v \in V(\vec{H}_2^{-1})$, $d_{\vec{H}_2^{-1}}(A, v) \leq 4$ and $d_{\vec{H}_2^{-1}}(v, B) \leq 4$. Since $d_{\vec{H}_1}(x, q) \leq 3$, $\forall x \in B$, we have $d_{\vec{H}_1 \cup \vec{H}_2^{-1}}(v, q) \leq 7$, $\forall v \in V(\vec{H}_2^{-1})$. Vertices in A can be from $S_{2,1}^c$, $S_{3,2}^c$ or $\{q\}$. By the definition of A there is an undirected path in G of length 3 from p to any vertex v_a in $(A \backslash \{q\})$, going through r. It can be verified that this undirected path is oriented from p to v_a by Algorithm ORIENTEDCORE. Hence $d_{\vec{H}_1}(p, v_a) \leq 3$, $\forall v_a \in (A \backslash \{q\})$ and hence $\forall v \in V(\vec{H}_2^{-1})$ with $d_{\vec{H}_2^{-1}}(A \backslash \{q\}, v) \leq 4$, $d_{\vec{H}_1 \cup \vec{H}_2^{-1}}(p, v) \leq 7$. But if a vertex $v \in V(\vec{H}_2^{-1})$ has $d_{\vec{H}_2^{-1}}(A \backslash \{q\}, v) > 4$, then $d_{\vec{H}_2^{-1}}(q, v) \leq 4$. In this case, since $d_{\vec{H}_1}(p, q) \leq 8$, we get

$d_{\vec{H}_1 \cup \vec{H}_2^{-1}}(p, v) \leq 12$. Notice that this is the only situation where $d_{\vec{H}_1 \cup \vec{H}_2^{-1}}(p, v) > 9$ and in this particular case $d_{\vec{H}_2^{-1}}(q, v) \leq 4$.

Now consider two vertices $x, y \in V(\vec{H}_1 \cup \vec{H}_2^{-1})$. We can see that $d_{\vec{H}_1 \cup \vec{H}_2^{-1}}(x, y) \leq d_{\vec{H}_1 \cup \vec{H}_2^{-1}}(x, q) + d_{\vec{H}_1 \cup \vec{H}_2^{-1}}(q, y)$. We have already proved that $d_{\vec{H}_1 \cup \vec{H}_2^{-1}}(x, q) \leq 7$. Now let us consider the $q - y$ path. If $y \in V(\vec{H}_1)$, from Table 1, we can see that $d_{\vec{H}_1}(p, y) \leq 9$ and therefore $d_{\vec{H}_1 \cup \vec{H}_2^{-1}}(x, y) \leq 17$. Now suppose if $y \in (V(\vec{H}_2^{-1}) \setminus V(\vec{H}_1))$. In this case we have already shown that $d_{\vec{H}_2^{-1}}(p, y) \leq 9$ or $d_{\vec{H}_2^{-1}}(q, y) \leq 4$. So, we either have a directed path of length 10 from q to y through p or a directed path of length 4 to y directly from q. Hence, $d_{\vec{H}_1 \cup \vec{H}_2^{-1}}(x, y) \leq 17$. Let $\vec{H}_2 = \vec{H}_1 \cup \vec{H}_2^{-1}$. By the above discussion, we get $d(\vec{H}_2) \leq 17$.

In both the cases we get an oriented subgraph $\vec{H}_2$ of G with $d(\vec{H}_2) \leq 17$. Moreover, it is clear from Conclusion (i) of Lemma 4 that $\vec{H}_2$ is a 1-step dominating subgraph of G. Hence the following Lemma.

Lemma 5. *For every 2-edge connected graph G with diameter 4 and $\eta(G) \geq 5$, there exists a 1-step dominating oriented subgraph $\vec{H}_2$ of G with $d(\vec{H}_2) \leq 17$.*

3.4   The Upper Bound

Now the main theorem of the section follows.

Theorem 5. $f(4) \leq 21$.

Proof. By Lemma 5, we get a 1-step dominating oriented subgraph $\vec{H}_2$ of G with $d(\vec{H}_2) \leq 17$. Let G_0 be a graph obtained by contracting the vertices in $V(\vec{H}_2)$ into a single vertex v_H. We can see that G_0 has radius at most 1. Thus by Theorem 1, G_0 has a strong orientation $\vec{G}_0$ with radius at most 2. Since $d \leq 2r$, we have $d(\vec{G}_0) \leq 4$. Notice that $\vec{G}_0$ and $\vec{H}_2$ do not have any common edges. Now we can see that G has an orientation with diameter at most 21 by combining the orientations in $\vec{H}_2$ and $\vec{G}_0$. ☐

References

1. Robbins, H.E.: A theorem on graphs, with an application to a problem of traffic control. Am. Math. Monthly **46**(5), 281–283 (1939)
2. Chvátal, V., Thomassen, C.: Distances in orientations of graphs. J. Comb. Theory Ser. B **24**(1), 61–75 (1978)
3. Chung, F.R.K., Garey, M.R., Tarjan, R.E.: Strongly connected orientations of mixed multigraphs. Networks **15**(4), 477–484 (1985)
4. Fomin, F.V., Matamala, M., Prisner, E., Rapaport, I.: AT-free graphs: linear bounds for the oriented diameter. Discrete Appl. Math. **141**(1–3), 135–148 (2004)
5. Fomin, F.V., Matamala, M., Rapaport, I.: Complexity of approximating the oriented diameter of chordal graphs. J. Graph Theory **45**(4), 255–269 (2004)

6. Eggemann, N., Noble, S.D.: Minimizing the oriented diameter of a planar graph. Electron. Notes Discrete Math. **34**, 267–271 (2009)
7. Bau, S., Dankelmann, P.: Diameter of orientations of graphs with given minimum degree. Eur. J. Comb. **49**, 126–133 (2015)
8. Surmacs, M.: Improved bound on the oriented diameter of graphs with given minimum degree. Eur. J. Comb. **59**, 187–191 (2017)
9. Kwok, P.K., Liu, Q., West, D.B.: Oriented diameter of graphs with diameter 3. J. Comb. Theory Ser. B **100**(3), 265–274 (2010)
10. Huang, X., Li, H., Li, X., Sun, Y.: Oriented diameter and rainbow connection number of a graph. Discret. Math. Theor. Comput. Sci. **16**(3), 51–60 (2014)

The Normalized Algorithmic Information Distance Can Not Be Approximated

Bruno Bauwens[(⊠)] and Ilya Blinnikov

National Research University Higher School of Economics, 11, Pokrovsky Boulevard,
109028 Moscow, Russia
brbauwens@gmail.com

Abstract. It is known that the normalized algorithmic information distance is not computable and not semicomputable. We show that for all $\varepsilon < 1/2$, there exist no semicomputable functions that differ from N by at most ε. Moreover, for any computable function f such that $|\lim_t f(x,y,t) - \mathrm{N}(x,y)| \leq \varepsilon$ and for all n, there exist strings x, y of length n such that $\sum_t |f(x,y,t+1) - f(x,y,t)| \geq \Omega(\log n)$. This is optimal up to constant factors.

We also show that the maximal number of oscillations of a limit approximation of N is $\Omega(n/\log n)$. This strengthens the $\omega(1)$ lower bound from [K. Ambos-Spies, W. Merkle, and S.A. Terwijn, 2019, *Normalized information distance and the oscillation hierarchy*].

Keywords: Algorithmic information distance · Kolmogorov complexity · Computability theory · Oscillation hierarchy

1 Introduction

The information distance defines a metric on bit strings that in some sense takes all "algorithmic regularities" into account. This distance was defined in [4] as $\mathrm{E}(x,y) = \max\{\mathrm{K}(x \mid y), \mathrm{K}(y \mid x)\}$, where $\mathrm{K}(\cdot|\cdot)$ denotes conditional prefix Kolmogorov complexity relative to some fixed optimal prefix-free Turing machine; we refer to appendix of the ArXiv version of this paper for the definition and basic properties, and to the books [6,8] for more background. After minor modifications, this distance satisfies the axioms of a metric, as explained in the appendix of the ArXiv version. We refer to [2] for an overview of many equivalent characterizations.

The distance is not computable. However, conditional Kolmogorov complexity is *upper semicomputable*, which means that there exists a computable function $f \colon \{0,1\}^* \times \{0,1\}^* \times \mathbb{N} \to \mathbb{Q}$ for which $\mathrm{K}(x \mid y) = \lim_t f(x,y,t)$, and that is non-increasing in its last argument t. Hence, also E is upper semicomputable.

The distance E is useful to compare strings of similar complexity. However, for strings of different complexity, a normalized variant is often preferable.

H. Fernau (Ed.): CSR 2020, LNCS 12159, pp. 130–141, 2020.
https://doi.org/10.1007/978-3-030-50026-9_9

Definition 1. *The* normalized *algorithmic information distance of strings* x *and* y *is*[1]

$$N(x,y) = \frac{\max\{K(x \mid y), K(y \mid x)\}}{\max\{K(x), K(y)\}}.$$

This normalized distance has inspired many applications in machine learning, where complexities are heuristically estimated using popular practical compression algorithms such as gzip, bzip2 and PPMZ, see [6, section 8.4]. Within small additive terms, the function N has values in the real interval $[0, 1]$ and satisfies the axioms of a metric:

- $0 \leq N(x,y) \leq 1 + O(1/K(x,y))$,
- $N(x,y) = N(y,x)$,
- $N(x,x) \leq O(1/K(x))$,
- $N(x,y) + N(y,z) \geq N(x,z) - O((\log K(y))/K(y))$.

See [6, Theorem 8.4.1].[2]

In this paper, we study the computability of N. Note that if Kolmogorov complexity were computable, then also N would be computable. But this is not the case, and in [9] it is proven that N is not upper semicomputable and not lower semicomputable, (i.e. $-N$ is not upper semicomputable). Below in Lemmas 2 and 4 we present simple proofs. In fact, in [9] it is proven that (i) there exists no lower semicomputable function that differs from N by at most some constant $\varepsilon < 1/2$, and (ii) there exists no upper semicomputable function that differs at most $\varepsilon = (\log n)/n$ from N on n-bit strings. Theorem 1 below implies that (ii) is also true for all $\varepsilon < 1/2$.

By definition, N is the ratio of two upper semicomputable functions, and hence it is *limit computable*, which means that there exists a computable function f such that $N(x,y) = \lim_t f(x,y,t)$. A function f that satisfies this property is called a *limit approximation* of N.

We define a *trivial limit approximation* f_{tr} of N where $f_{tr}(x,y,t)$ is obtained by replacing all appearances of $K(\cdot)$ and $K(\cdot \mid \cdot)$ in Definition 1 by upper approximations $K_t(\cdot)$ and $K_t(\cdot \mid \cdot)$, where $(x,t) \mapsto K_t(x)$ is a computable function satisfying $\lim K_t(x) = K(x)$ and $K_1(x) \geq K_2(x) \geq \ldots$; and similar for $K_t(\cdot \mid \cdot)$. We assume that $K_1(x \mid y)$ and $K_1(x)$ are bounded by $O(n)$ for all x of length n.

---

[1] The numerator is nonzero, even if $x = y$. $K_U(x) \geq 1$ holds for every choice of the optimal prefix-free Turing machine U, because such machines never halt on input the empty string. Indeed, if it halted, then it would be the only halting program by the prefix property, and hence, the machine can not be optimal.

[2] In [6, Exercise 8.4.3] it is claimed that for the prefix variant of the normalized information distance, one can improve the precision of the last item to $O(1/K(x,y,z))$. However, we do not know a proof of this. If this were true, then with minor modifications of N similar to those in the appendix of the ArXiv version, all axioms of a metric can be satisfied precisely.

Lemma 1. *For all n and strings x, y of length at most n:*

$$\sum_{t=1}^{\infty} |f_{\mathrm{tr}}(x, y, t+1) - f_{\mathrm{tr}}(x, y, t)| \leq 2 \ln n + O(1).$$

Definition 2. *An ε-approximation of a function g is a limit approximation of a function g' with $g - \varepsilon \leq g' \leq g + \varepsilon$.*

For a suitable choice of U, we have $0 \leq N \leq 1$, and the function defined by $f(x, y, t) = 1/2$ is a $(1/2)$-approximation.[3] We show that for $\varepsilon < 1/2$ and every ε-approximation, the sum in the above lemma is at least logarithmic.

Theorem 1. *Let f be an ε-approximation of N with $\varepsilon < 1/2$. For large n:*

$$\max_{x,y \in \{0,1\}^n} \sum_{t=1}^{\infty} |f(x, y, t+1) - f(x, y, t)| \geq \tfrac{1}{100} \cdot (1 - 2\varepsilon)^2 \cdot \log n.$$

This result implies that for each $\varepsilon < 1/2$, there exists no upper semicomputable function that differs from N by at most ε.

We now state the main result of [1].

Definition 3. *Let $k \geq 1$. A sequence $a_1, a_2, \ldots$ of real numbers has at most k oscillations if the sequence can be written as a concatenation of k sequences $(k-1$ finite and 1 infinite) such that each sequence is either monotonically non-increasing or non-decreasing. The sequence has 0 oscillations if $a_1 = a_2 = \ldots$*

The main result of [1] states that no 0-approximation f of N has at most a constant number of oscillations. More precisely, for each k, there exists a pair (x, y) such that $f(x, y, 1), f(x, y, 2), \ldots$ does not have at most k oscillations.

Let $k \colon \mathbb{N} \to \mathbb{N}$. We say that f *has at least $k(n)$ oscillations*, if for all n there exists a pair (x, y) of strings of length at most n, such that $f(x, y, 1), f(x, y, 2), \ldots$ does not have at most $k(n) - 1$ oscillations. (The proof of) Theorem 1 implies that if $\varepsilon < 1/2$, then any ε-approximation has at least $\Omega((1 - 2\varepsilon)^2 \log n)$ oscillations.

The trivial 0-approximation f_{tr} has at most $O(n)$ oscillations, because each upper-approximation of Kolmogorov complexity in its definition is bounded by $O(n)$ on n-bit strings, and hence, there can be at most this many updates. Can it be significantly less than n, for example at most $n/100$ for large n?

The answer is positive. For all constants c, there exist optimal machines U in the definition of complexity K for which the number of updates of K_t is at most $n/c + O(\log n)$. For example, one may select an optimal U whose halting programs all have length 0 modulo c. If N is defined relative to such a machine, than the total number of updates is $2n/c + O(\log n)$. Hence, for every constant e there exists a version of N and a 0-approximation that has at most n/e oscillations for large input sizes n. Our second main result provides an almost linear lower bound on the number of oscillations.

---

[3] For general optimal U, and for $\varepsilon > 1/2$, we can obtain an ε-approximation that is constant in t by choosing $f(x, y, t) = N(x, y)$ for some finite set of pairs (x, y), and by choosing $f(x, y, t) = 1/2$ otherwise.

Theorem 2. *Every* 0-*approximation of* N *has at least* $\Omega(n/\log n)$ *oscillations.*

In an extended version of this article, we plan to improve the $\Omega(n/\log n)$ lower bound to an $\Omega(n)$ bound. This requires a more involved variant of our proof.

Theorems 1 and 2 both imply that N and hence Kolmogorov complexity is not computable. In fact, they imply something stronger: $K(K(x \mid y) \mid x, y)$ can not be bounded by a constant.[4] It has been shown that $K(K(x) \mid x) \geq \log n - O(1)$, see [3,5], and our proofs are related. Like the proof in [3], we also use the game technique. This means that we present a game, present a winning strategy, and show that this implies the result. Using games one often obtains tight results with more intuitive proofs. (Moreover, the technique allows to easily involve students in research, because after the game is formulated, typically no specific background is needed to find a winning strategy.) For more examples of the game technique in computability theory and algorithmic information theory, we refer to [7].

N is Not Upper Nor Lower Semicomputable

For the sake of completeness, we present short proofs of the results in [9], obtained from Theorem 3.4 and Proposition 3.6 from [1] (presented in a form that is easily accessible to people with little background in the field). A function g is *lower semicomputable* if $-g$ is upper semicomputable.

Lemma 2. N *is not lower semicomputable.*

Proof. Note that for large n, there exist n-bit x and y such that

$$N(x, y) \geq 1/2 \,.$$

Indeed, for any y, there exists an n-bit x such that $K(x \mid y) \geq n$. The denominator of N is at most $n + O(\log n)$, and the inequality follows for large n.

Assume N was lower semicomputable. On input n, one could search for such a pair (x, y), and we denote the first such pair that appears by (x_n, y_n). We have $K(x_n) = K(n) + O(1)$ and $\max\{K(x_n \mid y_n), K(y_n \mid x_n)\} \leq O(1)$. Hence $N(x_n, y_n) \leq O(1/K(n))$. For large n this approaches 0, contradicting the equation above. $\square$

Remark. With the same argument, it follows that for any $\varepsilon < 1/2$, there exists no lower semicomputable function that differs from N by at most ε. Indeed, instead of $N(x, y) \geq 1/2$ we could as well use $N(x, y) \geq 1/2 + \varepsilon$, and search for (x_n, y_n) for which the estimate is at least $1/2$.

---

[4] Indeed, if this were bounded by c, there would exist an upper approximation f of $K(\cdot \mid \cdot)$ such that for each pair (x, y), the function $f(x, y, \cdot)$ has only finitely many values. (We modify any upper approximation of complexity by only outputting values k on input x, y, for which $K(k \mid x, y) \leq c$. There are at most 2^c such k.) Hence, there would exist an approximation f' of N such that for all x, y, the function $f'(x, y, \cdot)$ has only finitely many values. Such functions would have only finitely many oscillations, contradicting Theorem 2, and a finite total update, contradicting Theorem 1.

To prove that N is not upper semicomputable, we use the following well-known lemma.

Lemma 3. *The complexity function* $K(\cdot)$ *has no unbounded lower semicomputable lower bound.*

Proof. This is proven by the same argument as for the uncomputability of K (see appendix the appendix of the ArXiv version). Suppose such a bound $B(x) \leq K(x)$ exists. Then on input n, one can search for a string x_n with $n \leq B(x_n)$ and hence $n \leq K(x_n)$. But since there exists an algorithm to compute x_n given n, we have $K(x_n) \leq O(\log n)$. This is a contradiction for large n. Hence, no such B exists. □

Lemma 4. N *is not upper semicomputable.*

Proof. By optimality of the prefix-free machine in the definition of K, we have that $K(x \mid y) \geq 1$ for all x and y. Thus $1 \leq K(x \mid x) \leq O(1)$, and hence,

$$1/K(x) \leq N(x, x) \leq O(1/K(x)).$$

If N were upper semicomputable, we would obtain an unbounded lower semicomputable lower bound of K, which contradicts Lemma 3. □

2 Trivial Approximations Have at Most Logarithmic Total Update

Lemma 1 follows from the following lemma for $c \leq O(1)$ and the upper bound $m \leq O(n)$ on the upper approximations of Kolmogorov complexity.

Lemma 5. *Assume* $1 \leq a_1 \leq a_2 \leq \cdots \leq a_m \leq m$, $1 \leq b_1 \leq b_2 \leq \cdots \leq b_m \leq m$ *and* $a_i \leq b_i + c$. *Then,*

$$\sum_{i \leq m} \left| \frac{a_i}{b_i} - \frac{a_{i+1}}{b_{i+1}} \right| \leq 2 \ln m + O(c^2).$$

Proof. We first assume $c = 0$. We prove a continuous variant. Let $\alpha, \beta \colon [0, m] \to [1, m]$ be non-decreasing real functions with $\alpha(t) \leq \beta(t)$ and $1 \leq \alpha(0) \leq \beta(m) \leq m$. The sum in the lemma can be seen as a special case of

$$\int_{t=0}^{t=m} \left| d \frac{\alpha(t)}{\beta(t)} \right| = \int_{t=0}^{t=m} \frac{d\alpha(t)}{\beta(t)} + \int_{t=0}^{t=m} \frac{\alpha(t)}{\beta^2(t)} d\beta(t).$$

The left integral in the sum is maximized by setting $\beta(t)$ equal to its minimal possible value, which is $\alpha(t)$. The right one is maximized for the maximal value of $\alpha(t)$, which is $\beta(t)$. Thus,

$$\leq \int_{u=\alpha(0)}^{u=\alpha(m)} \frac{du}{u} + \int_{u=\beta(0)}^{u=\beta(m)} \frac{du}{u} \leq 2 \ln m.$$

For $c \geq 0$, the minimal value of β is $\max\{1, \alpha - c\}$ and the maximal value of α is $\min\{m, \beta + c\}$. The result follows after some calculation. □

3   Oscillations of 0-Approximations, the Game

For technical reasons, we first consider the *plain length conditional* variant of the normalized information distance N'. For notational convenience, we restrict the definition to pairs of strings of equal length.

Definition 4. *For all n and strings x and y of length n, let*

$$N'(x,y) = \frac{\max\{C(x \mid y), C(y \mid x)\}}{\max\{C(x \mid n), C(y \mid n)\}}.$$

If $C(x \mid n) = 0$, *let* $N'(x,x) = 0$.

Remarks

- For $x \neq y$, the denominator is at least 1, since at most 1 string can have complexity zero relative to n.
- The choice of the value of $N'(x,x)$ if $C(x \mid n) = 0$ is arbitrary, and does not affect Proposition 1 below.
- In the numerator, the length n is already included in the condition, since it equals the length of the strings.
- There exists a trivial approximation of N' with at most $2n + O(1)$ oscillations. Indeed, consider an approximation obtained by defining $C_t(\cdot \mid \cdot)$ with brute force searches among programs of length at most $n + O(1)$.
- Again, for every constant e, we can construct an optimal machine and a 0-approximation of N' for which the number of oscillations is at most n/e. We now present a matching lower bound.

Proposition 1. *Every 0-approximation of* N' *has at least* $\Omega(n)$ *oscillations.*

In this section, we show that the proposition is equivalent to the existence of a winning strategy for a player in a combinatorial (full information) game. In the last section of the paper, we present such a winning strategy.

Description of Game $\mathcal{G}_{n,c,k}$. The game has 3 integer parameters: $n \geq 1$, $c \geq 1$ and $k \geq 0$. It is played on two 2-dimensional grids X and Z. Grid X has size $n \times 2^n$. Its rows are indexed by integers $\{0, 1, \ldots, n-1\}$, and its columns are indexed by n-bit strings. Let X_u be the column indexed by the string u. See Fig. 1 for an example with $n = 3$. Grid Z has size $n \times \binom{2^n+1}{2}$. The rows are indexed by integers $\{0, \ldots, n-1\}$, and its columns are indexed by unordered pairs $\{u, v\}$, where u and v are n-bit strings, (that may be equal).[5] We sometimes denote unordered pairs $\{u, v\}$ of n-bit strings as uv, and write $Z_{\{u,v\}} = Z_{uv}$. Note that $Z_{uv} = Z_{vu}$. Let $u \in \{0,1\}^n$. The *slice* Z_u of Z is the 2-dimensional grid of size $n \times 2^n$ containing all columns Z_{uv} with $v \in \{0,1\}^n$. Additionally, Bob must generate a function f mapping unordered pairs of n-bit strings and natural numbers to real numbers.

---

[5] Formally, we associate sets $\{u, v\}$ with 2 elements to an unordered pair (u, v), and singleton sets $\{u\}$ to the pair (u, u).

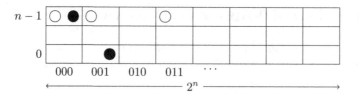

Fig. 1. Example of board X with $n = 3$. Alice has placed 2 tokens in row 2 (white), and Bob has placed 1 token in row 0 and 1 in row 2 (black). The row restrictions for both players are satisfied, since $\max\{1, 3\} \leq 2^2$ and $1 \leq 2^0$. $X_{000} = X_{011} = 2$, $X_{001} = 0$ and $X_{010} = 3$.

Two players, Alice and Bob, alternate turns. The rounds are numbered as $t = 1, 2, \ldots$ At each round, Alice plays first. At her turn, she places tokens on cells of the grids. She must place at least 1 token. Afterwards, Bob places zero or more tokens on the grids, and he declares all values $f(uv, t)$ for all unordered pairs $\{u, v\}$, where t is the number of the current round. This terminates round t, and the players start with round $t + 1$.

For each player, for each $i \in \{1, \ldots, n\}$, and for all grids $G \in \{X\} \cup \{Z_u : u \in \{0, 1\}^n\}$, the following *row restriction* should be satisfied: *The total number of tokens that the player has placed during the whole game in the i-th row of G, is at most 2^i.* If a player does not satisfy this restriction, the game terminates and the other player *wins*. See Fig. 1. Bob's moves should satisfy two additional requirements. If after his turn these requirements are not satisfied, the game terminates and Alice wins.

- Let X_u be the value of column X_u given by the minimal row-index of a cell in X_u containing a token. If X_u contains no tokens, then $X_u = n$. Similar for the value Z_{uv} of column Z_{uv}. For all u and v:

$$\frac{Z_{uv} - 1}{\max\{X_u, X_v\} + c} < f(uv, t) < \frac{Z_{uv} + c}{\max\{X_u, X_v\}}. \tag{c}$$

- For all u and v: $f(uv, 1), f(uv, 2), \ldots$ has at most k oscillations. (k)

Note that for decreasing c and k, it becomes easier for Alice to win.

Discussion. If Alice places a token in a row with small index, Bob has a dilemma: either he can change the function f, or he can place tokens on the other board to restore the ratios in (c). In the first case, he might increase the number of oscillations in (k) while in the second case, he exhausts his limited capacity to place tokens on rows of small indices, (by the row restriction, at most $1 + 2^1 + \ldots + 2^{i-1} = 2^i - 1$ tokens can be placed below row i in each grid G).

Remark. The game has at most $O(n2^{2n})$ rounds, because in each round, Alice must place at least 1 token, and by the row restriction, Alice can place at most $O(n2^{2n})$ tokens on all grids. Hence, the game above is finite and has full information. This implies that either Alice or Bob has a winning strategy.

Lemma 6. *Let $k \colon \mathbb{N} \times \mathbb{N} \to \mathbb{Z}$ be such that Alice has a winning strategy in the game $\mathcal{G}_{n,c,k(n,c)}$. Then for every 0-approximation of N' there exists a constant c such that for large n, the 0-approximation has more than $k(n,c)$ oscillations on n-bit inputs.*

Proof. The idea of the proof is to use any limit approximation f' to construct a strategy for Bob. By assumption there exists some winning strategy for Alice, and we let it play against this strategy for Bob. Then we show that Bob satisfies the row restrictions and requirement (c). Since Alice's strategy is winning, we conclude that requirement (k) must be violated. Our construction implies that f has fewer oscillations than f', thus also f' has more than $k(n,c)$ oscillations.

It suffices to prove the lemma for the largest function $k(n,c)$ for which Alice wins the game $\mathcal{G}_{n,c,k(n,c)}$. This function k is computable, since the game is finite, and for each value we can determine whether Alice has a winning strategy by brute force searching all strategies.

- Let $C_s(\cdot|\cdot)$ represent an upper approximation of $C(\cdot \mid \cdot)$.
- Let $C(u \leftrightarrow v) = \max\{C(u \mid v), C(v \mid u)\}$ and similar for $C_s(u \leftrightarrow v)$.
- Let f' be a 0-approximation of N'. Without loss of generality, we assume $f'(u,v,t) = f'(v,u,t)$.

For all c and n, we present a run of the game $\mathcal{G}_{n,c,k(n,c)}$. The mapping from c and n to a (transcript of) this run is computable. First, we fix a winning strategy of Alice in the game $\mathcal{G}_{n,c,k(n,c)}$ in a computable way. For example, we may brute force search all strategies and select the first winning strategy that appears. Let $r_0 = 1$. Consider the game in which Alice plays this strategy, and Bob replies as follows.

Bob's Strategy. At round t, Bob searches for a value s with $s > r_{t-1}$ such that for all u and v:

(i) $C_s(u \mid n) < X_u + c$ and $C_s(u \leftrightarrow v) < Z_{uv} + c$,
(ii) $f'(u,v,s) = \dfrac{C_s(u \leftrightarrow v)}{\max\{C_s(u|n), C_s(v|n)\}}$.

If such an s is found, he sets $r_t = s$ and $f(uv,t) = f'(u,v,s)$ for all u and v. For all u he places a token in column X_u at row $C_s(u \mid n)$. For all unordered pairs $\{u,v\}$, he places a token in column Z_{uv} at row $C_s(u \leftrightarrow v) + 1$. *End of Bob's strategy.*

We first show that if Bob does reply, he satisfies the row restriction. For $G = X$ this holds because there are at most 2^i programs of length i, and hence, at most 2^i strings u with $C_s(u) = i$ for some s. For $G = Z_u$, this holds because $C_s(u \leftrightarrow v) = i$ implies $C(v \mid u) \le i$, and there are less than 2^{i+1} such v.

Assuming that Bob plays in round t, requirement (c) holds. Indeed, after Bob's move and for $s = r_t$, condition (i) implies:

$$X_u \le C_s(u \mid n) < X_u + c \quad \text{and} \quad Z_{uv} - 1 \le C_s(u \leftrightarrow v) < Z_{uv} + c.$$

Together with (ii) and $f(t,u,v) = f'(s,u,v)$, this implies requirement (c).

We show that for large c, there always exists an s such that (i) and (ii) are satisfied, and hence, Bob plays in each round. Since f' is a 0-approximation, requirement (ii) is true for large s, and this does not depend on c. We show that (i) is also satisfied. To prove the left inequality, we first construct a Turing machine M. The idea is that the machine plays the game above, and each time Alice places a token in a cell of column X_u with row index i, it selects an unassigned i-bit string, and assigns to it the output u. Thus on input of a string p and integers c, n, it plays the game, waits until the p-th token is placed in the row with index equal to the length of p, and it outputs the column's index, (which is an n-bit string). The row restriction implies that enough programs are available for all tokens. Hence, $C_M(u \mid n, c) \leq i$, whenever Alice places a token in X_u at height i. By optimality of the Turing machine in $C(\cdot \mid \cdot)$, we have $C(u \mid n, c) \leq X_u + O(1)$ for all u, and hence,

$$C(u \mid n) \leq X_u + O(\log c).$$

For large c, this is less than $X_u + c$. By a similar reasoning, we have $C(u \leftrightarrow v) < Z_{uv} + c$, because each time Alice places a token in row i of column Z_{uv}, we assign two programs of length i: one that outputs u on input v, n, c, and one that outputs v on input u, n, c. Thus, for large s, also requirement (i) is satisfied, and Bob indeed plays at any given round, assuming he played in all previous rounds.

Recall that Alice plays a winning strategy, and that Bob satisfies the row restriction and requirement (c). Hence, requirement (k) must be violated, i.e., for some pair (u, v), the sequence $f(uv, 1)$, $f(uv, 2)$, ... has more than $k(n)$ oscillations. Since r_t is increasing in t, this sequence is a subsequence of $f'(u, v, 1)$, $f'(u, v, 2)$, ..., and the latter must also have more than $k(n)$ oscillations. This implies the lemma.   □

To prove Theorem 2 we need a version of the previous lemma for the prefix distance.

Lemma 7. *Under the assumption of Lemma 6, every 0-approximation of* N *has more than* $k(n, 5 \log n)$ *oscillations on* n-bit inputs for large n.

Proof. As a warm up, we observe that

$$K(x) \leq C(x \mid n) + 4 \log n + O(1).$$

Indeed, we can convert a program on a plain machine that has access to n, to a program on some prefix-free machine without access to n, by prepending prefix-free codes of the integers n and $C(x \mid n)$. Each such code requires $2 \log n + O(1)$ bits, and hence the inequality follows.

We modify the proof above by replacing all appearances of $C(x \mid n)$ by $K(x)$, of $C(x \mid y)$ by $K(x \mid y)$, and similarly for the approximations $C_s(\cdot \mid \cdot)$. We also set $c = 5 \log n$ and assume that f' is a 0-approximation of N. In Bob's strategy, no further changes are needed.

The row restriction for Bob is still satisfied, because the maximal number of halting programs of length i on a prefix-free machine is still at most 2^i. Requirement (c) follows in the same way from items (i) and (ii) in Bob's strategy. It remains to prove that for large c and s, these conditions (i) and (ii) are satisfied. Item (ii) follows directly, since f' is a 0-approximation of N.

For item (i), we need to construct a prefix-free machine M'. This is done in a similar way as above, by associating tokens in row i to programs of length i, but we also need to prepend 3 prefix-free codes: for the row index, for n, and for c. This implies

$$K(u) \leq X_u + 4\log n + O(\log c).$$

Recall that $c = 5\log n$. Hence, this is at most $X_u + c$ for large n. The lemma follows from the violation of requirement (k) in the same way as before.   □

4   Total Update of ε-Approximations, the Game

We adapt the game for the proof of Theorem 1.

Description of game $\mathcal{H}_{n,\varepsilon,a}$, where $\varepsilon > 0$ and $a \geq 0$ are real numbers. The game is the same as the game of the previous section, except that requirements (c) and (k) are replaced by:

– For all u and v with $\max\{X_u, X_v\} \geq \sqrt{n}$:

$$\left| f(u,v,t) - \frac{Z_{uv}}{\max\{X_u, X_v\}} \right| \leq \varepsilon. \qquad (\epsilon)$$

– For all u and v with $\max\{X_u, X_v\} \geq \sqrt{n}$:

$$\sum_{s=1}^{t-1} |f(u,v,s) - f(u,v,s+1)| \leq a. \qquad (a)$$

Remarks

– We call the sum in (a), the *total update* of f. Similar for the total update of an ε-approximation.
– The threshold $\sqrt{n}$ is chosen for convenience. Our proof also works with any computable threshold function that is at least super-logarithmic and at most n^α for some $\alpha < 1$.

Lemma 8. *Let* $a \colon \mathbb{N} \to \mathbb{R}$. *Suppose that for large n, Alice has a winning strategy in the game* $\mathcal{H}_{n,\varepsilon,a(n)}$. *Fix $\varepsilon' < \varepsilon$, and an ε'-approximation f' of either N' or N. Then, for large n, there exist n-bit inputs for which the total update of f' exceeds $a(n)$.*

Proof. We first consider an ε'-approximation f' of N', and at the end of the proof we explain the modifications for N. The proof has the same high-level structure as the proof of Lemma 6: from f' we obtain a strategy for Bob that is played against Alice's winning strategy. Then, from the violation of (a) we conclude that the total update of f' exceeds $a(n)$.

Let n be large such that Alice has a winning strategy in the game $\mathcal{H}_{n,\varepsilon,a(n)}$. We consider a run of the game where Alice plays a computably generated winning strategy and Bob's replies are as follows.

Bob's Strategy. He searches for an $s > r_{t-1}$ such that for all u and v with $\max\{C_s(u), C_s(v)\} \geq \sqrt{n}$:

(i) $C_s(u \mid n) \leq X_u + c$ and $C_s(u\leftrightarrow v) \leq Z_{uv} + c$,

(ii) $\left| f'(u,v,s) - \dfrac{C_s(u\leftrightarrow v)}{\max\{C_s(u\mid n), C_s(v\mid n)\}} \right| \leq \varepsilon'$,

If such an s is found, let $r_t = s$. Bob chooses $f(uv,t) = f'(u,v,s)$ for all u and v. For all u he places a token in column X_u at row $C_s(u \mid n)$. For all unordered pairs $\{u,v\}$, he places a token in column Z_{uv} at row $C_s(u\leftrightarrow v) + 1$. *End of Bob's strategy.*

For similar reasons as above, we have that for some c and for large s, requirements (i) and (ii) are satisfied. This implies that for some c, Bob always reacts.

We now verify that for large n, requirement (ϵ) holds. Recall that we need to check the inequality when the denominator is at least $\sqrt{n}$. After Bob's move we have again that

$$X_u \leq C_s(u \mid n) < X_u + c \qquad \text{and} \qquad Z_{uv} - 1 \leq C_s(u\leftrightarrow v) < Z_{uv} + c. \qquad (*)$$

Since $N' \leq e$ for some constant e, we may also assume that $f' \leq e$, because truncating f' can only decrease the number of oscillations. This and item (ii) imply that if n is large enough such that

$$(c+1)\frac{e+1}{\sqrt{n}} \leq \varepsilon - \varepsilon', \qquad\qquad (**)$$

inequality (ϵ) is indeed satisfied.

Because Bob loses, requirement (a) must be violated. Since the total update of f is at least the total update of f' as long as the $\sqrt{n}$-threshold is not reached, this implies that every ε'-approximation has total update more than $a(n)$. The statement for N' is proven.

The modifications for N are similar as in the previous section. Instead of choosing c to be a constant, we again choose it to be $5\log n$, and for the same reasons as above, this makes $(*)$ true if we replace conditional plain complexity by (conditional) prefix complexity. This increase from constant to logarithmic c increases the minimal value of n in $(**)$ only by a factor $O(\log^2 n)$. Otherwise, nothing changes in the above argument. The lemma is proven. $\square$

5   Conclusion

We have proven that statements about the incomputability of the normalized information distance are equivalent to the existence of winning strategies in a game. To prove the main results, we need to present these winning strategies. This is done in an extended version of the paper that is available on ArXiv.

References

1. Ambos-Spies, K., Merkle, W., Terwijn, S.A.: Normalized information distance and the oscillation hierarchy. arXiv preprint arXiv:1708.03583 (2017)
2. Bauwens, B.: Information distance revisited. In: Paul, C., Bläser, M. (eds.) 37th International Symposium on Theoretical Aspects of Computer Science (STACS 2020). Leibniz International Proceedings in Informatics (LIPIcs), vol. 154, pp. 46:1–46:14. Schloss Dagstuhl-Leibniz-Zentrum fuer Informatik, Dagstuhl (2020). https://doi.org/10.4230/LIPIcs.STACS.2020.46. https://drops.dagstuhl.de/opus/volltexte/2020/11907
3. Bauwens, B., Shen, A.: Complexity of complexity and maximal plain versus prefix-free Kolmogorov complexity. J. Symb. Log. **79**(2), 620–632 (2013)
4. Bennett, C.H., Gács, P., Li, M., Vitányi, P.M., Zurek, W.H.: Information distance. IEEE Trans. Inf. Theory **44**(4), 1407–1423 (1998)
5. Gács, P.: On the symmetry of algorithmic information. Soviet Math. Dokl. **15**, 1477–1480 (1974)
6. Li, M., Vitányi, P.M.: An Introduction to Kolmogorov Complexity and Its Applications, 4th edn. Springer, Cham (2019). https://doi.org/10.1007/978-3-030-11298-1
7. Muchnik, A.A., Mezhirov, I., Shen, A., Vereshchagin, N.: Game interpretation of Kolmogorov complexity, March 2010. Unpublished
8. Shen, A., Uspensky, V.A., Vereshchagin, N.: Kolmogorov Complexity and Algorithmic Randomness, vol. 220. American Mathematical Society, Providence (2017)
9. Terwijn, S., Torenvliet, L., Vitányi, P.M.: Nonapproximability of the normalized information distance. J. Comput. Syst. Sci. **77**, 738–742 (2011)

Definable Subsets of Polynomial-Time Algebraic Structures

Nikolay Bazhenov[1,2]($\boxtimes$) (iD)

[1] Sobolev Institute of Mathematics, 4 Acad. Koptyug Avenue,
Novosibirsk 630090, Russia
[2] Novosibirsk State University, 2 Pirogova St., Novosibirsk 630090, Russia
bazhenov@math.nsc.ru

Abstract. A structure S in a finite signature σ is polynomial-time if the domain of S, and the basic operations and relations of S are polynomial-time computable. Following the approach of semantic programming, for a given polynomial-time structure S, we consider the family $B(S)$ containing all subsets of $\mathrm{dom}(S)$, which are definable by a Δ_0 formula with parameters. It is known that each of these sets is polynomial-time computable; hence, $B(S)$, endowed with the standard set-theoretic operations, forms a natural Boolean algebra of polynomial-time languages, associated with S. We prove that up to isomorphism, the algebras $B(S)$, where S is a polynomial-time structure, are precisely computable atomic Boolean algebras.

Keywords: Polynomial-time structure · Δ_0 formula · Σ-definability · Hereditarily finite superstructure · Computable structure theory · List structure · Boolean algebra

1 Introduction

A common approach to *effective presentations* of infinite algebraic structures is provided by computable structure theory. The basic computation model of this theory is Turing machine: informally speaking, an effective presentation of a structure S is provided by Turing machines which compute the basic information about an isomorphic copy of S.

More formally, one gives the notion of *computable presentation*. Let τ be a finite signature. A τ-structure $\mathcal{A}$ is *computable* (or *constructive*) if the domain of $\mathcal{A}$ is a (Turing) computable subset of the set of natural numbers $\mathbb{N}$, and the basic operations and relations of $\mathcal{A}$ are (Turing) computable. A *computable*

The work is supported by Mathematical Center in Akademgorodok under agreement No. 075-15-2019-1613 with the Ministry of Science and Higher Education of the Russian Federation.

H. Fernau (Ed.): CSR 2020, LNCS 12159, pp. 142–154, 2020.
https://doi.org/10.1007/978-3-030-50026-9_10

presentation of a countable τ-structure $\mathcal{S}$ is a computable structure $\mathcal{A}$, which is isomorphic to $\mathcal{S}$.

The foundations of computable structure theory were developed in 1950s and early 1960s by Fröhlich and Shepherdson [17], Mal'tsev [29], and Rabin [36]. Nowadays, computable structure theory has become a flourishing area of mathematical logic. The reader is referred to, e.g., monographs [7,15,32] for a more detailed discussion.

From the computational point of view, a computable presentation $\mathcal{A}$ can be highly inefficient (in general, the basic $\mathcal{A}$-operations are not even primitive recursive). Hence, it is natural to ask when a given computable structure admits a presentation such that all its computations are resource bounded in a reasonable way. In order to attack this question, Nerode and Remmel [33] initiated the systematic studies of *polynomial-time structures*.

Let Σ be a finite alphabet. A τ-structure $\mathcal{S}$ is *polynomial-time* if the domain of $\mathcal{S}$ is a polynomial-time computable subset of Σ^*, and the basic operations and relations of $\mathcal{S}$ are computable in polynomial time (see Sect. 2.1 for more formal details).

It turned out that for many familiar classes of structures, the properties "being computably presentable" and "being polynomial-time presentable" coincide. Grigorieff [25] proved that every computable linear order has a polynomial-time isomorphic copy. Similar results hold for arbitrary relational structures [11], Boolean algebras [11], torsion abelian groups [13], etc. On the other hand, it is not difficult to build a computable structure in the signature $\{f^1; R^1; c\}$, which does not admit polynomial-time presentations (see Example 1.1 in [12]).

The reader is referred to [12] for a comprehensive survey of results that were obtained by the end of 1990s. In recent years, the interest in the field of polynomial-time structures has been rekindled: see, e.g., the papers [1–4,10]. We also note that the closely connected area of *fully primitive recursive* (or *punctual*) structures (see, e.g., [9,26,30,31]) has witnessed remarkable advances.

This paper is inspired by the recent applications of complexity theory in semantic programming: the articles [18,20,23,24,35] study various modifications of Δ_0 formulas (to be explained in Sect. 2) and their expressiveness over polynomial-time structures.

We give a brief overview of our paper's setting. Let $\mathcal{S}$ be a polynomial-time τ-structure, and let $\bar{b} := b_1, \ldots, b_n$ be a tuple from $\mathcal{S}$. It is not hard to show that for any quantifier-free τ-formula $\psi(x; y_1, \ldots, y_n)$, the set

$$\psi[\mathcal{S}; \bar{b}] := \{a : \mathcal{S} \models \psi(a; \bar{b})\}$$

is computable in polynomial time. Furthermore, a similar fact is true for an arbitrary Δ_0 formula ψ [35].

Therefore, one can associate with the structure $\mathcal{S}$ a natural class of polynomial-time languages: The family $\mathbb{P}[\mathcal{S}]$ contains all sets $A \subseteq \Sigma^*$ such that there are a Δ_0 formula $\psi(x; \bar{y})$ and a tuple $\bar{b} \in \mathcal{S}$ with $\psi[\mathcal{S}; \bar{b}] = A$.

Clearly, the definition of $\mathbb{P}[\mathcal{S}]$ is *not invariant* under isomorphisms of $\mathcal{S}$—it is easy to build two polynomial-time structures $\mathcal{S} \cong \mathcal{T}$ such that $\mathbb{P}[\mathcal{S}] \neq \mathbb{P}[\mathcal{T}]$

(e.g., for a given $\mathcal{S}$, construct $\mathcal{T}$ by replacing a fixed element $w \in \mathcal{S}$ with another element v not belonging to the domain of $\mathcal{S}$).

On the other hand, note that the set $\mathbb{P}[\mathcal{S}]$ forms a *Boolean algebra* under the standard set-theoretic operations:

$$\mathbb{B}[\mathcal{S}] := (\mathbb{P}[\mathcal{S}]; \cup, \cap, \overline{\cdot}; \emptyset, \mathrm{dom}(\mathcal{S})).$$

And this Boolean algebra can be treated as *invariant*: if $\mathcal{S} \cong \mathcal{T}$, then it is not hard to show that $\mathbb{B}[\mathcal{S}] \cong \mathbb{B}[\mathcal{T}]$, see Sect. 2 for details.

Recall that Selivanov and Konovalov gave a complete characterization for the isomorphism types of Boolean algebras induced by various families of languages: regular languages [37], regular ω-languages [38], piecewise testable languages [27], etc. In our setting, it is natural to follow their approach, and to consider the following question.

Problem $(\star)$. What are the isomorphism types of Boolean algebras $\mathbb{B}[\mathcal{S}]$ for polynomial-time structures $\mathcal{S}$?

The structure of the paper is as follows. Section 2 contains the necessary preliminaries. In particular, we give a detailed discussion of Δ_0 formulas. In the section, our working framework is based on the hereditarily finite list superstructure $\mathbb{HW}(\mathcal{S})$: the choice of this structure is motivated by the works [21,22,24] on the *semantic programming* paradigm, and the article [28] of Korovina, who studied various notions of computability over the ordered field of reals $\mathbb{R}$ (including Σ-definability over $\mathbb{HW}(\mathbb{R})$).

In Sect. 3, we give a solution of Problem $(\star)$: We prove that the isomorphism types of $\mathbb{B}[\mathcal{S}]$ for polynomial-time $\mathcal{S}$ are precisely the isomorphism types of computable atomic Boolean algebras (Theorem 3.1). The proof of the result is not difficult, but it uses some subtle techniques from computable structure theory and admissible set theory—hence, we aimed to make our paper as self-contained as possible, and added a lot of details to Sect. 2. Section 4 briefly discusses further directions.

2   Preliminaries

We mainly follow [21,22] and introduce our working framework for Δ_0 and Σ formulas. Note that starting from this point, the symbol Σ is always *reserved* for the term "Σ formula". A finite alphabet will be typically denoted by the letter Γ. By ω we denote the set of natural numbers.

The informal idea behind the framework is as follows. Given an arbitrary structure $\mathcal{S}$, one introduces a two-sorted structure $\mathbb{HW}(\mathcal{S})$: the first sort (called *atoms*) is the domain of $\mathcal{S}$, and the second sort (called *lists*) consists of finite lists of atoms, lists of lists of atoms, lists of lists of lists of atoms, etc. The structure $\mathbb{HW}(\mathcal{S})$ is endowed with standard list operations—informally, a list is treated as a "last in, first out" stack.

Now one can talk about formulas related to the structure $\mathbb{HW}(\mathcal{S})$. Roughly speaking, a Δ_0 formula is constructed as follows: our basic building blocks are

Boolean combinations of (standard) atomic formulas, but we are also allowed to use *bounded quantifiers* $\exists x \in t$ and $\forall x \in t$ in the construction (for now, one can think that "$\exists x \in t$" means "there is an atom x belonging to the list t," where t equals, say, $\langle y, z, w \rangle$). Starting with Δ_0 formulas as basic elements, a Σ formula is obtained by adding unbounded existential quantifiers $\exists x$ to the mix.

We emphasize that the discussion above is very informal—the formal details are given below. For a reader familiar with admissible set theory [8,14], we give another explanation: Essentially, $\mathbb{HW}(\mathcal{S})$ is a version of the hereditarily finite superstructure $\mathbb{HF}(\mathcal{S})$, where finite sets are replaced with finite lists. In this version, the classes of Δ_0 and Σ formulas are introduced in a standard way, see, e.g., [8, Chapter I].

Given a non-empty set A, by $S_0(A)$ we denote the set of all finite lists over A, i.e.,

$$S_0(A) := \{\text{nil}\} \cup \{\langle a_0, a_1, \ldots, a_n \rangle : n \in \omega, \ a_0, a_1, \ldots, a_n \in A\},$$

where nil is the empty list.

We consider a two-sorted structure $(A; S_0(A))$, where the first sort atom is A, and the second sort list is $S_0(A)$. For convenience, we will slightly abuse our notations and treat the empty list as an element simultaneously belonging to both sorts.

Let τ be a finite signature. For simplicity, we assume that τ is purely relational. We introduce new symbols nil, *cons*, *head*, *tail*, *conc*, $\in$, $\sqsubseteq$, which do not belong to τ.

The new symbols admit a natural interpretation in the structure $(A; S_0(A))$:

1. nil is a constant, which is interpreted as the empty list.
2. *cons* is a binary function of sort list$\times$atom$\rightarrow$list. It is interpreted as appending an atom a to a list x: e.g., if $x = \langle 0, 2 \rangle$ and $a = 1$, then $cons(\langle 0, 2 \rangle, 1) = \langle 0, 2, 1 \rangle$.
3. *head* is a unary function of sort list$\rightarrow$atom. The function produces the last element of a given list: e.g., $head(\langle 0, 2 \rangle) = 2$. We assume that $head(\text{nil}) = \text{nil}$.
4. *tail* is a unary function of sort list$\rightarrow$list. It outputs the list, which is obtained by deleting the head of a given list x: e.g., $tail(\langle 0, 2 \rangle) = \langle 0 \rangle$. Again, we set $tail(\text{nil}) = \text{nil}$.
5. *conc* is a binary function of sort list$\times$list$\rightarrow$list. It is interpreted as the concatenation of two given lists: e.g., $conc(\langle 0, 2 \rangle, \langle 1, 3 \rangle) = \langle 0, 2, 1, 3 \rangle$.
6. $\in$ is a binary predicate of sort atom$\times$list. We have $a \in x$ if and only if the atom a occurs in the list x: e.g., $0 \in \langle 0, 2 \rangle$ and $1 \notin \langle 0, 2 \rangle$.
7. $\sqsubseteq$ is a binary predicate of sort list$\times$list. We have $x \sqsubseteq y$ iff x is an initial segment of y: e.g., $\langle 0, 2 \rangle \sqsubseteq \langle 0, 2, 1 \rangle$ and $\langle 2, 1 \rangle \not\sqsubseteq \langle 0, 2, 1 \rangle$.

Let $\mathcal{M}$ be a τ-structure. Note that all the definitions below are given for arbitrary $\mathcal{M}$, but this paper is focused on polynomial-time $\mathcal{M}$.

Consider a new signature

$$\tau^+ := \tau \cup \{U\} \cup \{\text{nil}, cons, head, tail, conc, \in, \sqsubseteq\},$$

where U is a unary predicate. Then the *hereditarily finite list superstructure* (or *hfl superstructure* for short) $\mathbb{HW}(\mathcal{M})$ is a τ^+-structure, which is defined as follows.

The domain of $\mathbb{HW}(\mathcal{M})$ is given by:

$$HW_0(M) := S_0(\mathrm{dom}(\mathcal{M})), \ HW_{n+1}(M) := S_0(\mathrm{dom}(\mathcal{M}) \cup HW_n(M));$$

$$\mathrm{dom}(\mathbb{HW}(\mathcal{M})) = \mathrm{dom}(\mathcal{M}) \cup \bigcup_{n \in \omega} HW_n(M).$$

The unary predicate U distinguishes the set $\mathrm{dom}(\mathcal{M})$. If R is an n-ary relation from τ, then $\mathbb{HW}(\mathcal{M}) \models R(a_1, \ldots, a_n)$ if and only if $a_1, \ldots, a_n \in \mathrm{dom}(\mathcal{M})$ and $\mathcal{M} \models R(a_1, \ldots, a_n)$.

The interpretations of the remaining τ^+-symbols are essentially the same as those given above for $(A; S_0(A))$: For example, $\mathbb{HW}(\mathcal{M}) \models (x \in y)$ if and only if x occurs as an element of the list y. Note that x itself can also be a list—e.g., $x = \langle 0, 1 \rangle$ occurs in $y = \langle \mathrm{nil}, \langle 0, 1 \rangle, 2, \langle 2 \rangle \rangle$.

Terms in the signature τ^+ are defined in a standard way. The notion of Δ_0 *formula* is introduced according to the following rules:

- If ψ is an atomic formula in the signature τ^+, then ψ is a Δ_0 formula.
- If ψ and ξ are Δ_0 formulas, then $(\psi \& \xi)$, $(\psi \vee \xi)$, $(\psi \rightarrow \xi)$, and $\neg\psi$ are also Δ_0 formulas.
- If ψ is a Δ_0 formula, x is a variable, and t is a term, then $(\exists x \in t)\psi$, $(\forall x \in t)\psi$, $(\exists x \sqsubseteq t)\psi$, and $(\forall x \sqsubseteq t)\psi$ are Δ_0 formulas.

The corresponding $\mathbb{HW}$-semantics (i.e. evaluation of truth values in the structure $\mathbb{HW}(\mathcal{M})$) is pretty straightforward—the only non-trivial detail is the following: The formula $(\exists x \in t)\psi$ is treated as an abbreviation for $\exists x(x \in t \,\& \, \psi)$; the formula $(\forall x \in t)\psi$ abbreviates $\forall x(x \in t \rightarrow \psi)$. The formulas $(\exists x \sqsubseteq t)\psi$ and $(\forall x \sqsubseteq t)\psi$ are interpreted in a similar way.

The class of Σ *formulas* is defined as follows:

- Every Δ_0 formula is a Σ formula.
- If ψ and ξ are Σ formulas, then $(\psi \& \xi)$ and $(\psi \vee \xi)$ are also Σ formulas.
- If ψ is a Σ formula, x is a variable, and t is a term, then $(\exists x \in t)\psi$, $(\forall x \in t)\psi$, $(\exists x \sqsubseteq t)\psi$, and $(\forall x \sqsubseteq t)\psi$ are Σ formulas.
- If ψ is a Σ formula and x is a variable, then $(\exists x)\psi$ is a Σ formula.

Let $n \geq 1$. We say that a set $A \subseteq (\mathrm{dom}(\mathcal{M}))^n$ is Δ_0-*definable* (with parameters) if there are a Δ_0 formula $\psi(x_1, \ldots, x_n; y_1, \ldots, y_k)$ and a tuple $\bar{p} = p_1, \ldots, p_k$ from $\mathbb{HW}(\mathcal{M})$ such that for all $\bar{a} = a_1, \ldots, a_n$ from $\mathbb{HW}(\mathcal{M})$, the following holds:

$$\bar{a} \in A \ \Leftrightarrow \ \mathbb{HW}(\mathcal{M}) \models \psi(\bar{a}; \bar{p}).$$

The notion of a Σ-*definable* set is introduced in a similar way.

We will heavily exploit the key property of Σ-definable sets given below (Theorem 2.1). We note that Theorem 2.1 is essentially a modification of the result of Vajtsenavichyus [39] obtained for hereditarily finite superstructures (the proof of this result can be found, e.g., in Proposition 6.12 of [16]). A proof sketch for Theorem 2.1 is given in Proposition 2.1 of [5].

Theorem 2.1. *Let $\mathcal{M}$ be a τ-structure. Suppose that $A \subseteq \mathrm{dom}(\mathcal{M})^n$ is a Σ-definable set. Then there are a tuple $\bar{p} \in \mathrm{dom}(\mathcal{M})^k$, for some $k \in \omega$, and a computable sequence $\{\psi_\ell(x_1, \ldots, x_n; y_1, \ldots, y_k)\}_{\ell \in \omega}$ of existential τ-formulas such that for any tuple $\bar{a} = a_1, \ldots, a_n$ from $\mathrm{dom}(\mathcal{M})$,*

$$\bar{a} \in A \;\Leftrightarrow\; \mathcal{M} \models \bigvee_{\ell \in \omega} \psi_\ell(\bar{a}; \bar{p}).$$

In other words, for a Σ-definable set $A \subseteq \mathrm{dom}(\mathcal{M})^n$, there is the following trade-off: One can omit all the $\mathbb{HW}$-semantic details—and just work with truth values inside the structure $\mathcal{M}$ itself; *but* in order to do this, we have to go beyond the first-order logic, and consider *computable infinite disjunctions* of $\exists$-formulas.

Δ_0- and Σ-definable sets satisfy the following simple properties:

- The set $\emptyset \subset \mathrm{dom}(\mathcal{M})$ is definable by a Δ_0 formula $(x \neq x)$.
- The set $\mathrm{dom}(\mathcal{M})$ is definable by a Δ_0 formula $U(x)$.
- Δ_0-definable subsets of $\mathrm{dom}(\mathcal{M})^n$ are closed under set-theoretic operations: union, intersection, and complement (i.e. $\overline{A} := \mathrm{dom}(\mathcal{M})^n \backslash A$).
- Σ-definable subsets of $\mathrm{dom}(\mathcal{M})^n$ are closed under $\cup$ and $\cap$.

Therefore, the following notions are well-defined:

Definition 2.1. *By $\mathbb{P}[\mathcal{M}]$ we denote the family of all Δ_0-definable subsets of $\mathrm{dom}(\mathcal{M})$. Let $\mathbb{P}_\Sigma[\mathcal{M}]$ denote the family of all Σ-definable subsets of $\mathrm{dom}(\mathcal{M})$. Then:*

(a) $\mathbb{B}[\mathcal{M}] := (\mathbb{P}[\mathcal{M}]; \cup, \cap, \overline{}; \emptyset, \mathrm{dom}(\mathcal{M}))$ *is a Boolean algebra (under standard set-theoretic operations).*

(b) $\mathbb{D}_\Sigma[\mathcal{M}] := (\mathbb{P}_\Sigma[\mathcal{M}]; \cup, \cap)$ *is a distributive lattice.*

Clearly, if $\mathcal{M}$ is isomorphic to $\mathcal{N}$, then $\mathbb{B}[\mathcal{M}] \cong \mathbb{B}[\mathcal{N}]$ and $\mathbb{D}_\Sigma[\mathcal{M}] \cong \mathbb{D}_\Sigma[\mathcal{N}]$.

2.1   Polynomial-Time Structures

Let Γ be a finite alphabet. In this subsection, we consider strings from Γ^*.

Following [12], our basic computational model is multitape Turing machine. Let $t(n)$ be a function acting from ω to ω. A Turing machine M is $t(n)$-*time bounded* if each computation of M on inputs of length n, where $n \geq 2$, requires at most $t(n)$ steps. A function $f(x)$ on strings belongs to $\mathrm{DTIME}(t)$ if there is a $t(n)$-bounded deterministic Turing machine M, which computes $f(x)$. If f is a function of several variables, then the length of a tuple $(x_1, x_2, \ldots, x_n)$ can be defined as the sum of lengths of $x_1, x_2, \ldots, x_n$. A set of strings or a relation on strings is in $\mathrm{DTIME}(t)$ if its characteristic function is in $\mathrm{DTIME}(t)$.

Let $\mathbf{P} = \bigcup_{1 \leq i < \omega} \mathrm{DTIME}(n^i)$. For a finite signature τ, a τ-structure $\mathcal{S}$ is *polynomial-time* if:

- the domain of $\mathcal{S}$ is a subset of Γ^*, which belongs to $\mathbf{P}$;
- if R is a k-ary predicate from τ, then $R^{\mathcal{S}}$ is a subset of $\mathrm{dom}(\mathcal{S})^k$, which belongs to $\mathbf{P}$;

– if f is an ℓ-ary function from τ, then $f^{\mathcal{S}}: \mathrm{dom}(\mathcal{S})^{\ell} \to \mathrm{dom}(\mathcal{S})$ can be computed in **P**-time.

We recall the following fact, which was already mentioned in the introduction:

Proposition 2.1 (Lemma 2 of [35]; see also [24]). *If a τ-structure $\mathcal{M}$ is polynomial-time, then every Δ_0-definable subset of $\mathrm{dom}(\mathcal{M})^n$ is also polynomial-time.*

2.2 Boolean Algebras

The reader is referred to the monograph [19] for the background on Boolean algebras. We treat Boolean algebras as structures in the signature $\tau_{BA} = \{\vee, \wedge, \overline{\cdot\,}; 0, 1\}$.

Let $\mathcal{B}$ be a Boolean algebra. An element $a \in \mathcal{B}$ is an *atom* if a is a minimal non-zero element. The algebra $\mathcal{B}$ is *atomic* if for every non-zero $b \in \mathcal{B}$, there is an atom a with $a \leq_{\mathcal{B}} b$. The *Fréchet ideal* of $\mathcal{B}$ is the set $F(\mathcal{B}) = \{b \in \mathcal{B} : b$ is a finite sum of atoms$\}$. The binary relation $\sim_F$, which is defined as $(x \sim_F y) \Leftrightarrow (x \wedge \overline{y}) \vee (\overline{x} \wedge y) \in F(\mathcal{B})$, is a congruence of the structure $\mathcal{B}$.

Proposition 2.2 (see, e.g., Proposition 1.5.2 of [19]). *Let $\mathcal{A}$ and $\mathcal{B}$ be countable atomic Boolean algebras. If the quotients $\mathcal{A}/\sim_F$ and $\mathcal{B}/\sim_F$ are isomorphic, then the structures $\mathcal{A}$ and $\mathcal{B}$ are isomorphic.*

Let $\mathcal{L}$ be a linear order. The *interval Boolean algebra* $Int(\mathcal{L})$ is defined as follows. The domain of $Int(\mathcal{L})$ contains all finite unions of the intervals:

– $(-\infty; a) = \{x : x <_{\mathcal{L}} a\}$, where $a \in \mathcal{L}$;
– $[b; c) = \{x : b \leq_{\mathcal{L}} x <_{\mathcal{L}} c\}$, where $b <_{\mathcal{L}} c$; and
– $[d; +\infty) = \{x : d \leq_{\mathcal{L}} x\}$, where $d \in \mathcal{L}$.

The τ_{BA}-operations are interpreted as standard set-theoretic operations on elements from $Int(\mathcal{L})$. The following fact is well-known (see, e.g., Chap. 3.2 of [19]):

Proposition 2.3. *(a) If $\mathcal{L}$ is a computable linear order, then the Boolean algebra $Int(\mathcal{L})$ has a computable presentation.*
(b) If $\mathcal{B}$ is a computable Boolean algebra, then there is a computable linear order $\mathcal{L}$ such that $Int(\mathcal{L}) \cong \mathcal{B}$.

3 Main Result

Theorem 3.1. *Let $\mathcal{B}$ be a countable Boolean algebra. Then the following conditions are equivalent:*

(i) There is a polynomial-time structure $\mathcal{S}$ such that the algebra $\mathbb{B}[\mathcal{S}]$ is isomorphic to $\mathcal{B}$.
(ii) $\mathcal{B}$ is atomic, and $\mathcal{B}$ has a computable presentation.

The proof of Theorem 3.1 is given in the next two subsections.

3.1   From (i) To (ii)

Let $\mathcal{S}$ be a polynomial-time structure. First, we show that the Boolean algebra $\mathbb{B}[\mathcal{S}]$ is atomic. Let A be a non-empty Δ_0-definable subset of $\mathrm{dom}(\mathcal{S})$. Choose an element $a \in A$. Then it is clear that the set $\{a\}$ is definable by a Δ_0 formula $(x = a)$, and $\{a\}$ is a minimal non-empty subset of $\mathrm{dom}(\mathcal{S})$. Therefore, we deduce that $\{a\}$ is an atom inside $\mathbb{B}[\mathcal{S}]$, and $\mathbb{B}[\mathcal{S}]$ is an atomic algebra.

Recall that a Boolean algebra $\mathcal{A}$ is Π_1^0 if there are a computable τ_{BA}-structure $\mathcal{C}$ and a co-c.e. equivalence relation E such that E is a congruence of the structure $\mathcal{C}$, and the quotient $\mathcal{C}/E$ is isomorphic to $\mathcal{A}$. Odintsov and Selivanov [34, Theorem 1] proved that every Π_1^0 Boolean algebra is isomorphic to a computable one.

For our structure $\mathcal{S}$, one can define a natural computable τ_{BA}-structure $\mathcal{C}$ associated with $\mathcal{S}$:

- The domain of $\mathcal{C}$ contains (Gödel numbers of) all Δ_0 formulas $\psi(x; \bar{p})$, where $\bar{p}$ is a finite tuple of parameters from $\mathcal{S}$.
- The τ_{BA}-operations on these formulas are interpreted as disjunction, conjunction and negation. The least element 0 is interpreted as the formula $(x \neq x)$. The greatest element 1 is interpreted as $U(x)$.

Consider an equivalence relation E on $\mathrm{dom}(\mathcal{C})$:

$$E := \{(\psi(x; \bar{p}), \xi(x; \bar{q})) : \psi[\mathcal{S}; \bar{p}] = \xi[\mathcal{S}; \bar{q}]\}.$$

Clearly, E is a congruence of $\mathcal{C}$, and the quotient $\mathcal{C}/E$ is isomorphic to the Boolean algebra $\mathbb{B}[\mathcal{S}]$.

Hence, by the result of Odintsov and Selivanov, it is sufficient to show that the relation E is Π_1^0. But this fact is obvious, since

$$\psi[\mathcal{S}; \bar{p}] = \xi[\mathcal{S}; \bar{q}] \iff (\forall w \in \mathrm{dom}(\mathcal{S}))\, \mathcal{S} \models \psi(w; \bar{p}) \leftrightarrow \xi(w; \bar{q}),$$

and the truth-checking of $\psi(\cdot; \bar{p})$ is realized by a computable procedure, which is uniform in ψ and $\bar{p}$. Therefore, we deduce that $\mathbb{B}[\mathcal{S}]$ is isomorphic to a computable Boolean algebra.

3.2   From (ii) To (i)

Let $\mathcal{B}$ be a computable atomic Boolean algebra. First, assume that $\mathcal{B}$ is finite, and $\mathcal{B}$ contains precisely n atoms, where $n \geq 1$. Then the structure $\mathcal{T} := (\{1, 2, \ldots, n\}; f^1)$, where $f(x) = x$ for all x, is polynomial-time. Each of the subsets of $\mathrm{dom}(\mathcal{T})$ is Δ_0-definable, and this fact easily implies $\mathbb{B}[\mathcal{T}] \cong \mathcal{B}$.

Hence, without loss of generality, we may assume that $\mathcal{B}$ is infinite. By ζ we denote the order type of the integers. We prove the following auxiliary result:

Lemma 3.1 (folklore). *Let $\mathcal{B}$ be a computable atomic infinite Boolean algebra. Then there is a computable linear order $\mathcal{L}$ such that:*

1. $\mathcal{L}$ is isomorphic to $\zeta \cdot \mathcal{L}_1$ for some Δ_3^0-computable linear order $\mathcal{L}_1$, and

2. *the interval algebra $Int(\mathcal{L})$ is isomorphic to $\mathcal{B}$.*

Proof (sketch). Consider the quotient algebra $\mathcal{B}_1 := \mathcal{B}/\sim_F$. Since the structure $\mathcal{B}$ is computable, the ideal $F(\mathcal{B})$ is Σ_2^0, and hence $\mathcal{B}_1$ has a Δ_3^0-computable presentation. Furthermore, since the algebra $\mathcal{B}$ is infinite, the structure $\mathcal{B}_1$ contains at least two elements.

By a relativized version of Proposition 2.3, there is a Δ_3^0-computable linear order $\mathcal{L}_1$ such that $Int(\mathcal{L}_1) \cong \mathcal{B}_1$. By Theorem 2a of [6], the order $\mathcal{L} := \zeta \cdot \mathcal{L}_1$ has a computable presentation.

It is not hard to show that the algebra $Int(\mathcal{L})$ is atomic, and the quotient $Int(\mathcal{L})/\sim_F$ is isomorphic to $Int(\mathcal{L}_1)$. Since $Int(\mathcal{L}_1) \cong \mathcal{B}/\sim_F$, by Proposition 2.2, we deduce that $Int(\mathcal{L})$ is isomorphic to $\mathcal{B}$. $\qquad\square$

Fix a computable linear order $\mathcal{L}$ from Lemma 3.1. By the result of Grigorieff [25], $\mathcal{L}$ has a polynomial-time presentation. Without loss of generality, we may assume that $\mathcal{L}$ itself is polynomial-time.

We will show that the Boolean algebra $\mathbb{B}[\mathcal{L}]$ is isomorphic to our structure $\mathcal{B}$. In order to obtain this, we prove the following:

Lemma 3.2. *Let $\psi(x; y_1, \ldots, y_k)$ be a Σ formula, and let $\bar{b} := b_1, \ldots, b_k$ be a tuple from $\mathcal{L}$. Then the set $\psi[\mathcal{L}; \bar{b}]$ is a finite union of intervals in $\mathcal{L}$. In other words, every Σ-definable subset of $\mathrm{dom}(\mathcal{L})$ is an element of the interval Boolean algebra $Int(\mathcal{L})$.*

Proof. Let $A := \psi[\mathcal{L}; \bar{b}]$. Without loss of generality, we may assume that $b_1 <_{\mathcal{L}} b_2 <_{\mathcal{L}} \cdots <_{\mathcal{L}} b_k$. For the sake of simplicity, we consider the case when $k = 2$.

It is sufficient to establish the following three facts:

(a) If $A \cap (-\infty; b_1)$ is non-empty, then there is an element c_0 such that $A \cap (-\infty; b_1)$ is equal to "$(-\infty; c_0)$ plus finitely many elements," i.e. there is a finite set F such that $F \cap (-\infty; c_0) = \emptyset$ and $A \cap (-\infty; b_1) = (-\infty; c_0) \cup F$.

(b) If $A \cap [b_1; b_2) \neq \emptyset$, then there are $c_1 <_{\mathcal{L}} d_1$ such that $A \cap [b_1; b_2)$ equals "$[c_1; d_1)$ plus finitely many elements".

(c) If $A \cap [b_2; +\infty) \neq \emptyset$, then there is c_2 such that $A \cap [b_2; +\infty)$ equals "$[c_2; +\infty)$ plus finitely many elements".

We give a detailed proof only for (b), since the other two claims can be obtained in a similar way.

Suppose that the set $A \cap [b_1; b_2)$ is non-empty. Choose an arbitrary element $d \in A \cap [b_1; b_2)$.

Recall that by Theorem 2.1, there is a computable sequence $\{\psi_\ell(x; y_1, y_2)\}_{\ell \in \omega}$ of $\exists$-formulas in the signature $\{\le\}$ such that inside $\mathcal{L}$, the original Σ formula $\psi(x; y_1, y_2)$ is equivalent to the infinite disjunction $\bigvee_{\ell \in \omega} \psi_\ell(x; y_1, y_2)$.

Without loss of generality, we may assume that $\mathcal{L} \models \psi_0(d; b_1, b_2)$, and $\psi_0 = \exists z_1 \ldots \exists z_t \xi(x; y_1, y_2; \bar{z})$, where the formula ξ is quantifier-free. Fix the elements $e_1, \ldots, e_t$ from $\mathcal{L}$ with $\xi(d; b_1, b_2; e_1, \ldots, e_t)$.

In order to avoid bulky technicalities, we consider a particular case, which is illustrative enough to recover all formal details: Suppose that $t = 6$ and $e_1 <_{\mathcal{L}} b_1 <_{\mathcal{L}} e_2 <_{\mathcal{L}} e_3 <_{\mathcal{L}} d <_{\mathcal{L}} e_4 <_{\mathcal{L}} e_5 <_{\mathcal{L}} b_2 <_{\mathcal{L}} e_6$.

Consider two sequences of $\leq_{\mathcal{L}}$-successive elements $b_1 <_{\mathcal{L}} f_2 <_{\mathcal{L}} f_3 <_{\mathcal{L}} f_3^+$ and $f_4 <_{\mathcal{L}} f_5 <_{\mathcal{L}} b_2$. Since the formula ξ is quantifier-free, it is clear that every element d' satisfies the following:

$$f_3^+ \leq_{\mathcal{L}} d' <_{\mathcal{L}} f_4 \ \Rightarrow \ \mathcal{L} \models \xi(d'; b_1, b_2; e_1, f_2, f_3, f_4, f_5, e_6).$$

Therefore, we deduce that $[f_3^+; f_4) \subseteq A$ and $A \cap [b_1; b_2)$ equals $[f_3^+; f_4)$ plus finitely many elements—these elements can be taken only from the finite set $\{b_1, f_2, f_3, f_4, f_5\}$.

It is not hard to see that the proof extends to the case of arbitrary t and $e_1, \ldots, e_t$. Lemma 3.2 is proved.   □

Lemma 3.2 shows that every Σ-definable subset of $\operatorname{dom}(\mathcal{L})$ is an element of $Int(\mathcal{L})$. On the other hand, it is easy to see that every set $A \in Int(\mathcal{L})$ is definable by a quantifier-free formula (with parameters) inside $\mathcal{L}$. Hence, we deduce that $\operatorname{dom}(Int(\mathcal{L})) = \mathbb{P}[\mathcal{L}] = \mathbb{P}_{\Sigma}(\mathcal{L})$, and both structures $\mathbb{B}[\mathcal{L}]$ and $\mathbb{D}_{\Sigma}(\mathcal{L})$ are isomorphic to the interval algebra $Int(\mathcal{L})$. Recall that this algebra is isomorphic to our $\mathcal{B}$. Theorem 3.1 is proved.   □

4   Further Discussion

We leave open the following questions:

Problem 4.1. What are the isomorphism types of lattices $\mathbb{P}_{\Sigma}[\mathcal{S}]$ for polynomial-time structures $\mathcal{S}$? Note that the proof of Theorem 3.1 shows that all computable atomic Boolean algebras can be realized as $\mathbb{P}_{\Sigma}[\mathcal{S}]$ for appropriate $\mathcal{S}$.

One can modify the setting of Sect. 2, and consider only sets $A \subseteq \operatorname{dom}(\mathcal{S})$, which are Δ_0-definable *without parameters* (i.e. we explicitly forbid to use parameter tuples $\bar{p} \in \mathcal{S}$ in our Δ_0 definitions). Again, these sets form a "parameter-free" Boolean algebra $\mathbb{B}_{pf}[\mathcal{S}]$ under usual set-theoretic operations.

Problem 4.2. What are the isomorphism types of Boolean algebras $\mathbb{B}_{pf}[\mathcal{S}]$ for polynomial-time structures $\mathcal{S}$?

In general, the algebras $\mathbb{B}_{pf}[\mathcal{S}]$ can be *non-atomic*. Indeed, consider a structure $\mathcal{M}$ which is defined as follows.

The signature of $\mathcal{M}$ contains a unary functional symbol f and a unary relational symbol R. The domain of $\mathcal{M}$ is equal to $\{a_{\sigma,k} : \sigma \in \{0,1\}^{<\omega}, \sigma \neq \Lambda, k \in \omega\}$. We put $f(a_{\sigma,k}) := a_{\sigma,k+1}$, and

$$\mathcal{M} \models R(a_{\sigma,k}) \ \Leftrightarrow \ k < \operatorname{length}(\sigma) \text{ and } \sigma(k) = 1.$$

It is not hard to build a polynomial-time presentation of $\mathcal{M}$.

Proposition 4.1. *The structure $\mathbb{B}_{pf}[\mathcal{M}]$ is a countable atomless Boolean algebra.*

Proof (sketch). In order to obtain the desired result, it is sufficient to establish the following: Suppose that $\psi(x)$ is a Σ formula without parameters such that $\psi[\mathcal{M}] \neq \emptyset$. Then there is a parameter-free Δ_0 formula $\xi(x)$ such that both sets $\psi[\mathcal{M}] \cap \xi[\mathcal{M}]$ and $\psi[\mathcal{M}] \backslash \xi[\mathcal{M}]$ are not empty.

By Theorem 2.1, there is a sequence $\{\psi_\ell(x)\}_{\ell \in \omega}$ of $\exists$-formulas in the signature $\{f, R\}$ such that inside $\mathcal{M}$, our Σ formula $\psi(x)$ is equivalent to the disjunction $\bigvee_{\ell \in \omega} \psi_\ell(x)$. Without loss of generality, we assume that $\psi_0[\mathcal{M}] \neq \emptyset$.

Suppose that $\psi_0(x) = \exists y_1 \ldots \exists y_n \theta(x, y_1, \ldots, y_n)$, where θ is quantifier-free. Let p be the greatest natural number such that θ contains term $f^p(z)$ for some $z \in \{x, y_1, \ldots, y_n\}$.

Choose a tuple $a_{\sigma,k}, a_{\tau_1,m_1}, \ldots, a_{\tau_n,m_n}$ satisfying the formula $\theta(x, \bar{y})$. Set $q = p + \max\{k, m_1, \ldots, m_n\}$. Then it is not difficult to show that the tuples

$$a_{\sigma^\frown 0^q 1, k}, a_{\tau_1^\frown 0^q 1, m_1}, \ldots, a_{\tau_n^\frown 0^q 1, m_n} \text{ and } a_{\sigma^\frown 0^{q+1} 1, k}, a_{\tau_1^\frown 0^{q+1} 1, m_1}, \ldots, a_{\tau_n^\frown 0^{q+1} 1, m_n}$$

both satisfy θ. On the other hand, consider $N = \text{length}(\sigma) + q - k$. Then we have

$$\mathcal{M} \models R(f^N(a_{\sigma^\frown 0^q 1, k})) \,\&\, \neg R(f^N(a_{\sigma^\frown 0^{q+1} 1, k})).$$

Therefore, the formula $\xi(x) := R(f^N(x))$ has the desired properties. $\square$

Acknowledgements. The author is grateful to the anonymous reviewers for their helpful suggestions.

References

1. Alaev, P., Selivanov, V.: Polynomial-time presentations of algebraic number fields. In: Manea, F., Miller, R.G., Nowotka, D. (eds.) CiE 2018. LNCS, vol. 10936, pp. 20–29. Springer, Cham (2018). https://doi.org/10.1007/978-3-319-94418-0_2
2. Alaev, P.E.: Structures computable in polynomial time. I. Algebra Log. **55**(6), 421–435 (2017). https://doi.org/10.1007/s10469-017-9416-y
3. Alaev, P.E.: Structures computable in polynomial time. II. Algebra Log. **56**(6), 429–442 (2018). https://doi.org/10.1007/s10469-018-9465-x
4. Alaev, P.E., Selivanov, V.L.: Polynomial computability of fields of algebraic numbers. Dokl. Math. **98**(1), 341–343 (2018). https://doi.org/10.1134/S1064562418050137
5. Aleksandrova, S.A.: The uniformization problem for Σ-predicates in a hereditarily finite list superstructure over the real exponential field. Algebra and Logic **53**(1), 1–8 (2014). https://doi.org/10.1007/s10469-014-9266-9
6. Ash, C.J.: A construction for recursive linear orderings. J. Symb. Log. **56**(2), 673–683 (1991). https://doi.org/10.2307/2274709
7. Ash, C.J., Knight, J.F.: Computable Structures and the Hyperarithmetical Hierarchy, Studies in Logic and the Foundations of Mathematics, vol. 144. Elsevier Science B.V, Amsterdam (2000)
8. Barwise, J.: Admissible Sets And Structures. Springer, Berlin (1975)
9. Bazhenov, N., Downey, R., Kalimullin, I., Melnikov, A.: Foundations of online structure theory. Bull. Symb. Log. **25**(2), 141–181 (2019). https://doi.org/10.1017/bsl.2019.20

10. Bazhenov, N., Harrison-Trainor, M., Kalimullin, I., Melnikov, A., Ng, K.M.: Automatic and polynomial-time algebraic structures. J. Symb. Log. **84**(4), 1630–1669 (2019). https://doi.org/10.1017/jsl.2019.26

11. Cenzer, D., Remmel, J.: Polynomial-time versus recursive models. Ann. Pure Appl. Log. **54**(1), 17–58 (1991). https://doi.org/10.1016/0168-0072(91)90008-A

12. Cenzer, D., Remmel, J.B.: Complexity theoretic model theory and algebra. In: Ershov, Y.L., Goncharov, S.S., Nerode, A., Remmel, J.B. (eds.) Handbook of Recursive Mathematics: Volume 1: Studies in Logic and the Foundations of Mathematics, vol. 138, pp. 381–513. North-Holland, Amsterdam (1998). https://doi.org/10.1016/S0049-237X(98)80011-6

13. Cenzer, D.A., Remmel, J.B.: Polynomial-time Abelian groups. Ann. Pure Appl. Log. **56**(1–3), 313–363 (1992). https://doi.org/10.1016/0168-0072(92)90076-C

14. Ershov, Y.L.: Definability and Computability. Consultants Bureau, New York (1996)

15. Ershov, Y.L., Goncharov, S.S.: Constructive Models. Kluwer Academic/Plenum Publishers, New York (2000)

16. Ershov, Y.L., Puzarenko, V.G., Stukachev, A.I.: HF-computability. In: Cooper, S.B., Sorbi, A. (eds.) Computability in Context, pp. 169–242. Imperial College Press, London (2011). https://doi.org/10.1142/9781848162778_0006

17. Fröhlich, A., Shepherdson, J.C.: Effective procedures in field theory. Philos. Trans. Roy. Soc. Lond. Ser. A **248**(950), 407–432 (1956). https://doi.org/10.1098/rsta.1956.0003

18. Goncharov, S., Ospichev, S., Ponomaryov, D., Sviridenko, D.: The expressiveness of looping terms in the semantic programming. Sib. Elektron. Mat. Izv. **17**, 380–394 (2020). https://doi.org/10.33048/semi.2020.17.024

19. Goncharov, S.S.: Countable Boolean Algebras and Decidability. Consultants Bureau, New York (1997)

20. Goncharov, S.S.: Conditional terms in semantic programming. Sib. Math. J. **58**(5), 794–800 (2017). https://doi.org/10.1134/S0037446617050068

21. Goncharov, S.S., Sviridenko, D.I.: Theoretical aspects of Σ-programming. In: Bibel, W., Jantke, K.P. (eds.) MMSSS 1985. LNCS, vol. 215, pp. 169–179. Springer, Heidelberg (1986). https://doi.org/10.1007/3-540-16444-8_13

22. Goncharov, S.S., Sviridenko, D.I.: Σ-programming. Am. Math. Soc. Transl. Ser. 2 **142**, 101–121 (1989). https://doi.org/10.1090/trans2/142/10

23. Goncharov, S.S., Sviridenko, D.I.: Recursive terms in semantic programming. Sib. Math. J. **59**(6), 1014–1023 (2018). https://doi.org/10.1134/S0037446618060058

24. Goncharov, S.S., Sviridenko, D.I.: Logical language of description of polynomial computing. Dokl. Math. **99**(2), 121–124 (2019). https://doi.org/10.1134/S1064562419020030

25. Grigorieff, S.: Every recursive linear ordering has a copy in DTIME-SPACE(n, log(n)). J. Symb. Log. **55**(1), 260–276 (1990). https://doi.org/10.2307/2274966

26. Kalimullin, I., Melnikov, A., Ng, K.M.: Algebraic structures computable without delay. Theor. Comput. Sci. **674**, 73–98 (2017). https://doi.org/10.1016/j.tcs.2017.01.029

27. Konovalov, A., Selivanov, V.: The Boolean algebra of piecewise testable languages. In: Beckmann, A., Bienvenu, L., Jonoska, N. (eds.) CiE 2016. LNCS, vol. 9709, pp. 292–301. Springer, Cham (2016). https://doi.org/10.1007/978-3-319-40189-8_30

28. Korovina, M.V.: Generalized computability of real functions. Sib. Adv. Math. **2**(4), 85–103 (1992)

29. Mal'tsev, A.I.: Constructive algebras. I. Russ. Math. Surv. **16**(3), 77–129 (1961). https://doi.org/10.1070/RM1961v016n03ABEH001120

30. Melnikov, A.G.: Eliminating unbounded search in computable algebra. In: Kari, J., Manea, F., Petre, I. (eds.) CiE 2017. LNCS, vol. 10307, pp. 77–87. Springer, Cham (2017). https://doi.org/10.1007/978-3-319-58741-7_8

31. Melnikov, A.G., Ng, K.M.: The back-and-forth method and computability without delay. Isr. J. Math. **234**(2), 959–1000 (2019). https://doi.org/10.1007/s11856-019-1948-5

32. Montalbán, A.: Computable structure theory: within the arithmetic. https://math.berkeley.edu/~antonio/CSTpart1.pdf

33. Nerode, A., Remmel, J.B.: Polynomial time equivalence types. In: Sieg, W. (ed.) Logic and Computation, Contemporary Mathematics, vol. 106, pp. 221–249. American Mathematical Society, Providence (1990). https://doi.org/10.1090/conm/106/1057825

34. Odintsov, S.P., Selivanov, V.L.: Arithmetic hierarchy and ideals of enumerated Boolean algebras. Sib. Math. J. **30**(6), 952–960 (1989). https://doi.org/10.1007/BF00970918

35. Ospichev, S., Ponomarev, D.: On the complexity of formulas in semantic programming. Sib. Elektron. Mat. Izv. **15**, 987–995 (2018). https://doi.org/10.17377/semi.2018.15.083

36. Rabin, M.O.: Computable algebra, general theory and theory of computable fields. Trans. Am. Math. Soc. **95**(2), 341–360 (1960). https://doi.org/10.2307/1993295

37. Selivanov, V., Konovalov, A.: Boolean algebras of regular languages. In: Mauri, G., Leporati, A. (eds.) DLT 2011. LNCS, vol. 6795, pp. 386–396. Springer, Heidelberg (2011). https://doi.org/10.1007/978-3-642-22321-1_33

38. Selivanov, V., Konovalov, A.: Boolean algebras of regular ω-languages. In: Dediu, A.-H., Martín-Vide, C., Truthe, B. (eds.) LATA 2013. LNCS, vol. 7810, pp. 504–515. Springer, Heidelberg (2013). https://doi.org/10.1007/978-3-642-37064-9_44

39. Vajtsenavichyus, R.: On necessary conditions for the existence of a universal function on an admissible set. Mat. Logika Primen. **6**, 21–37 (1989). in Russian

Families of Monotonic Trees:
Combinatorial Enumeration
and Asymptotics

Olivier Bodini[1], Antoine Genitrini[2]($\boxtimes$), Mehdi Naima[1]($\boxtimes$),
and Alexandros Singh[1]

[1] Université Sorbonne Paris Nord, Laboratoire d'Informatique de Paris Nord,
CNRS, UMR 7030, 93430 Villetaneuse, France
{Olivier.Bodini,Mehdi.Naima,Alexandros.Singh}@lipn.univ-paris13.fr
[2] Sorbonne Université, CNRS, Laboratoire d'Informatique de Paris 6 -LIP6- UMR
7606, 75005 Paris, France
Antoine.Genitrini@lip6.fr

Abstract. There exists a wealth of literature concerning families of
increasing trees, particularly suitable for representing the evolution of
either data structures in computer science, or probabilistic urns in mathematics, but are also adapted to model evolutionary trees in biology. The
classical notion of increasing trees corresponds to labeled trees such that,
along paths from the root to any leaf, node labels are strictly increasing;
in addition nodes have distinct labels. In this paper we introduce new
families of increasingly labeled trees relaxing the constraint of unicity of
each label. Such models are especially useful to characterize processes
evolving in discrete time whose nodes evolve simultaneously. In particular, we obtain growth processes for biology much more adequate than the
previous increasing models. The families of monotonic trees we introduce
are much more delicate to deal with, since they are not decomposable in
the sense of Analytic Combinatorics. New tools are required to study the
quantitative statistics of such families. In this paper, we first present a
way to combinatorially specify such families through evolution processes,
then, we study the tree enumerations.

Keywords: Analytic Combinatorics · Asymptotic enumeration ·
Increasing trees · Monotonic trees · Borel transform · Evolution process

1 Introduction

An increasing tree is a rooted tree whose nodes are labeled by integers in
$\{1, \ldots, n\}$, n being the number of nodes in the tree. Furthermore, each label
appears exactly once and, along each branch, the sequence of labels is strictly
increasing. Families of such increasing trees have been the subject of many

This work was also supported by the ANR projects METACONC ANR-15-CE40-0014.

H. Fernau (Ed.): CSR 2020, LNCS 12159, pp. 155–168, 2020.
https://doi.org/10.1007/978-3-030-50026-9_11

studies, owing to their wide applicability to representing data structures in computer science, probabilistic urn models in mathematics, and evolutionary trees in biology.

For example, in the analysis of algorithms and data structures, the study of increasing trees is useful in understanding the typical behavior of heaps and search trees (see [7]). In the study of permutations, increasing trees found usage in illuminating the behavior of local order patterns in permutations (see, for example, [6]). In biology, increasing trees find application as models of phylogenetic trees which, apart from encoding the relations between species, also encode temporal information in a way such that it encodes the history of some evolutionary process (see [5]). For a detailed and generic analysis of families of increasing trees, see [1]. A study of combinatorial differential equations related to various enumerative aspects of increasing trees, including path length enumeration for general increasing trees and enumeration of enriched increasing trees with respect to node height is presented in [11].

Increasingly labeled tree structures have also been studied in [15] under the guise of *monotone functions of tree structures*. These are mappings f from the nodes of a tree t to the set $\{1, \ldots, k\}$ such that if a_i is the child of a_j, then $f(a_i) \geq f(a_j)$. The authors studied this labelling on t-ary, plane and non-plane trees. Other authors studied this scheme on different tree models like Motzkin trees in [2]. The typical shapes of these trees have been studied in [10,12]. A good summary can be found in the thesis presented in [13]. Note that, unlike the case of increasing trees, this model allows labels of $\{1, \ldots, k\}$ to appear any number of times, including zero times.

A related model, *rooted increasing m-ary trees with label repetitions*, also allowing multiple nodes to have the same label, appeared in [4] and can be seen as the foundations of our following new study that widely extends the latter model. These models of increasing trees are related to *evolution processes* in discrete time: starting from a single leaf, a tree is grown by selecting at each time-step a leaf and replacing it by an internal node to which new leaves are attached. By allowing at each step multiple leaves to be expanded in parallel, the authors thus obtain trees with label repetitions. Interpreting the evolution process they establish functional equations satisfied by the enumerating series of their model. But these series are purely formal: their radius of convergence is 0, thus the use of direct analytic methods to solve the equations is non-viable. For the analysis of the series, an approximate Borel transform is used and then arguments based on the asymptotics of certain differential equations give the asymptotic behavior for the tree enumeration.

In our work we extend the study of [4], presenting a generic framework for an even more general class of combinatorial structures; in particular by relaxing the restrictions on node degrees (more precisely node out-degree in graph theory) and by allowing also for weakly increasing labeling sequences along branches of the trees. We study the following broad classes:

– *Strictly monotonic trees*: rooted trees T whose internal nodes are labeled with integers such that the root is labeled by 1 and along each branch the sequence of labels is strictly increasing. We also require that if ℓ is the greatest label in T, then all integers from 1 to ℓ also appear as labels of some internal nodes. Finally, we take the size of a tree to be its number of leaves.

- *Monotonic trees*: these are as above, except that in this case we allow for weakly increasing sequences of labels along each branch.

The plan of the paper is as follows. We conclude this section with the formalization of our problem and the exact statement of the results of the paper. Section 2 is then dedicated to the presentation of a number of applications in our framework. In Sect. 3 we present a detailed discussion of the combinatorial and asymptotic properties of the model we have quickly described above. This includes the derivation of a recurrence relation for such families of trees and a general asymptotic analysis of the recurrence relation. We then conclude the paper with a discussion of open problems and potential future directions.

To formalize the previous description, we recall the notion of a *degree function*, (following [1]), which in our case describes the tree evolution.

Degree Function. We define a degree function to be a power series of the form[1] $\phi(z) = \sum_{i \geq 1} \phi_i z^i$. Combinatorially they are interpreted in one of two following ways, depending on how we see the integer non-negative values ϕ_i. Firstly, we can interpret ϕ_i, $i \geq 2$, as the number of possible colors of a node of degree i. In this context the objects of study will correspond to the aforementioned strictly monotonic trees (these also include the so-called *weakly increasing trees* of [4,5]). Alternatively, the coefficients ϕ_i, $i \geq 2$, can be seen as the number of trees with i leaves belonging to some class of plane rooted unlabeled trees (in the sequel, we will refer to elements of such classes as *tree-shapes*). In this second context the objects that we will construct are monotonic trees as defined above. In both contexts, ϕ_1 must be interpreted carefully, owing to the definition of the evolutionary process below: $(\phi_1 - 1)$ corresponds either to the number of colors for unary nodes or to the number of unlabeled trees of size 1 in the corresponding plane rooted unlabeled tree class. Using this notion of a degree function we can now define the following evolution process.

Evolution Process. Given some degree function ϕ with $\phi_1 = 1$, the following evolution process generates a strictly monotonic tree. The process starts at time-step 0 with a single leaf and at each time-step $i \geq 1$ is as follows:

1. Choose a non-empty subset L of leaves of the so-far built tree.
2. For each leaf $\ell \in L$ choose an admissible degree and color (r, c), $r > 1$, $\phi_r > 0$ and one of its colors $1 \leq c \leq \phi_r$.
3. Replace each leaf ℓ with an internal node labeled by i with color c and having r new leaves attached to it.

In order to generate monotonic trees, in which case the coefficients of $\phi(z)$ are alternatively interpreted as enumerating tree-shapes rather than node colors, a slight modification of the above process is required: at each iteration step i, each

---

[1] We take $\phi_0 = 0$ in anticipation of our model.

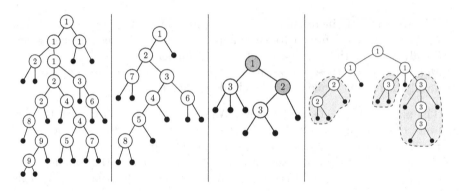

Fig. 1. (first) A monotonic binary tree of size 17; (second) a strictly monotonic binary tree of size 9; (third) a strictly monotonic tree of size 6, with $\phi(z) = z + 2z^2 + 2z^3$; (fourth) a monotonic 2-3 tree of size 11: highlighted sub-trees shapes are the substitution of some step in the evolution process

selected leaf is replaced by a tree-shape, rather than a colored internal node, and all internal nodes of this tree-shape are labeled by i.

Remark 1. If unary nodes are allowed, i.e., if we have $\phi_1 > 1$, at the end of each iteration step we can choose any subset of the unselected leaves (the ones that were present at the previous step and not newly-added during the current step) and expand each into a unary node with the desired color/shape.

Translating the above process using the framework of the *symbolic method* (see [7]), we obtain the following functional relation for the ordinary generating function B enumerating trees built via the evolution process based on a degree function ϕ:

$$B(z) = z + B\left(\phi(z)\right) - B\left(\phi_1 z\right). \tag{1}$$

Alternatively, the aforementioned evolution process may be expressed in terms of a function equation for $B(z)$ as

$$B(z) = z + \sum_{n=1}^{\infty} \frac{z^n}{n!} B^{(n)}(z) \; (\phi(z) - z)^n \, .$$

where $B^{(k)}(z)$ is the k-th derivative of $B(z)$. The n-th term of the sum represents the process of pointing at n leaves and substituting each by an element of the class represented by ϕ. Note that the order in which we choose the leaves is irrelevant and so we divide by $n!$.

Remark 2. This last formulation with a sum works when $\phi_1 = 1$ (i.e there are no unary nodes allowed).

The process is defined for $\phi_1 \geq 1$, but it is worth noting that when $\phi_1 = 0$ we get families of balanced trees (all the leaves are at the same level). Meanwhile when $\phi_1 = 1$ we have trees with no unary nodes. Finally when $\phi_1 > 1$ we have $(\phi_1 - 1)$ colors for unary nodes. More details will be given in Sect. 3. The generating series solution of the above functional relation are invariably purely formal, having radius of convergence equal to zero.

See Fig. 1 (first) and (second) for examples of, respectively a monotonic and a strictly monotonic binary tree.

We partition our analysis in two parts according to the value ϕ_1 of the degree function. Let B_n^ϕ be the number of trees of size n built via the evolution process.

Theorem 1. *Let $\phi(z)$ be such that $\phi_1 = 1, \phi_2 \geq 1$ and $\forall i \geq 3, \phi_i \leq i\phi_{i-1}$. Then*

$$B_n^\phi \sim \kappa_\phi \, (n-1)! \, \left(\frac{\phi_2}{\ln 2}\right)^n \, n^{\left(-1+\frac{\phi_3}{\phi_2^2}\right)\ln 2}.$$

Theorem 2. *Let $\phi(z)$ be such that $\phi_1 > 1, \phi_2 \geq 1$ and $\forall i \geq 3, \phi_i \leq i\phi_{i-1}$. Then*

$$B_n^\phi \sim \kappa_\phi \, \phi_1^{\frac{(n-1)(n-2)}{2}} \, (n-1)! \, \phi_2^{\,n}.$$

In both cases κ_ϕ is a positive constant defined through an implicit equation.

It is of interest to note that the presence (or not) of unary nodes radically affects the asymptotic regime and that the first-order behavior of the asymptotics depends only on the first few terms of the degree function $\phi(z)$ (ϕ_2 and ϕ_3 for the first case and ϕ_1, ϕ_2 for the second). Finally the technical conditions $\phi_i \leq i! \, \phi_2$ are not sharp but they are good enough for all practical applications.

As mentioned above, our scheme encompasses some classical tree models. For instance the models of balanced 2-3 trees and their generalization, the 2-3-4 trees, are obtained with $\phi(z) = z^2 + z^3$ and $\phi(z) = z^2 + z^3 + z^4$ respectively. Note in both cases we have $\phi_1 = 0$. The model of 2-3 trees has been introduced by Hopcroft as an efficient data structure and their asymptotics enumeration was given by Odlyzko [14]. In fact when $\phi_1 = 0$ it is possible to obtain the exponential growth of these structures by computing the fixed point of $\phi(z)$, after which the asymptotic enumeration can usually be obtained by means of singularity analysis. In this paper we are interested in the cases where $\phi_1 > 0$, which leads to different asymptotic regimes, see Fig. 2.

	ϕ_1	degree function	Asymptotics	Ref.
2-3 Trees	0	$\phi(z) = z^2 + z^3$	$\Omega(\ln n)\frac{\phi^n}{n}$	[14], [7]
S. M. Binary Trees	1	$\phi(z) = z + z^2$	$\alpha(n-1)! \, n^{-\ln 2} \left(\frac{1}{\ln 2}\right)^n$	Thm. 1, [4]
S. M. General Schröder Trees	2	$\phi(z) = \frac{z}{1-z} + z$	$\beta(n-1)! \, 2^{\frac{(n-1)(n-2)}{2}}$	Thm. 2

Fig. 2. Some known combinatorial classes specifiable via Eq. (1), with different ϕ_1 which results in different asymptotic scales. The function Ω is a periodic function with mean $(\phi \ln(4 - \phi))^{-1}$ and $\phi = \frac{1+\sqrt{5}}{2}$. Here "S. M." stands for "Strictly Monotonic".

2 Applications

We now exhibit examples of asymptotic enumeration for a number of interesting combinatorial structures. These are a direct application of the results presented in this paper.

Strictly Monotonic Trees

Example 1. Consider the class $\mathcal{T}$ with $\phi(z) = z + z^2 + z^3$. The first few values of $\mathcal{T}_n$ are:

$$0, 1, 1, 3, 12, 68, 482, 4122, 41253, 472795, \ldots$$

Asymptotically by Theorem 1, we have $\mathcal{T}_n \sim \kappa\,(n-1)!\,\left(\dfrac{2}{\ln 2}\right)^n$ with $\kappa \approx 0.41$.

	$\phi(z)$	Asymptotics	Ref.
S. M. Binary Trees	$z + z^2$	$\alpha(n-1)!\,n^{-\ln 2}\,(\frac{1}{\ln 2})^n$	Thm.1,[4]
S. M. Binary-Ternary Trees	$z + z^2 + z^3$	$\kappa(n-1)!\,(\frac{1}{\ln 2})^n$	Thm.1
S. M. Schröder trees	$\frac{z}{1-z}$	$\beta(n-1)!\,(\frac{1}{\ln 2})^n$	Thm.1,[5]

Fig. 3. An example of the change in behavior of the asymptotics for different classes of our model. Here "S. M." stands for "Strictly Monotonic".

This example is of interest as it indicates where the change of asymptotic behavior occurs when one varies the allowed node arities, going from binary trees with label repetitions (see [4]) to weakly increasing Schröder trees (see [5]) (with arbitrary arity of nodes).

The above example shows that the addition of just z^3 to $\phi(z) = z + z^2$ already results in a vanishing of the polynomial factor. Therefore adding higher powers of z to $\phi(z)$ only affects the constant term. See also Fig. 3. As a further example, consider the class of such trees having binary nodes of two colors and ternary nodes again of two colors (see Fig. 1 (third)). By straightforward application of Theorem 2 we obtain the following.

Example 2. Consider the class $\mathcal{T}$ with $\phi = z + 2z^2 + 2z^3$. The first few values of $\mathcal{T}_n$ are:

$$0, 1, 2, 10, 76, 804, 10800, 176240, 3384176, 74744016, 1866432032 \ldots$$

Asymptotically, we have that, $\mathcal{T}_{\phi,n} \sim \kappa\,(n-1)!\,\left(\dfrac{2}{\ln 2}\right)^n\, n^{\frac{-\ln 2}{2}}$ with $\kappa \approx 0.27$.

Monotonic Trees

Let $\mathcal{T}$ be some family of unlabeled rooted plane trees. We will denoted by $\mathcal{MT}$ the corresponding family of monotonic trees, i.e trees in $\mathcal{T}$ that have been labeled according to the rules for monotonic trees.

For example, consider the class of monotonic binary-ternary trees (see Fig. 1 (fourth)).

Example 3. Consider the class $\mathcal{BT}$ of rooted plane binary-ternary unlabeled trees (whose size is their number of leaves). The specification of this class is

$$\mathcal{BT} = \mathcal{Z} + Seq_{\{2,3\}}\mathcal{BT},$$

where the first terms are $BT(z) = z + z^2 + 3z^3 + 10z^4 + 38z^5 + 154z^6 + \ldots$. Then the first few values of $\mathcal{MBT}_n$, i.e., the number of monotonic binary-ternary trees with n leaves, are

$$0, 1, 1, 5, 32, 252, 2340, 25048, 303862, 4121730, \ldots$$

By applying Theorem 1, $\mathcal{MBT}_n \sim \kappa\,(n-1)!\left(\dfrac{1}{\ln 2}\right)^n n^{2\ln 2}$ with $\kappa \approx 0.17$ (Fig. 4).

Example 4. Let S be the class of Schröder trees (all arities except unary are allowed) which has the following specification,

$$\mathcal{S} = \mathcal{Z} + Seq_{\geq 2}\mathcal{S}.$$

By solving the above equation, we have $S(z) = \frac{1}{4}(1 + z - \sqrt{1 - 6z + z^2})$. The first terms of $S(z)$ are $z + z^2 + 3z^3 + 11z^4 + 45z^5 + 197z^6 + \ldots$. Hence, the first values of $\mathcal{MS}_n$, i.e., the number of monotonic Schröder trees with n leaves, are

$$0, 1, 1, 5, 33, 265, 2497, 27017, 330409, 4510065, \ldots$$

By Theorem 1 we have $\mathcal{MS}_n \sim \kappa\,(n-1)!\left(\dfrac{1}{\ln 2}\right)^n n^{2\ln 2}$ with $\kappa \approx 0.19$.

	ϕ	Asymptotics.	Refs.
Strongly Increasing Schröder	—	$\frac{n!}{2}$	[5]
Strictly Monotonic Schröder	$\frac{z}{1-z}$	$\alpha(n-1)!\,(\frac{1}{\ln 2})^n$	Thm. 1, [5]
Monotonic Schröder	$S(z)$	$\beta(n-1)!\,(\frac{1}{\ln 2})^n\,n^{2\ln 2}$	Thm. 1

Fig. 4. Comparison of the asymptotic behavior of increasing Schröder trees (where strongly increasing Schröder trees are increasing Schröder without label repetitions).

Monotonic M-ary Trees

It is a fact that our specification, by construction, enumerates families of trees by number of leaves. However there exists a special case, that of *monotonic m-ary trees* ($\mathcal{MT}$ where $\mathcal{T}$ is a variety of rooted plane m-ary trees), where our specification also allows for enumeration by number of internal nodes. In this specific case then, we are also able to enumerate by number of internal nodes since any m-ary tree with k leaves has $(k-1)/(m-1)$ internal nodes.

As an example, we consider the case of monotonic binary trees (see for example Fig. 1). In this case we obtain the following.

Example 5. Let C be the class of plane binary trees with size equal to the number of leaves, given by $C = Z + C^2$. These are counted by shifted Catalan numbers. By solving the above equation we find that $C(z) = \frac{1-\sqrt{1-4z}}{2}$. Then the first few values of $\mathcal{MC}_n$, i.e the number of monotonic binary trees with $(n-1)$ internal nodes and n leaves, are

$$0, 1, 1, 4, 22, 152, 1264, 12304, 137332, 1729584, \ldots$$

By Theorem 1, we have that $\mathcal{MS}_n \sim \kappa \, (n-1)! \left(\frac{1}{\ln 2}\right)^n n^{\ln 2}$ with $\kappa \approx 0.34$ (Fig. 5).

	ϕ	Asymptotics.	Ref.
Increasing Binary Trees	—	$(n-1)!$	[7]
Strictly Monotonic Binary Trees	$z + z^2$	$\alpha(n-1)! \left(\frac{1}{\ln 2}\right)^n n^{-\ln 2}$	Thm. 1
Monotonic Binary Trees	$C(z)$	$\beta(n-1)! \left(\frac{1}{\ln 2}\right)^n n^{\ln 2}$	Thm. 1

Fig. 5. Comparison of the asymptotic behavior between 3 classes of increasing binary trees.

3   Combinatorial Model and Asymptotic Analysis

Using the above functional equation we can directly obtain the following recurrence:

$$B_1 = 1,$$
$$B_n = \sum_{k=1}^{n-1} \left(\sum_{i=1}^{n-k} \phi_1^{n-k-i} \binom{n-k}{i} [z^{k+i}] \left(\phi(z) - \phi_1 z\right)^i \right) B_{n-k}. \tag{2}$$

Let us define $T_n(n-k) = \sum_{i=1}^{n-k} \phi_1^{n-k-i} \binom{n-k}{i} [z^{k+i}] \left(\phi(z) - \phi_1 z\right)^i$. We can then rewrite $B_n = \sum_{k=1}^{n-1} T_n(n-k)B_{n-k}$. In essence the coefficients of $T_n(n-k)$ represent the number of different combinations for a tree of size $n-k$ to be made into a tree of size n by a subset of leaves into internal nodes that contains new leaves. The recurrence in Eq. (2) can be used to iterate on the specification and get the first few coefficients for a combinatorial class. It is also possible to write another recurrence for B_n which involves sums over integer partitions.

Proof Sketch of Theorem 1

The evolution process, cf. Eq. (1), that we study translates to generating functions which has a null convergence radius. Therefore its study needs a more elaborate approach. We present here a summary of our approach for the proof in 6 points:

1. Performing a Borel transform by rescaling coefficients by $n!$.
2. Exhibit the dominant coefficients in the rescaled recurrence.

3. Find two new recurrences for the upper bound and the lower bound of b_n.
4. Write a new recurrence for the rescaled coefficients with a remainder term.
5. Deduce from it a linear differential equation.
6. Deduce the asymptotic behavior of the differential equation and determine the growth conditions on $\phi(z)$ necessary for the asymptotic to hold.

Let us carry on with the aforementioned plan, by first defining the following rescaled version of B_n:

$$b_n = \frac{B_n}{n!}$$

From this we see that $B_n = n!b_n$. Formally, this gives the following,

$$b_n = \sum_{k=1}^{n-1} t_n(n-k)b_{n-k} \tag{3}$$

where $t_n(i)$ is a sum of terms representing the different ways for a tree of size i to be made into a tree of size n by expanding some leaves into internal nodes with new leaves which all get the same label. These terms are each multiplied by $\frac{k!}{n!}$ due to the transform we have just performed. We notice also that the result are of power $j \leq 0$ in terms of n. To wit, the first few terms $T_n(n-1)$, $T_n(n-2)$ transform (under Borel transforms) to $t_n(n-1)$, $t_n(n-2)$, as follows:

$$T_n(n-1)B_{n-1} = \phi_2(n-1)B_{n-1} \xrightarrow{Borel} \phi_2(n-1)\frac{(n-1)!}{n!}b_{n-1} = t_n(n-1)b_{n-1}$$

$$T_n(n-2)B_{n-2} = \left[\frac{\phi_2^2(n-2)\,(n-3)}{2} + \phi_3\,(n-2)\right]B_{n-2}$$

$$\xrightarrow{Borel} \left[\frac{\phi_2^2(n-2)(n-3)}{2n(n-1)} + \frac{\phi_3(n-2)}{n(n-1)}\right]b_{n-2} = t_n(n-2)b_{n-2}$$

The coefficients $t_n(n-k)$ are sums whose terms look like $f_n(\phi, k)\frac{p_k(n)}{n(n-1)...(k+1)}$ where p is a polynomial in n of order at most k and f is a function which includes a ratio between the product of of elements of $\phi(z)$ divided by some factorial of k.

The highest order polynomial is of order k and appears in the case where all leaves were replaced with binary nodes. In this case the corresponding term of t_{n-k} is $\frac{\phi_2^k(n-k)...(n-2k+1)}{k!\,n(n-1)...(n-k+1)}$ which is of power 0 in n, as can be seen in the above example.

Our method of determining the dominant coefficients is based on a combinatorial argument. Nodes of lowest degree are the ones that count most, as the tree will have many permutations to create a lot of other trees of the same size.

Proposition 1. *For $k \in \{1,\ldots,n-1\}$ the terms in $t_n(n-k)$ of order 0 in n are*

$$\frac{\phi_2^k(n-k)}{k!\,n}.$$

Proof. From the discussion above, the term of highest order in n is $t_n(n-k)$ is (for $k \in \{\lceil \frac{n}{2} \rceil, \ldots, n-1\}$):

$$\frac{\phi_2^k(n-k)\ldots(n-2k+1)}{k!\, n(n-1)\ldots(n-k+1)}.$$

It is then possible to factor out the desired term by making a polynomial division with two polynomials having the same order. The quotient is equal to 1 and we have a remainder term. □

Proposition 2. *For $k \in \{2, \ldots, n-1\}$ the terms in $t_n(n-k)$ of order n^{-1} are the following*

$$\frac{\phi_2^{k-2}\phi_3}{(k-1)!(n-k+1)} - \frac{\phi_2^k}{(k-2)!\,(n-k+1)}.$$

Proof. The term of second highest order in n is $t_n(n-k)$ is:

$$\frac{\phi_2^{k-1}\phi_3(n-k)\ldots(n-2k+2)}{(k-2)!\, n(n-1)\ldots(n-k+1)} \quad (\text{for } k \in \{\lfloor \frac{n}{2} \rfloor, \ldots, n-2\}).$$

We should also take the second order term of $\frac{\phi_2^k(n-k)\ldots(n-2k+1)}{k!\, n(n-1)\ldots(n-k+1)}$ into account, since it involves a term of order -1 in n. □

From here by reasoning on the recurrence relation it is possible to conclude:

Proposition 3. *Under the conditions of Theorem 1 on $\phi(z)$, the following holds:*

$$b_n = \Theta\left(\left(\frac{\phi_2}{\ln 2}\right)^n n^{\left(-1+\frac{\phi_3}{\phi_2^2}\right)\ln 2}\right).$$

Proof. The proof can be made upon finding upper and lower bounds on the coefficients $t_n(k)$ and translating the results to linear differential equations from which asymptotic behavior can be determined. □

The last result does not give the asymptotic equivalent of the first order. For now we do not know if the function oscillates or not. We can write a new recurrence for b_n

$$b_n = \sum_{k=2}^{n-1} \left(\frac{\phi_2^k}{k!} + \frac{\phi_2^{k-2}\phi_3}{(k-1)!(n-k+1)} - \frac{\phi_2^k}{(k-2)!(n-k+1)}\right) b_{n-k}$$
$$+ \phi_2 b_{n-1} + a_n.$$

where a_n groups all omitted terms.

We can now determine the linear differential equation satisfied by $b(z)$, using the above recurrence, in which the coefficients transform into corresponding terms of the differential equation by simple manipulations as follows

$$\frac{\phi_2^k (n-k)}{k! \, n} b_n \rightarrow \frac{\phi_2^k}{k!} \int_0^z z^k b'(z) \, dz;$$

$$\frac{\phi_2^{k-2}\phi_3}{(k-1)!(n-k+1)} b_n \rightarrow \frac{\phi_2^{k-2}\phi_3 \, z^{k-1}}{(k-2)!} \int_0^z b(z) \, dz;$$

$$\frac{-\phi_2^k}{(k-2)!(n-k+1)} b_n \rightarrow \frac{-\phi_2^k \, z^{k-1}}{(k-2)!} \int_0^z b(z) \, dz; \quad \text{and } a_n \text{ translates into } a(z).$$

From the above we can derive the following integral form for $b(z)$

$$b(z) = \int_0^z \left(e^{\phi_2 z} - 1\right) b'(z) \, dz + (z\phi_3 - \phi_2^2 z)e^{\phi_2 z} \int_0^z b(z) \, dz + a(z).$$

By differentiating the latter equation once we obtain

$$\left(e^{\phi_2 z} - 2\right) b'(z) + (z\phi_3 - \phi_2^2 z)e^{\phi_2 z} b(z) + \tilde{a}(z) \tag{4}$$

where $\tilde{a}(z) = a'(z) + \left((z\phi_3 - \phi_2^2 z)e^{\phi_2 z}\right)' \int_0^z b(z) \, dz$. Note that the second term in $\tilde{a}(z)$ is of smaller order than the first two terms of Eq. (4).

Our problem has been transformed now into a more classical one where we have a linear differential equation with a regular singularity. The asymptotic behavior can be found by applying some classical theorems with some additional computations. The generic solution to the related homogeneous differential equation

$$\left(e^{\phi_2 z} - 2\right) y'(z) + (z\phi_3 - \phi_2^2 z)e^{\phi_2 z} y(z) = 0$$

is $y(z) = Cg(z)$ with $g(z)$ as follows

$$g(z) = C \cdot \left(2 - e^{\phi_2 z}\right)^{\ln 2\left(\frac{\phi_2^2 - \phi_3}{\phi_2^2}\right)} e^{-\left(\frac{\phi_2^2 - \phi_3}{\phi_2^2}\right)\left((\ln 2)^2 + Li_2\left(\frac{\exp(\phi_2 z)}{2}\right)\right)},$$

where the function $Li_2(z)$ stands for the dilogarithm function. Then, by variation of constants we obtain $C'(z) \cdot \exp(\phi_2 z - 2)g(z) = \tilde{a}(z)$ and hence, as b_0 is 0,

$$b(z) = g(z) \int_0^z \frac{\tilde{a}(t)}{(e^{\phi_2 t} - 2)g(t)} dt.$$

In the following all the constants are positive. By the theory of complex linear differential equations studied in [9] and [16] and a good summary of the theorems in [7] we can deduce that $y(z)$ has a regular singular point at $z = \frac{\ln 2}{a}$, around it can be expanded as

$$y(z) \underset{z \to \frac{\ln 2}{\phi_2}}{\sim} \kappa' \left(z - \frac{\ln 2}{\phi_2}\right)^{\left(1 - \frac{\phi_3}{\phi_2^2}\right)\ln 2}, \quad \text{for some constant } \kappa'.$$

Therefore the expansion of $g(z)$ satisfies

$$g(z) \underset{z \to \frac{\ln 2}{\phi_2}}{\sim} \kappa \left(z - \frac{\ln 2}{\phi_2} \right)^{\left(1 - \frac{\phi_3}{\phi_2^2}\right) \ln 2}, \qquad \text{for some constant } \kappa.$$

Now that we have the singular expansion of $g(z)$. We need to understand how the integral $\int_0^z \frac{\tilde{a}(t)}{(e^{\phi_2 t} - 2)g(t)} dt$ affects the main order asymptotic. And it will turn out that these integral is bounded depending on the growth of the coefficients of $\phi(z)$ as stated in the Theorem 1.

Proposition 4. *If the integral $\int_0^z \frac{\tilde{a}(t)}{(e^{\phi_2 t} - 2)g(t)} dt$ is bounded as $z \to \frac{\ln(2)}{\phi_2}$, then*

$$b_n \sim \alpha \left(\frac{\phi_2}{\ln 2} \right)^n n^{-1 + \left(-1 + \frac{\phi_3}{\phi_2^2}\right) \ln 2} \qquad \text{with} \qquad \alpha = \kappa \int_0^{\frac{\ln 2}{\phi_2}} \frac{\tilde{a}(t)}{(e^{\phi_2 t} - 2)g(t)} dt.$$

The proof is based on the coefficients of the Cauchy product of $g(z)$ and the integral. We end the proof with a lemma that shows under which conditions on $\phi(z)$ the integral is bounded.

Lemma 1. *If $\phi(z)$ fulfills the conditions of Theorem 1, then $\tilde{a}_n = \mathcal{O}\left(\frac{g_n}{n^\epsilon}\right)$, which in turn implies that the integral $\int_0^{\frac{\ln 2}{\phi_2}} \frac{\tilde{a}(t)}{(e^{at} - 2)g(t)} dt$ is bounded.*

Proof. For the first implication, the result follows from Proposition 3 and a subsequent estimate on a_n. For the second implication we use the following argument. Let us denote $\beta = \left(-1 + \frac{\phi_3}{\phi_2^2}\right) \ln 2$. We notice that

$$\frac{1}{(e^{at} - 2)g(t)} \underset{z \to \frac{\ln 2}{a}}{\sim} c \left(z - \frac{\ln 2}{a} \right)^{-\beta},$$

for some constant c. Furthermore, the coefficients of $\tilde{a}(z)$ are bounded above by some $\bar{a}_n = [z^n]c'\left(z - \frac{\ln 2}{a}\right)^{\beta - \epsilon}$, with some constant c'. Finally, we get

$$[z^n] \frac{\tilde{a}(t)}{(e^{\phi_2 t} - 2)g(t)} \underset{z \to \frac{\ln 2}{a}}{\sim} \mathcal{O}\left(n^{-\epsilon}\right)$$

and therefore

$$[z^n] \int_0^{\frac{\ln 2}{\phi_2}} \frac{\tilde{a}(t)}{(e^{\phi_2 t} - 2)g(t)} \underset{z \to \frac{\ln 2}{a}}{\sim} \mathcal{O}\left(n^{-\epsilon-1}\right).$$

$\square$

From Lemma 1, if $\phi(z)$ fulfills the conditions, then the integral is bounded and the result holds.

4   Conclusion

We have presented an evolution-process-based framework for specifying and counting families of increasing trees allowing for label repetitions and weakly-increasing sequences of labels along branches. Specifically, we have shown that under most interesting cases only binary and ternary do count in the main order asymptotic (unary and binary if unary nodes are allowed).

In this paper we discuss the case, where the coefficients of the degree function grow less rapidly than the process itself. Therefore, an interesting direction to pursue would be to have a full characterization of the asymptotic behavior of these processes depending on $\phi(z)$ for the other two cases. This analysis naturally follows from further applying the notions we have presented in this work.

Furthermore, the tools developed in this work may provide a novel way to approach the study of other structures of interest, such as linear λ-terms (see [3]), whose specifications also make critical use of the composition operation.

Acknowledgment. We thank Stephan Wagner for a fruitful discussion about the relationship of an involved proof of this paper and the article [8]. Furthermore we are grateful for the anonymous reviewers whose comments and suggestions helped improving and clarifying this manuscript.

References

1. Bergeron, F., Flajolet, P., Salvy, B.: Varieties of increasing trees. In: Raoult, J.-C. (ed.) CAAP 1992. LNCS, vol. 581, pp. 24–48. Springer, Heidelberg (1992). https://doi.org/10.1007/3-540-55251-0_2
2. Blieberger, J.: Monotonically labelled Motzkin trees. Discrete Appl. Math. **18**(1), 9–24 (1987)
3. Bodini, O., Gardy, D., Gittenberger, B., Jacquot, A.: Enumeration of generalized BCI lambda-terms. Electr. J. Comb. **20**(4), P30 (2013)
4. Bodini, O., Genitrini, A., Gittenberger, B., Wagner, S.: On the number of increasing trees with label repetitions. Discrete Math. (2019, in press). https://doi.org/10.1016/j.disc.2019.111722
5. Bodini, O., Genitrini, A., Naima, M.: Ranked Schröder trees. In: 2019 Proceedings of the Sixteenth Workshop on Analytic Algorithmics and Combinatorics (ANALCO), pp. 13–26. SIAM (2019)
6. Flajolet, P., Gourdon, X., Martínez, C.: Patterns in random binary search trees. Random Struct. Algorithms **11**(3), 223–244 (1997)
7. Flajolet, P., Sedgewick, R.: Analytic Combinatorics. Cambridge University Press, Cambridge (2009)
8. Genitrini, A., Gittenberger, B., Kauers, M., Wallner, M.: Asymptotic enumeration of compacted binary trees of bounded right height. J. Comb. Theory Ser. A **172** (2020). https://doi.org/10.1016/j.jcta.2019.105177
9. Henrici, P.: Applied and Computational Complex Analysis, Volume 2. Pure and Applied Mathematics. A Wiley-Interscience Series of Texts, Monographs, and Tracts. Wiley, Hoboken (1974)
10. Kirschenhofer, P.: On the average shape of monotonically labelled tree structures. Discrete Appl. Math. **7**(2), 161–181 (1984)

11. Mendez, M.A.: Combinatorial differential operators in: Faa di Bruno formula, enumeration of ballot paths, enriched rooted trees and increasing rooted trees. Technical report. arXiv:1610.03602 [math.CO], Cornell University (2016)

12. Morris, K.: On parameters in monotonically labelled trees. In: Drmota, M., Flajolet, P., Gardy, D., Gittenberger, B. (eds.) Mathematics and Computer Science III, pp. 261–263. Birkhäuser Basel (2004)

13. Morris, K.: Contributions to the analysis of increasing trees and other families of trees. Ph.D. thesis, University of the Witwatersrand, Johannesburg, South Africa (2005)

14. Odlyzko, A.M.: Periodic oscillations of coefficients of power series that satisfy functional equations. Adv. Math. **44**(2), 180–205 (1982)

15. Prodinger, H., Urbanek, F.J.: On monotone functions of tree structures. Discrete Appl. Math. **5**(2), 223–239 (1983)

16. Wasow, W.: Asymptotic Expansions for Ordinary Differential Equations. Pure and Applied Mathematics, Vol. XIV, Interscience Publishers Wiley, New York (1965)

Nested Regular Expressions Can Be Compiled to Small Deterministic Nested Word Automata

Iovka Boneva[1] , Joachim Niehren[2] , and Momar Sakho[1,2]([✉])

[1] Université de Lille, Lille, France
`momar.sakho@univ-lille.fr`
[2] Inria Lille, Lille, France

Abstract. We study the problem of whether regular expressions for nested words can be compiled to small deterministic *nested word automata* (NWAs). In theory, we obtain a positive answer for small deterministic regular expressions for nested words. In practice of navigational path queries, nondeterministic NWAs are obtained for which NWA determinization explodes. We show that practical good solutions can be obtained by using *stepwise hedge automata* as intermediates.

Keywords: Automata · Regular expressions · Nested words · XPATH

1 Introduction

Nested words are nested structures omnipresent in computer science. They were used in particular to represent data trees or XML documents, or to analyze the call structure of recursive programs. The idea of nested words is to generalize words and unranked trees at the same time. Nested words can be obtained by enriching Dyck words with internal letters, besides opening and closing parentheses. Nested words can also be defined recursively as the elements of the least set that contains internal letters from a given alphabet, triples consisting of an opening parenthesis, a nested word, and a closing parenthesis, and all sequences of nested words. Alternatively, nested words can be specified as finite sequences of internal letters, opening parentheses and closing parentheses. Only well-nested sequences are permitted in which every opening parenthesis is properly closed and every closing parenthesis is properly opened. Or else, nested words can be identified with sequences of unranked trees, which are often called hedges.

From the viewpoint of formal language theory, the natural question is how to lift and relate the notions of finite automata and regular expressions for words and trees to the case of nested words. Automata for nested words (NWAs) are well studied [1,3,23] and also known as visibly pushdown automata. While having the same expressiveness as hedge automata [10,26], which generalize tree automata from ranked to unranked trees, they are often defined as pushdown automata with visible stacks, meaning that exactly one symbol is pushed when

H. Fernau (Ed.): CSR 2020, LNCS 12159, pp. 169–183, 2020.
https://doi.org/10.1007/978-3-030-50026-9_12

reading an opening parenthesis, and exactly one symbol is popped when reading a closing parenthesis, while the stack is not used otherwise. Their main advantage is a powerful notion of determinism, generalizing both over bottom-up and top-down determinism of tree automata for ranked trees [1]. In contrast to more general pushdown automata, NWAs permit determinization, basically since they are so closely related to tree automata.

Regular expressions for nested words were first introduced under the name of *regular expression types* by Hosoya et al. in the context of the XML programming language XDuce [19]. We will call them *nested regular expressions* (NREs) instead. Independently, more complex notions of regular expressions were proposed [21,25] that can also deal to some extent with generalizations of nested words, in which dangling opening and closing parentheses are permitted. It was already claimed in [19], that NREs have the same expressiveness as hedge automata [10,26], which in turn have the same expressiveness as NWAs [1]. However, the question under which conditions nested words can be compiled to small deterministic NWAs has not been studied. Whenever possible, one can decide language inclusion or equivalence in P. Otherwise, these problems may not be feasible since EXP-complete for general NWAs or NREs.

Our concrete interest in the universality of deterministic NWAs is motivated by XML stream processing: we want to compute the certain answers of a CoreXPath query on an XML stream [15,24], i.e., those elements that are selected in all possible futures of the stream. Whether an answer is certain is computationally hard for tiny syntactic fragments of CoreXPath [4,15], but can be done in polynomial time for queries defined by deterministic NWAs [16]. A natural question is therefore, whether it is possible to compile CoreXPath queries as in the usual benchmark [14] to deterministic NWAs of reasonable size. Unfortunately, the existing compilers fail to do so [12], since they are based on NWA determinization for dealing with disjunction, negation, and recursive steps. Thereby they produce huge deterministic automata even for very simple CoreXPath queries from the benchmark, or do not terminate after some hours.

In this paper, we consider NREs for defining queries on nested words, since there exist compilers that can map the CoreXPath queries from the usual benchmark to NREs of reasonable size, under the condition that the path query contains only forwards steps. We then distinguish a subclass of "deterministic" NREs that can be complied in polynomial time to deterministic NWAs by generalizing on Glushkov's construction of deterministic finite-state automata (DFAs) from "deterministic" regular expressions [6,7]. However, the NREs obtained by compilation from CoreXPath queries are rarely deterministic, so neither are the NWAs obtained from them by direct compilation. Neither can we apply NWA determinization to them as argued above. We show that deterministic NWAs can be obtained nevertheless based on stepwise hedge automata (SHAs), that we introduce. SHAs combine stepwise tree automata [8] for unranked trees with finite state automata on words (NFAs). They can be determinized in a bottom-up and left-to-right manner, simply by combining the determinization procedures for tree automata and for NFAs. Furthermore, we can compile deterministic SHAs to deterministic NWAs in polynomial time. Conversely, NWAs can be compiled to SHAs in polynomial time too, but at the cost of introducing nondeterminism.

By composing these compilers and determinization algorithms, NREs can be compiled to deterministic NWAs in the following two manners. The first method is to compile the NRE to an SHA, from there to an NWA, which is then determinized. The second way consists of compiling the NRE to an SHA, determinize it, and convert the result to a deterministic NWA. In an experimental study, we consider a collection of NREs that we constructed automatically from CoreXPath queries in the XMark benchmark [14]. It turns out a little surprisingly that both above algorithms yield a satisfactory solution: they produce small deterministic NWAs for all NREs in our collection. The sizes of the deterministic may differ, sometimes in favor of the one or the other algorithm. We also discuss, why the NWA determinization behaves reasonably for the NWAs obtained from SHAs, while it behaved so badly for NWAs obtained directly from NREs. The reason seems to be that the former NWAs in contrast to the latter have the single entry property, which basically states that the NWA performs all its work in a bottom-up and left-to-right manner, and none when moving top-down. This conjecture is supported by practical evidence rather than some formal statement.

Related Work. CoreXPath [17] is a fragment of nested regular path queries on data trees, in which recursion is restricted to basic steps up, down, left and right. Nested regular path queries were introduced in the seventies [13] under the name of the propositional dynamic logic (PDL). There they were applied to general labeled graphs, rather than being restricted to data trees.

Since certain query answering for CoreXPath was considered as difficult, the currently existing approaches to CoreXPath evaluation on XML streams [12,24] either approximate certain query answers based on nondeterministic machines or restrict the queries so that answers certainty can be decided without latency [4,22]. This also holds for recent streaming algorithms on words without nesting in the context of complex event processing [18].

2   Nested Words

Nested words are words with parentheses that are well-nested. They can be identified with hedges, that is sequences of internal symbols and unranked trees.

Nested words are constructed with an opening and a closing parentheses, respectively $\langle$ and $\rangle$. An unranked alphabet Σ is a possibly infinite set of so called "internal" symbols, that does not contain the two parentheses. Nested words over Σ then have the following abstract syntax:

$$h, h' ::= \varepsilon \quad | \quad a \quad | \quad \langle h \rangle \quad | \quad h \cdot h' \qquad \text{where } a \in \Sigma$$

The empty word is denoted by ε and assumed to satisfy
$\varepsilon \cdot h = h = h \cdot \varepsilon$. Nested words can be identified with hedges,
i.e., words of trees and internal symbols. Seen as a graph,
the inner nodes are labeled by the tree constructor $\langle\rangle$ and
the leaves by symbols in Σ or the tree constructor. For
instance $\langle a \cdot \langle b \rangle \cdot \varepsilon \rangle \cdot c \cdot \langle d \cdot \langle \varepsilon \rangle \rangle$ corresponds to the hedge
on the right. A nested word of type *tree* has the form $\langle h \rangle$.

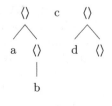

Variants. Our notion of nested words accepts only well-nested words without
dangling opening or closing parentheses in contrast to others [1,3]. This will lead
to simpler notion of regular expressions, avoiding the more complex operators
as with visibly rational expressions [5,25]. A less important difference is that we
do not support labeled parentheses.

Labeled Trees. Labeled parentheses can be simulated by using internal letters.
For instance, the labeled tree $a(b(), c())$ can be represented by the nested word
of type tree $\langle a \cdot \langle b \rangle \cdot \langle c \rangle \rangle$. In this way, the labeled tree $a()$ is represented by the
nested word $\langle a \rangle$ which is of type tree (while the internal letter a alone is not).
XML documents are particular labeled trees, such as for instance: $\langle a \ name =$
$"uff" \rangle \langle b \rangle isgaga \langle d/ \rangle \langle /b \rangle \langle c/ \rangle \langle /a \rangle$. Labeled trees satisfying the XML data model
can be represented as nested words over a signature that contains the XML node-
types $(elem, attr, text, \ldots)$, the XML names of the document $(a, \ldots, d, name)$,
and the characters of the data values, say UTF8. For the above example, we get
the nested word $\langle elem \cdot a \cdot \langle attr \cdot name \cdot u \cdot f \cdot f \rangle \langle elem \cdot b \cdot \langle text \cdot i \cdot s \cdot g \cdot a \cdot g \cdot$
$a \rangle \langle elem \cdot d \rangle \rangle \langle elem \cdot c \rangle \rangle$.

3   Nested Regular Expressions

We present nested regular expressions (NREs), that were introduced under the
name *regular expression types* in the context of XDuce [19] up to minor details.
A NRE over alphabet Σ has the following abstract syntax:

$$E, E' ::= \varepsilon \mid a \mid \neg \Sigma' \mid \emptyset \mid E \cdot E' \mid E + E' \mid E \& E' \mid E^* \mid \langle E \rangle \mid \mu a.E$$

where $a \in \Sigma$ and $\Sigma' \subseteq \Sigma$ is finite. We restrict the recursive expressions $\mu a.E$
such that all occurrences of a in E are nested below parentheses. The sets of
free and bound symbols $fn(E)$ and $bn(E)$ are defined as usual where $\mu a.E$ binds
symbol a with scope E and there is no other binder.

There are two differences with the regular expression types from [19]. First,
our NREs treat labels as internal symbols instead of labels of parentheses. Second, they provide recursion through the μ-operator instead of using recursive
equation systems. Even though not needed from the view point of expressiveness, we allow conjunctions $E \& E'$ to simplify the compilation of CoreXPath
expressions with filters to NREs. NREs having no subexpressions $E \& E'$ are
called conjunction-free (CF-NREs). Any NRE describes a language of nested

words that is defined by structural induction as follows:

$$L(\varepsilon) = \{\varepsilon\} \qquad L(a) = \{a\} \qquad L(\neg\Sigma') = \Sigma\backslash\Sigma' \qquad L(\emptyset) = \emptyset$$
$$L(E \cdot E') = L(E) \cdot L(E') \qquad L(E^*) = L(E)^*$$
$$L(E + E') = L(E) \cup L(E') \qquad L(E\&E') = L(E) \cap L(E')$$
$$L(\langle E \rangle) = \{\langle h \rangle \mid h \in L(E)\} \qquad L(\mu a.E) = \cup_{n \geq 0} L(\mu^n a.E)$$

A negation $\neg\Sigma'$ stands for $\Sigma\backslash\Sigma'$. This is useful for dealing with infinite alphabets and with large finite alphabets. For all expressions E, E_1 and E_2, the notation $E[E_1/E_2]$ stands for the expression E where all the occurrences of E_1 have been replaced by E_2. The semantics of a μ-operator is then defined using the shortcuts $\mu^0 a.E = E[a/\emptyset]$ and $\mu^n a.E = E[a/\mu^{n-1}a.E]$ for all $n \geq 1$. Note that $\mu a. \ b \cdot a \cdot c + \varepsilon$ would define the string language $\{b^n \cdot c^n \mid n \geq 0\}$ which is not regular. But this expression is ruled out since the μ-bound name a is not nested below parentheses.

In the context of XML queries, we can express the child and descendant-or-self axes of XPATH expressions by using the following NREs:

$$ch(E) \ =_{df} T \cdot \langle E \rangle \cdot T \qquad T =_{df} \mu x. \ (\langle x \rangle + \neg\emptyset)^*$$
$$ch^*(E) =_{df} \mu x. \ (E + ch(x)) \qquad \text{where } x \notin fn(E)$$
$$ch^+(E) =_{df} \mu x. \ (ch(E) + ch(x)) \qquad \text{where } x \notin fn(E)$$

Thereby, the XPath expression $a[following\text{-}sibling::b]/descendant::c$ can be expressed as a NRE, in which $x \in \Sigma$ serves as the selection variable, while the negation $\neg\{x\}$ expresses nonselection.

$$\langle elem \cdot a \cdot \neg\{x\} \cdot ch^+(\langle elem \cdot c \cdot x \cdot T \rangle) \rangle \cdot T \cdot \langle elem \cdot b \cdot \neg\{x\} \cdot T \rangle \cdot T$$

Our next objective is to distinguish NREs that can be evaluated deterministically in polynomial time, for instance by compilation to deterministic NWAs. For this, we consider the language of NREs $nregexp(ch, T)$ extended by the constant T and the unary constructor ch.

Definition 1. *An expression of $nregexp(ch, T)$ is deterministic if it does not contain a subexpression of any of the forms: $E_1 + E_2$, E^*, $T \cdot E$, $\mu a.E$.*

Note in particular that $ch(a)$ is a deterministic expression of $nregexp(ch, T)$. In contrast, the semantically equivalent expression $T.\langle a \rangle.T$ is not deterministic. Similarly, T is deterministic while the equivalent expression $\mu x.(\langle x \rangle + \neg\emptyset)^*$ is not. The expression $ch^*(E)$ is not deterministic since its definition relies on the μ-operator.

4 Nested Word Automata

Nested word automata (NWAs) are pushdown automata reading nested words, whose stacks are visible: they push a single stack symbol when reading an opening parenthesis, pop a single stack symbol when reading a closing parenthesis, and do not alter or inspect the stack otherwise.

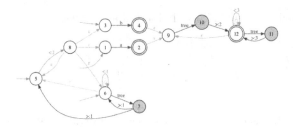

Fig. 1. Nested word automaton $nwa(ch^*(a+b))$.

Definition 2. *An NWA is a tuple* $A = (Q_h, Q_t, \Sigma, \Gamma, \Delta, I, F)$ *consisting of a possibly infinite set* Σ *of internal symbols, finite sets* Q_h *and* Q_t *of states of type hedge and tree respectively, sets of initial and final states* $I, F \subseteq Q_h$, *a finite set* Γ *of stack symbols, and a finite set* Δ *of transition rules of the forms:*

hedge rules	$a^\Delta, \_^\Delta, \varepsilon^\Delta \subseteq Q_h \times Q_h$	*where* $a \in \Sigma$
opening rules	$\langle^\Delta_\gamma \subseteq Q_h \times Q_h$	*where* $\gamma \in \Gamma$
hedge ending rules	$tree^\Delta \subseteq Q_h \times Q_t$	
closing rules	$\rangle^\Delta_\gamma \subseteq Q_t \times Q_h$	

Our NWAs are symbolic, in that they come with else rules, i.e., elements of $(q, q') \in \_^\Delta$ that we will denote by $q \overset{\_}{\to} q'$, for dealing with large or infinite alphabets. An example for an NWA is given in a graphical syntax in Figurefig1. Tree states are drawn in circles that are filled in light gray (q), while hedge states are in unfilled circles (q). Initial states are drawn as $\to(q)$ and final states as $(\!(q)\!)$. Hedge rules that have the form $(q_1, q_2) \in o^\Delta$ where $o \in \Sigma \cup \{\_, \varepsilon, tree\}$ are denoted by $q_1 \overset{o}{\to} q_2$. They are either label, else, epsilon, or tree rules depending of the type of letter o. Opening rules $(q_1, q_2) \in \langle^\Delta_\gamma$ are represented as $q_1 \overset{\langle\downarrow\gamma}{\longrightarrow} q_2$ and closing rules $(q_1, q_2) \in \rangle^\Delta_\gamma$ as $q_1 \overset{\rangle\uparrow\gamma}{\longrightarrow} q_2$.

Our notion of NWAs supports factorization in the spirit of [9]. It is obtained by distinguishing two types of states $q \in Q_h$ and $p \in Q_t$, and adding explicit type coercion rules $q \overset{tree}{\longrightarrow} p$. Semantically, both kinds of states could be merged when replacing type the coercion rules by the epsilon rule $q \overset{\varepsilon}{\to} p$, but at the cost of introducing additional nondeterminism. This may lead to quadratically larger deterministic automata, as we will illustrate at the NWA in Fig. 5.

The language of nested words between two states $q_1, q_2 \in Q_h$ is defined as the least language such that:

$$L_{q_1,q_2}(\Delta) = \{\varepsilon \mid \text{if } q_1 = q_2 \text{ or } q_1 \overset{\varepsilon}{\to} q_2 \text{ wrt.} \Delta\} \cup \bigcup_{q_3 \in Q_h} L_{q_1,q_3}(\Delta) \cdot L_{q_3,q_2}(\Delta)$$
$$\cup \{a \mid \text{if } q_1 \overset{a}{\to} q_2 \in \Delta \text{ or } (q_1 \overset{\_}{\to} q_2 \in \Delta' \text{ and } \neg\exists q_2'. q_1 \overset{a}{\to} q_2' \in \Delta)\}$$
$$\cup \{\langle h \rangle \mid \exists q_1', q_2' \in Q_h. \exists q_3 \in Q_t. \exists \gamma \in \Gamma. \ q_1 \overset{\langle\downarrow\gamma}{\longrightarrow} q_1', h \in L_{q_1',q_2'}(\Delta),$$
$$q_2' \overset{tree}{\longrightarrow} q_3 \in \Delta \text{ and } q_3 \overset{\rangle\uparrow\gamma}{\longrightarrow} q_2 \in \Delta\}.$$

The language of the NWA then is $L(A) = \bigcup_{q_1 \in I, q_2 \in F} L_{q_1, q_2}(\Delta)$. NWAs can be determinized by adapting the usual determinization procedure for NWAs [1,12] so that it can account for hedge ending and else rules.

As argued earlier, NREs have the same expressiveness as NWAs and thus as deterministic NWAs. But in practice, the sizes may blow-up surprisingly by determinization [12]. We next discuss on a compiler from expression an E of $nregexp(T, ch)$ to an NWA $nwa(E)$ that preserves determinism. For instance, the NWA for the expression $ch^*(a + b)$ is shown in Fig. 1. For regular expressions without nesting, the compiler is based on Glushkov's construction recursively on the structure of the expression while eliminating ε-edges on the fly. Such construction is known to preserve determinism [7]. For deterministic expressions $ch(E)$, we adapt ideas from [12]. As for conjunctions, product of automata are used. Special care has to be given to $\mu a.E$ expressions in order to preserve the recognized language and also have the following complexity result:

Theorem 1. *For any CF-NRE E, we can construct in time $O(|E|^2)$ an NWA A while preserving determinism such that $L(A) = L(E)$.*

This quadratic time result generalizes on a previous result for the Glushkov construction [6]. Because of automata products used to build them, NREs having conjunctions may in the worst case yield NWAs with an exponential size. As a consequence of Theorem 1, small deterministic CF-NREs can be compiled to small deterministic NWAs. This gives a first positive answer to the motivating question of the present paper.

As for nondeterministic expressions, the NWA determinization procedure is not a solution to the problem at hand, due to huge size increase. For instance, the NWA $det(nwa(ch^*(a + b)))$ obtained by determinization of the NWA in Fig. 1 has size 271, which may seem way too large. Even worse cases can be found in the experimental section. The problem is not solved by factorization, and actually confirms a size increase reported earlier for NWAs obtained from XPath by a different compiler [12]. So the question is, whether there do not exist better methods to obtain smaller deterministic NWAs for nested regular expressions.

5   Stepwise Hedge Automata

We propose SHAs as an extension of stepwise tree automata [8] to recognize not only unranked trees but also hedges. The problematic notion of determinism of the hedge automata from [10,20,26] is avoided.

Our notion of SHAs will be symbolic in using else rules, and factorized in the sense of [9]: there are two types of states for hedges and trees and an operator for explicit type coercion. We also propose a novel treatment of internal letters inspired by nested word automata, so that SHAs generalize both on stepwise tree automata and on NFAs.

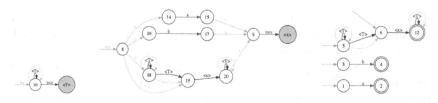

Fig. 2. Stepwise hedge automaton $sha(ch^*(a + b))$: the part with the stepwise tree automaton is on the left and middle, and the NFA part on the right.

Definition 3. *A SHA is a tuple $S = (Q_h, Q_t, \Sigma, \Delta, I, F)$ such that Q_t and Q_h are finite set of states of two types t for tree and respectively h for hedge, Σ an alphabet of internal letters (that may be infinite), $I, F \subseteq Q_h$ subsets of initial and final states respectively, and Δ a finite set of transition rules such that for all $q \in Q_t$ and $a \in \Sigma$:*

$$\begin{aligned}
&\text{hedge rules} && q^\Delta, a^\Delta, \_^\Delta, \varepsilon^\Delta \subseteq Q_h \times Q_h \\
&\text{tree final rules} && tree^\Delta \subseteq Q_h \times Q_t \\
&\text{tree initial states} && \langle\rangle^\Delta \subseteq Q_h
\end{aligned}$$

An example for a SHA is given in graphical syntax in Fig. 2. It recognizes all hedges which are either just a or b or contain some tree node that contains either just a or b. In the graphical syntax, the states of type tree $q \in Q_t$ are drawn in circles filled in light gray $\boxed{q}$, while the states of type hedge $q' \in Q_h$ are drawn in unfilled circles $\boxed{q'}$. The right part of the graph is an NFA which uses tree states as additional edge labels, while the left part is a stepwise tree automaton, that defines the tree languages of these tree states.

Let Δ_h be the restriction of Δ to the hedge rules. Then, $(Q_h, \Sigma \uplus Q_t, \Delta_h, I, F)$ is a standard NFA with ε-rules, which is symbolic [11] in providing else rules for dealing with large or infinite alphabets in addition. Therefore, we denote the initial states $q \in I$ by $\rightarrow\!\boxed{q}$ and the final states $q \in F$ by $\boxed{\boxed{q}}$. A rule with an internal letter $(q_1, q_2) \in a^\Delta$ is denoted by $q_1 \xrightarrow{a} q_2$ wrt. Δ stating that a hedge in state q_1 can be extended by the internal letter a leading to a hedge in state q_2. Similarly, an epsilon rule $(q_1, q_2) \in \varepsilon^\Delta$ is denoted by $q_1 \xrightarrow{\varepsilon} q_2$, and an else rule $(q_1, q_2) \in \_^\Delta$ is denoted by $q_1 \xrightarrow{} q_2$. In the same spirit, a hedge rule $(q_1, q_2) \in q^\Delta$ is denoted by $q_1 \xrightarrow{q} q_2$ wrt. Δ, stating that a hedge in state q_1 can be extended by a tree in state q leading to a hedge in state q_2.

A tree initial state $q \in \langle\rangle^\Delta$ is graphically denoted by $\xrightarrow{\langle\rangle} q$ and a tree final rule $(q_1, q_2) \in tree^\Delta$ by $q_1 \xrightarrow{tree} q_2$. Intuitively, a tree $\langle h \rangle$ can be evaluated to state q if h can be evaluated starting with some tree initial state $\xrightarrow{\langle\rangle} q_1$ to some state q_2 such that $q_2 \xrightarrow{tree} q$. More formally, the hedge languages $L_{q_1, q_2}(S)$ between any two hedge states $q_1, q_2 \in Q_h$ are defined as follows:

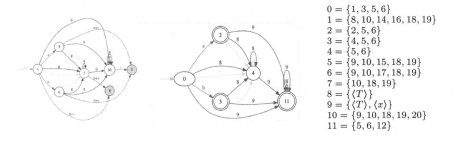

$$0 = \{1, 3, 5, 6\}$$
$$1 = \{8, 10, 14, 16, 18, 19\}$$
$$2 = \{2, 5, 6\}$$
$$3 = \{4, 5, 6\}$$
$$4 = \{5, 6\}$$
$$5 = \{9, 10, 15, 18, 19\}$$
$$6 = \{9, 10, 17, 18, 19\}$$
$$7 = \{10, 18, 19\}$$
$$8 = \{\langle T \rangle\}$$
$$9 = \{\langle T \rangle, \langle x \rangle\}$$
$$10 = \{9, 10, 18, 19, 20\}$$
$$11 = \{5, 6, 12\}$$

Fig. 3. The determinized SHA $det(sha(ch^*(a + b)))$.

$$L_{q_1,q_2}(S) = \{\varepsilon \mid \text{if } q_1 = q_2 \text{ or } q_1 \xrightarrow{\varepsilon} q_2 \text{ wrt. } \Delta\} \cup \bigcup_{q_3 \in Q_h} L_{q_1,q_3}(S) \cdot L_{q_3,q_2}(S)$$
$$\cup \{a \mid \text{if } q_1 \xrightarrow{a} q_2 \in \Delta \text{ or } (q_1 \xrightarrow{\cdot} q_2 \in \Delta \text{ and } \neg \exists q_2'. \, q_1 \xrightarrow{a} q_2' \in \Delta)\}$$
$$\cup \bigcup_{q_1 \xrightarrow{q} q_2 \in \Delta} L_q(S)$$

This definition is mutually recursive with the definition of the tree languages $L_q(S)$ of all tree states $q \in Q_t$:

$$L_q(S) = \{\langle h \rangle \mid \xrightarrow{\langle \rangle} q_1, \ h \in L_{q_1,q_2}(S), \ q_2 \xrightarrow{tree} q\}$$

The hedge language $L(S)$ that is recognized by automaton is $\bigcup_{q_1 \in I, q_2 \in F} L_{q_1,q_2}(S)$. The rules of standard bottom-up tree automata have the form $a(q_1, \ldots, q_n) \to q$ where a is a symbol of arity n. With SHAs, this rule can be encoded by the sequence $\xrightarrow{\langle \rangle} \textcircled{p_0} \xrightarrow{a} \textcircled{p_1} \xrightarrow{q_1} \ldots \xrightarrow{q_n} \textcircled{p_n} \xrightarrow{tree} q$ where the states $q_1, \ldots, q_n, q$ are all tree states, and $p_0, \ldots, p_n$ fresh hedge states. Stepwise hedge automata have a natural notion of determinism, generalizing both on that of stepwise tree automata and on NFAs, in contrast to the earlier notion of hedge automata in [10,26]. For instance, the SHA in Fig. 3 is obtained by determinization of the automaton in Fig. 2. It consists of a DFA on the right and a deterministic stepwise tree automaton on the left. We show that

Proposition 1. *Any SHA can be made deterministic in at most exponential time such that the hedge language is preserved.*

Any expression E can be compiled to a SHA $sha(E) = (Q_h, Q_t, \Sigma, \Delta, I, F)$ such that $Q_t = \{E' \mid E' = \langle E'' \rangle \text{ subexpression of } E\}$ and $L_t(E') = L(E')$ for all tree states $E' \in Q_t$. The SHA $sha(E)$ can be partitioned into disjoint SHAs $sha(E) = A^{top} \cup \bigcup_{E' \in Q_t} A^{E'}$ such that $A^{top} = (Q_h^{top}, Q_t, \Sigma, \Delta^{top}, I, F)$ and $A^{E'} = (Q_h^{E'}, Q_t, \Sigma, \Delta^{E'}, \emptyset, \emptyset)$ for all $E' \in Q_t$ and $\langle \overset{\Delta^{top}}{} = \emptyset$. Note that the transitions relation Δ is decomposed thereby into independent connected components.

Proposition 2. *For any CF-NRE E we can construct in time $O(|E|^2)$ a SHA $sha(E)$ such that $L(sha(E)) = L(E)$.*

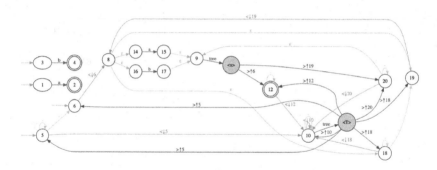

Fig. 4. The NWA from the SHA $nwa(sha(ch^*(a+b)))$.

However, the construction does not preserve determinism. For the deterministic NRE $\langle a_1 \cdot \langle a_2 \cdot \ldots \cdot \langle a_n \rangle \ldots \rangle \rangle$, one would have an SHA having a tree initial state for each of the $\langle a_i \ldots \rangle$ subtree, implying nondeterminism. This is in contrast to the compiler to NWAs, which can rely on top-down determinism that is unavailable for SHAs though. Furthermore, as for NWAs, conjunctions may cause an exponential blow-up of the produced SHA.

6   NWAs Versus SHAs

We next show how to compile SHAs to NWAs such that determinism is preserved, and back while introducing nondeterminism. Thereby we can obtain small NWAs for NREs such as $E = ch^*(a+b)$ for which $det(nwa(E))$ blows up in size in a surprising manner.

SHAs to NWAs. Any SHA $S = (Q_h, Q_t, \Sigma, \Delta, I, F)$ can be compiled to an NWA $nwa(S) = (Q_h, Q_t, \Sigma, \Gamma, \Delta', I, F)$ such that $L_{q_1,q_2}(S) = L_{q_1,q_2}(nwa(S))$. We set $\Gamma = Q_h$, $\_^{\Delta'} = \_^{\Delta}$, $a^{\Delta'} = a^{\Delta}$ for all $a \in \Sigma$, $\varepsilon^{\Delta'} = \varepsilon^{\Delta}$, $tree^{\Delta'} = tree^{\Delta}$:

$$\frac{q_1 \xrightarrow{q} q_2 \in \Delta \quad p \in \langle \rangle^{\Delta}}{q_1 \xrightarrow{\langle \downarrow q_1} p \in \Delta' \text{ and } q \xrightarrow{\rangle \uparrow q_1} q_2 \in \Delta'}$$

Clearly, if S is deterministic then so is $nwa(S)$, since p is unique in this case. Furthermore, one might be tempted to restrict the above construction rule to states p such that $L_q(S[\langle \rangle^{\Delta}/\{p\}]) \neq \emptyset$ where the set of tree initial states $\langle \rangle^{\Delta}$ is replaced by $\{p\}$. However, this would lead to huge blow-up when determinizing these NWAs, basically since this change spoils the single-entry property discussed in Definition 4.

The conversion of $sha(ch^*(a+b))$ in Fig. 2 yields the NWA in Fig. 4. Note that the opening rules are deterministic (but not the whole NWA), since for all tree states q there is at most one hedge state $p \in \langle \rangle^{\Delta}$ such that q is accessible from p. The NWA has size 64, while its determinization has size 159, which yields a size increase of $95 = 159 - 64$. The size increase for determinization

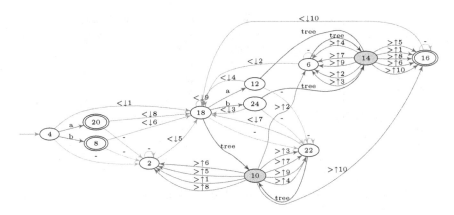

Fig. 5. Deterministic NWA: $nwa(det(sha(ch^*(a + b))))$.

$$\frac{o \in \Sigma \cup \{tree, \_, \varepsilon\} \quad q_1 \xrightarrow{o} q_2 \in \Delta \quad q \in Q_h}{(q, q_1) \xrightarrow{o} (q, q_2) \in \Delta^s} \qquad \frac{q_1 \xrightarrow{\langle\downarrow\gamma} q_2 \in \Delta}{\xrightarrow{\langle\rangle} (q_2, q_2) \in \Delta^s}$$

$$\frac{q_1 \xrightarrow{\langle\downarrow\gamma} q_2 \in \Delta \quad q_3 \in Q_t \quad q_3 \xrightarrow{\rangle\uparrow\gamma} q_4 \in \Delta \quad q \in Q_h}{(q, q_1) \xrightarrow{(q_2, q_3)} (q, q_4) \in \Delta^s}$$

Fig. 6. NWA to SHA conversion.

is considerably smaller for the NWA obtained from the regular expressions by indirection via a SHA, than for the NWA obtained by direct compilation. Indeed, the determinization of $nwa(ch^*(a + b))$ blows the size from 39 to 271. The size increase for the determinization of $nwa(ch^*(a + b))$ is thus $242 = 271 - 39$ and while for $nwa(sha(ch^*(a + b)))$ is only $95 = 159 - 64$.

The experiments will show that this is not an exception but the general rule. Intuitively, the reason is that NWAs obtained from stepwise hedge automata do all work bottom-up, where NWAs obtained directly from the regular expression do a considerable amount of work top-down. In terms of [2] this restriction can be characterized syntactically by the single-entry property:

Definition 4. *An NWA A has the single-entry property, if there exists a single state $q_{entry} \in Q_h$ such that all opening rules of A have the form $q \xrightarrow{\langle\downarrow q} q_{entry}$.*

It can be shown that $nwa(S)$ has the single-entry property for all SHAs S for which the p's are unique in the above construction rule, i.e., such that $\langle\rangle \to p$. Note that this was not the case for $sha(ch * (a + b))$ in Fig. 2 but could have been imposed w.l.o.g., leading to a slightly different NWA than in Fig. 4.

The conversion of the determinization $det(sha(ch^*(a+b)))$ in Fig. 3 yields the deterministic NWA in Fig. 5. The size goes up slightly from 53 to 73. It should be noticed, that factorization avoids a quadratic blow up in this case. This can be observed at state 14, which has 3 incoming tree-edges and 10 outgoing

	$det(nwa(.))$	$nwa(det(sha(.)))$	$det(nwa(sha(.)))$	$nwa(det(sha(nwa(.))))$	$det(nwa(sha(nwa(.))))$
A1		398 (37)	302 (62)	398 (37)	398 (37)
A2	362600 (6782)	668 (57)	4889 (221)	1648 (127)	4105 (148)
A3	318704 (8216)	469 (44)	542 (66)	625 (56)	907 (62)
A4		487 (42)	335 (67)	487 (42)	487 (42)
A5		676 (55)	1054 (110)	856 (67)	1192 (73)
A6		548 (45)	332 (62)	548 (45)	548 (45)
A7		468 (41)	285 (54)	468 (41)	468 (41)
A8		2520 (124)	1236 (137)	1804 (118)	

Fig. 7. Deterministic NWAs for XPath benchmark: size (#states).

	$det(nwa(.))$	$nwa(det(sha(.)))$	$det(nwa(sha(.)))$	$nwa(det(sha(nwa(.))))$	$det(nwa(sha(nwa(.))))$
$ch^3[a]$	19828 (1281)	85 (13)	157 (30)	192 (24)	352 (32)
$ch^4[a]$		177 (21)	206 (39)	664 (56)	2200 (88)
$ch^5[a]$		457 (37)	255 (48)	3336 (168)	
$ch^7[a]$		4825 (133)	353 (66)		
$ch^9[a]$			451 (84)		

Fig. 8. Deterministic NWAs queries $ch^n[a]$ for $n = 3, 4, 5, 7, 9$: size (#states).

closing edges. Without factorization, the 3 tree edges could be replaced by 3 ε-edges whose elimination would produce 30 closing edges. This would increase the number $3 + 10$ edges to $3 * 10$ edges.

NWAs to SHAs. Conversely, NWAs can be compiled to stepwise hedge automata, but at the cost of introducing nondeterminism, since an NWA may traverse the branches of a tree top-down, while a stepwise must traverse them bottom-up. For this, the stepwise guesses the state in which the NWA will arrive from above and then evaluates the subtree starting with this state, while verifying the correctness of the guess later on. Let $A = (Q_h, Q_t, \Sigma, \Delta', I, F)$ be an NWA. We build a SHA $sha(A) = (Q_h^s, Q_t^s, \Sigma, \Delta^s, I^s, F^s)$ where $Q_h^s = Q_h \times Q_h$, $Q_t^s = Q_h \times Q_t$, $I^s = \{(q, q) \mid q \in I\}$, $F^s = I \times F$ and Δ^s is the smallest satisfying the rule schemas in Fig. 6. The construction is such that $L(A) = L(sha(A))$.

7   Experimental Results and Discussion

We now compare the sizes of deterministic NWAs that we can obtain by composing the various compilers in different orders.

We test the $A1, \dots, A8$ XPATH queries in the usual XPATH benchmark [14], which contain not only forward child, descendant and following-sibling axes, but also filters and path compositions. Note that the queries A4 until A8 contain filters, which are mapped to NREs with conjunctions. We compiled these queries automatically to nested regular expressions, then compiled these expressions to deterministic NWAs, by composing the various compilers presented earlier in all reasonable manners. A1 is the only query for which we obtain a deterministic regular expression. But since we replaced $ch(E)$ systematically by $T \cdot \langle E \rangle \cdot T$ in our experiments, all nested regular expression become nondeterministic.

The overall size of the resulting automata and the number of their rules are given in Fig. 7. We can see that determinization applied to the NWAs for these expressions fails. Only 2 out of 8 automata have a size less than 400000, and for the others, the determinization ran out of time. In contrast, 3 of the 4 other methods – that use stepwise hedge automaton intermediately – produce reasonable small deterministic NWAs. For the fourth method in the last column, NWA-determinization did not terminate on $nwa(sha(nwa(A8)))$ after a few hours (Fig. 8).

We also tested our algorithms on collections of XPath queries with a scalable parameter, such as the queries $ch^n(a)$ for increasing n. This series is known to require many states for deterministic bottom-up evaluation. Indeed, the determinization for stepwise hedge automata $nwa(det(sha))$ leads to a size explosion. The method $det(nwa(sha(.)))$, however, still yields small deterministic automata! Generally this method produced satisfactory results in all our experiments. In quite some cases, however, $nwa(det(sha(.)))$ still behaves better.

We conjecture that these differences are related to the lack of minimization in our current implementation. The main problem here is that minimal deterministic NWAs do not exist for all regular languages of nested words [2]. This is in strict contrast to the cases of word automata, tree automata, and SHAs.

We point out that SHAs are determinized in bottom-up and left-to-right manner by combining the usual bottom-up determinization algorithms for tree automata and the usual left-to-right determinization algorithm for NFAs. In contrast to deterministic NWAs, they cannot support top-down determinism in combination with bottom-up and left-to-right determinism though. The NWAs obtained by compilation from SHAs are special in that they perform all their work in a bottom-up and left-to-right manner, and nothing top-down. Such NWAs were characterized syntactically as single-entry NWAs, and deterministic single-entry NWAs are shown to admit a unique minimization in [2]. Our experiments show that NWA determinization often works nicely for single-entry NWAs, while it explodes quickly without the single-entry restriction. The intuition is that single-entry NWAs behave like SHAs.

References

1. Alur, R.: Marrying words and trees. In: 26th ACM SIGMOD-SIGACT-SIGART Symposium on Principles of Database Systems, pp. 233–242. ACM Press (2007). https://doi.org/10.1145/1265530.1265564
2. Alur, R., Kumar, V., Madhusudan, P., Viswanathan, M.: Congruences for visibly pushdown languages. In: Caires, L., Italiano, G.F., Monteiro, L., Palamidessi, C., Yung, M. (eds.) ICALP 2005. LNCS, vol. 3580, pp. 1102–1114. Springer, Heidelberg (2005). https://doi.org/10.1007/11523468_89

3. Alur, R., Madhusudan, P.: Visibly pushdown languages. In: 36th ACM Symposium on Theory of Computing, pp. 202–211. ACM Press (2004). http://portal.acm.org/citation.cfm?coll=GUIDE&dl=GUIDE&id=1007390
4. Benedikt, M., Jeffrey, A., Ley-Wild, R.: Stream firewalling of XML constraints. In: ACM SIGMOD International Conference on Management of Data, pp. 487–498. ACM Press (2008)
5. Bozzelli, L., Sánchez, C.: Visibly rational expressions. Acta Inf. **51**(1), 25–49 (2014). https://doi.org/10.1007/s00236-013-0190-6
6. Brüggemann-Klein, A.: Regular expressions into finite automata. Theor. Comput. Sci. **120**(2), 197–213 (1993). https://doi.org/10.1016/0304-3975(93)90287-4
7. Brüggemann-Klein, A., Wood, D.: One-unambiguous regular languages. Inf. Comput. **142**(2), 182–206 (1998)
8. Carme, J., Niehren, J., Tommasi, M.: Querying unranked trees with stepwise tree automata. In: van Oostrom, V. (ed.) RTA 2004. LNCS, vol. 3091, pp. 105–118. Springer, Heidelberg (2004). https://doi.org/10.1007/978-3-540-25979-4_8. http://www.ps.uni-sb.de/Papers/abstracts/stepwise.html
9. Champavère, J., Gilleron, R., Lemay, A., Niehren, J.: Efficient inclusion checking for deterministic tree automata and XML schemas. Inf. Comput. **207**(11), 1181–1208 (2009). https://doi.org/10.1016/j.ic.2009.03.003
10. Comon, H., et al.: Tree automata techniques and applications, October 2007. http://tata.gforge.inria.fr. Accessed 1997
11. D'Antoni, L., Alur, R.: Symbolic visibly pushdown automata. In: Biere, A., Bloem, R. (eds.) CAV 2014. LNCS, vol. 8559, pp. 209–225. Springer, Cham (2014). https://doi.org/10.1007/978-3-319-08867-9_14
12. Debarbieux, D., Gauwin, O., Niehren, J., Sebastian, T., Zergaoui, M.: Early nested word automata for XPath query answering on XML streams. Theor. Comput. Sci. **578**, 100–125 (2015). https://doi.org/10.1016/j.tcs.2015.01.017
13. Fischer, M.J., Ladner, R.E.: Propositional dynamic logic of regular programs. J. Comput. Syst. Sci. **18**(2), 194–211 (1979). https://doi.org/10.1016/0022-0000(79)90046-1
14. Franceschet, M.: XPathMark performance test. https://users.dimi.uniud.it/~massimo.franceschet/xpathmark/PTbench.html. Accessed 30 Mar 2020
15. Gauwin, O., Niehren, J.: Streamable fragments of forward XPath. In: Bouchou-Markhoff, B., Caron, P., Champarnaud, J.-M., Maurel, D. (eds.) CIAA 2011. LNCS, vol. 6807, pp. 3–15. Springer, Heidelberg (2011). https://doi.org/10.1007/978-3-642-22256-6_2
16. Gauwin, O., Niehren, J., Tison, S.: Earliest query answering for deterministic nested word automata. In: Kutyłowski, M., Charatonik, W., Gębala, M. (eds.) FCT 2009. LNCS, vol. 5699, pp. 121–132. Springer, Heidelberg (2009). https://doi.org/10.1007/978-3-642-03409-1_12. http://hal.inria.fr/inria-00390236/en
17. Gottlob, G., Koch, C., Pichler, R.: The complexity of XPath query evaluation. In: 22nd ACM SIGMOD-SIGACT-SIGART Symposium on Principles of Database Systems, pp. 179–190 (2003)
18. Grez, A., Riveros, C., Ugarte, M.: A formal framework for complex event processing. In: Barceló, P., Calautti, M. (eds.) 22nd International Conference on Database Theory, ICDT 2019. LIPIcs, Lisbon, Portugal, 26–28 March 2019, vol. 127, pp. 5:1–5:18. Schloss Dagstuhl - Leibniz-Zentrum fuer Informatik (2019). https://doi.org/10.4230/LIPIcs.ICDT.2019.5
19. Hosoya, H., Pierce, B.C.: XDuce: a statically typed XML processing language. ACM Trans. Internet Technol. **3**(2), 117–148 (2003). https://doi.org/10.1145/767193.767195

20. Martens, W., Niehren, J.: On the minimization of XML-schemas and tree automata for unranked trees. J. Comput. Syst. Sci. **73**(4), 550–583 (2007). https://doi. org/10.1016/j.jcss.2006.10.021. https://hal.inria.fr/inria-00088406. Special issue of DBPL 05

21. Mozafari, B., Zeng, K., Zaniolo, C.: From regular expressions to nested words: unifying languages and query execution for relational and XML sequences. PVLDB **3**(1), 150–161 (2010). https://doi.org/10.14778/1920841.1920865. http://www. vldb.org/pvldb/vldb2010/pvldb_vol3/R13.pdf

22. Mozafari, B., Zeng, K., Zaniolo, C.: High-performance complex event processing over XML streams. In: Candan, K.S., et al. (eds.) SIGMOD Conference, pp. 253–264. ACM (2012). https://doi.org/10.1145/2213836.2213866

23. Neumann, A., Seidl, H.: Locating matches of tree patterns in forests. In: Arvind, V., Ramanujam, S. (eds.) FSTTCS 1998. LNCS, vol. 1530, pp. 134–145. Springer, Heidelberg (1998). https://doi.org/10.1007/978-3-540-49382-2_12

24. Olteanu, D.: SPEX: streamed and progressive evaluation of XPath. IEEE Trans. Know. Data Eng. **19**(7), 934–949 (2007). https://doi.org/10.1109/TKDE.2007. 1063

25. Pitcher, C.: Visibly pushdown expression effects for XML stream processing. In: PlanX (2005)

26. Thatcher, J.W.: Characterizing derivation trees of context-free grammars through a generalization of automata theory. J. Comput. Syst. Sci. **1**, 317–322 (1967)

On Embeddability of Unit Disk Graphs
onto Straight Lines

Onur Çağırıcı[✉]📷

Masaryk University, Brno, Czech Republic
onur@mail.muni.cz

Abstract. Unit disk graphs are the intersection graphs of unit radius
disks in the Euclidean plane. Deciding whether there exists an embedding
of a given unit disk graph, i.e., unit disk graph recognition, is an important geometric problem, and has many application areas. In general, this
problem is known to be $\exists\mathbb{R}$-complete. In some applications, the objects
that correspond to unit disks have predefined (geometrical) structures to
be placed on. Hence, many researchers attacked this problem by restricting the domain of the disk centers. One example to such applications is
wireless sensor networks, where each disk corresponds to a wireless sensor node, and a pair of intersecting disks corresponds to a pair of sensors
being able to communicate with one another. It is usually assumed that
the nodes have identical sensing ranges, and thus a unit disk graph model
is used to model problems concerning wireless sensor networks. We consider the unit disk graph realization problem on a restricted domain, by
assuming a scenario where the wireless sensor nodes are deployed on the
corridors of a building. Based on this scenario, we impose a geometric
constraint such that the unit disks must be centered onto given straight
lines. In this paper, we first describe a polynomial-time reduction which
shows that deciding whether a graph can be realized as unit disks onto
given straight lines is NP-hard, when the given lines are parallel to either
the x-axis or y-axis. Using the reduction we described, we also show that
this problem is NP-complete when the given lines are only parallel to the
x-axis (and one another). We obtain these results using the idea of the
logic engine introduced by Bhatt and Cosmadakis in 1987.

1 Introduction

An *intersection graph* is a graph that models the intersections among geometric objects. In an intersection graph, each vertex corresponds to a geometric
object, and each edge corresponds to a pair of intersecting geometric objects. A
unit disk graph is the intersection graph of a set of unit disks in the Euclidean
plane. Some well-known NP-hard problems, such as chromatic number, independent set, and dominating set, remain hard on unit disk graphs [4,6,11]. We are
particularly interested in the unit disk recognition problem, i.e., given a simple

This work is supported by the Czech Science Foundation, project no. 20-04567S.

H. Fernau (Ed.): CSR 2020, LNCS 12159, pp. 184–197, 2020.
https://doi.org/10.1007/978-3-030-50026-9_13

graph, deciding whether there exists an embedding of disks onto the plane which corresponds to the given graph. This problem is known to be NP-hard [9], and even $\exists \mathbb{R}$-complete [16] in general.

A major application area of unit disk graphs is wireless sensor networks, since it is an accurate model (in an ideal setting) of communicating wireless sensor nodes with identical range [3,12]. In a wireless sensor network, the sensor nodes are deployed on bounded areas [2,10,19]. Thus, it becomes more interesting to observe the behavior of the unit disk graph recognition problem when the domain is restricted [1,8,13,15].

We assume that the sensor nodes are deployed onto the corridors in a building, and the floor plans are available. We model the corridors on a floor as straight lines, and consider the recognition problem where the unit disks are centered on the given lines. We show that this problem is NP-hard, even when the given straight lines are either vertical or horizontal, i.e., any pair of lines is either parallel, or perpendicular to each other. In addition, we show that if there are no pairs of perpendicular lines i.e., all lines are parallel to x-axis, then the recognition problem is NP-complete.

Due to space restrictions, the proofs of some statements are omitted, and those statements are marked with (*). The full version of this paper is available online at http://arxiv.org/abs/1811.09881.

Related Work

Breu and Kirkpatrick showed that the unit disk graph recognition problem is NP-hard in general [9]. Later on, this result was extended, and it was proved that the problem is also $\exists \mathbb{R}$-complete [16,18]. Kuhn et al. showed that finding a "good" embedding is not approximable when the problem is parameterized by the maximum distance between any pair of disk centers [17]. In the very same paper, they also give a short reduction that the realization problem and the recognition problem on unit disk graphs are polynomially equivalent [17].

Intuitively, the most restricted domain for unit disk graphs is when the disks are centered on a single straight line in the Euclidean plane. In this case, the unit disks become *unit intervals* on the line, and they yield a *unit interval graph* [14]. To recognize or realize whether a given graph is a unit interval graph is a linear-time task [7]. Our domain is restricted to not only one straight line, but to a set of straight lines given by their equations. Given a simple graph, and a set of straight lines, we ask the question "can this graph be realized as unit disks on the given set of straight lines?" We show that even though these lines are restricted to be parallel to either the x-axis or y-axis, it is NP-hard to determine whether the given graph can be embedded onto the given lines (Theorem 1). We, however, do not know whether this variant belongs to the class NP, or is possibly $\exists \mathbb{R}$-complete. If, on the other hand, the lines are restricted to be parallel only to the x-axis, then we show that the problem belongs to NP and thus is NP-complete.

2 Basic Terminology and Notations

A *unit disk* around a point p is the set of points in the plane whose distance from p is one unit. Two unit disks, centered at two points p and q, intersect when the Euclidean distance between p and q is less than or equal to two units. A graph $G = (V, E)$ is called a *unit disk graph* when every vertex $v \in V$ corresponds to a disk $\mathcal{D}_v$ in the Euclidean plane, and an edge $uv \in E$ exists when $\mathcal{D}_u$ and $\mathcal{D}_v$ intersect.

The *unit disk recognition problem* is deciding whether a given graph $G = (V, E)$ is a unit disk graph. That is, determining whether there exists a mapping $\Sigma : V \to (\mathbb{R} \times \mathbb{R})$, such that each vertex is the center of a unit disk without violating the intersection property. The mapping Σ is also called the *embedding of G by unit disks*. We use the domain of *axes-parallel straight lines* which is a set of lines in 2D, where the angle between a pair of lines is either 0 or $\pi/2$. This implies that the equation of a straight line is either $y = a$ if it is a horizontal line, or $x = b$ if it is a vertical line, where $a, b \in \mathbb{R}$. The input for axes-parallel straight lines recognition problem contains two sets, $\mathcal{H}, \mathcal{V} \subset \mathbb{R}$, where $\mathcal{H}$ contains the Euclidean distance of each horizontal line from the x-axis, and $\mathcal{V}$ contains the Euclidean distance of each vertical line from the y-axis. Thereby in the domain that we use, each vertex is mapped either onto a vertical line, or onto a horizontal line. We denote the class of axes-parallel unit disk graphs as $\mathrm{APUD}(k, m)$ where k is the number of horizontal lines, and m is the number of vertical lines. Formally, we define the problem as follows.

Definition 1 (Axes-parallel unit disk graph recognition on k horizontal and m vertical lines). *The input is a graph $G = (V, E)$, where $V = \{1, 2, \ldots, n\}$, and two sets $\mathcal{H}, \mathcal{V} \subset \mathbb{Q}$ of rational numbers with $|\mathcal{H}| = k$ and $|\mathcal{V}| = m$. The task is to determine whether there exists a mapping $\Sigma : V \to (\mathbb{R} \times \mathcal{H}) \cup (\mathcal{V} \times \mathbb{R})$ such that there is a unit disk realization of G in which $u \in \ell_{\Sigma(u)}$ for each $u \in V$.*

3 APUD(k, m) Recognition Is NP-Hard

We prove that axes-parallel unit disk recognition ($\mathrm{APUD}(k, m)$ recognition with k and m given as input) is NP-hard by giving a reduction from the *Monotone not-all-equal 3-satisfiability* (NAE3SAT) problem[1]. NAE3SAT is a variation of 3SAT where three values in each clause are not all equal to each other, and due to Schaefer's dichotomy theory, the problem remains NP-complete when all clauses are monotone (i.e., none of the literals are negated) [20]. Our main theorem is as follows.

Theorem 1. *There is a polynomial-time reduction of any instance Φ of Monotone NAE3SAT to some instance Ψ of $\mathrm{APUD}(k, m)$ such that Φ is a YES-instance if, and only if Ψ is a YES-instance.*

---

[1] This problem is equivalent to the 2-coloring of 3-uniform hypergraphs. We choose to give the reduction from Monotone NAE3SAT as it is more intuitive to construct for our problem.

We construct our hardness proof using the scheme called a *logic engine*, which is used to prove the hardness of several geometric problems [5]. For a given instance Φ of Monotone NAE3SAT, there are two main components in our reduction. First, we construct a backbone for our gadget. The backbone models only the number of clauses and the number of literals. Next, we model the relationship between the clauses and literals, i.e., which literal appears in which clause.

Let us begin by describing the input graph. For the sake of simplicity, we assume that the given formula has 3 clauses, A, B, C, and 4 literals, q, r, s, t for the moment. In general, we denote the clauses by $C_1, \ldots, C_k$, and the literals by $x_1, \ldots, x_m$. Later on, we explain how to generalize the input graph according to any given instance of Monotone NAE3SAT formula. For the following part, we describe the input graph given in Fig. 1a. Throughout the manuscript, we index the vertices from left to right, and from bottom to top, in ascending order.

Three essential components of the input graph are the following induced paths $P_\alpha = (\alpha_1, \alpha_2, \ldots, \alpha_{11})$, $P_L = (L_1, L_2, \ldots, L_{15})$, and $P_R = (R_1, R_2, \ldots, R_{15})$. The length of P_α is $2m + 3$ for m literals. In our case, $(2 \times 4) + 3 = 11$. The lengths of P_L and P_R are the same, equal to $3 + 4k$ for k clauses. In our case, $3 + (4 \times 3) = 15$.

The middle vertices of P_L and P_R are the end vertices of P_α. That is, $\alpha_1 = L_8$, and $\alpha_{11} = R_8$. The paths P_L and P_R define the left and the right boundary for our gadget, respectively.

For $i = q, r, s, t$, there is an induced path $P_i = (i_1, \ldots, i_{15})$ for each literal, with 15 vertices. In general, we denote those paths by $P^1, P^2, \ldots, P^m$ for m literals. The vertices of these paths are denoted by blue circles in Fig. 1a, they are mutually disjoint, but each of them shares one vertex with P_α. The shared vertices are precisely the middle vertices, which are indicated by green rectangles in the figure. That is, $\alpha_3 = q_8$, $\alpha_5 = r_8$, $\alpha_7 = s_8$, and $\alpha_9 = t_8$. Moreover, i_1 is a vertex of an induced 4-cycle, and i_{15} is a vertex of another induced 4-cycle for $i = q, r, s, t$. The three vertices in a 4-cycle, except the one in one of the induced paths, are indicated by the red color in the figure. Precisely two of them, that are adjacent to a blue vertex (either i_1 or i_{15}) are indicated by squares, and the remaining is indicated by a triangle.

Starting from the second edge of P_L (respectively P_R), every second edge is a chord of a 4-cycle (C_4). Throughout the paper, we refer to such 4-cycles with a chord as a *diamond*. Two vertices of these diamonds are of P_L (respectively P_R), and remaining two are denoted by red triangles in Fig. 1a.

Remember that the problem takes two inputs: a graph, and a set of lines determined by their equations (or rather by two sets of rational numbers, since every line is parallel to either the x- or y- axis). For a Monotone NAE3SAT formula with 3 clauses and 4 literals, we have described the input graph above. Now, let us discuss the input lines of our gadget. The input graph is given in Fig. 1a, and the corresponding lines are given in Fig. 1b. We claim that the given graph can be embedded onto the given lines with ε flexibility, and the resulting realization looks like the set of unit disks given in Fig. 1c.

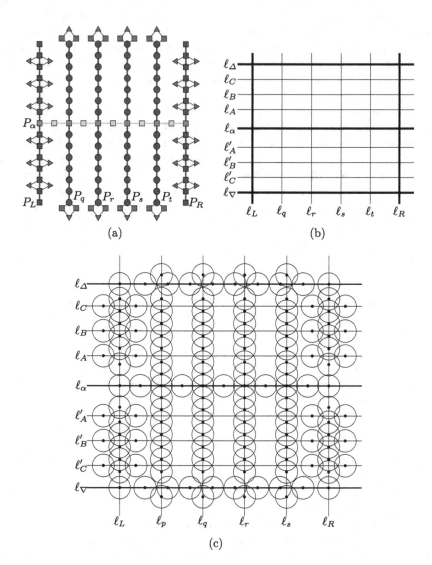

(a)

(b)

(c)

Fig. 1. (a) Skeleton of the input graph for Φ. The consecutive induced paths, labeled as P_q, P_r, P_s, P_t, are to be embedded on the literal lines $\ell_q, \ell_r, \ell_s, \ell_t$ in Fig. 1b, respectively. The vertices in the long induced paths P_L and P_R in 1a (indicated by rectangles) must be embedded on the lines ℓ_L and ℓ_R given in 1b. Similarly, the vertices in P_α (indicated by blue and green rectangles) must be embedded on the line ℓ_α given in 1b.
(b) The line set of the configuration for a Monotone NAE3SAT formula Φ with 4 literals (q, r, s, t) and 3 clauses (A, B, C).
(c) Realization of the graph given in 1a onto the lines given in 1b. (Color figure online)

In order to force such an embedding, we adjust the Euclidean distance between each pair of parallel lines carefully. We start by defining the horizontal line ℓ_α. This line is the axis of horizontal symmetry for our line configuration. Thus, it is safe to assume that ℓ_α is the x-axis. On the positive side of the y-axis, for each clause A, B, and C, there is a straight line parallel to ℓ_α, and another horizontal line acting as the top boundary of the configuration. These lines are denoted by ℓ_A, ℓ_B, ℓ_C, and ℓ_Δ, and their equations are $y = a$, $y = b$, $y = c$, and $y = \Delta$, respectively, where $a < b < c < \Delta$. For every pair of consecutive horizontal lines, the Euclidean distance between them is precisely 2.01 units. That is, $a = 2.01$, $b = 4.02$ and $c = 6.03$, and $\Delta = 8.04$. For every horizontal line described above, there is another horizontal line symmetric to it about the x-axis. These lines are $\ell'_A, \ell'_B, \ell'_C$, and ℓ_∇ (see Fig. 1b).

The leftmost vertical line is ℓ_L, which is the left boundary of our configuration. We can safely assume that ℓ_L is the y-axis for the sake of simplicity. For each literal q, r, s and t, there exists a vertical line parallel to ℓ_L, and another vertical line that defines the right boundary of our configuration. These lines are denoted by $\ell_q, \ell_r, \ell_s, \ell_t$, and ℓ_R, and their equations are $x = q$, $x = r$, $x = s$, $x = t$ and $x = R$, respectively, where $q < r < s < t < R$. The Euclidean distance between each pair of consecutive vertical lines is precisely 3.8 units. That is $q = 3.8$, $r = 7.6$, $s = 11.4$, $t = 15.2$, and $R = 19$.

Up to this point, we have described the input graph, and the input lines for a given Monotone NAE3SAT formula with 3 clauses and 4 literals. In general, for a given Monotone NAE3SAT formula Φ with k clauses $C_1, C_2, \ldots, C_k$, and m literals $x_1, x_2, \ldots, x_m$, our gadget has the following components.

1. An induced path $P_\alpha = (\alpha_1, \alpha_2, \ldots, \alpha_{2m+3})$ with $2m + 3$ vertices.
2. m induced paths $P^1 = (P_1^1, P_2^1, \ldots, P_{4k+3}^1), \ldots, P^m = (P_1^m, P_2^m, \ldots, P_{4k+3}^m)$, each with $4k+3$ vertices, where $\alpha_3 = P_{2k+2}^1, \alpha_5 = P_{2k+2}^2, \ldots, \alpha_{2k+1} = P_{2k+2}^m$, and induced 4-cycles containing the first and the last vertices of each of these paths.
3. Two induced paths $P_L = (L_1, \ldots, L_{4k+3})$ and $P_R = (R_1, \ldots, R_{4k+3})$, each with $4k+3$ vertices, where the edges $L_2L_3, L_4L_5, \ldots, L_{2k}L_{2k+1}, L_{2k+3}L_{2k+4}, \ldots, L_{4k+1}L_{4k+2}$, and $R_2R_3, R_4R_5, \ldots, R_{2k}R_{2k+1}, R_{2k+3}R_{2k+4}, \ldots, R_{4k+1}R_{4k+2}$ are chords of disjoint 4-cycles.
4. $2k + 3$ horizontal lines $\ell_\nabla, \ell_k'^C, \ell_{k-1}'^C, \ldots, \ell_\alpha, \ell_1^C, \ell_2^C, \ldots, \ell_k^C, \ell_\Delta$, with equations $\ell_\nabla : y = -2.01(k+1)$, $\ell_\Delta : -\ell_\nabla$, $\ell_\alpha : y = 0$, $\ell_i'^C = -2.01i$, and $\ell_i^C = 2.01i$ for $i = 1, 2, \ldots, k$.
5. $m + 2$ vertical lines $\ell_L, \ell_1^x, \ell_2^x, \ldots, \ell_m^x, \ell_R$, with equations $\ell_L : x = 0$, $\ell_R : x = 3.8(m+1)$, and $\ell_i^x : 3.8i$ for $i = 1, 2, \ldots, m$.

In total, for the given formula Φ with k clauses and m literals, our gadget is an instance of APUD($2k + 3, m + 2$). Here, we conclude the proof of Theorem 1.

Now, let us show that the given graph has a unique embedding onto the given lines, up to ε flexibility.

Claim. The vertices indicated by rectangles in Fig. 1a can only be embedded on the bold lines in Fig. 1b.

Let us start by discussing the embedding of P_L onto ℓ_L (and respectively P_R onto ℓ_R). We give the following two trivial lemmas as preliminaries for the proof of our claim.

Lemma 1 (*). *Consider two disks A and B, centered on $(a,0)$ and $(b,0)$ with $0 < |a| < |b|$. Another disk, C that is centered on $(0,c)$ cannot intersect B without intersecting A.*

Lemma 2 (*). *An induced 4-star $(K_{1,4})$ can be realized as a unit disk graph on two perpendicular lines, but not on two parallel lines.*

Now, with the help of Lemmas 1 and 2, we state the following lemmas, and prove our claim.

Lemma 3. *The induced paths P_L, P_R and P_α in the input graph (Fig. 1a) can only be embedded onto ℓ_L, ℓ_R, and ℓ_α, respectively (Fig. 1b).*

Proof (Sketch). The diamonds on the left and the right side of the figure should be embedded around an intersection. There are a total of six diamonds, and thus six intersections are required.

P_α has induced 4-stars, and those 4-stars are vertices of long induced paths, P_α cannot be embedded on multiple lines (via bending etc.). The middle vertices of P_L and P_R are the two ends of P_α. Since P_α is realized on a single line, another intersection is required to realize $K_{1,3}$ consists of $L_8; L_7, L_9, \alpha_2$.

In total, P_L requires seven intersections. Those seven intersections are between a vertical line and seven horizontal lines, excluding ℓ_Δ and ℓ_∇. The same argument applies to P_R up to symmetry. □

Claim. For the given input graph for 3 clauses and 4 literals, the following hold:

i) The induced paths $P_q = (q_1, \ldots, q_{15})$, $P_r = (r_1, \ldots, r_{15})$, $P_s = (s_1, \ldots, s_{15})$ and $P_t = (t_1, \ldots, t_{15})$ in the input graph given in Fig. 1a can only be embedded onto ℓ_q, ℓ_r, ℓ_s, and ℓ_t, respectively.

ii) The center of each disk that correspond to a vertex of those induced paths must be between ℓ_Δ and ℓ_∇.

iii) A pair of non-intersecting disks that are included in an induced 4-cycle, but not included in any of P_q, P_r, P_s, P_t, must lie on either ℓ_Δ or ℓ_∇ (red rectangles in Fig. 1a).

Proof (Sketch). Due to Lemma 3, we know that P_α is realized on ℓ_α, thus (i) holds. 4-cycles require at least two lines, and those two lines cannot be two parallel lines, as the Euclidean distance between each pair of consecutive parallel lines is larger than 2. Thus, (ii) holds. The induced paths P_q, P_r, P_s, P_t can be squeezed enough to be realized between ℓ_α and ℓ_C because $\ell_C : y = 6.03$, but then the 4-cycles cannot be realized. Therefore, the disks that correspond to two vertices of these 4-cycles must be centered on ℓ_Δ (and symmetrically on ℓ_∇). Thus, (iii) holds. □

Lemma 4 (*). *For the given input graph for k clauses and m literals, the following hold:*

i) *The induced paths $P_1 = (P_1^1, \ldots, P_{4k+3}^1)$, $P^2 = (P_1^2, \ldots, P_{4k+3}^2)$, $\ldots$, $P^n = (P_1^n, \ldots, P_{4k+3}^n)$ in the input graph can only be embedded onto ℓ_1^x, ℓ_2^x, $\ldots$ ℓ_m^x, respectively.*

ii) *The center of each disk that correspond to a vertex of those induced paths must be between ℓ_Δ and ℓ_∇.*

iii) *A pair of non-intersecting disks that are included in an induced 4-cycle, but not included in any of $P^1, P^2, \ldots, P^n$, must lie on either ℓ_Δ or ℓ_∇ (red rectangles in Fig. 1a).*

With Lemmas 3 and 4, we have shown that the vertices denoted by rectangles in Fig. 1a must be embedded onto the bold lines in Fig. 1b.

Using the backbone we have described, we now show how to model the relationship between the clauses and the literals. To make it easier to follow, we also refer to Fig. 1a in parentheses in the following description. Consider a sub-path $(P_{2k+3}^i, P_{2k+4}^i, \ldots, P_{4k+3}^i)$ of the induced path P^i. This part corresponds to the literal x_i of the given Monotone NAE3SAT formula (corresponding to $(q_9, \ldots, q_{15})$ of P_q in our example). The edges $P_{2k+3}^i P_{2k+4}^i$, $P_{2k+5}^i P_{2k+6}^i$, $\ldots$, $P_{4k+1}^i P_{4k+2}^i$ (corresponding to $q_9 q_{10}$, $q_{11} q_{12}$, and $q_{13} q_{14}$ in P_q in our example) are used to model membership of x_i in the clauses $C_1, C_2, \ldots, C_k$ (corresponding to the clauses A, B, and C in our example), respectively.

If x_i appears in a clause C_j, then we do nothing for the edges correspond do those clauses. Otherwise, if x_i does not appear in C_j, then we introduce a *flag vertex* in the graph, which is adjacent to $P_{2(k+j)+1}^i$ and $P_{2(k+j)+2}^i$. Due to the rigidity of the backbone (up to ε flexibility), this flag vertex lies on ℓ_j^C. Similarly, in our example, if q appears in B, then $q_{11} q_{12}$ stays as is, but otherwise, a flag vertex is introduced, adjacent to both q_{11} and q_{12}.

Every clause has 3 literals. Thus, on each horizontal line, 3 out of m possible flag vertices will be missing. That sums up to a total of $k(m-3)$ flag vertices for this part of the graph. For the remaining sub-path $(P_1^i, \ldots, P_{2k+2}^i)$ of P^i (corresponding to $(q_1, \ldots, q_8)$ of P_q in our example), we introduce the flag vertices for the pairs (P_2^i, P_3^i), (P_4^i, P_5^i), $\ldots$, P_{2k}^i, P_{2k+1}^i (corresponding to (q_6, q_7) (q_2, q_3), (q_4, q_5), (q_6, q_7) in our example). That is a total number of km flag vertices for this part of the graph. In the whole graph, there are precisely $2\,km$–$3\,km$ flag vertices.

Realize that the embeddings on some vertical lines must be flipped upside-down to create space for the flag vertices. This operation corresponds to the truth assignment of the literal that corresponds to that vertical line. The configuration forces at least one literal to have a different truth assignment, because for a pair of symmetrical horizontal lines, say ℓ_A and ℓ_A', there must be at least one missing flag, and at most two missing flags for the disks to fit between ℓ_L and ℓ_R.

The input graph, a YES-instance, and the realization of the YES-instance of the Monotone NAE3SAT formula $\Phi = (q \vee s \vee t) \wedge (q \vee r \vee t) \wedge (q \vee r \vee s)$ is given in Fig. 2.

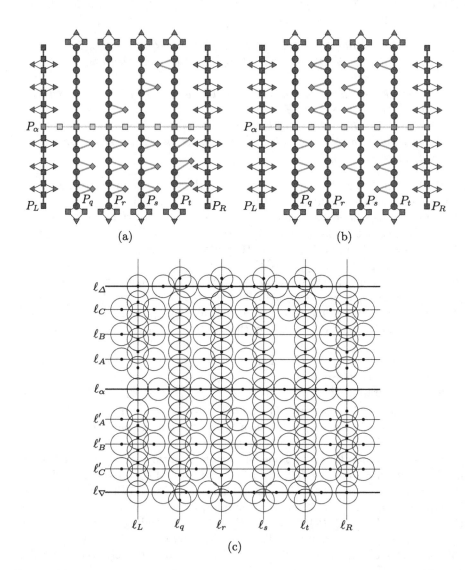

(a)

(b)

(c)

Fig. 2. (a) The input graph for the Monotone NAE3SAT formula Φ with literals q, r, s, t, and clauses $A, B, C.\Phi = A \wedge B \wedge C$ where $A = (q \vee s \vee t)$, $B = (q \vee r \vee t)$, and $C = (q \vee r \vee s)$. The flag vertices, indicated by orange diamonds, are adjacent to the vertices that correspond to a clause on an induced path, if the literal does not appear in that clause.
(b) A truth assignment that satisfies the formula given in 2a: $q =$ TRUE, $r =$ FALSE, $s =$ FALSE, and $t =$ TRUE.
(c) Realization of the graph given in 2b. (Color figure online)

Corollary 1. *Given a graph $G = (V, E)$, deciding whether G is a unit disk graph is an NP-hard problem when the size of the largest induced cycle in G is of length 4.*

4 APUD$(k, 0)$ Recognition Is NP-Complete

In this section, we show that the recognition of axes-parallel unit disk graphs is NP-complete when all the given lines are parallel to each other. This version of the problem is referred to as APUD$(k, 0)$, as there are k horizontal lines given as input, but no vertical lines. We use the reduction given in Sect. 3.

Theorem 2. APUD$(k, 0)$ *recognition is NP-hard.*

Proof. Consider the realization given in Fig. 1c. Notice that the length of the paths $P^1, P^2, \ldots, P^m$ (P_q, P_r, P_s, P_t in our example), and thus the number of disks on vertical lines, is equal. Lemma 3 (ii) implies that those disks must be centered between ℓ_∇ and ℓ_Δ. Thus, for the disks that correspond to the vertices on these paths, we do not need any vertical line. We can simply remove the vertical lines, and add an extra horizontal line for each clause. For the disks that are adjacent to, but not on P_L and P_R, we can simply add another horizontal line. That is an extra horizontal line for each clause. As a result, for a given instance Φ of Monotone NAE3SAT formula with k clauses and m literals, we have an instance Ψ of APUD$(2k + 3, m + 2)$ to prove NP-hardness with vertical lines, and an instance Ψ' of APUD$(4k + 3, 0)$ to prove NP-completeness without vertical lines. For each clause, we have 3 horizontal lines.

In Ψ', only disks that can "jump" from one horizontal line to another are the ones that are on the top line of ℓ_1^C and bottom line of $\ell_1'^C$. And those jumps do not change the overall configuration.   □

To show that APUD$(k, 0)$ recognition is in NP, we need to prove that a given solution can be verified in polynomial time and additionally that any feasible input will have a solution that takes up polynomial space, with respect to the input size. Thus, we show that for any graph $G \in$ APUD$(0, k)$, there exists an embedding where the disk centers are represented using polynomially many decimals with respect to the input size.

Below, we describe a procedure to show the existence of an embedding with polynomially many digits for every feasible output. Let $h_0, h_1, h_2 \ldots, h_k$ denote the rational numbers that correspond to the horizontal lines given as input.

1. Let $G_0, G_1, \ldots, G_k$ denote the disjoint induced subgraphs of G, such that the vertices of G_i correspond to the disks centered on line $y = h_i$.
2. Embed G_0 on x-axis with small perturbations which results in all disks on $y = 0$ having polynomially many decimals.
3. For each $1 \le i \le k$, find an embedding of G_i onto the $y = h_i$ line by only considering neighbors from $\bigcup_{j<i} G_j$.

Therefore, APUD$(k, 0)$ recognition is NP-complete.

5 APUD(1, 1) Recognition Is Open

In this section, we discuss a natural basis for the APUD(k, m) recognition problem, that is, $k = 1$ and $m = 1$. For the sake of simplicity, we can assume that our given two lines are the x-axis and the y-axis. First, we give some forbidden induced subgraphs for APUD(1, 1). Namely, those subgraphs are the 5-cycle (C_5), the 4-sun (S_4), and the 5-star ($K_{1,5}$).

Lemma 5 (*). $C_5, S_4, K_{1,5} \notin$ APUD(1, 1).

Lemma 6. *A given graph G can be embedded on x-axis and y-axis as a unit disk intersection graph, without using negative coordinates for the disk centers if, and only if, G is a unit interval graph.*

Proof. In this proof, let us denote the class of graphs that can be embedded on the x- and y- axes as unit disks, using positive coordinates only by $(xy)^+$. We show that the disks on the y-axis can be rotated by $\pi/2$ degrees counterclockwise, and the intersection relationships can be preserved as given in G.

$\mathbf{G} \in \mathbf{(xy)^+} \Rightarrow \mathbf{G} \in \mathbf{UIG}$: Consider two disks, A and B, whose centers are $(a, 0)$ and $(0, b)$, respectively, where a and b are both positive numbers. If A and B do not intersect, then $\sqrt{a^2 + b^2} > 2$. After the rotation, the center of B will be on $(-b, 0)$. The new distance between the centers is $a + b$. Since $(a + b)^2 > a^2 + b^2 > 4$, the inequality $a + b > 2$ holds.

If A and B intersect, then $\sqrt{a^2 + b^2} \leq 2$. After the rotation, it might still be the case that $a + b > 2$. However, now we can safely move the center of B and the other centers that have negative coordinates closer to the center of A, recovering the intersection. Note that if a disk C is centered in between the centers of A and B after the rotation, then both A and B must intersect C by Lemma 1.

$\mathbf{G} \in \mathbf{UIG} \Rightarrow \mathbf{G} \in \mathbf{(xy)^+}$: Since G is a unit interval graph, we can assume that every interval is a unit disk, and the graph is embedded on x-axis. Consider two disks, A and B, whose centers are $(-a, 0)$ and $(b, 0)$, respectively, where a and b are both positive numbers. If A and B are intersecting, then $a + b \leq 2$ Then, after the rotation, since $a + b \leq 2$ holds, then $\sqrt{a^2 + b^2} \leq 2$ also holds.

If A and B are not intersecting, then $a + b > 2$. After the rotation $\sqrt{a^2 + b^2} \leq 2$ might hold, creating an intersection between A and B. However, we can simply shift the center of A (along with the other centers that are on y-axis) far away from the center of B, separating A and B.   □

Lemma 6 shows APUD$^+$(1, 1) = APUD$^+$(1, 0) = UIG if we use only non-negative coordinates. This also applies if we use only non-positive coordinates. Thus, a given APUD(1, 1) can always be partitioned into two unit interval graphs. Considering the embedding, one of these two partitions contains the disks that are centered on the positive sides of x- and y-axes, and the other partition contains the disks centered on the negative sides of x- and y-axes.

Lemma 7. *A graph $G \in \text{APUD}(1,1)$ can be vertex-partitioned into four parts, such that any two form a unit interval graph.*

Proof. Let $\Sigma(G)$ be an embedding of G onto x- and y- axes as unit disks. Denote the set of unit disks in $\Sigma(G)$ that are centered on the positive side of the x-axis, and the positive side of the y-axis by $\Sigma^+(G)$. Similarly, by $\Sigma^-(G)$ denote the set of unit disks that are centered on the negative side of the x-axis, and the negative side of the y-axis. By Lemma 6, both $\Sigma^+(G)$ and $\Sigma^-(G)$ yield separate interval graphs. The vertices corresponding to the disks in $\Sigma^+(G)$ yield a unit interval graph, as do the vertices corresponding to the disks in $\Sigma^-(G)$. If there exists a disk centered at $(0,0)$, then it can be included in one of the partitions arbitrarily. Hence, we can vertex-partition G into two unit interval graphs. $\square$

Up to this point, we showed that if a unit disk graph can be embedded onto two orthogonal lines, then it can be partitioned into two interval graphs. However, this implication obviously does not hold the other way around. Thus, we now identify some structural properties of $\text{APUD}(1,1)$.

Remark 1. Consider four unit disks A, B, C, D, embedded onto x-axis and y-axis. If they induce a 4-cycle, then the centers of those disks will be at $(a, 0)$, $(0, b)$, $(-c, 0)$, $(0, -d)$, respectively, where a, b, c, d are non-negative numbers.

For the upcoming lemma, we will utilize Remark 1. The lemma is an important step towards describing a characterization of $\text{APUD}(1,1)$.

Lemma 8 (*). *Consider eight unit disks embedded onto x-axis and y-axis, around the origin, whose intersection graph contains two induced 4-cycles. Then, this intersection contains at least four 4-cycles, each with a chord, not necessarily as induced subgraphs. Moreover, those 4-cycles are formed by pairs of disks on the same direction ($+x$, $+y$, $-x$, $-y$) with respect to the origin.*

The lemmas imply that for a connected graph $G \in \text{APUD}(1,1)$, we can deduce:

  (i) G does not contain either of the 4-sun (S_4) or the 5-star($K_{1,5}$) as an induced subgraph, and the largest induced cycle in G is of length 4 (by Lemma 5).
 (ii) G can be vertex partitioned into into four parts, such that any two form a unit interval graph (by Lemma 7).
(iii) Given two 4-cycles, (a, b, c, d) and (u, v, w, x) in G, each one of the quadruplets $\{a, b, u, v\}$, $\{b, c, v, w\}$, $\{c, d, w, x\}$, and $\{d, a, x, u\}$ is either a diamond or a K_4 (by Lemma 8).

Although this characterization gives a rough idea regarding the structure of a graph $G \in \text{APUD}(1,1)$, it is not clear if the recognition can be done in polynomial time. Our characterization is a necessary step through recognition, but it is not yet known whether it is sufficient. Hence, we conjecture that given a graph G, it can be determined whether $G \in \text{APUD}(1,1)$ in polynomial time.

Acknowledgments. The author wants to thank Petr Hliněný for his insight on the hardness proof. In addition, he thanks Deniz Ağaoğlu and Michał Dębski for their extensive comments and generous help during the preparation of this manuscript.

References

1. Alber, J., Fiala, J.: Geometric separation and exact solutions for the parameterized independent set problem on disk graphs. J. Algorithms **52**(2), 134–151 (2004)
2. Alomari, A., Aslam, N., Phillips, W., Comeau, F.: Three-dimensional path planning model for mobile anchor-assisted localization in Wireless Sensor Networks. In: 30th IEEE Canadian Conference on Electrical and Computer Engineering, CCECE, pp. 1–5 (2017)
3. Aspnes, J., et al.: A theory of network localization. IEEE Trans. Mob. Comput. **5**(12), 1663–1678 (2006)
4. Balasundaram, B., Butenko, S.: Optimization problems in unit-disk graphs. In: Floudas, C., Pardalos, P. (eds.) Encyclopedia of Optimization, pp. 2832–2844. Springer, Boston (2009). https://doi.org/10.1007/978-0-387-74759-0
5. Bhatt, S.N., Cosmadakis, S.S.: The complexity of minimizing wire lengths in VLSI layouts. Inf. Process. Lett. **25**(4), 263–267 (1987)
6. Bonnet, É., Giannopoulos, P., Kim, E.J., Rzążewski, P., Sikora, F.: QPTAS and subexponential algorithm for maximum clique on disk graphs. In: Speckmann, B., Tóth, C.D. (eds.) 34th International Symposium on Computational Geometry, SoCG. LIPIcs, vol. 99, pp. 12:1–12:15. Schloss Dagstuhl - Leibniz-Zentrum für Informatik (2018)
7. Booth, K.S., Lueker, G.S.: Testing for the consecutive ones property, interval graphs, and graph planarity using PQ-tree algorithms. J. Comput. Syst. Sci. **13**(3), 335–379 (1976)
8. Breu, H.: Algorithmic aspects of constrained unit disk graphs. Ph.D. thesis, University of British Columbia (1996)
9. Breu, H., Kirkpatrick, D.G.: Unit disk graph recognition is NP-hard. Comput. Geom. **9**(1–2), 3–24 (1998)
10. Çağırıcı, O.: Exploiting coplanar clusters to enhance 3D localization in wireless sensor networks. Master's thesis, Izmir University of Economics (2015). http://arxiv.org/abs/1502.07790
11. Clark, B.N., Colbourn, C.J., Johnson, D.S.: Unit disk graphs. Discrete Math. **86**(1–3), 165–177 (1990)
12. Dil, B., Dulman, S., Havinga, P.: Range-based localization in mobile sensor networks. In: Römer, K., Karl, H., Mattern, F. (eds.) EWSN 2006. LNCS, vol. 3868, pp. 164–179. Springer, Heidelberg (2006). https://doi.org/10.1007/11669463_14
13. Evans, W., van Garderen, M., Löffler, M., Polishchuk, V.: Recognizing a DOG is hard, but not when it is thin and unit. In: Demaine, E.D., Grandoni, F. (eds.) 8th International Conference on Fun with Algorithms, FUN. LIPIcs, vol. 49, pp. 16:1–16:12. Schloss Dagstuhl - Leibniz-Zentrum für Informatik (2016)
14. Fishburn, P.C.: Interval Orders and Interval Graphs – A Study on Partially Ordered Sets. Wiley, Hoboken (1985)
15. Ito, H., Kadoshita, M.: Tractability and intractability of problems on unit disk graphs parameterized by domain area. In: Zhang, X.S., Liu, D.G., Wu, L.Y., Wang, Y. (eds.) Operations Research and Its Applications, 9th International Symposium, ISORA. Lecture Notes in Operations Research, vol. 12, pp. 120–127 (2010)
16. Kang, R.J., Müller, T.: Sphere and dot product representations of graphs. Discrete Comput. Geom. **47**(3), 548–568 (2012). https://doi.org/10.1007/s00454-012-9394-8
17. Kuhn, F., Moscibroda, T., Wattenhofer, R.: Unit disk graph approximation. In: Basagni, S., Phillips, C.A. (eds.) Proceedings of the DIALM-POMC Joint Workshop on Foundations of Mobile Computing, pp. 17–23. ACM (2004)

18. McDiarmid, C., Müller, T.: Integer realizations of disk and segment graphs. J. Comb. Theory Ser. B **103**(1), 114–143 (2013)
19. Neto, M.F., Goussevskaia, O., dos Santos, V.F.: Connectivity with backbone structures in obstructed wireless networks. Comput. Netw. **127**, 266–281 (2017)
20. Schaefer, T.J.: The complexity of satisfiability problems. In: Lipton, R.J., Burkhard, W.A., Savitch, W.J., Friedman, E.P., Aho, A.V. (eds.) Proceedings of the 10th Annual ACM Symposium on Theory of Computing, STOC, pp. 216–226. ACM (1978)

On the Decision Tree Complexity
of Threshold Functions

Anastasiya Chistopolskaya[1] and Vladimir V. Podolskii[1,2]([⊠]) [iD]

[1] National Research University Higher School of Economics, Moscow, Russia
achistopolskaya@hse.ru
[2] Steklov Mathematical Institute, Russian Academy of Sciences, Moscow, Russia
podolskii@mi-ras.ru

Abstract. In this paper we study decision tree models with various types of queries. For a given function it is usually not hard to determine the complexity in the standard decision tree model (each query evaluates a variable). However in more general settings showing tight lower bounds is substantially harder. Threshold functions often have non-trivial complexity in such models and can be used to provide interesting examples.

Standard decision trees can be viewed as a computational model in which each query depends on only one input bit. In the first part of the paper we consider natural generalization of standard decision tree model: we address decision trees that are allowed to query any function depending on two input bits. We show the first lower bound of the form $n - o(n)$ for an explicit function (namely, the majority function) in this model. We also show that in the decision tree model with AND and OR queries of arbitrary fan-in the complexity of the majority function is $n - 1$.

In the second part of the paper we address parity decision trees that are allowed to query arbitrary parities of input bits. There are various lower bound techniques for parity decision trees complexity including analytical techniques (degree over $\mathbb{F}_2$, Fourier sparsity, granularity) and combinatorial techniques (generalizations of block sensitivity and certificate complexity). These techniques give tight lower bounds for many natural functions. We give a new inductive argument tailored specifically for threshold functions. A combination of this argument with granularity lower bound allows us to provide a simple example of a function for which all previously known lower bounds are not tight.

Keywords: Decision tree · Parity decision tree · Granularity · Threshold function · Lower bound

1 Introduction

Decision trees are a computational model in which we compute a known Boolean function $f\colon \{0,1\}^n \to \{0,1\}$ on an unknown input $x \in \{0,1\}^n$ and in one step we

The paper is partially supported by RFBR Grant 18-01-00822 and by HSE University Basic Research Program.

can query $q(x)$ for $q\colon \{0,1\}^n \to \{0,1\}$ from a fixed set of queries. In the standard and the most studied decision tree model we can query only individual variables of the input x [2,8] (the complexity of f is denoted by $\mathsf{D}(f)$). The studies of this model among other things are related to the well-known sensitivity conjecture [6] that was recently resolved [7]. Among other cases studied in the literature are parity decision trees that can query any parity of input bits [14], linear decision trees in which the queries are linear threshold functions [4,8] and decision trees with AND and OR queries [1].

In this paper we will mainly deal with threshold functions. The threshold function THR_n^k on n bits outputs 1 iff there are at least k ones in the input. The majority function MAJ_n is simply $\mathrm{THR}_n^{\lceil n/2 \rceil}$.

The main goal of the first part of the paper is to study a natural generalization of standard decision tree model: we address decision trees that are allowed to query any function depending on two input bits. We denote the complexity in this model by $\mathsf{D}_{\mathrm{B}_2}(f)$. More generally, we can consider decision trees that can query arbitrary functions depending on at most r inputs, where r is a parameter. The standard decision tree model corresponds to the case $r = 1$.

This model can be viewed as a uniform version of multi-party Communication Complexity (see the book by Kushilevitz and Nisan [10] for details on Communication Complexity). In this model k players are trying to compute the function $f\colon \{0,1\}^{nk} \to \{0,1\}$ and the input is shared by the players. Each player is associated with a piece of input of size n. In the Number in Hand model (NIH) players see only the input bits associated with them. In the Number on the Forehead model (NOF) players see all input bits except those that are associated with them (thus the inputs visible to players have large overlaps). Decision tree model with $r = n$ for the computation of f can be viewed as a version of communication model where for each n bits there is a player seeing exactly these n bits. The decision tree model with $r = (k-1)n$ can be viewed as a generalization of NOF communication model.

A special case of this model was considered by Posobin [13], where the computation of MAJ_n with MAJ_k-queries was studied for $k < n$. There are results on a related model with non-Boolean counting queries [3,9]. Related settings with non-Boolean domain also arise in algebraic decision tree models (see, e. g., [8, Section 14.8]).

We initiate (to the best of our knowledge) the study of strong lower bounds for decision trees with queries of bounded fan-in considering the case of queries of fan-in 2. It is easy to see that the complexity $\mathsf{D}_{\mathrm{B}_2}(f)$ in this model is lower bounded by $\mathsf{D}(f)/2$ (each binary query can be simulated by two unary queries). However, in the view of generalization to larger r it is interesting to obtain lower bounds greater than $n/2$. We show that

$$\mathsf{D}_{\mathrm{B}_2}(\mathrm{MAJ}_n) \geq n - o(n).$$

We also show that if we allow additionally to query parities of three bits, the complexity of majority (as well as any symmetric function) drops to at most $2n/3$. Thus to obtain strong lower bounds for $r = 3$ more complicated functions need to be considered.

Also in this part of the paper we address the complexity of majority function MAJ in decision tree model with AND and OR queries (of arbitrary fan-in). We denote the complexity of a function f in this model by $D_{\wedge,\vee}(f)$. The complexity of threshold functions in this model was studied by Ben-Asher and Newman [1] with the relation to a certain PRAM model. It was shown there that THR_n^k functions have complexity $\Theta(k/\log(n/k))$. In this paper we are interested in the precise complexity of functions in this model. We show that $D_{\wedge,\vee}(\text{MAJ}) = n-1$.

In the second part of the paper we turn to parity decision tree model $D_\oplus(f)$. Apart from being natural and interesting on its own parity decision tree model was studied mainly in connection with Communication Complexity and more specifically, with Log-rank Conjecture. It is known that the two-party deterministic communication complexity $D^{cc}(F)$ of the function $F\colon \{0,1\}^n \times \{0,1\}^n \to \{-1,1\}$ is lower bounded by $\log \text{rank}(M_F)$, where M_F is a communication matrix of F [10]. It is a long standing conjecture and one of the key open problems in Communication Complexity, called Log-rank Conjecture [11], to prove that $D^{cc}(F)$ is upper bounded by a polynomial of $\log \text{rank}(M_F)$.

An important special case of Log-rank Conjecture addresses the case of XOR-functions $F(x,y) = f(x \oplus y)$ for some f, where $x \oplus y$ is a bit-wise XOR of Boolean vectors x and y. On one hand, this class of functions is wide and captures many important functions (including equality and Hamming distance), and on the other hand the structure of XOR-functions allows to use analytic tools. For such functions $\text{rank}(M_F)$ is equal to the Fourier sparsity $\text{spar} f$, the number of non-zero Fourier coefficients of f. Thus, the Log-rank Conjecture for XOR-functions can be restated: is it true that $D^{cc}(F)$ is bounded by a polynomial of $\log \text{spar} f$?

In a recent paper [5] it was shown that actually $D^{cc}(F)$ and $D_\oplus(f)$ are polynomially related. This reduces the Log-rank Conjecture for XOR-functions to studies of parity decision trees.

Known techniques for lower bounds for parity decision trees fall into one of the two categories: of analytical and combinatorial flavor. Analytical techniques include lower bounds on $D_\oplus(f)$ through sparsity $\text{spar}(f)$, granularity $\text{gran}(f)$ and degree $\deg_2(f)$ over $\mathbb{F}_2$. The strongest lower bound among these is $D_\oplus(f) \geq \text{gran}(f) + 1$.

Regarding combinatorial techniques, for standard decision trees there are several combinatorial measures known that lower bound decision tree complexity. Among them the most common are certificate complexity and block sensitivity. Zhang and Shi [14] generalized these measures to the setting of parity decision tree complexity.

Parity decision tree complexity versions of combinatorial measures are actually known to be polynomially related to parity decision tree complexity [14]. For analytical techniques it is known that existence of polynomial relation between $D_\oplus(f)$ and $\text{gran}(f)$ (or $\text{spar}(f)$) is equivalent to Log-rank Conjecture for XOR-functions [5].

In view of this it is interesting to further study lower bounds for parity decision trees.

In this paper we prove a new lower bound for parity decision tree complexity of threshold functions. We show that

$$\mathsf{D}_\oplus(\mathrm{THR}_{n+2}^{k+1}) \geq \mathsf{D}_\oplus(\mathrm{THR}_n^k) + 1$$

for any k, n.

The combination of this result with granularity lower bound allows to show that for $n = 8k + 2$, $k > 0$ we have $\mathsf{D}_\oplus(\mathrm{THR}_n^3) = n - 1$, whereas all previous techniques give at most $n - 2$ lower bound. Thus, we give an example of a function, for which all known general techniques are not tight.

The rest of the paper is organized as follows. In Sect. 2 we provide necessary definitions, preliminary information and review lower bounds for parity decision trees. In Sect. 3 we study decision trees with binary queries as well as decision trees with AND and OR queries. In Sect. 4 we study parity decision tree complexity of threshold functions. Due to the space constraints many of the technical proofs are omitted.

2 Preliminaries

In many parts of the paper we assume that Boolean functions are functions of the form $f \colon \{0,1\}^n \to \{-1,1\}$, for $n \in \mathbb{N}$. That is, input bits are treated as 0 and 1 and to them we will usually apply operations over $\mathbb{F}_2$. Output bits are treated as -1 and 1 and the arithmetic will be over $\mathbb{R}$. The value -1 corresponds to 'true' and 1 corresponds to 'false'. In other parts of the paper it is more convenient to consider Boolean functions in the form $f \colon \{-1,1\}^n \to \{-1,1\}$ with the same semantics of -1 and 1.

We denote the variables of functions by $x = (x_1, \ldots, x_n)$. We use the notation $[n] = \{1, \ldots, n\}$.

We briefly review the notation and needed facts from Boolean Fourier analysis. For extensive introduction see [12].

For functions $f, g \colon \{0,1\}^n \to \mathbb{R}$ consider an inner product

$$\langle f, g \rangle = \mathop{\mathbf{E}}_{x} f(x)g(x),$$

where the expectation is taken over uniform distribution of x on $\{0,1\}^n$.

For a subset $S \subseteq [n]$ we denote by $\chi_S(x) = \prod_{i \in S}(-1)^{x_i}$ the Fourier character corresponding to S. We denote by $\widehat{f}(S) = \langle f, \chi_S \rangle$ the corresponding Fourier coefficient of f.

It is well-known that for any $x \in \{0,1\}^n$ we have $f(x) = \sum_{S \subseteq [n]} \widehat{f}(S)\chi_S(x)$.

Consider a binary fraction α, that is α is a rational number that can be written in a form that its denominator is a power of 2. By the *granularity* $\mathsf{gran}(\alpha)$ of α we denote the minimal integer $k \geq 0$ such that $\alpha \cdot 2^k$ is an integer.

Note that for Boolean f the Fourier coefficients of f are binary fractions. By the *granularity* of $f \colon \{0,1\}^n \to \mathbb{Z}$ we call the following value

$$\mathsf{gran}(f) = \max_{S \subseteq [n]} \mathsf{gran}(\widehat{f}(S)).$$

It is easy to see that for any $f\colon \{0,1\}^n \to \{-1,1\}$ it is true that $0 \leq \mathsf{gran}(f) \leq n-1$ and both of these bounds are achievable (for example, for $f(x) = \bigoplus_i x_i$ and $f(x) = \bigwedge_i x_i$ respectively).

2.1 Decision Trees

A decision tree T is a rooted directed binary tree. Each of its leaves is labeled by -1 or 1, each internal vertex v is labeled by some function $q_v\colon \{0,1\}^n \to \{-1,1\}$. Each internal node has two outgoing edges, one labeled by -1 and another by 1. A computation of T on input $x \in \{0,1\}^n$ is the path from the root to one of the leaves that in each of the internal vertices v follows the edge, that has label equal to the value of $q_v(x)$. Label of the leaf that is reached by the path is the output of the computation. The tree T *computes* the function $f\colon \{0,1\}^n \to \{-1,1\}$ iff on each input $x \in \{0,1\}^n$ the output of T is equal to $f(x)$.

Decision tree models differ by the types of functions q_v that are allowed in the vertices of the tree. For any set $\mathcal{Q}$ of functions the decision tree complexity of the function f is the minimal depth of a tree (that is, the number of edges in the longest path from the root to a leaf) using functions from $\mathcal{Q}$ and computing f. We denote this value by $\mathsf{D}_\mathcal{Q}(f)$.

The standard decision tree model allows to query individual variables in the vertices of the tree. The complexity in this model is denoted simply by $\mathsf{D}(f)$. In the paper we also consider $\mathsf{D}_\oplus(f)$, $\mathsf{D}_{\wedge,\vee}(f)$, $\mathsf{D}_{B_2}(f)$ standing for $\mathcal{Q}$ equal to the set of all parities, the set of all AND and OR functions and the set B_2 of all binary functions respectively.

2.2 Parity Decision Trees

There are various techniques known for parity decision trees (that is, decision trees with parity queries). Here we list only the most relevant ones. A detailed exposition will be provided in the full version of the paper.

The strongest known analytical lower bound is through granularity.

Lemma 1. *For any non-constant function $f\colon \{0,1\}^n \to \{-1,1\}$ we find that* $\mathsf{D}_\oplus(f) \geq \mathsf{gran}(f) + 1$.

It is not hard to deduce this lemma from [12, Exercise 3.26]. We omit the proof here.

Another more combinatorial approach goes through analogs of certificate complexity and block sensitivity for parity decision trees [14]. Since parity block sensitivity is always less or equal than parity certificate complexity and we are interested in lower bounds, we will introduce only certificate complexity here.

For a function $f\colon \{0,1\}^n \to \{-1,1\}$ and $x \in \{0,1\}^n$ denote by $C_\oplus(f,x)$ the minimal co-dimension of an affine subspace in $\{0,1\}^n$ containing x, on which f is constant. The *parity certificate complexity* of f is $C_\oplus(f) = \max_x C_\oplus(f,x)$.

Lemma 2. ([14]) *For any function $f\colon \{0,1\}^n \to \{-1,1\}$ we have* $\mathsf{D}_\oplus(f) \geq C_\oplus(f)$.

3   Decision Trees with B_2-Queries

In this section we show a $n - o(n)$ lower bound for the complexity of MAJ_n function for B_2-queries. As a warm-up we start with the analysis of the complexity of MAJ_n in decision tree model with AND and OR queries of arbitrary fan-in. This model was studied in [1] with the relation to certain PRAM model.

In this section it will be convenient to switch to $\{-1, 1\}$ variables, that is we will consider $MAJ_n \colon \{-1, 1\}^n \to \{-1, 1\}$ that is equal to 1 iff $\sum_{i=1}^{n} x_i \geq 0$.

First we observe that the complexity of all monotone functions cannot be maximal.

Lemma 3. *For any monotone function $f \colon \{-1, 1\}^n \to \{-1, 1\}$ we find that* $D_{\wedge, \vee}(f) \leq n - 1$.

Proof. We can query all variables one by one until two variables are left. Now, observe that a monotone function of two remaining variables is either a constant, or a variable, or AND_2, or OR_2. We can compute this function in at most one query. $\square$

This upper bound is tight for MAJ_n.

Theorem 4. $D_{\wedge, \vee}(MAJ_n) = n - 1$.

Proof. The upper bound follows from Lemma 3.

For the lower bound we will argue by adversary argument, that is we will describe the strategy of query answering forcing the decision tree to make at least $n - 1$ queries.

During the computation we will fix the values of some of the variables. We will maintain an undirected graph G on the variables that are not yet fixed. Each vertex in this graph will have degree either 0 or 1 (that is, our graph is a matching). Each edge in the graph is labeled by either 1 or -1. The intuition behind the edges is the following. We connect x_i and x_j by an edge labeled by a iff we add the restriction that at least one of the variables x_i and x_j is equal to a. That is, we are not allowed to fix both variables to $-a$ in the future.

In the beginning the set of vertices of G consists of all variables and there are no edges. In one query the number of connected components will reduce by at most 1, with only one exception (when we remove one connected component without making queries). We will answer the queries in such a way that to know the value of the function the decision tree should reduce our graph to an empty graph. From this it follows that at least $n - 1$ queries are needed.

Along with the graph we maintain the parameter t that is equal to the sum of the values of already fixed variables. During most of the process we will have that $t \in \{-1, 0\}$.

Next we describe how to answer the queries. If the query asks the value of one of the variables x_i, there are two cases. If this variable is isolated in G, we fix the value of the variable in such a way that the new value of t still lies in

$\{-1, 0\}$. If x_i was connected by an edge to x_j, we fix $x_i = 1$ and $x_j = -1$. The value of t does not change. In both cases we remove one connected component from G.

Next suppose the query asks AND or OR of several variables. Without loss of generality consider a query $\bigwedge_{i \in S} x_i$ for some $S \subseteq [n]$ with $|S| \geq 2$. The case of OR-query is symmetric. We can assume that none of the variables in S are already fixed, since otherwise we can either answer the query without fixing new variables, or simplify the query. Suppose there is an edge $\{x_i, x_j\}$ in G, such that $i \in S$. In this case we fix $x_i = 1$ and $x_j = -1$. The answer to the query is 1 (that is, 'false'). The number of connected components has reduced by 1 and t does not change. Next suppose that all vertices x_i with $i \in S$ are isolated. Since $|S| \geq 2$ we can consider two distinct variables x_i and x_j with $i, j \in S$. We connect these vertices by an edge with the label 1. Thus, we promise that at least one of the variables is 1 and the answer to the query is 1. Since we introduce one edge, the number of connected components reduces by 1. The value of t does not change since we do not fix any variables.

We might need to interrupt this process at one point of the computation. At an arbitrary point of computation denote by A the number of -1-edges, by B the number of 1-edges and by C the number of isolated vertices. Note that $A + B + C$ is the number of connected components in G. Suppose that at some point we have $A + C = 1$ or $B + C = 1$. These cases are symmetric, without loss of generality suppose we have $A + C = 1$. If at this point of the computation we have $t = 0$ this might be a potential problem: note that none of 1-edges can change the balance to the negative side. If after fixing the last isolated vertex or the last -1-edge the balance does not decrease, it must be non-negative for all assignments of variables consistent with the current restrictions. So, to keep the function non-constant we will fix the last isolated vertex or -1-edge as soon as $A + C = 1$. If $C = 1$, we set the isolated vertex to -1 and we have $t = -1$ or $t = -2$. If $A = 1$ and $t = 0$, we set both of the variables connected by the edge to -1 and we have $t = -2$. If $A = 1$ and $t = -1$ we set one of the variables to 1 and the other to -1 and we have $t = -1$.

In the rest of the process we have that all the remaining vertices are connected by 1-edges. Answering the queries as before we keep t the same. Thus, there is an input consistent with our answers such that MAJ_n is -1 on this input. On the other hand, if at least one of 1-edges is still present in the graph we can set both of its vertices to 1 and make the balance t non-negative. Thus, in this case there is also an assignment on which the value of the function is equal to 1. Thus, to make the function to be constant we should remove all connected components from G. $\qquad\square$

We now proceed to the proof of the lower bound for binary queries.

Theorem 5. $\mathrm{D}_{\mathrm{B}_2}(\mathrm{MAJ}_n) \geq n - O(\sqrt{n})$.

Proof. Let us first consider what type of queries can be made by functions in B_2. First note, that the queries q and $-q$ are equivalent. Then note that there are queries of the form of one variable, $(x_i^a \vee x_j^b)$ for $a, b \in \{0, 1\}$, where $x_i^1 = x_i$

and $x_i^0 = -x_i$, and $x_i \oplus x_j$. Note that the last query basically asks whether variables x_i and x_j are equal.

The proof strategy is similar to the one in the previous proof. During the computation we will maintain the graph G. But now the vertices of G are new fresh variables that we denote by $y_1, \ldots, y_k$ (here k is the number of vertices in G). To each of the vertices y_i we assign some integer weight c_i. Some of the vertices are connected by edges, labeled by 1 or -1. The edges form a matching. We will maintain that the weights of connected vertices are equal. We will maintain that $1 \leq c_i \leq \sqrt{n}$.

The intuition behind the graph is the following. Each original input variable x_i at each point of the computation is fixed either to a constant or to some variable y_j, or to its negation $-y_j$. We will maintain the following relation

$$\sum_{i: \, x_i \text{ is unfixed}} x_i = \sum_{j=1}^{k} c_j y_j.$$

Initially $k = n$, for all i we set $x_i = y_i$, $c_i = 1$ and there are no edges in the graph.

We will answer queries in such a way that the number of connected components of G will reduce as slowly as possible. We will show how to answer queries in such a way that to know the value of the function the decision tree must reduce the number of connected components to a small number.

We also maintain a parameter t that is equal to the sum of the values of already fixed variables x_i.

The computation will proceed in two phases. In the first phase we will maintain that $-\sqrt{n} \leq t \leq \sqrt{n}$.

We now explain how to answer the queries in the first phase. Note that each query to variables of x can be restated as a query to variables of y (since each x_i is fixed either to a constant or to some variable y_j). First we consider the case that the query addresses the variables of y that are isolated.

Queries to Isolated Vertices. Suppose the query asks the value of one of the variables y_i. We then fix the value of the variable in such a way that $c_i y_i$ and t have opposite signs. We remove the variable y_i from the graph. Since $c_i \leq \sqrt{n}$ the balance t is still at most $\sqrt{n}$ in absolute value.

Suppose the query asks whether $y_i = y_j$. Suppose first that $c_i \neq c_j$, suppose without loss of generality that $c_i > c_j$. Then the adversary reply with $y_i \neq y_j$, so we identify $y_j = -y_i$, remove the vertex y_j from G and subtract c_j from c_i. It is easy to see that all properties are maintained. The number of connected components reduces by 1.

If on the other hand $c_i = c_j$, then if $c_i > \sqrt{n}/2$, we fix $y_i = 1$ and $y_j = -1$, and remove both vertices from G. The number of connected components in this case reduces by 2. If on the other hand $c_i \leq \sqrt{n}/2$, we set $y_j = y_i$, remove y_j from G and add c_j to c_i. The number of connected components reduces by 1.

Suppose next that the query asks the function $y_i \vee \neg y_j$. In this case if $t \geq 0$ we set $y_i = -1$, otherwise we fix $y_j = 1$. In both cases we remove y_i from G. The answer to the query in both cases is -1. The number of connected components reduces by 1 and since $c_i, c_j \leq \sqrt{n}$ the balance t is still at most $\sqrt{n}$ in absolute value.

Finally, suppose the query asks $y_i \vee y_j$ or $y_i \wedge y_j$. Suppose first that $c_i \neq c_j$, suppose without loss of generality that $c_i > c_j$. Then again we set $y_j = -y_i$, remove the vertex y_j from G and subtract c_j from c_i. It is easy to see that all properties are maintained. The number of connected components reduces by 1.

If on the other hand $c_i = c_j$ we connect y_i and y_j by an edge. We label the edge by -1 for the case of $y_i \vee y_j$ query and by 1 for the case of $y_i \wedge y_j$ query. The number of connected components reduces by 1.

Next we proceed to queries to non-isolated vertices.

Queries to Non-isolated Vertices. First consider arbitrary queries of the form y_i, $y_i \vee \neg y_j$, $y_i \vee y_j$ or $y_i \wedge y_j$ and suppose y_i is connected by an edge to some other vertex y_l (possibly $l = j$). For all these types of queries we can fix the answer to the query by fixing y_i to some constant. We also fix y_l to the opposite constant and remove both vertices from G. Since $c_i = c_l$ the balance t does not change. The number of connected components reduces by 1.

The only remaining case is the query of the form $y_i = y_j$ for the case when y_i is connected to some other vertex y_l by an edge. If $l = j$ we simply set $y_i = 1$, $y_j = -1$ and remove both vertices from G. The balance t does not change and the number of connected components reduces by 1. If $l \neq j$ we let $y_i = y_j$ and $y_l = -y_j$. We remove vertices y_i and y_l from the graph. Since c_i and c_l are equal the weight of y_j does not change. The number of connected components reduce by 1.

We have described how to answer queries in the first phase. Next we describe at which point this phase ends. For this denote by A the sum of weights of vertices connected by -1-edges, by B the sum of weights of vertices connected by 1-edges and by C the sum of weights of isolated vertices. The first phase ends once either $A + C \leq 3\sqrt{n}$ or $B + C \leq 3\sqrt{n}$. Without loss of generality assume that $A + C \leq 3\sqrt{n}$ (the other case is symmetric). Note that we can claim that $A + C > \sqrt{n}$. Indeed, note that in one step of the first phase at most two vertices are removed and the weight of each vertex is at most $\sqrt{n}$, so if $A + C \leq \sqrt{n}$, then on the previous step we already had $A + C \leq 3\sqrt{n}$.

At this step of the computation we fix all isolated vertices and all vertices connected by -1-edges to -1. Before that we had $-\sqrt{n} \leq t \leq \sqrt{n}$. Thus, since $\sqrt{n} < A + C \leq 3\sqrt{n}$, after this step we have $-4\sqrt{n} \leq t < 0$ (we could be more careful here, but this only results in a multiplicative constant factor in $O(\sqrt{n})$ in the theorem). After this the second phase of the computation starts. There are only vertices connected by 1-edges remained. We answer the queries as in the first phase. Note that the balance t does not change anymore. Thus if the sum of the weights of the remaining variables is at least $4\sqrt{n}$, then the function is non-constant: on one hand setting one vertex in each pair to 1 and the other to -1 we set function to -1 and setting all variables to 1 we set the function to 1. Thus the function becomes constant only once the total weight of the remaining vertices is below $4\sqrt{n}$, that is there are less than $2\sqrt{n}$ connected components.

Let us now calculate how many queries the decision tree needs to make to set the function to a constant. In the beginning G has n connected components and in the end it has at most $2\sqrt{n}$ connected components. On each step the number of connected components reduces by 1 with some exceptions that we consider below.

On the first phase there is the case when the number of connected components reduces by 2. Note that in this case the total weight of all vertices reduces by at least $\sqrt{n}$. Since originally the total weight is n and the total weight never increases, this step can occur at most $\sqrt{n}$ times.

Between the two phases we fix a lot of variables without answering any queries. Note that their total weight is at most $3\sqrt{n}$, thus the number of connected components reduces by at most $3\sqrt{n}$.

Thus, in total the tree needs to make at least $n-2\sqrt{n}-\sqrt{n}-3\sqrt{n} = n-O(\sqrt{n})$ queries to fix the function to a constant. □

We observe that the complexity of MAJ_n drops substantially if we allow to query parities of three variables.

Lemma 6. *Suppose $f\colon \{-1,1\}^n \to \{-1,1\}$ is symmetric function. Then there is a decision tree of depth $\lceil \frac{2n}{3} \rceil$ making queries only of the form AND_2, OR_2 and XOR_3 and computing f.*

Proof. Split the variables in blocks of size 3. In each block query the parity of its variables. If the answer is -1, query AND_2 of any two variables in the block. If the answer to the first query is 1, query OR_2 of any two variables in the block. It is easy to see that after these two queries we know the number of -1 variables in the block. In the case when n is not divisible by 3, if there is a small block of size 1, it requires one query to handle. If there is block of size 2 we will handle it with two queries. Knowing the number of -1 variables in all blocks is enough to output the value of the symmetric function. □

4 Parity Decision Tree Complexity of Threshold Functions

In this section we show a new lower bound for parity decision tree complexity of threshold functions.

To show that all previous techniques are not tight for some threshold functions we need an approach to prove even better lower bounds. We will do this via the following theorem.

Theorem 7. *For any s, k, n if $\mathsf{D}_\oplus(\text{THR}_n^k) \geq s$, then $\mathsf{D}_\oplus(\text{THR}_{n+2}^{k+1}) \geq s + 1$.*

Proof. We argue by a contradiction. Assume that $\mathsf{D}_\oplus(\text{THR}_{n+2}^{k+1}) \leq s$. We will construct a parity decision tree for THR_n^k making no more than $s - 1$ queries.

Denote the input variables to THR_n^k by $x = (x_1, \ldots, x_n) \in \{0, 1\}^n$. We introduce one more variable y (which we will fix later) and consider $x_1, \ldots, x_n, y, \neg y$ as inputs to the algorithm for THR_{n+2}^{k+1}. Note that $\mathrm{THR}_n^k(x) = \mathrm{THR}_{n+2}^{k+1}(x, y, \neg y)$. Our plan is to simulate the algorithm for THR_{n+2}^{k+1} on $(x, y, \neg y)$ (possibly reordered) and save one query on our way.

Consider the first query that the algorithm for THR_{n+2}^{k+1} makes. Suppose first that the query does not ask the parity of all input variables (we will deal with the other case later). Since the function THR_{n+2}^{k+1} is symmetric we can reorder the inputs in such a way that the query contains input y and does not contain $\neg y$, that is the query asks the parity $(\bigoplus_{i \in S} x_i) \oplus y$ for some $S \subseteq [n]$. Now it is time for us to fix the value of y. We let $y = \bigoplus_{i \in S} x_i$. Then the answer to the first query is 0, we can skip it and proceed to the second query. For each next query of the algorithm for THR_{n+2}^{k+1} if it contains y or $\neg y$ (or both) we substitute them by $\bigoplus_{i \in S} x_i$ and $(\bigoplus_{i \in S} x_i) \oplus 1$ respectively. The result is the parity of some variables among $x_1, \ldots, x_n$ and we make this query to our original input x. Clearly the answer to the query to x is the same as the answer to the original query to $(x, y, \neg y)$. Thus, making at most $s - 1$ queries we reach the leaf of the tree for THR_{n+2}^{k+1} and thus compute $\mathrm{THR}_{n+2}^{k+1}(x, y, \neg y) = \mathrm{THR}_n^k(x)$.

It remains to consider the case when the first query to THR_{n+2}^{k+1} is $(\bigoplus_{i=1}^n x_i) \oplus y \oplus \neg y$. This parity is equal to $\bigoplus_{i=1}^n x_i$ and we make this query to x. Now we proceed to the second query in the computation of THR_{n+2}^{k+1} and this query does not query the parity of all input variables. We perform the same analysis as above for this query: rename the inputs, fix y to the parity of subset of x to make the answer to the query to be equal to 0, simulate further queries to $(x, y, \neg y)$. Again we save one query in this case and compute $\mathrm{THR}_n^k(x)$ in at most $s - 1$ queries. $\qquad\square$

Next we analyze the decision tree complexity of THR_n^2 functions. For them the lower bound through granularity is tight. We need this analysis to use in combination with Theorem 7 to prove lower bound for THR_n^3.

Lemma 8. *For even n, we have $\mathsf{D}_\oplus(\mathrm{THR}_n^2) = n$, and for odd n, we have $\mathsf{D}_\oplus(\mathrm{THR}_n^2) = n - 1$.*

The proof of this lemma is omitted.
Next we compute the granularity for threshold functions with threshold three.

Lemma 9. *For $n = 8m + 2$ for any integer m, we have $\mathsf{gran}(\mathrm{THR}_n^3) = n - 3$.*

To prove this lemma we need to analyse Fourier coefficients of THR_n^2. It turns out that the maximal granularity is achieved on one of the Fourier coefficients $\widehat{\mathrm{THR}}_n^3(\emptyset)$ and $\widehat{\mathrm{THR}}_n^3([n])$. We omit the details of the proof.

We now show that for functions in Lemma 9 their decision tree complexity is greater than their granularity plus one. Note, that since granularity lower bound is not worse than the lower bounds through the sparsity and the degree, they also do not give tight lower bounds. Also it is easy to see that the certificate complexity does not give optimal lower bound as well (note that each input x lies in an affine subspace of dimension 2 on which the function is constant).

Theorem 10. *For $n = 8m + 2$ for integer $m > 0$ we have* $D_\oplus(\text{THR}_n^3) = n - 1$.

Proof. For the lower bound we note that $n - 2$ is even and thus by Lemma 8 we have $D_\oplus(\text{THR}_{n-2}^2) \geq n - 2$. Then by Theorem 7 we have $D_\oplus(\text{THR}_n^3) \geq n - 1$.

For the upper bound we construct an algorithm. Our algorithm will maintain splitting of input variables into blocks of two types with the following properties:

- all variables in each block of type 1 are equal;
- blocks of type 2 are balanced, that is they have equal number of ones and zeros.

In the beginning of the computation each variable forms a separate block of size one.

During each step the algorithm will merge two blocks into a new one. Thus, after k steps the number of blocks is $n - k$.

More specifically, on each step we pick two blocks of type 1 of equal size. We pick one variable from each block and query the parity of these two variables. If the variables are equal, we merge the blocks into a new block of type 1. If the variables are not equal, the new block is of type 2.

The algorithm works as follows. We first combine input variables into blocks of size 2 and then combine all unbalanced blocks, except possibly one, into blocks of size 4. If in the end we have at least one balanced block we just query one variable from all other blocks thus learning the number of ones in the input in at most $n - 1$ queries. If all blocks are of type 1, then there is one block of size 2. We observe that two variables in this block do not affect the value of the function. Indeed, $\text{THR}_n^3(x) = 1$ iff $\sum_i x_i \geq 3$ iff there is a block of size 4 containing variables equal to 1. Again, we can query one variable from each block except one to compute the output.   □

Thus, we have shown that previously known lower bounds are not tight for THR_{8m+2}^3. However, the gap between the lower bound and the actual complexity is 1.

Remark 11. We note that from our analysis it is straightforward to determine the complexity of THR_n^3 for all n. If $n = 4m$ or $4m + 3$ for some m, then $D_\oplus(\text{THR}_n^3) = n$ and if $n = 4m + 1$ or $n = 4m + 2$, then $D_\oplus(\text{THR}_n^3) = n - 1$. The lower bounds (apart from the case covered by Theorem 10) follows from the consideration of $\widehat{\text{THR}}_n^3(\emptyset)$ and $\widehat{\text{THR}}_n^3([n])$ as in the proof of Lemma 9. The upper bound follows the same analysis as in the proof of Theorem 10.

Acknowledgments. We would like to thank Alexander Kulikov for useful discussions on parity decision tree complexity.

References

1. Ben-Asher, Y., Newman, I.: Decision trees with Boolean threshold queries. J. Comput. Syst. Sci. **51**(3), 495–502 (1995). https://doi.org/10.1006/jcss.1995.1085

2. Buhrman, H., de Wolf, R.: Complexity measures and decision tree complexity: a survey. Theor. Comput. Sci. **288**(1), 21–43 (2002). https://doi.org/10.1016/S0304-3975(01)00144-X

3. Eppstein, D., Hirschberg, D.S.: From discrepancy to majority. Algorithmica **80**(4), 1278–1297 (2018). https://doi.org/10.1007/s00453-017-0303-7

4. Gröger, H.D., Turán, G.: On linear decision trees computing Boolean functions. In: Albert, J.L., Monien, B., Artalejo, M.R. (eds.) ICALP 1991. LNCS, vol. 510, pp. 707–718. Springer, Heidelberg (1991). https://doi.org/10.1007/3-540-54233-7_176

5. Hatami, H., Hosseini, K., Lovett, S.: Structure of protocols for XOR functions. In: IEEE 57th Annual Symposium on Foundations of Computer Science, FOCS 2016, Hyatt Regency, New Brunswick, New Jersey, USA, 9–11 October 2016, pp. 282–288 (2016). https://doi.org/10.1109/FOCS.2016.38

6. Hatami, P., Kulkarni, R., Pankratov, D.: Variations on the sensitivity conjecture. Theory Comput. Grad. Surv. **4**, 1–27 (2011). https://doi.org/10.4086/toc.gs.2011.004

7. Huang, H.: Induced subgraphs of hypercubes and a proof of the sensitivity conjecture. CoRR abs/1907.00847 (2019). http://arxiv.org/abs/1907.00847

8. Jukna, S.: Boolean Function Complexity - Advances and Frontiers. Algorithms and Combinatorics, vol. 27. Springer, Heidelberg (2012). https://doi.org/10.1007/978-3-642-24508-4

9. Knop, D., Pilipczuk, M., Wrochna, M.: Tight complexity lower bounds for integer linear programming with few constraints. In: 36th International Symposium on Theoretical Aspects of Computer Science, STACS 2019, Berlin, Germany, 13–16 March 2019, pp. 44:1–44:15 (2019). https://doi.org/10.4230/LIPIcs.STACS.2019.44

10. Kushilevitz, E., Nisan, N.: Communication Complexity. Cambridge University Press, Cambridge (1997)

11. Lovász, L., Saks, M.E.: Lattices, Möbius functions and communication complexity. In: 29th Annual Symposium on Foundations of Computer Science, White Plains, New York, USA, 24–26 October 1988, pp. 81–90 (1988). https://doi.org/10.1109/SFCS.1988.21924

12. O'Donnell, R.: Analysis of Boolean Functions. Cambridge University Press (2014). http://www.cambridge.org/de/academic/subjects/computer-science/algorithmics-complexity-computer-algebra-and-computational-g/analysis-boolean-functions

13. Posobin, G.: Computing majority with low-fan-in majority queries. CoRR abs/1711.10176 (2017). http://arxiv.org/abs/1711.10176

14. Zhang, Z., Shi, Y.: On the parity complexity measures of Boolean functions. Theor. Comput. Sci. **411**(26–28), 2612–2618 (2010). https://doi.org/10.1016/j.tcs.2010.03.027

Randomized and Symmetric Catalytic Computation

Samir Datta[1]([⊠]), Chetan Gupta[2], Rahul Jain[2][iD], Vimal Raj Sharma[2], and Raghunath Tewari[2]

[1] Chennai Mathematical Institute and UMI ReLaX, Chennai, India
sdatta@cmi.ac.in
[2] Indian Institute of Technology Kanpur, Kanpur, India
{gchetan,jain,vimalraj,rtewari}@cse.iitk.ac.in

Abstract. A catalytic Turing machine is a model of computation that is created by equipping a Turing machine with an additional *auxiliary* tape which is initially filled with arbitrary content; the machine can read or write on auxiliary tape during the computation but when it halts auxiliary tape's initial content must be restored. In this paper, we study the power of catalytic Turing machines with $O(\log n)$-sized clean tape and a polynomial-sized auxiliary tape.

We introduce the notion of randomized catalytic Turing machine and show that the resulting complexity class CBPL is contained in the class ZPP. We also introduce the notion of symmetricity in the context of catalytic computation and prove that, under a widely believed assumption, in the logspace setting the power of a randomized catalytic Turing machine and a symmetric catalytic Turing machine is equal to a deterministic catalytic Turing machine which runs in polynomial time.

Keywords: Catalytic computation · Logspace · Randomized computation

1 Introduction

Buhrman et al. [1] first introduced the catalytic computational model. This model of computation has an auxiliary tape filled with arbitrary content in addition to the clean tape of a standard Turing machine. The machine during the computation can use this auxiliary tape to read or write, but at the end of the computation, it is constrained to have the same content in the auxiliary tape as initial. The central question here is, whether catalytic computational model is more powerful than the traditional Turing machine model or not. It seems intuitive that the content of auxiliary tape must be stored in one form or another at each step of the computation, making the auxiliary tape useless

The first author was partially funded by a grant from Infosys foundation and SERB-MATRICS grant MTR/2017/000480. The second and fourth author were supported by Visvesvaraya PhD Scheme.

H. Fernau (Ed.): CSR 2020, LNCS 12159, pp. 211–223, 2020.
https://doi.org/10.1007/978-3-030-50026-9_15

if the original auxiliary tape content is incompressible. However, Buhrman et al. [1] showed that problems that are not known to be solvable by a standard Turing machine using $O(\log n)$ space (Logspace, L) can be solved by a catalytic Turing machine with $O(\log n)$ clean space and $n^{O(1)}$ auxiliary space (Catalytic logspace, CL). Specifically, they showed that the circuit class uniform TC_1, which contains L is contained in CL. This result gives evidence that the auxiliary tape might not be useless.

Since its introduction, researchers have tried to understand the power and limitation of catalytic Turing machine. Buhrman et al. [2] also introduced a nondeterministic version of the catalytic Turing machine and proved that non-deterministic catalytic logspace class CNL is closed under complement. They also showed that CNL is contained in ZPP. Girard et al. [3] studied catalytic computation in a nonuniform setting. More recently, Gupta et al. [4] studied the notion of unambiguity in catalytic logspace and proved that unambiguous catalytic Turing machines are as powerful as nondeterministic catalytic Turing machines in the logspace setting.

In this paper, we study the notion of randomized computation and symmetric computation in the context of catalytic Turing machines. Following the earlier results in the field of catalytic computation, we define the classes of problems by limiting the catalytic Turing machine to $O(\log n)$-size clean tape and $n^{O(1)}$-sized auxiliary tape. We thus get the classes CBPL and CSL for randomized and symmetric logspace catalytic Turing machine respectively (see Sect. 2 for complete definitions). We show that CBPL $\subseteq$ ZPP. We also prove that under a widely believed assumption, not only CBPL is equal to CSL, but they are also equal to the class of problems that can be solved by a *deterministic* catalytic Turing machine running in *polynomial time* with $O(\log n)$-size clean (or work) tape and $n^{O(1)}$-sized auxiliary tape (CSC$_1$). Formally, we prove the following.

Theorem 1 (Main Theorem). *If there exists a constant $\epsilon > 0$ such that we have* DSPACE$(n) \not\subseteq$ SIZE$(2^{\epsilon n})$, *then* CBPL = CL = CSL = CSC$_1$.

Our result requires (i) a pseudorandom generator to get a small size configuration graph of a catalytic machine, and (ii) universal exploration sequence to traverse those small size configuration graphs. The required pseudorandom generator was used in [2] and [4] as well. Universal exploration sequence was first introduced by Koucky [7]. Reingold [9] presented a logspace algorithm to construct a polynomially-long universal exploration sequence for undirected graphs. Since the catalytic Turing machines we study have $O(\log n)$ size clean space, we can use Reingold's algorithm to construct those sequences in catalytic machines as well.

1.1   Outline of the Paper

In Sect. 2, we give preliminary definitions of various catalytic classes and state the lemmas on the pseudorandom generator and universal exploration sequences used by us. In Sect. 3, we prove CBPL $\subseteq$ ZPP. In Sect. 4, we prove our main result

Theorem 1. Finally, in Sect. 5, without using the class CSL we give an alternative proof of $CL = CSC_1$ under the same assumption as in Theorem 1.

2 Preliminaries

We start with the brief definitions of a few well-known complexity classes.

ZPP, DSPACE(n), SIZE(k): ZPP denotes the set of the languages which are decidable in expected polynomial time. DSPACE(n) denotes the set of the languages which are decidable in linear space. SIZE(k) denotes the set of the languages which are decidable by circuits of size k.

The deterministic catalytic Turing machine was formally defined by Buhrman et al. [2] in the following way.

Definition 2. Let $\mathcal{M}$ be a deterministic Turing machine with four tapes: one input and one output tape, one work-tape, and one *auxiliary tape* (or *aux-tape*).

$\mathcal{M}$ is said to be a *deterministic catalytic Turing machine* using workspace $s(n)$ and auxiliary space $s_a(n)$ if for all inputs $x \in \{0,1\}^n$ and auxiliary tape contents $w \in \{0,1\}^{s_a(n)}$, the following three properties hold.

1. **Space bound.** The machine $\mathcal{M}$ uses space $s(n)$ on its work tape and space $s_a(n)$ on its auxiliary tape.
2. **Catalytic condition.** $\mathcal{M}$ halts with w on its auxiliary tape.
3. **Consistency.** $\mathcal{M}$ either accepts x for all choices of w or it rejects for all choices of w.

Definition 3. CSPACE($s(n)$) is the set of languages that can be solved by a deterministic catalytic Turing machine that uses at most $s(n)$ size workspace and $2^{s(n)}$ size auxiliary space on all inputs $x \in \{0,1\}^n$. CL denotes the class CSPACE($O(\log n)$).

Definition 4. CTISP($t(n), s(n)$) is the set of languages that can be solved by a deterministic catalytic Turing machine that halts in at most $t(n)$ steps and uses at most $s(n)$ size workspace and $2^{s(n)}$ size auxiliary space on all inputs $x \in \{0,1\}^n$. CSC_i denotes the class CTISP($poly(n), O((\log n)^i)$).

A *configuration* of a catalytic machine $\mathcal{M}$ with $s(n)$ workspace and $s_a(n)$ auxiliary space consists of the state, at most $s(n)$ size work tape content, at most $s_a(n)$ size auxiliary tape content, and the head positions of all the three tapes. We will use the notion of *configuration graph* in our results, which is often used in proving space-bounded computation results for traditional Turing machines. In the context of catalytic Turing machines, the configuration graph was defined in [1,2] in a slightly different manner than traditional Turing machines.

Definition 5. For a deterministic catalytic Turing machine $\mathcal{M}$, input x, and initial auxiliary content w, the configuration graph denoted by $\mathcal{G}_{\mathcal{M},x,w}$ is a directed acyclic graph in which every vertex is a configuration which is reachable when $\mathcal{M}$ runs on (x, w). $\mathcal{G}_{\mathcal{M},x,w}$ has a directed edge from a vertex u to a vertex v if $\mathcal{M}$ in one step can move to v from u.

$|\mathcal{G}_{\mathcal{M},x,w}|$ denotes the number of the vertices in $\mathcal{G}_{\mathcal{M},x,w}$. We call a configuration in which a machine accepts (rejects) the input an accepting (rejecting) configuration. We note that the configuration graph of a deterministic catalytic Turing machine is a line graph.

Motivated by the symmetric Turing machines defined in [8], we study the notion of symmetricity in catalytic computation. We define the symmetric catalytic Turing machine below.

Definition 6. A *symmetric catalytic Turing machine* is a catalytic Turing machine with two sets of transitions δ_0 and δ_1. At each step, the machine uses either δ_0 or δ_1 arbitrarily. δ_0 and δ_1 are the finite set of transitions of the following form. (For simplicity, we have described these transitions for a single tape machine.)

– $(p, a, 0, b, q)$: If machine's current state is p, the head is on a cell containing a, then in one step machine changes the state to q, a is changed to b, and the head doesn't move.
– (p, ab, L, cd, q): If machine's current state is p, the head is on a cell containing b and the cell left to it contains a, then in one step machine changes the state to q, the head moves to the left, and both a and b are changed to c and d respectively.
– (p, ab, R, cd, q): If machine's current state is p, the head is on a cell containing a and the cell right to it contains b, then in one step machine changes the state to q, the head moves to the right, and both a and b are changed to c and d respectively.

Additionally, the following two properties hold:

– Every transition has its inverse, i.e. each of δ_0 and δ_1 has (p, ab, L, cd, q) if and only if it has (q, cd, R, ab, p) and $(p, a, 0, b, q)$ if and only if it has $(q, b, 0, a, p)$.
– The machine has two special states q_{start} and q_{accept}. The machine in the beginning is in the state q_{start}. During the run, at every configuration where the state is q_{start} or q_{accept}, the machine is constrained to have the same auxiliary content as initial.

The notion of the configuration graph extends to symmetric catalytic machines as well. Due to inverse transitions, configuration graphs of a symmetric catalytic machine are bidirectional, i.e. for any two vertices in a configuration graph, say u and v, an edge goes from u to v if and only if an edge goes from v to u.

We say a symmetric catalytic Turing machine $\mathcal{M}$ decides or solves a language L if on every input x and every initial auxiliary content w, an accepting configuration (i.e., configuration with q_{accept}) is reachable when $\mathcal{M}$ runs on (x, w) if and only if $x \in L$.

Definition 7. CSSPACE($s(n)$) is the set of languages that can be solved by a symmetric catalytic Turing machine that uses at most $s(n)$ size workspace and $2^{s(n)}$ size auxiliary space on all inputs $x \in \{0,1\}^n$. CSL denotes the class CSSPACE($O(\log n)$).

The following lemma follows from Theorem 1 of [8].

Lemma 8. $\mathsf{CL} \subseteq \mathsf{CSL}$.

In this paper, we also study randomized catalytic computation. We define the randomized catalytic Turing machine as follows.

Definition 9. A *randomized catalytic Turing machine* is a catalytic Turing machine with two transition functions δ_0 and δ_1. At each step the machine applies δ_0 with $\frac{1}{2}$ probability and δ_1 with $\frac{1}{2}$ probability, independent of the previous choices. On all possible choices of transition functions δ_0 and δ_1, the machine is constrained to have the same auxiliary content as initial when it halts.

We say a randomized catalytic Turing machine $\mathcal{M}$ decides or solves a language L if for every input x and initial auxiliary content w, $\mathcal{M}$ accepts x with probability at least $\frac{2}{3}$ if $x \in L$ and rejects x with probability at least $\frac{2}{3}$ if $x \notin L$.

Definition 10. $\mathsf{CBPSPACE}(s(n))$ is the set of languages that can be solved by a randomized catalytic Turing machine that uses at most $s(n)$ size workspace and $2^{s(n)}$ size auxiliary space on all inputs $x \in \{0, 1\}^n$. CBPL denotes the class $\mathsf{CBPSPACE}(O(\log n))$.

The class CBPL is the catalytic equivalent of the well-known complexity class BPL. Configuration graph for a randomized catalytic Turing machine is defined in the same way it was defined for a deterministic catalytic Turing machine. Although note here that non-halting configurations have out-degree two in a configuration graph of a randomized catalytic machine.

For a deterministic catalytic machine $\mathcal{M}$ with $c \log n$ size workspace and n^c size auxiliary space, an input x and initial auxiliary content w, $|\mathcal{G}_{\mathcal{M},x,w}|$ can be as large as exponential in $|x|$. But in [1,2], authors showed that the average size of the configuration graphs over all possible initial auxiliary contents for a particular x and $\mathcal{M}$ is only polynomial in $|x|$. This observation holds for symmetric and randomized catalytic Turing machines as well. The following lemma is a direct adaption of Lemma 8 from [2] for symmetric and randomized catalytic machines.

Lemma 11. *Let $\mathcal{M}$ be a symmetric or randomized catalytic Turing machine with $c \log n$ size workspace and n^c size auxiliary space. Then for all x,*

$$\mathop{\mathbb{E}}_{w \in_R \{0,1\}^{n^c}} [|\mathcal{G}_{\mathcal{M},x,w}|] \leq O(n^{2c+2}),$$

where $\mathbb{E}$ is the expectation symbol.

In Sect. 4, we will prove $\mathsf{CBPL} = \mathsf{CL} = \mathsf{CSL} = \mathsf{CSC}_1$ under the same assumption the following standard derandomization result holds.

Lemma 12. *[5, 6] If there exists a constant $\epsilon > 0$ such that* $\mathsf{DSPACE}(n) \not\subseteq$ $\mathsf{SIZE}(2^{\epsilon n})$ *then for all constants c there exists a constant c' and a function G :* $\{0, 1\}^{c' \log n} \to \{0, 1\}^n$ *computable in $O(\log n)$ space, such that for any circuit C of size n^c*

$$\left| \Pr_{r \in \{0,1\}^n}[C(r) = 1] - \Pr_{s \in \{0,1\}^{c' \log n}}[C(G(s)) = 1] \right| < \frac{1}{n}.$$

We will use a pseudorandom generator to produce small size configuration graphs of symmetric and randomized catalytic machines. From [2] we know that such a pseudorandom generator exists for nondeterministic catalytic Turing machines under the same assumption as that of Lemma 12. Their result trivially implies the following lemma.

Lemma 13. *Let $\mathcal{M}$ be a symmetric or randomized catalytic Turing machine using $c \log n$ size workspace and n^c size auxiliary space. If there exists a constant $\epsilon > 0$ such that* $\mathsf{DSPACE}(n) \not\subseteq \mathsf{SIZE}(2^{\epsilon n})$, *then there exists a function* $G : \{0, 1\}^{O(\log n)} \to \{0, 1\}^{n^c}$, *such that on every input x and initial auxiliary content w, for more than half of the seeds $s \in \{0, 1\}^{O(\log n)}$, $|\mathcal{G}_{\mathcal{M}, x, w \oplus G(s)}| \leq n^{2c+3}$. Moreover, G is logspace computable. ($w \oplus G(s)$ denotes the bitwise XOR of w and $G(s)$.)*

We will also need *universal exploration sequences*. Let $\mathcal{G}$ be an undirected graph, then *labelling* is a function where every edge uv leaving a vertex u is mapped to an integer $\{0, 1, \ldots, degree(u) - 1\}$ in such a way that any two distinct edges leaving a common vertex get different labels. Note that, in such a labelling an undirected edge, say uv, gets two labels, one with respect to u and another with respect to v.

An (n, d)-universal exploration sequence of length m is a sequence of integers $(s_1, s_2, \ldots, s_m)$ where each $s_i \in \{0, 1, \ldots, d - 1\}$, which can be used to visit all the vertices of any connected undirected graph $\mathcal{G}$ of n vertices and maximum degree d in the following way. Let $\mathcal{G}$ has a labelling l, in the first step we pick a vertex u and take an edge e leaving u labeled by s_1 mod $degree(u)$ to move to the next vertex, after this, in the ith step if we arrived at a vertex, say v, through an edge labeled with p with respect to v then we take an edge with label $(p + s_i)$ mod $degree(v)$ with respect to v to move to the next vertex. Reingold [9] proved that an (n, d)-universal exploration sequence can be constructed in $O(\log n)$ space.

An essential property of universal exploration sequences that we will use in our result is that at any point during the traversal using a universal exploration sequence we can stop and traverse back the vertices visited so far in the exact reverse order that they were visited.

3   CBPL $\subseteq$ ZPP

In this section, we will prove that CBPL is contained in ZPP. Our proof, similar to the proof of CNL $\subseteq$ ZPP, uses the observation that the average size of the configuration graphs over all possible auxiliary content is polynomial in the length of the input.

Theorem 14. $\mathsf{CBPL} \subseteq \mathsf{ZPP}$.

Proof. Let $\mathcal{M}$ be a CBPL machine with $c \log n$ size workspace and n^c size auxiliary space. We construct a ZPP machine $\mathcal{M}'$ such that $L(\mathcal{M}) = L(\mathcal{M}')$. On input x, $\mathcal{M}'$ first randomly generates a string w of size $|x|^c$ and constructs the configuration graph $\mathcal{G}_{\mathcal{M},x,w}$.

For every $v \in \mathcal{G}_{\mathcal{M},x,w}$, let $\mathsf{prob}(v)$ denote the probability of reaching an accepting configuration from v. $\mathcal{M}'$ computes the $\mathsf{prob}(v)$ for every vertex in the following way.

1. Set $\mathsf{prob}(v) = 1$ if v is an accepting configuration and $\mathsf{prob}(v) = 0$ if v is a rejecting configuration.
2. For every vertex v whose $\mathsf{prob}(v)$ is still not computed, if $\mathsf{prob}(v_1)$ and $\mathsf{prob}(v_2)$ are already computed and there is an edge from v to both v_1 and v_2, set $\mathsf{prob}(v) = \frac{1}{2} \cdot \mathsf{prob}(v_1) + \frac{1}{2} \cdot \mathsf{prob}(v_2)$.
3. Repeat 2 until $\mathsf{prob}(v)$ is computed for all $v \in \mathcal{G}_{\mathcal{M},x,w}$.

In the end, $\mathcal{M}'$ accepts x if and only if $\mathsf{prob}(v_{init}) \geq \frac{2}{3}$, where v_{init} is the initial configuration. The procedure to compute $\mathsf{prob}(v)$ can easily be done by $\mathcal{M}'$ in time polynomial in $|\mathcal{G}_{\mathcal{M},x,w}|$. Since from Lemma 11 we know that $\mathbb{E}_{w \in_R \{0,1\}^{n^c}}[|\mathcal{G}_{\mathcal{M},x,w}|] \leq O(n^{2c+2})$, the machine runs in expected polytime.   □

4   Proof of Main Theorem

Since we know $\mathsf{CL} \subseteq \mathsf{CSL}$ from Lemma 8 and $\mathsf{CSC}_1 \subseteq \mathsf{CBPL}$ follows from the definition, it is enough to prove $\mathsf{CBPL} \subseteq \mathsf{CL}$ and $\mathsf{CSL} \subseteq \mathsf{CSC}_1$.

Proof of $\mathsf{CBPL} \subseteq \mathsf{CL}$:

Let $\mathcal{M}$ be a CBPL machine with $c \log n$ size workspace and n^c size auxiliary space. We will construct a CL machine $\mathcal{M}'$ such that $L(\mathcal{M}) = L(\mathcal{M}')$.

From Lemma 13, we know that there exists a logspace computable function $G : \{0,1\}^{O(\log n)} \to \{0,1\}^{n^c}$, such that on every input x and initial auxiliary content w, for more than half of the seeds $s \in \{0,1\}^{O(\log n)}$, $|\mathcal{G}_{\mathcal{M},x,w \oplus G(s)}| \leq n^{2c+3}$. We call a seed s *good*, if $|\mathcal{G}_{\mathcal{M},x,w \oplus G(s)}| \leq n^{2c+3}$.

We first prove the existence of another pseudorandom generator which $\mathcal{M}'$ will use to deterministically find $\mathcal{M}$'s output in case of a good seed. Let $\tilde{s}$ be a good seed and $C_{x,w \oplus G(\tilde{s})}$ be a polynomial size boolean circuit which on input $r \in \{0,1\}^{n^{2c+3}}$ traverses $\mathcal{G}_{\mathcal{M},x,w \oplus G(\tilde{s})}$ using r in the following way. Assume a label on every edge of $\mathcal{G}_{\mathcal{M},x,w \oplus G(\tilde{s})}$, such that an edge uv is labeled by 0 if u changes to v using δ_0 or 1 if u changes to v using δ_1. $C_{x,w \oplus G(\tilde{s})}$ starts from the initial vertex and in the ith step moves to the next vertex using the outgoing edge with label same as the ith bit of r. $C_{x,w \oplus G(\tilde{s})}$ outputs 1 if it reaches an accepting vertex while traversing $\mathcal{G}_{\mathcal{M},x,w \oplus G(\tilde{s})}$, else it outputs 0.

From Lemma 12, we know that there exists a logspace computable function $F : \{0,1\}^{(O\log n)} \rightarrow \{0,1\}^{n^{2c+3}}$ such that,

$$\left| \Pr_{r\in\{0,1\}^{n^{2c+3}}}[C_{x,w\oplus G(\tilde{s})}(r) = 0] - \Pr_{s'\in\{0,1\}^{O(\log n)}}[C_{x,w\oplus G(\tilde{s})}(F(s')) = 0] \right| < \frac{1}{n}.$$

For sufficiently large n, if $x \in L(\mathcal{M})$, then

$$\left| \Pr_{s'\in\{0,1\}^{O(\log n)}}[C_{x,w\oplus G(\tilde{s})}(F(s')) = 0] \right| < \frac{1}{3} + \frac{1}{n} < \frac{1}{2} \tag{1}$$

Similarly, we can prove that, if $x \notin L(\mathcal{M})$, then

$$\left| \Pr_{s'\in\{0,1\}^{O(\log n)}}[C_{x,w\oplus G(\tilde{s})}(F(s')) = 1] \right| < \frac{1}{3} + \frac{1}{n} < \frac{1}{2} \tag{2}$$

Equations (1) and (2) together prove that on less than half of the seeds $s' \in \{0,1\}^{O(\log n)}$, the simulation of $\mathcal{M}$ on $(x, w \oplus G(\tilde{s}))$ by picking δ_0 or δ_1 according to $F(s')$ gives the wrong answer.

We now present the algorithm of $\mathcal{M}'$.

Algorithm 1. Algorithm of $\mathcal{M}'$

G and F are the above described pseudorandom generators. S and S' are the set of seeds for G and F respectively.

```
1: procedure DETERMINISTICSIMULATION(Input x, Auxiliary Content w)
2:     cnt_final = 0
3:     for s ∈ S do
4:         w ← w ⊕ G(s)
5:         cnt_acc = 0
6:         for s' ∈ S' do
7:             Simulate M on (x,w) by picking δ0 and δ1 according to F(s').
8:             if M halts with an accepting state during the simulation then
9:                 cnt_acc = cnt_acc + 1
10:            end if
11:            if M doesn't halt during the simulation using F(s') then
12:                Continue the simulation using either δ0 or δ1 until M halts
13:            end if
14:        end for
15:        if cnt_acc > |S'|/2 then
16:            cnt_final = cnt_final + 1
17:        end if
18:        w ← w ⊕ G(s)
19:    end for
20:    if cnt_final > |S|/2 then
21:        Accept
22:    else
23:        Reject
24:    end if
25: end procedure
```

If $x \in L(\mathcal{M})$, then on every good seed s of G, $cnt_{acc} > \frac{|S'|}{2}$. Since more than half of G's seeds are good, cnt_{final} is incremented in line 16 more than $\frac{|S|}{2}$ times. Hence, in line 21 $\mathcal{M}'$ will **Accept** after checking $cnt_{final} > \frac{|S|}{2}$. On the other hand, if $x \notin L(\mathcal{M})$, then on every good seed s, $cnt_{acc} < \frac{|S'|}{2}$. So cnt_{final} is not incremented in line 16 more than $\frac{|S|}{2}$ times. Hence, $\mathcal{M}'$ will **Reject** in line 23.

Proof of $\mathsf{CSL} \subseteq \mathsf{CSC_1}$:

Let $\mathcal{M}$ be a CSL machine with $c \log n$ size workspace and n^c size auxiliary space. We will construct a $\mathsf{CSC_1}$ machine $\mathcal{M}'$ such that $L(\mathcal{M}) = L(\mathcal{M}')$.

We will again use the pseudorandom generator G of Lemma 13 with the property that on every input x and initial auxiliary content w, for more than half of the seeds $s \in \{0,1\}^{O(\log n)}$, $|\mathcal{G}_{\mathcal{M},x,w \oplus G(s)}| \leq n^{2c+3}$.

We will also use universal exploration sequence to traverse all the vertices of $\mathcal{G}_{\mathcal{M},x,w \oplus G(s)}$ on good seeds s. Let seq denote a (n^{2c+3}, d)-universal exploration sequence, where d is a constant upper bound on the maximum degree of $\mathcal{G}_{\mathcal{M},x,w \oplus G(s)}$. We now present the algorithm of $\mathcal{M}'$.

Algorithm 2. Algorithm of $\mathcal{M}'$

---

seq is a (n^{2c+3}, d)-universal exploration sequence. G is the above described pseudorandom generator and S is the set of seeds for G.

1: **procedure** POLYTIME-DETERMINISTICSIMULATION(Input x, Auxiliary Content w)
2:     $accept = \text{FALSE}$
3:     **for** $s \in S$ **do**
4:         $w \leftarrow w \oplus G(s)$
5:         Traverse $\mathcal{G}_{\mathcal{M},x,w}$ by simulating $\mathcal{M}$ on (x,w) using seq.
6:         Set $accept = \text{TRUE}$ if an accepting config. is reached during the simulation.
7:         Reverse simulate $\mathcal{M}$ on (x,w) using seq.   ▷ Restoring the aux. content.
8:         $w \leftarrow w \oplus G(s)$
9:         **if** $accept = \text{TRUE}$ **then**
10:             **Accept**
11:         **end if**
12:     **end for**
13:     **Reject**
14: **end procedure**

$\mathcal{M}'$ uses a flag variable $accept$ which it sets to TRUE when it finds an accepting configuration while traversing $\mathcal{G}_{\mathcal{M},x,w}$ using seq. If $x \in L(\mathcal{M})$, then $\mathcal{M}'$ on a good seed s must visit all the vertices of $\mathcal{G}_{\mathcal{M},x,w}$ in the simulation of line 5, and hence also visit an accepting configuration. In which case, it sets $accept = \text{TRUE}$ in line 6 and later **Accepts** in line 10. If $x \notin L(\mathcal{M})$, then clearly $\mathcal{M}'$ can never reach an accepting configuration during any simulation. Therefore, $\mathcal{M}'$ never sets $accept$ to TRUE and finally, **Rejects** in line 13.

$\mathcal{M}'$ takes polynomial time because there are only polynomially many seeds of G, and for every seed of G, it runs two simulations using polynomially-long seq.

We note here that our proof works even for a relaxed definition of CSL, in which a CSL machine is constrained to have the original auxiliary content only when it enters a configuration with q_{start}, not q_{accept}.

5   An Alternative Proof of $\mathsf{CL} = \mathsf{CSC}_1$

Under the assumption that $\mathsf{DSPACE}(n) \not\subseteq \mathsf{SIZE}(2^{\epsilon n})$, we provide an alternative proof of $\mathsf{CL} = \mathsf{CSC}_1$, without using the class CSL. For this we need to define the notion of *undirected configuration graph* for the deterministic catalytic machines.

Definition 15. For a deterministic catalytic Turing machine $\mathcal{M}$, input x, and initial auxiliary content w, the undirected configuration graph denoted by $\widetilde{\mathcal{G}}_{\mathcal{M},x,w}$ contains the two types of vertices.

- **Type 1**: A vertex for every configuration which is reachable when $\mathcal{M}$ runs on (x,w).
- **Type 2**: A vertex for every configuration which is not reachable when $\mathcal{M}$ runs on (x,w) but which can reach some configuration which is reachable when $\mathcal{M}$ runs on (x,w) by applying the transition function of $\mathcal{M}$.

$\widetilde{\mathcal{G}}_{\mathcal{M},x,w}$ has an undirected edge between a vertex v_1 and a vertex v_2 if $\mathcal{M}$ in one step can move to v_2 from v_1 or to v_1 from v_2.

In the following lemma, we prove a result similar to Lemma 11 for undirected configuration graphs of a CL machine.

Lemma 16. *Let $\mathcal{M}$ be a deterministic catalytic Turing machine with $c \log n$ size workspace and n^c size auxiliary space. Then for all x,*

$$\mathop{\mathbb{E}}_{w \in_R \{0,1\}^{n^c}} [|\widetilde{\mathcal{G}}_{\mathcal{M},x,w}|] \leq O(n^{2c+2}).$$

Proof. We first show that for an input x and any two different initial auxiliary contents w and w', $\widetilde{\mathcal{G}}_{\mathcal{M},x,w}$ and $\widetilde{\mathcal{G}}_{\mathcal{M},x,w'}$ cannot have a common vertex (or configuration). Let's assume for the sake of contradiction that $\widetilde{\mathcal{G}}_{\mathcal{M},x,w}$ and $\widetilde{\mathcal{G}}_{\mathcal{M},x,w'}$ have a common vertex v. Then, the following two cases are possible for v:

Case 1: v is a Type 1 vertex in both $\widetilde{\mathcal{G}}_{\mathcal{M},x,w}$ and $\widetilde{\mathcal{G}}_{\mathcal{M},x,w'}$.

First note that if v is a Type 1 vertex in both $\widetilde{\mathcal{G}}_{\mathcal{M},x,w}$ and $\widetilde{\mathcal{G}}_{\mathcal{M},x,w}$, then v is also a common vertex of $\mathcal{G}_{\mathcal{M},x,w}$ and $\mathcal{G}_{\mathcal{M},x,w'}$. Buhrman et al. [1] proved that two different configuration graphs $\mathcal{G}_{\mathcal{M},x,w}$ and $\mathcal{G}_{\mathcal{M},x,w'}$ cannot have a common vertex. We present their argument here for the sake of completion.

If v is a common vertex of $\mathcal{G}_{\mathcal{M},x,w}$ and $\mathcal{G}_{\mathcal{M},x,w'}$, then v is reachable both the times when $\mathcal{M}$ runs on (x, w) and when $\mathcal{M}$ runs on (x, w'). Since $\mathcal{M}$ is a deterministic machine, its run on (x, w) and (x, w') must go through the same sequence of configurations after reaching v. This implies that $\mathcal{M}$ on (x, w) has the same halting configuration as $\mathcal{M}$ on (x, w'), which is not possible because in such a halting configuration auxiliary content can either be w or w' violating the property that $\mathcal{M}$ restores the initial auxiliary content when it halts. This proves that v cannot be a common vertex of $\mathcal{G}_{\mathcal{M},x,w}$ and $\mathcal{G}_{\mathcal{M},x,w'}$, hence, v can also not be a Type 1 vertex in both $\widetilde{\mathcal{G}}_{\mathcal{M},x,w}$ and $\widetilde{\mathcal{G}}_{\mathcal{M},x,w'}$.

Case 2: v is either a Type 2 vertex in $\widetilde{\mathcal{G}}_{\mathcal{M},x,w}$ or a Type 2 vertex in $\widetilde{\mathcal{G}}_{\mathcal{M},x,w'}$.

For simplicity we only consider the case where v is a Type 2 vertex in both $\widetilde{\mathcal{G}}_{\mathcal{M},x,w}$ and $\widetilde{\mathcal{G}}_{\mathcal{M},x,w'}$, the other cases can be analysed similarly.

If v is a Type 2 vertex in $\widetilde{\mathcal{G}}_{\mathcal{M},x,w}$, then there must be a sequence of configurations, say $S_1 = v \to C_1 \to C_2 \cdots \to C_{k_1}$, where every configuration in the sequence yields the next configuration in the sequence, and C_{k_1} is reachable when $\mathcal{M}$ runs on (x, w). Similarly, since v is also a Type 2 vertex in $\widetilde{\mathcal{G}}_{\mathcal{M},x,w'}$, there must also be a sequence of configurations, say $S_2 = v \to C_1' \to C_2' \cdots \to C_{k_2}$, where every configuration in the sequence yields the next configuration in the sequence, and C_{k_2} is reachable when $\mathcal{M}$ runs on (x, w'). Existence of S_1 and S_2 follows from Definition 15.

Without loss of generality, assume that $k_1 < k_2$. Since $\mathcal{M}$ is a deterministic machine where a configuration can yield at most one configuration, $C_i = C_i'$ for $i = 1$ to k_1. This implies that C_{k_1} is present in S_2, and therefore, C_{k_2} is also reachable when $\mathcal{M}$ runs on (x, w). Therefore, C_{k_2} must be a common Type 1 vertex of $\mathcal{G}_{\mathcal{M},x,w}$ and $\mathcal{G}_{\mathcal{M},x,w'}$, which is not possible as we proved in Case 1.

A configuration of $\mathcal{M}$ can be described with at most $c \log n + n^c + \log n + \log(c \log n) + \log n^c + O(1)$ bits, where we need $c \log n + n^c$ bits for work and auxiliary tape content, $\log n + \log(c \log n) + \log n^c$ bits for the tape heads, and $O(1)$ bits for the state. Since no two different undirected configuration graphs for $\mathcal{M}$ and x can have a common vertex, the total number of possible configurations bounds the sum of the size of all the undirected configuration graphs for $\mathcal{M}$ and x.

$$\sum_{w \in \{0,1\}^{n^c}} |\widetilde{\mathcal{G}}_{\mathcal{M},x,w}| \leq O(2^{c \log n} . 2^{n^c} . n . c \log n . n^c).$$

This implies:

$$\mathbb{E}_{w \in_R \{0,1\}^{n^c}} [|\widetilde{\mathcal{G}}_{\mathcal{M},x,w}|] \leq O(n^{2c+2}).$$

$\square$

Here again, we will use a pseudorandom generator to create an auxiliary content on which a CL machine produces a small size undirected configuration graph. Lemma 16 and the assumption of Lemma 12 gives us such a pseudorandom generator. We are omitting the proof here as it is similar to the proof of Lemma 10 of [2].

Lemma 17. *Let $\mathcal{M}$ be a deterministic catalytic Turing machine using $c \log n$ size workspace and n^c size auxiliary space. If there exists a constant $\epsilon > 0$ such that $\mathsf{DSPACE}(n) \not\subseteq \mathsf{SIZE}(2^{\epsilon n})$, then there exists a function $G : \{0,1\}^{O(\log n)} \to \{0,1\}^{n^c}$, such that on every input x and initial auxiliary content w, for more than half of the seeds $s \in \{0,1\}^{O(\log n)}$, $|\tilde{\mathcal{G}}_{\mathcal{M},x,w \oplus G(s)}| \le n^{2c+3}$. Moreover, G is logspace computable. ($w \oplus G(s)$ represents the bitwise XOR of w and $G(s)$).*

Now to complete the proof we will construct a CSC_1 machine $\mathcal{M}'$ for a deterministic catalytic machine $\mathcal{M}$ with $c \log n$ size workspace and n^c size auxiliary space, such that $L(\mathcal{M}) = L(\mathcal{M}')$.

On an input x and initial auxiliary content w, $\mathcal{M}'$ uses the pseudorandom generator G of Lemma 17 and a universal exploration sequence to traverse the vertices of $\tilde{\mathcal{G}}_{\mathcal{M},x,w \oplus G(s)}$. Let seq denote a logspace computable $\left(n^{2c+3}, d\right)$-universal exploration sequence, where d is a constant upper bound on the degree of every vertex in $\tilde{\mathcal{G}}_{\mathcal{M},x,w \oplus G(s)}$.

The algorithm of $\mathcal{M}'$ is same as Algorithm 2, except in line 5 instead of traversing $\mathcal{G}_{\mathcal{M},x,w \oplus G(s)}$ it traverses the vertices of $\tilde{\mathcal{G}}_{\mathcal{M},x,w \oplus G(s)}$ using seq.

If $x \in L(\mathcal{M})$, then $\mathcal{M}'$ on a good seed s must reach an accepting configuration while simulating $\mathcal{M}$ using seq. In this case it will set $accept = \mathrm{TRUE}$ in line 6 and finally **Accept** in line 10.

If $x \notin L(\mathcal{M})$, then clearly a Type 1 vertex of $\tilde{\mathcal{G}}_{\mathcal{M},x,w \oplus G(s)}$ cannot be an accepting configuration for any seed s of G. Observe that a Type 2 vertex of $\tilde{\mathcal{G}}_{\mathcal{M},x,w \oplus G(s)}$ can also not be an accepting configuration because configurations corresponding to Type 2 vertices are non-halting by definition. Therefore, $\mathcal{M}'$ cannot reach an accepting configuration during any simulation if $x \notin L(\mathcal{M})$, due to which $\mathcal{M}'$ never sets $accept$ to TRUE and finally **Rejects** in line 13.

References

1. Buhrman, H., Cleve, R., Koucký, M., Loff, B., Speelman, F.: Computing with a full memory: catalytic space. In: Proceedings of the Forty-Sixth Annual ACM Symposium on Theory of Computing, STOC 2014, pp. 857–866. ACM, New York (2014). https://doi.org/10.1145/2591796.2591874
2. Buhrman, H., Koucký, M., Loff, B., Speelman, F.: Catalytic space: non-determinism and hierarchy. Theory Comput. Syst. **62**(1), 116–135 (2018). https://doi.org/10.1007/s00224-017-9784-7
3. Girard, V., Koucký, M., McKenzie, P.: Nonuniform catalytic space and the direct sum for space. Electron. Colloq. Comput. Complex. (ECCC) **22**, 138 (2015). http://eccc.hpi-web.de/report/2015/138
4. Gupta, C., Jain, R., Sharma, V.R., Tewari, R.: Unambiguous catalytic computation. In: FSTTCS 2019. Leibniz International Proceedings in Informatics (LIPIcs), vol. 150, pp. 16:1–16:13 (2019). https://doi.org/10.4230/LIPIcs.FSTTCS.2019.16. https://drops.dagstuhl.de/opus/volltexte/2019/11578
5. Impagliazzo, R., Wigderson, A.: P = BPP if E requires exponential circuits: derandomizing the XOR lemma. In: Proceedings of the Twenty-Ninth Annual ACM Symposium on Theory of Computing, STOC 1997, pp. 220–229. ACM, New York (1997). https://doi.org/10.1145/258533.258590

6. Klivans, A.R., van Melkebeek, D.: Graph nonisomorphism has subexponential size proofs unless the polynomial-time hierarchy collapses. SIAM J. Comput. **31**(5), 1501–1526 (2002). https://doi.org/10.1137/S0097539700389652
7. Koucký, M.: Universal traversal sequences with backtracking. J. Comput. Syst. Sci. **65**(4), 717–726 (2002). http://www.sciencedirect.com/science/article/pii/S0022000002000235
8. Lewis, H.R., Papadimitriou, C.H.: Symmetric space-bounded computation. Theoret. Comput. Sci. **19**(2), 161–187 (1982). http://www.sciencedirect.com/science/article/pii/0304397582900585
9. Reingold, O.: Undirected connectivity in log-space. J. ACM **55**(4) (2008). https://doi.org/10.1145/1391289.1391291

On the Parameterized Complexity
of the Expected Coverage Problem

Fedor V. Fomin[1] and Vijayaragunathan Ramamoorthi[2]([⊠])

[1] Department of Informatics, University of Bergen, 5020 Bergen, Norway
fedor.fomin@ii.uib.no
[2] Department of Computer Science and Engineering, IIT Madras, Chennai, India
vijayr@cse.iitm.ac.in

Abstract. The *Maximum covering location problem* (MCLP) is a well-studied problem in the field of operations research. Given a network with demands (demands can be positive or negative) on the nodes, an integer budget k, the MCLP seeks to find k potential facility centers in the network such that the neighborhood coverage is maximized. We study the variant of MCLP where edges of the network are subject to random failures due to some disruptive events. One of the popular models capturing the unreliable nature of the facility location is the linear reliable ordering (LRO) model. In this model, with every edge e of the network, we associate its survival probability $0 \le p_e \le 1$, or equivalently, its failure probability $1 - p_e$. The failure correlation in LRO is the following: If an edge e fails then every edge e' with $p_{e'} \le p_e$ surely fails. The task is to identify the positions of k facilities that maximize the *expected* coverage. We refer to this problem as EXPECTED COVERAGE problem. We study the EXPECTED COVERAGE problem from the parameterized complexity perspective and obtain the following results.

1. For the parameter treewidth, we show that the EXPECTED COV-ERAGE problem is W[1]-hard. We find this result a bit surprising, because the variant of the problem with non-negative demands is fixed-parameter tractable (FPT) parameterized by the treewidth of a graph.

2. We complement the lower bound by the proof that EXPECTED COV-ERAGE is FPT being parameterized by the treewidth and the maximum vertex degree. We give an algorithm that solves the problem in time $2^{\mathcal{O}(\mathrm{tw} \log \Delta)} n^{\mathcal{O}(1)}$, where tw is the treewidth, Δ is the maximum vertex degree, and n the number of vertices of the input graph. In particular, since $\Delta \le n$, it means the problem is solvable in time $n^{\mathcal{O}(\mathrm{tw})}$, that is, is in XP parameterized by treewidth.

Keywords: Facility location · Treewidth · W[1]-hard · Subexponential parameterized algorithm · Apex-minor-free graph

This work was done while the second author was visiting University of Bergen, Bergen, Norway supported by the Norwegian Research Council (NFR) MULTIVAL project.

H. Fernau (Ed.): CSR 2020, LNCS 12159, pp. 224–236, 2020.
https://doi.org/10.1007/978-3-030-50026-9_16

1   Introduction

Maximum covering location problem (MCLP) is a well-studied problem in the field of operations research [8]. Given a network with demands on the nodes, an integer budget k, the MCLP asks to find k potential facility centers in the network such that the neighborhood coverage is maximized.

We are interested in investigating the unreliable nature of the MCLP. Unreliability is introduced by associating survival probabilities on the edges of the input network. The notion of unreliability is used in disaster management, surviving network design and influence maximization. Assume that the network is subjected to a disaster event. During the course of disaster, some link may become non-functional. This yield a structural change in the underlying graph of the network. The resulting graph is an edge-induced subgraph of the original graph. In certain case, the resulting graph can have multiple connected components. The real challenge is to place a limited number of potential facility centers a priori such that the expected coverage after an event of disaster is maximized. See [9,13–15,19,32] for further references on unreliable MCLP.

In this paper, we study the following model of the MCLP with edge failures. Let $G = (V, E, w)$ be a vertex weighted underlying graph of the MCLP. On each edge $e \in E$, let p_e be the survival probability associated with e such that the edge e can survive in the network with probability p_e. Under the assumption that edges fail independently, the input graph can be rendered into one of 2^m edge subgraphs called *realization*, where m is number of edges in the graph. Each realization can have a non-zero probability of occurrence. Since the number of realizations is exponential and many of them occur with close to zero probability, Hassin, Ravi and Salman [26,27] formulated a dependency model for edge failure in unreliable facility networks called *linear reliable ordering* (LRO). In LRO model, for each pair of edges $e \neq e' \in E$, $p(e) \neq p(e')$, and for any pair of edges e_i and e_j with $p_{e_i} > p_{e_j}$, the $\Pr[e_j$ fails $\mid e_i$ fails $] = 1$. More precisely, if an edge e fails then every edge e' with $p_{e'} < p_e$ surely fails. It is clear that, in this model, we have exactly $m + 1$ edge subgraphs that can be rendered and they can be linearly ordered by the subgraph relationship.

While in most articles dealing with maximum coverage problems the weights assumed to be positive, there are situations when the weights can be negative. Such mixed-weight coverage problems are useful for modeling situations when some of the demand nodes are obnoxious and their inclusion in the coverage area may be detrimental [4,5]. Nodes with a negative demand are nodes we do not wish to cover. If a node has negative demand, then we wish to cover as little as possible. For example, opening a new facility (grocery store) close to many positive weighted modes (customers) seems as an excellent opportunity but the proximity of a big supermarket (a neighbor with negative weight) could decrease the expected profit.

Problem Statement. Let $G = (V, E, w, p)$ be a vertex-weighted undirected graph where $w : V \to \mathbb{R}$ and $p : E \to [0, 1]$, and k be an integer. In the LRO model, let $G_0 \preceq G_1 \preceq \cdots \preceq G_\ell$ be the linear ordering of the realizations of G, that occur with probability $P(G_i)$ for $0 \leq i \leq m$. The EXPECTED-COVERAGE problem asks to find a k sized vertex set S such that the expected coverage by S on the distribution $\{G_i \mid 0 \leq i \leq m\}$ is maximized. We use the expected coverage function $\mathcal{C}$ defined by Narayanaswamy et al. [33]. Given a pair of sets $S, T \subseteq V$, the expected coverage of T by S is

$$\mathcal{C}(T, S) = \sum_{i=0}^{m} \left(P(G_i) \sum_{v \in N_{G_i}[S] \cap T} w(v) \right).$$

The decision version of the problem is defined as follows.

EXPECTED COVERAGE
Instance: A graph $G = (V, E)$, $w : V \rightarrow \mathbb{R}$, $p : E \rightarrow [0, 1]$, an integer k and value of coverage $t \in \mathbb{R}$
Decide: Is there a set $F \subseteq V$ of size at most k such that $\mathcal{C}(V, F) \geq t$.

Related Works. The facility location problems can take many forms, depending on the objective function. In the most facility location problems, the objective function focuses on comforting the clients. For example, in the k-center problem, the goal is minimizing the maximum distance of each client from its nearest facility center [7]. The facility location problem has received a good deal of attention in the parameterized perspective [1,6,20,21].

The MCLP with edge failure has been studied with various constraints. Eiselt et al. [19] considered the problem with a single edge failure. In this case, exactly one edge would have failed after a disaster and the objective is to place k facility centers such that the expected weight of non-covered vertices is minimized. If the number of facility centers is $k = 1$, and the facility center can cover all the vertices in the connected component, then the problem is studied as *most reliable source* (MRS) problem. In this problem, the edges are failed independently. The MRS problem has received a good deal of attention in the literature [9,13,15,32]. Hassin et al. [26] studied the problem with edge failure follows LRO failure model. The problem is referred to as the MAX-EXP-COVER-R problem. An additional input radius of coverage R is also given such that any facility center can cover a vertex at distance at most R. The MAX-EXP-COVER-R problem is shown to be NP-hard even when $R = 1$. When $R = \infty$, the problem is polynomial time solvable [26].

In *budgeted dominating set* problem, we are given a graph G and an integer k, and asked to find a set of at most k vertices S maximizing the value $w(N[S])$ in G. Set theoretic version of the BDS problem is studied as budgeted maximum coverage in [28,29]. The EXPECTED COVERAGE problem can be viewed as a generalization of the BDS problem. When we have probability 1 on all the edges, then both these problems are the same. The BDS problem generalizes *partial dominating set* (PDS) problem, where one seeks a set of size at most k vertices dominating at least t vertices [31]. Of course, all these problems also generalize the fundamental *dominating set* problem, where the task is to find a set of at most k vertices dominating all remaining vertices of the graph.

The dominating set problem on general graphs is W[2]-hard [16]. However, on planar graphs it is FPT [25]. Moreover, on planar, and more generally on H-minor-free graphs it is solvable in subexponential time [3,11]. It also admits a linear kernel on planar graphs, H-minor-free graphs and graphs of bounded expansion [2,18,23,24,34]. Subexponential parameterized algorithm for PDS on planar, and more generally, apex-minor-free graphs, was given in [22].

On graphs of bounded treewidth, the classical dynamic programming, see, e.g., [10], shows that DOMINATING SET is FPT parameterized by the treewidth of an input graph. The FPT algorithm for dominating set can be adapted to solve the BDS problem in FPT time. Further, when we have mixed vertex weights on the BDS problem, it remains FPT parameterized by the treewidth of the input graph. Narayanaswamy et al. [33] gave an FPT algorithm to solve the EXPECTED COVERAGE problem with *non-negative* weights.[1]

Our Results. Since the EXPECTED COVERAGE problem (with mixed-weights) generalizes both the BDS problem and the EXPECTED COVERAGE problem with non-negative weights, it is also natural to ask what algorithmic results for these problems can be extended to the EXPECTED COVERAGE problem. We obtain the following results.

1. For the parameter treewidth, we show that the EXPECTED COVERAGE problem is W[1]-hard. Moreover, the problem remains W[1]-hard for any combination of parameters treewidth, budget and value of coverage. This is interesting because as it was shown by Narayanaswamy et al. [33], the variant of the problem with only non-negative weight is FPT parameterized by the treewidth. Thus the results for non-negative weights cannot be (unless FPT = W[1]) extended to the mixed-weight model.
2. We complement the lower bound by the proof that EXPECTED COVERAGE is FPT being parameterized by the treewidth and the maximum vertex degree. We give an algorithm that solves the problem in time $2^{\mathcal{O}(\text{tw} \log \Delta)} n^{\mathcal{O}(1)}$, where tw is the treewidth, Δ is the maximum vertex degree, and n the number of vertices of the input graph. In particular, since $\Delta \leq n$, it means the problem is solvable in time $n^{\mathcal{O}(\text{tw})}$, that is, is in XP parameterized by treewidth.

We refer to the recent books of Cygan et al. [10] and Downey and Fellows [17] for detailed introductions to parameterized complexity. We use standard graph theoretic notations based on Diestel [12].

2 Parameterized Intractability: The EXPECTED COVERAGE Problem is W[1]-hard for the Parameter Treewidth

In this section, we show that the EXPECTED COVERAGE problem is W[1]-hard parameterized by the treewidth. We reduce from the MULTICOLOR CLIQUE problem which is defined as follows. Given a k-partite graph $G = (V, E)$ where $V = V_1 \cup V_2 \cup \cdots \cup V_k$, and an integer k, is there a k-clique with exactly one

---

[1] Narayanaswamy et al. [33] called this problem MAX-EXP-COVER-1-LRO.

vertex from each partition. By the classical work of Downey and Fellows [16], MULTICOLOR CLIQUE problem is W[1]-complete.

Theorem 1 ([16]). MULTICOLOR CLIQUE *is W[1]-complete for the parameter* k.

For each $1 \leq i \neq j \leq k$, let $E_{i,j} \subseteq E$ be the set of all edges where one end vertex is in V_i and another one is in V_j. That is, $E_{i,j} = \{uv \in E \mid u \in V_i \wedge v \in V_j\}$. Then, $E = \bigcup_{1 \leq i < j \leq k} E_{i,j}$ is a $\binom{k}{2}$ partition of the edge set.

2.1 Construction

We will show how given an instance (G, k) of MULTICOLOR CLIQUE problem, we can construct an instance (H, w, p, k', t') of the EXPECTED COVERAGE problem where $k' = k + \binom{k}{2}$ and $t' = k^4 + k^3 - k^2 + k$ such that both instances are equivalent and treewidth of the graph H is $\mathcal{O}(k^2)$. Now we describe the construction of the graph H, and the functions $w : V(H) \to \mathbb{R}$ and $p : E(H) \to [0, 1]$.

For each $1 \leq i \leq k$, we construct a *vertex-partition gadget* H_i corresponding to the vertex partition V_i as follows. For each vertex $v \in V_i$, we add a vertex a_v with $w(a_v) = 0$ in the gadget H_i. We add two more vertices t_i with $w(t_i) = k^2$, and q_i with $w(q_i) = k^2$ to the gadget H_i. For each vertex $v \in V_i$, the vertex a_v is made adjacent to the vertices t_i and q_i. For each edge $e \in E(H_i)$, we define the survival probability $p(e) = 1$. Thus, the gadget H_i has $|V_i| + 2$ vertices and $2|V_i|$ edges.

For each $1 \leq i < j \leq k$, we construct an *edge-partition gadget* $H_{i,j}$ corresponding to the edge partition $E_{i,j}$. For each edge $e \in E_{i,j}$, we add a vertex a_e with $w(a_e) = 0$ in the gadget $H_{i,j}$. We add two more vertices $t_{i,j}$ and $q_{i,j}$ with $w(t_{i,j}) = k^2 = w(q_{i,j}) = k^2$ to the gadget $H_{i,j}$. For each edge $e \in E_{i,j}$, the vertex a_v is made adjacent to the vertices $t_{i,j}$ and $q_{i,j}$. For each edge $e \in E(H_{i,j}$, we define the survival probability $p(e) = 1$. Thus, the gadget H_i has $|E_{i,j}| + 2$ vertices and $2|E_{i,j}|$ edges.

Finally, we introduce the *connector vertices* between the edge-partition gadgets and vertex-partition gadgets. Let $R = \{s_{i,j}^i, s_{i,j}^j, r_{i,j}^i, r_{i,j}^j \mid 1 \leq i < j \leq k\}$ be the connector vertices. For each vertex $x \in R$, we define $w(x) = -1$. To establish the edges between the gadgets and the connector vertices, we define a probability function $z : V \to [0, 1]$ such that for any two vertices $u \neq v$, $z(v) \neq z(v)$. For $1 \leq i \leq k$, the gadget H_i is connected to the set R as follows. For each vertex $v \in V_i$, the vertex $a_v \in H_i$ is made adjacent to the vertices $s_{i,j}^i$ and $r_{i,j}^i$ with survival probabilities $z(v)$ and $1 - z(v)$, respectively. For $1 \leq i < j \leq k$, the gadget $H_{i,j}$ is connected to the set R as follows. For each edge $e = uv \in E_{i,j}$ with $u \in V_i$ and $v \in V_j$, the vertex a_e is made adjacent to the vertices $s_{i,j}^i, r_{i,j}^i, s_{i,j}^j$ and $r_{i,j}^j$ with survival probabilities $z(u)$, $1 - z(u)$, $z(v)$ and $1 - z(v)$, respectively. An illustration of a vertex-partition gadget and an edge-partition gadget connected to the connector vertices is given in Fig. 1. For clarity, we denote the vertices a_v and a_e in $V(H)$ for each $v \in V$ and $e \in E$, as *real* vertices, and the remaining vertices in $V(H)$ are denoted by *non-real* vertices. Thus, the graph H is constructed with $N = n + m + 3k^3 - 3k$ vertices and $M = (2k+2)n + 6m$ edges.

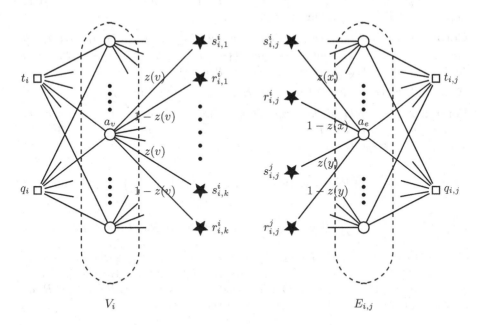

$$V_i \qquad\qquad\qquad E_{i,j}$$

Fig. 1. Gadgets for a partition V_i and $E_{i,j}$ for some $i \neq j$ are given. Star shaped vertices are connector vertices. Real vertices are represented by circle shape. Square shaped vertices are part of gadgets. Let $a_e \in V(H_{i,j})$ be the vertex illustrated for some edge $e = xy \in E_{i,j}$ such that $x \in V_i$ and $y \in V_j$.

For each vertex $v \in V$, $\mathcal{C}(V(H), a_v) \leq 2k^2$ since a_v has some neighbors with negative weight. Similarly, for each edge $e \in E$, $\mathcal{C}(V(H), a_e) \leq 2k^2$. For each $1 \leq i \leq k$, $\mathcal{C}(V(H), t_i) = \mathcal{C}(V(H), q_i) = k^2$ since all neighbors of t_i are vertices with weight 0. Similarly, for each $1 \leq i < j \leq k$, $\mathcal{C}(V(H), t_{i,j}) = \mathcal{C}(V(H), q_{i,j}) = k^2$.

Lemma 1. *For each $1 \leq i < j \leq k$, let a_v and a_{xy} be a pair of vertices for some $v, x \in V_i$ and $y \in V_j$. Then, the expected coverage of the vertices $s_{i,j}^i$ and $r_{i,j}^i$ by a_v and a_{xy} is given as follows.*

1. $\mathcal{C}(s_{i,j}^i, \{a_v, a_{xy}\}) = -(\max(z(v), z(x)))$.
2. $\mathcal{C}(r_{i,j}^i, \{a_v, a_{xy}\}) = -(\max(1 - z(v), 1 - z(x)))$.

Proof. The weights of the vertices $s_{i,j}i$ and $r_{i,j}^i$ are -1. The probabilities of the edges $a_v s_{i,j}^i$, $a_v r_{i,j}^i$, $a_{xy} s_{i,j}^i$ and $a_{xy} r_{i,j}^i$ are $z(v), 1 - z(v), z(x)$ and $1 - z(x)$, respectively. Then we have $\mathcal{C}(s_{i,j}^i, \{a_v, a_{xy}\}) = -1 \times (\max(z(v), z(x))) = -(\max(z(v), z(x)))$. Similarly, we have $\mathcal{C}(r_{i,j}^i, \{a_v, a_{xy}\}) = -(\max(1 - z(v), 1 - z(x)))$. Hence the lemma is proved. $\qquad\square$

Any solution that contains either $s_{i,j}^i$, $r_{i,j}^i$ or both, we can switch the selection of these vertices by one of their neighbor such that the expected coverage value is not decreased. To maximize the expected coverage, we need the coverage of $r_{i,j}^i$ and $s_{i,j}^i$ by the pair of vertices a_v and a_{xy} is as minimum as possible. This implies the following corollary from Lemma 1.

Corollary 1. *The expected coverage of vertices $s_{i,j}^i$ and $r_{i,j}^i$ by a_v and a_{xy} is maximum when $v = x$. That is, we have $\mathcal{C}(\{s_{i,j}^i, r_{i,j}^i\}, \{a_v, a_{xy}\}) = -1$. Otherwise ($v \neq x$), the expected coverage value is less than -1.*

We bound the treewidth of the graph H using the following lemma.

Lemma 2. *Treewidth of the graph H is at most $5\binom{k}{2} + k + 1$. Moreover, the vertex cover number and feedback vertex set number of the graph H is $3k^2 - k$ and $5\binom{k}{2} + k$, respectively.*

Proof. Consider the following two sets $T = \bigcup_{i=1}^{k}\{t_i\} \cup \bigcup_{1 \leq i < j \leq k}\{t_{i,j}\}$ and $Q = \bigcup_{i=1}^{k}\{q_i\} \cup \bigcup_{1 \leq i < j \leq k}\{q_{i,j}\}$. If we remove the vertices in $R \cup T$ from H, then we obtain a forest. Treewidth of a graph can decrease at most one when we remove a vertex. Then, treewidth of H is at most $|R \cup T| + 1$. Thus, treewidth of H is at most $5\binom{k}{2} + k + 1 = \mathcal{O}(k^2)$. Observe that the sets $T \cup Q \cup R$ and $T \cup R$ are the vertex cover and feedback vertex set of the graph H. $\qquad\square$

2.2 Equivalence

Now we show the equivalence of both the problems. More precisely, the graph G has a k-clique if and only if H has k' vertices that achieve the expected coverage of value at least t'.

Lemma 3. *If (G, k) is an yes instance of the MULTICOLOR CLIQUE problem, then (H, k', t') is also an yes instance the EXPECTED COVERAGE problem.*

Proof. Let $K = \{v_1, v_2, \ldots v_k\}$ be a k-clique in G such that for $1 \leq i \leq k$, $v_i \in V_i$. Now we construct a feasible solution $S \subseteq V(H)$ for the instance (H, k', t') of the EXPECTED COVERAGE problem. Let $S = \{a_{v_i} \mid 1 \leq i \leq k\} \cup \{a_{v_i v_j} \mid 1 \leq i < j \leq k\}$. Clearly, the size of S is exactly the budget k'. The expected coverage by the set S is given as:

$$\mathcal{C}(V(H), S) = \sum_{1 \leq i \leq k}\left(\mathcal{C}(\{t_i, a_{v_i}, q_i\}, a_{v_i})\right) + \sum_{1 \leq i < j \leq k}\left(\mathcal{C}(\{t_{i,j}, a_{v_i v_j}, q_{i,j}\}, a_{v_i v_j})\right)$$

$$+ \sum_{1 \leq i < j \leq k, l \in \{i,j\}}\left(\mathcal{C}(s_{i,j}^l, \{a_{v_i}, a_{v_i v_j}\}) + \mathcal{C}(r_{i,j}^l, \{a_{v_i}, a_{v_i v_j}\})\right)$$

$$= \sum_{1 \leq i \leq k} 2k^2 + \sum_{1 \leq i < j \leq k} 2k^2 + \sum_{1 \leq i < j \leq k} -2 = \left(k + \binom{k}{2}\right)(2k^2) - 2\binom{k}{2}$$

$$= (k^2 + k)k^2 - (k^2 - k) = k^4 + k^3 - k^2 + k = t'.$$

We apply Lemma 1 in the second step to replace the exact value of expected coverage. Thus, we showed that $\mathcal{C}(V(H), S) = k^4 + k^3 - k^2 + k = t'$.   □

Now we prove the other direction of equivalence. Let $S \subseteq V(H)$ be a feasible solution that achieves an expected coverage of at least $t' = k^4 + k^3 - k^2 + k$. Observe that any vertex can achieve an expected coverage of value at most $2k^2$. Specifically, any vertex a_v or a_e for some $v \in V$ or $e \in E$ can achieve an expected coverage of value at most $2k^2$, whereas, other vertices can achieve at most k^2.

Lemma 4. *Any set $S \subseteq V(H)$ that contains a non-real vertex is infeasible.*

Proof. Assume that the set S contains a non-real vertex. The maximum value of the expected coverage by the set S is achieved when S has exactly on non-real vertex. That is, other vertices in the set S are real vertices. A non-real vertex can cover at most a value of k^2. Then the expected coverage of S is given as follows:

$$\mathcal{C}(V(H), S) \leq (k' - 1)(2k^2) + k^2 = (k^2 + k - 2)(k^2) + k^2$$
$$\leq k^4 + k^3 - 2k^2 + k^2 = k^4 + k^3 - k^2 < t'.$$

Thus, the solution S is not a feasible solution.   □

Lemma 4 states that every feasible solution is a set of k' real vertices from $V(H)$. Further, we claim that a feasible solution must intersect with every gadget.

Lemma 5. *Every feasible solution $S \subseteq V(H)$ must have*

- *for each $1 \leq i \leq k$, $S \cap \{a_v \mid v \in V_i\} \neq \emptyset$, and*
- *for each $1 \leq i < j \leq k$, $S \cap \{a_e \mid e \in E_{i,j}\} \neq \emptyset$.*

Proof. As we stated early in the construction, the gadgets for the vertex-partition and edge-partition are disjoint and connected through vertices with negative weights. By contradiction, assume that there exists solution $S \subseteq V(H)$ such that there exists a gadget with no vertex from the gadget is in S. Without loss of generality assume that a vertex-partition gadget corresponding to a partition V_i is having no intersection with the set S. Since there are $\binom{k}{2} + k$ gadgets, exactly one of the gadget will have two vertices in S and all other gadgets will have one vertex from S. For any gadget, the expected coverage contribution by the vertices in the gadget is at most $2k^2$ even the gadget has more than one vertex from S. Then we have the following:

$$\mathcal{C}(V(H) \setminus R, S) \leq (k' - 1)(2k^2)$$
$$= (k^2 + k)(k^2) - 2k^2 = k^4 + k^3 - 2k^2 < t'.$$

Then we have, $\mathcal{C}(V(H), S) \leq \mathcal{C}(V(H) \setminus R_E, S) \leq (k^4 + k^3 - 2k^2) < t'$. Thus, the solution S is infeasible.   □

Lemmas 4 and 5 together state that every feasible solution S must be a subset of real vertices and for each $1 \leq i \leq k$, there exists a vertex $v \in V_i$ such that $a_v \in S$, and for each $1 \leq i < j \leq k$, there exists an edge $e \in E_{i,j}$ such that $a_e \in S$. Now we prove the other direction of the equivalence.

Lemma 6. *If (H, k', t') is an* yes *instance of the* EXPECTED COVERAGE *problem then (G, k) is also an yes instance of the* MULTICOLOR CLIQUE *problem.*

Proof. Let S be a feasible solution for the instance (H, k', t') of the EXPECTED COVERAGE problem. The feasibility of S ensures that every gadget has non-zero coverage. More specifically, each gadget contributes an expected coverage of $2k^2$. Then $\mathcal{C}(V(H) \setminus R, S) = k'(2k^2) = k^4 + k^3$. Since we claim that S is a feasible solution, we have to show that $\mathcal{C}(R, S) = k - k^2$.

There are $2\binom{k}{2}$ pairs of $s^i_{i,j}$ and $r^i_{i,j}$ connector vertices in H. By Lemma 1, each pair can contribute at most -1. Then, the value $k - k^2$ can be achieved only when each pair is contributing exactly -1. From Corollary 1, for each $1 \leq i < j \leq k$, the pair $r^i_{i,j}$ and $s^i_{i,j}$ together can contribute exactly -1 when $a_v \in S$ and $a_{vx} \in S$ for some $v \in V_i$ and $x \in V_j$.

Let $K = \{v \in V_i \mid a_v \in S\}$ be a k sized vertex set from V of G, and form a k-clique because there is an edge between every pair of vertices in K. □

Thus, we state the following theorem using the Lemmas 2, 3 and 6.

Theorem 2. *The* EXPECTED COVERAGE *problem is W[1]-hard for the parameter treewidth.*

Moreover, the parameterized reduction preserves the parameters k', t' and the treewidth of the constructed graph as a functions of k. That is, $k' = k + \binom{k}{2}$, $t' = k^4 + k^3 - k^2 + k$ and treewidth of the graph H is $\mathcal{O}(k^2)$. We conclude the section with the following corollary.

Corollary 2. *The* EXPECTED COVERAGE *problem is W[1]-hard for any combination of parameters, budget k, treewidth tw, feedback vertex set number and vertex cover number.*

3 FPT Algorithm for the EXPECTED COVERAGE Problem Parameterized by Treewidth on Bounded Degree Graphs

While, as we have seen, the EXPECTED COVERAGE problem is W[1]-hard for the parameter treewidth, we give an FPT algorithm for the combined treewidth and maximum vertex degree parameter.

To describe the algorithm, we first define the tree decomposition of a graph. A tree decomposition of an undirected graph $G = (V, E)$ is a pair $(X, \mathcal{T})$ where $\mathcal{T}$ is a tree whose vertices are called *nodes* and $X = \{X_i \subseteq V \mid i \in V(\mathcal{T})\}$ such that

1. for each vertex $u \in E$, there is a node $i \in V(\mathcal{T})$ such that $u \in X_i$,
2. for each edge $uv \in E$, there is a node $i \in V(\mathcal{T})$ such that $u, v \in X_i$, and
3. for each vertex $v \in V$, the set $\{i \in V(\mathcal{T}) \mid v \in X_i\}$ forms a subtree of T.

The *width* of a tree decomposition $(X, \mathcal{T})$ equals $max_{i \in V(\mathcal{T})} |X_i| - 1$. The treewidth of a graph G is the minimum width over all tree decompositions of G. We give a dynamic programming algorithm working on a so-called *nice tree decomposition* of the input graph G. A tree decomposition $(X, \mathcal{T})$ is a nice tree decomposition if $\mathcal{T}$ is rooted by a node r with $X_r = \emptyset$ and every node in $\mathcal{T}$ is either an *insert node*, *forget node*, *join node* or *leaf node*. Thereby, a node $i \in V(\mathcal{T})$ is an insert node if i has exactly one child j such that $X_i = X_j \cup \{v\}$ for some $v \notin X_j$; it is a forget node if i has exactly one child j such that $X_i = X_j \setminus \{v\}$ for some $v \in X_j$; it is a join node if i has exactly two children j and k such that $X_i = X_j = X_k$; and it is a leaf node if $X_i = \emptyset$. Given a tree decomposition of width tw, a nice tree decomposition of width tw can be obtained in linear time [30]. For a node $i \in V(\mathcal{T})$, let $\mathcal{T}_i$ be a subtree rooted at i and $X_i^+ = \cup_{j \in V(\mathcal{T}_i)} \{X_j\}$.

3.1   Solution Structure

Let $\langle G, w, p, k \rangle$ be an input to the optimization version of the EXPECTED COVERAGE problem. Let $(\mathcal{X}, \mathcal{T})$ be a nice tree decomposition of G with treewidth tw. Narayanaswamy et al. [33] introduced the notion of best neighbor to solve the EXPECTED COVERAGE problem with non-negative weights on bounded treewidth graphs. Consider any feasible solution $S \subseteq V$ of size k with expected coverage value t. Since the failure model follows the LRO distribution, then for each vertex u (with $u \in N(S)$) that contributes to t, there exists a unique vertex in S called *best neighbor* of u in S, denoted by $bn(u, S)$. We use the fact that the graph G has bounded degree. We define a structural ordering called *neighborhood indexing* on the neighborhood of each vertex. This LRO specific intuitions *"best neighbor"* and *"neighborhood indexing"*, help us to solve the problem efficiently in tree decomposition.

Neighborhood Indexing - We define an ordered indexing on the neighborhood of each vertex v of G. Let $v \in V$ be a vertex. Let $D = N(v)$ be the open neighborhood of v. We order the vertices in D based on the survival probability of the edge connected to v in non-increasing order. Let $D_v = \{u_1, u_2, \ldots, u_l\}$ with $l = deg(v) \leq \Delta$ be the ordering of the vertices described above. For each vertex v, let $\mathfrak{N}_v : \{1, 2, \ldots, deg(v)\} \to N(v)$ be a function on input an integer $r \leq deg(v)$ outputs the rth-vertex u from the ordered set D_v.

Now we show the structure of a solution in the tree decomposition. Let i be a node in $\mathcal{T}$. We define two labels c and d as follows. Let $c : X_i \to \{0, 1\}$ be a function decides for each vertex $v \in X_i$ whether v is covered at the current stage. Let $d : X_i \to \{-1, 0, 1, \ldots, \Delta\}$ be a function that assigns $\Delta + 2$ different values to the vertices in X_i as follows.

- If $d(v) = -1$, then the vertex v has no neighbor from the solution at current stage.
- If $d(v) = 0$, then the vertex v is part of the solution at current stage.
- If $d(v) > 0$, then the vertex $\mathfrak{N}_v(d(v))$ be the best neighbor of v in the solution at current stage.

Intuitively, the label d determines the index of best neighbor in the solution at current stage. Let ℓ be an integer with $\ell \leq k$. A triple (ℓ, c, d) is said to be *valid* if it satisfies the following conditions.

1. $d^{-1}(0) \cap c^{-1}(0) = \emptyset$.
2. $d^{-1}(-1) \cap c^{-1}(1) = \emptyset$.
3. For any vertex $v \in X_i$ with $d(v) > 0$, $\mathfrak{N}_v(d(v)) \in X_i^+$.
4. Let $A = d^{-1}(0) \cup \bigcup\limits_{v \in X_i | d(v) > 0} \{\mathfrak{N}_v(d(v))\}$. Then, $|A| \leq \ell$ and for each vertex $v \in X_i$ with $d(v) > 0$, $bn(v, A) = \mathfrak{N}_v(d(v))$.

Otherwise, the triple (ℓ, c, d) is *invalid*.

3.2 Dynamic Programming

We present a dynamic programming formulation for the EXPECTED COVERAGE problem on bounded degree graphs parameterized by treewidth. Our dynamic programming maintains a table T_i for every node $i \in \mathcal{T}$ that has three dimensions, an integer ℓ with $\ell \leq k$ and a pair of labels $c : X_i \to \{0, 1\}$ and $d : X_i \to \{-1, 0, 1, \ldots, \Delta\}$. For every valid triple (ℓ, c, d), the table entry $T_i[\ell, c, d]$ is a pair (**Solution, Value**) that consists of a set of size ℓ and a value of expected coverage. Let $A = d^{-1}(0) \cup \bigcup\limits_{v \in X_i | d(v) > 0} \{\mathfrak{N}_v(d(v))\}$. Let $S_i = T_i[\ell, c, d].$**Solution**. Then the set S_i is the optimal solution for the instance $(G[X_i^+ \setminus c^{-1}(0)], \ell)$ of the EXPECTED COVERAGE problem conditioned on

1. $A \subseteq S_i$ and $|S_i| \leq \ell$,
2. for each vertex $v \in X_i \setminus (d^{-1}(0) \cup d^{-1}(-1))$, $\mathfrak{N}_v^{-1}(bn(v, S_i)) = d(v)$,
3. $\mathcal{C}(X_i^+ \setminus c^{-1}(0), S_i)$ is maximized, and
4. $T_i[\ell, c, d].$**Value** $= \mathcal{C}(X_i^+ \setminus c^{-1}(0), S_i)$

For each invalid entry (ℓ, c, d), $T_i[\ell, c, d].$**Solution** $= \emptyset$ and $T_i[\ell, c, d].$**Value** $= -\infty$. Clearly, the table contained in the root node r has the solution for the EXPECTED COVERAGE problem. Note that the root node r has $X_r = \emptyset$. For $c : \emptyset \to \{0, 1\}$, $d : \emptyset \to \{-1, 0, 1, \ldots, \Delta\}$ and $\ell = k$, the table entry $T_r[\ell, c, d]$ gives an optimal solution for the EXPECTED COVERAGE problem.

Due to lack of space, the update operation of the dynamic programming are deferred to the full version of this paper.

Theorem 3. *The* EXPECTED COVERAGE *problem can be solved optimally in time* $2^{\mathcal{O}(\text{tw} \log \Delta)} n^{\mathcal{O}(1)}$ *where tw and Δ are the treewidth and max-degree of the input graph.*

Proof. We have shown that every feasible solution follows the labeling c and d as described in solution structure. We enumerate all such labeling on the bags of tree decomposition and output the optimal one on the root of tree decomposition. Note that every node has a table of size $(k+1)(2\Delta+4)^{tw}$ and each entry can be updated in time $\mathcal{O}(2\Delta+4)^{tw}$. Since the nice tree decomposition with $\mathcal{O}(n \cdot tw)$ nodes can be computed in polynomial time [30], the EXPECTED COVERAGE problem can be solved in time $2^{\mathcal{O}(\text{tw} \log \Delta)} n^{\mathcal{O}(1)}$. $\qquad \square$

References

1. Ageev, A.A.: A criterion of polynomial-time solvability for the network location problem. In: Proceedings of the 2nd Integer Programming and Combinatorial Optimization Conference, Pittsburgh, PA, USA, May 1992, pp. 237–245 (1992)
2. Alber, J., Fellows, M.R., Niedermeier, R.: Polynomial-time data reduction for dominating set. J. ACM **51**(3), 363–384 (2004)
3. Alber, J., Fernau, H., Niedermeier, R.: Parameterized complexity: exponential speed-up for planar graph problems. J. Algorithms **52**(1), 26–56 (2004)
4. Berman, O., Drezner, Z., Krass, D.: Generalized coverage: new developments in covering location models. Comput. Oper. Res **37**(10), 1675–1687 (2010)
5. Berman, O., Drezner, Z., Wesolowsky, G.O.: The maximal covering problem with some negative weights. Geograph. Anal. **41**(1), 30–42 (2009)
6. van Bevern, R., Tsidulko, O.Y., Zschoche, P.: Fixed-parameter algorithms for maximum-profit facility location under matroid constraints. In: Heggernes, P. (ed.) CIAC 2019. LNCS, vol. 11485, pp. 62–74. Springer, Cham (2019). https://doi.org/10.1007/978-3-030-17402-6_6
7. Charikar, M., Guha, S.: Improved combinatorial algorithms for the facility location and k-median problems. In: 40th Annual Symposium on Foundations of Computer Science, FOCS 1999, New York, NY, USA, 17–18 October 1999, pp. 378–388 (1999)
8. Church, R., Velle, C.R.: The maximal covering location problem. Pap. Reg. Sci. **32**(1), 101–118 (1974)
9. Colbourn, C.J., Xue, G.: A linear time algorithm for computing the most reliable source on a series-parallel graph with unreliable edges. Theor. Comput. Sci. **209**(1), 331–345 (1998)
10. Cygan, M., et al.: Parameterized Algorithms. Springer, Cham (2015). https://doi.org/10.1007/978-3-319-21275-3
11. Demaine, E.D., Fomin, F.V., Hajiaghayi, M.T., Thilikos, D.M.: Subexponential parameterized algorithms on graphs of bounded genus and H-minor-free graphs. J. ACM **52**(6), 866–893 (2005)
12. Diestel, R.: Graph Theory. GTM, vol. 173. Springer, Heidelberg (2017). https://doi.org/10.1007/978-3-662-53622-3
13. Ding, W.: Computing the most reliable source on stochastic ring networks. In: 2009 WRI World Congress on Software Engineering, vol. 1, pp. 345–347, May 2009
14. Ding, W.: Extended most reliable source on an unreliable general network. In: 2011 International Conference on Internet Computing and Information Services, pp. 529–533, September 2011
15. Ding, W., Xue, G.: A linear time algorithm for computing a most reliable source on a tree network with faulty nodes. Theor. Comput. Sci. **412**(3), 225–232 (2011). Combinatorial Optimization and Applications
16. Downey, R.G., Fellows, M.R.: Fixed parameter tractability and completeness. In: Complexity Theory: Current Research, Dagstuhl Workshop, 2–8 February 1992, pp. 191–225 (1992)
17. Downey, R.G., Fellows, M.R.: Fundamentals of Parameterized Complexity. Texts in Computer Science. Springer, London (2013). https://doi.org/10.1007/978-1-4471-5559-1
18. Drange, P.G., et al.: Kernelization and sparseness: the case of dominating set. In: Proceedings of the 33rd International Symposium on Theoretical Aspects of Computer Science (STACS). LIPIcs, vol. 47, pp. 31:1–31:14. Schloss Dagstuhl - Leibniz-Zentrum fuer Informatik (2016)

19. Eiselt, H.A., Gendreau, M., Laporte, G.: Location of facilities on a network subject to a single-edge failure. Networks **22**(3), 231–246 (1992)
20. Feldmann, A.E., Marx, D.: The parameterized hardness of the k-center problem in transportation networks. In: 16th Scandinavian Symposium and Workshops on Algorithm Theory, SWAT 2018, Malmö, Sweden, 18–20 June 2018, vol. 101, pp. 19:1–19:13 (2018)
21. Fellows, M.R., Fernau, H.: Facility location problems: a parameterized view. Discrete Appl. Math. **159**(11), 1118–1130 (2011)
22. Fomin, F.V., Lokshtanov, D., Raman, V., Saurabh, S.: Subexponential algorithms for partial cover problems. Inf. Process. Lett. **111**(16), 814–818 (2011)
23. Fomin, F.V., Lokshtanov, D., Saurabh, S., Thilikos, D.M.: Bidimensionality and kernels. In: Proceedings of the 21st Annual ACM-SIAM Symposium on Discrete Algorithms (SODA), pp. 503–510. SIAM (2010)
24. Fomin, F.V., Lokshtanov, D., Saurabh, S., Thilikos, D.M.: Kernels for (connected) dominating set on graphs with excluded topological minors. ACM Trans. Algorithms **14**(1), 6:1–6:31 (2018)
25. Frick, M., Grohe, M.: Deciding first-order properties of locally tree-decomposable graphs. In: Wiedermann, J., van Emde Boas, P., Nielsen, M. (eds.) ICALP 1999. LNCS, vol. 1644, pp. 331–340. Springer, Heidelberg (1999). https://doi.org/10.1007/3-540-48523-6_30
26. Hassin, R., Ravi, R., Salman, F.S.: Tractable cases of facility location on a network with a linear reliability order of links. In: Fiat, A., Sanders, P. (eds.) ESA 2009. LNCS, vol. 5757, pp. 275–276. Springer, Heidelberg (2009). https://doi.org/10.1007/978-3-642-04128-0_24
27. Hassin, R., Ravi, R., Salman, F.S.: Multiple facility location on a network with linear reliability order of edges. J. Comb. Optim. **34**, 931–955 (2017). https://doi.org/10.1007/s10878-017-0121-5
28. Hochbaum, D.S. (ed.): Approximation Algorithms for NP-Hard Problems. PWS Publishing Co., Boston (1997)
29. Khuller, S., Moss, A., Naor, J.: The budgeted maximum coverage problem. Inf. Process. Lett. **70**(1), 39–45 (1999)
30. Kloks, T. (ed.): Treewidth. LNCS, vol. 842. Springer, Heidelberg (1994). https://doi.org/10.1007/BFb0045375
31. Kneis, J., Mölle, D., Rossmanith, P.: Partial vs. complete domination: t-dominating set. In: van Leeuwen, J., Italiano, G.F., van der Hoek, W., Meinel, C., Sack, H., Plášil, F. (eds.) SOFSEM 2007. LNCS, vol. 4362, pp. 367–376. Springer, Heidelberg (2007). https://doi.org/10.1007/978-3-540-69507-3_31
32. Melachrinoudis, E., Helander, M.E.: A single facility location problem on a tree with unreliable edges. Networks **27**(4), 219–237 (1996)
33. Narayanaswamy, N.S., Nasre, M., Vijayaragunathan, R.: Facility location on planar graphs with unreliable links. In: Fomin, F.V., Podolskii, V.V. (eds.) CSR 2018. LNCS, vol. 10846, pp. 269–281. Springer, Cham (2018). https://doi.org/10.1007/978-3-319-90530-3_23
34. Philip, G., Raman, V., Sikdar, S.: Polynomial kernels for dominating set in graphs of bounded degeneracy and beyond. ACM Trans. Algorithms **9**(1), 11 (2012)

Computational Hardness
of Multidimensional Subtraction Games

Vladimir Gurvich[1,4] and Mikhail Vyalyi[1,2,3(✉)]

[1] National Research University Higher School of Economics, Moscow, Russia
vladimir.gurvich@gmail.com, vyalyi@gmail.com
[2] Moscow Institute of Physics and Technology, Moscow, Russia
[3] Dorodnicyn Computing Centre, FRC CSC RAS, Moscow, Russia
[4] Rutgers University, New Brunswick, NJ 08901-8554, USA

Abstract. We study the algorithmic complexity of solving subtraction games in a fixed dimension with a finite difference set. We prove that there exists a game in this class such that solving the game is **EXP**-complete and requires time $2^{\Omega(n)}$, where n is the input size. This bound is optimal up to a polynomial speed-up.

The results are based on the construction introduced by Larsson and Wästlund. It relates subtraction games and cellular automata.

Keywords: Subtraction games · Cellular automata · Computational hardness

1 Introduction

The algorithmic complexity of solving combinatorial games is an important area of research. Some famous games can be solved efficiently. For example, the ancient game of nim was solved by Bouton in [8].

A position of nim is determined by n heaps of pebbles. By one move it is allowed to reduce (strictly) exactly one heap. Two players move alternately. One who is out of moves loses. In fact, in his paper Bouton obtained an explicit formula for the Sprague-Grundy (SG) function of the disjunctive compound (or, for brevity, the sum) of impartial games (see [2,9,15] and Sect. 2.1 for definitions.)

Later efficient algorithms were developed for several versions and/or generalizations of nim: Wythoff's nim [13,25], Fraenkel's nim [12,13], nim(a, b) game [3], Moore's (n, k)-nim with $k = n - 1$ [7,19,21], and the exact (n, k)-nim with $2k \geqslant n$ [5,7].

In all these versions it is allowed to reduce by one move several heaps, not necessarily only one. These versions differ in rules of choosing heaps and the numbers of pebbles that can be taken. For example, in the exact (n, k)-nim

The article was prepared within the framework of the HSE University Basic Research Program. The second author was supported in part by RFBR grant 20-01-00645 and the state assignment topic no. 0063-2016-0003.

© Springer Nature Switzerland AG 2020
H. Fernau (Ed.): CSR 2020, LNCS 12159, pp. 237–249, 2020.
https://doi.org/10.1007/978-3-030-50026-9_17

a player by one move reduces exactly k from n heaps, strictly but otherwise arbitrarily.

A move is called *slow* if at most one pebble is taken from each heap. The so-called *slow versions* of the Moore and exact (n, k)-nim were introduced in [16,17], respectively. In these games the players are restricted to their slow moves.

An explicit formula for the SG function of the exact slow (n, k)-nim was obtained for $(n, k) = (3, 2)$ in [17] and then extended to $(n, k) = (4, 2)$ in [16]. However, for many values of parameters n and k, no explicit formula for the SG function of the exact slow (n, k)-nim is known and even the set of the P-positions looks rather chaotic.

Is it possible that there are no efficient algorithms solving these variants of nim? Right now we have no answer to this question and even no clue.

Looking for hardness results in solving combinatorial games, we see numerous examples of **PSPACE**-complete games; see e.g. [10,23]. As for generalizations of nim, there are hardness results for the *hypergraph nim*.

Given an arbitrary hypergraph $\mathcal{H} \subseteq 2^{[n]} \setminus \{\varnothing\}$ on the ground set $[n] = \{1, \ldots, n\}$, the game *hypergraph nim*, $\text{NIM}_{\mathcal{H}}$, is played as follows. By one move a player chooses an edge $H \in \mathcal{H}$ and reduces (strictly) all heaps of H. The games of standard (not slow) exact and Moore's nim considered above provide examples of the hypergraph nim. For a position $x = (x_1, \ldots, x_n)$ of $\text{NIM}_{\mathcal{H}}$ its *height* $h(x) = h_{\mathcal{H}}(x)$ is defined as the maximum number of successive moves that players can make beginning in x. (Clearly, they can restrict themselves by their slow moves.) A hypergraph $\mathcal{H}$ is called *intersecting* if $H' \cap H'' \neq \varnothing$ for any two edges $H', H'' \in \mathcal{H}$. The following two statements were proven in [4,6]. For any intersecting hypergraph $\mathcal{H}$, its height and SG functions are equal. Computing the height $h_{\mathcal{H}}(x)$ is **NP**-hard already for the intersecting hypergraphs with edges of size at most 4 (see [4]). It follows from these two statements that, already for the above family of hypergraphs, computing the SG function is **NP**-hard.

For the case of fixed number of heaps Larsson and Wästlund [20] obtained an important result. Actually, they studied a wider class of games, so-called *vector subtraction games* introduced by Golomb [14]. Later they were studied under the name of *invariant games* [11]. The *subtraction games* include all versions of nim mentioned above. In these games, the positions are d-dimensional vectors with nonnegative integer coordinates. The game is specified by a set of d-dimensional integer vectors (the difference set), and a move is subtraction of any vector from the difference set. Larsson and Wästlund considered subtraction games of finite dimension with a finite difference set (FDG for brevity: fixed dimension and fixed difference set).

The P-positions of a 1-dimensional FDG form a periodic structure [1]. This provides an efficient algorithm of solving such games.

Yet, the FDG of higher dimensions may behave in a very complicated way. Larsson and Wästlund proved in [20] that in some fixed dimension the equivalence problem for FDG is undecidable.

Nevertheless, this remarkable result does not answer the main question; whether efficient algorithms solving FDG exist. For example, there are polynomial algorithms solving the membership problem for context-free languages (CFLs) but the equivalence problem for CFLs is undecidable [18].

In this paper we extend arguments of Larsson and Wästlund and prove the existence of a FDG such that solving the game is **EXP**-complete and requires time $2^{\Omega(n)}$, where n is the input size. Furthermore, this bound is optimal up to a polynomial speed-up.

The rest of the paper is organized as follows. In Sect. 2 we introduce the concepts that we will need and outline our contribution. In Sect. 3 we sketch the proof of the main theorem. The following sections contain some more detailed exposition of main steps of the proof: in Sect. 4 we describe a simulation of a binary cellular automaton by a subtraction game; in Sect. 5 we discuss converting a Turing machine to a binary cellular automaton; in Sect. 6 we present a way to launch a Turing machine on all inputs simultaneously. Finally, Sect. 7 contains the proof itself. Due to space limitation, we skip several technical proofs.

2 Concepts and Results

2.1 Impartial Games

An *impartial* game is defined by a directed acyclic graph (DAG), which vertices and arcs are interpreted as positions and moves. Such DAG may be infinite, but the set of positions reachable (by one or several moves) from any given position is finite. An initial position is fixed. Two players move alternately. One who has to move but is out of moves loses.

Recall the standard classification of positions of an impartial game. If a player who moves at a position x has a winning strategy then x is called an N-*position* and it is called a P-*position* otherwise (from words **N**ext and **P**revious). In graph theory, the set of P-positions of a DAG is called its *kernel*; it can be found in time linear in the size of the DAG [22].

A refinement of this concept is given by the so-called Sprague-Grundy (SG) function. For a set S of nonnegative integers the *minimum excluded value* of S is defined as the smallest nonnegative integer that is not in S and denoted by $\text{mex}(S)$. In particular, $\text{mex}(\varnothing) = 0$. The SG value of a position x is defined recursively as

$$G(x) = \text{mex}\{G(y) : (x, y) \in E(G)\}, \tag{1}$$

where G is the corresponding DAG. A position of SG value t is called a t-position. Then, P-positions are exactly 0-positions, while N-positions have positive SG values.

Taking in mind the relation with the Sprague-Grundy function, we assign to P- and N-positions the (Boolean) values 0 and 1, respectively. The basic relation between values of positions is

$$p(v) = [p(v_1), \ldots, p(v_k)] = \neg \bigwedge_{i=1}^{k} p(v_i) = \bigvee_{i=1}^{k} \neg p(v_i), \tag{2}$$

where all the possible moves from the position v are to the positions $v_1, \ldots, v_k$. If v is a sink then $p(v) = 0$. We will use notation $[\ldots]$ introduced in [20] for Boolean functions in Eq. (2).

Yet, we will assume that a game is given by the succinct description, in size of which its DAG, as well as the above algorithm, become exponential.

2.2 Subtraction Games and Modular Games

Now we introduce the class FDG of subtraction games. Subtraction games generalize naturally all versions of nim mentioned above. Note that a position in a version of nim with d heaps is specified by a d-dimensional vector $x = (x_1, \ldots, x_d)$ with nonnegative integer coordinates which are just the numbers of pebbles in a heap. A move in the game decreases some coordinates of this vector. Thus possible moves are specified by a set $D(x) \subseteq \mathbb{N}^d$ of d-dimensional vectors with nonnegative integer coordinates (the *difference set*). A move from x to y is possible if $x - y \in D(x)$.

The set $\mathcal{D}$ is defined by rules of a game.

Example 1. The exact slow (n, k)-nim (see Sect. 1 and [17] for more details) is an n-dimensional FDG with the difference set consisting of all $(0, 1)$-vectors with exactly k coordinates equal 1.

A subtraction game is defined by the same rules. The requirements of the difference set are changes In a subtraction game the difference set $\mathcal{D}$ is the same for all positions but may contain integer vectors with negative coordinates. In other words, a player is allowed to add pebbles to heaps. To guarantee that each play terminates after finite number of moves, we put a restriction on coordinates of difference vectors: if $a \in \mathcal{D}$ then

$$\sum_{i=1}^{d} a_i > 0. \tag{3}$$

This restriction implies that the total number of pebbles is strictly reduced by each move.

Finally, a game from the class FDG is defined by a *finite* difference set D of d-dimensional vectors satisfying (3).

Example 2. Let $\mathcal{D} = \{(2, -1), (-1, 2)\}$. Then possible moves from position $(3, 3)$ are to positions $(1, 4)$ and $(4, 1)$. It is an easy exercise to compute the value of the position: $p(3, 3) = 0$, i.e. it is a P-position.

In a FDG starting at a position $(x_1, \ldots, x_d)$, the total number of positions is $O((M + 1)^d)$, where $M = \max_i x_i$. Therefore, due to the kernel construction algorithm mentioned above, solving FDG belongs to the class **EXP** if x_i are binary.

If the difference set is a part of the input, then clearly solving FDG is **PSPACE**-hard. To show this, we reduce solving of the game called NODE

KAYLES to solving a FDG. Recall the rules of NODE KAYLES. It is played on a graph G. At each move a player puts a pebble on an unoccupied vertex of G that is non-adjacent to any occupied vertex. As usual, the player unable to make a move loses. It is known that solving NODE KAYLES is **PSPACE**-complete [23]. So, **PSPACE**-hardness of solving FDG is an immediate corollary of the following claim.

Proposition 1. *Solving NODE KAYLES is polynomially reducible to solving FDG.*

In the sequel we will solve a particular FDG (the difference set is fixed). In other words, we are going to determine algorithmic complexity of the language $\mathcal{P}(D)$ that consists of the binary representations of all P-positions $(x_1, \ldots, x_d)$ of the FDG with the difference set D.

Our main result is unconditional hardness of this problem.

Theorem 1. *There exist a constant d and a finite set $\mathcal{D} \subset \mathbb{N}^d$ such that the language $\mathcal{P}(\mathcal{D})$ is **EXP**-complete and $\mathcal{P}(\mathcal{D}) \notin \mathbf{DTIME}(2^{n/11})$, where n is the input size.*

Note that Theorem 1 implies that for some languages $\mathcal{P}(\mathcal{D})$ the kernel construction algorithm is optimal up to polynomial speed-up.

In the proofs we will need a generalization of FDG—the so-called k-modular FDG introduced in [20]. A k-modular d-dimensional FDG is determined by k finite sets $D_0, \ldots, D_{k-1}$ of vectors from $\mathbb{Z}^d$. The rules are similar to those of FDG, but the set of possible moves at a position x is D_r, where r is the residue of $\sum_i x_i$ modulo k.

Example 3. Let $D_0 = \{(1,0),(0,1)\}$, $D_1 = \{(2,-1),(-1,2)\}$. Then possible moves from position $(3,3)$ in 2-modular game D_0, D_1 are to positions $(3,2)$ and $(2,3)$ (since $3+3 = 6$ is even). Possible moves from $(2,3)$ are to $(0,4)$ and $(3,1)$ (since $2 + 3$ is odd).

2.3 Turing Machines and Cellular Automata

The notion of a Turing machine is well-known. We will use the definition from Sipser's book [24].

Although the cellular automata are also well-known, for reader's convenience we provide the definition. Formally, a cellular automaton (CA) C is a pair (A, δ), where A is a finite set (called the *alphabet*), and $\delta \colon A^{2r+1} \to A$ is the *transition function*. The number r is called *the size of a neighborhood*. The automaton operates on an infinite tape consisting of *cells*. Each cell carries a symbol from the alphabet. Thus, a *configuration* of C is a function $c \colon \mathbb{Z} \to A$.

At each step a CA changes the content of the tape using the transition function. If a configuration before the step is c, then the configuration after the step is c', where

$$c'(u) = \delta\big(c(u-r), c(u-r+1), \ldots, c(u), \ldots, c(u+r-1), c(u+r)\big).$$

Note that the changes are local: the content of a cell depends only on the content of $2r + 1$ neighbor cells.

We assume that there exists a blank symbol Λ in the alphabet and the transition function satisfies the condition $\delta(\Lambda, \ldots, \Lambda) = \Lambda$ ("nothing generates nothing"). This convention guarantees that the configurations containing only a finite number of non-blank symbols produce configurations with the same property.

A 2CA (a binary CA) is a CA with the binary alphabet $\{0, 1\}$. It will be convenient to assume that 1 is the blank symbol in 2CAs, because of their connections with games.

It is well-known that Turing machines can be simulated by CA with $r = 1$ and any CA can be simulated by a 2CA (with a neighborhood of a larger size).

We will need some specific requirements on these simulations; see Sect. 5 for more details.

3   Sketch of the Proof

The proof of Theorem 1 consists of the following steps.

1. Choose an **EXP**-complete language L such that $L \notin \mathbf{DTIME}(2^{n/2})$ and fix a Turing machine M recognizing it.
2. Construct another machine U that simulates an operation of M *on all inputs in parallel* (see Sect. 6 fore more details).
3. Machine U is simulated by a CA C_U. The cellular automaton C_U is simulated in its turn by a 2CA $C_U^{(2)}$ (see Sect. 5 for more details) and $C_U^{(2)}$ is simulated by a d-dimensional FDG $\mathcal{D}_U$ (see Sect. 4), where d depends on $C_U^{(2)}$.
4. It is important to note that the result of operating M on an input w is completely determined by the value of a specific position of $\mathcal{D}_U$ that is computed in polynomial time. Thus, we obtain a polynomial reduction of L to $\mathcal{P}(D_U)$.
5. Now Theorem follows from the assumption of hardness of language L.

4   From Cellular Automata to Subtraction Games

We will follow Larsson and Wästlund's [20] construction with minor modifications.

4.1   First Step: Simulation of a 2CA by a 2-Dimensional Modular Game

Let $C = (\{0, 1\}, \delta)$ be a 2CA. Symbol 1 is assumed to be blank: $\delta(1, \ldots, 1) = 1$. We will relate the evolution of C beginning with the configuration $c_0 = (\ldots 11011 \ldots)$ with the values $p(x_1, x_2)$ of positions of a 2-dimensional $2N$-modular FDG $\mathcal{D}'_C$. The value of N depends on C and we will choose it greater than r.

The exact form of the relation is as follows. The time arrow is a direction $(1, 1)$ in the space of positions, while the coordinate along the automaton tape is in the direction $(1, -1)$.

The configuration c_t of C at moment t corresponds to the positions on a line $x_1 + x_2 = 2Nt$. The cell coordinate is $u = (x_1 - x_2)/2$, as shown in Fig. 1 ($N = 1$). For the configuration $(\ldots 11011 \ldots)$ we assume that 0 has coordinate 0 on the automaton tape.

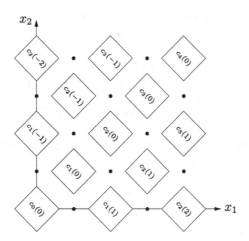

Fig. 1. Encoding configurations of 2CA by positions of a modular FDG

The content of an automaton tape and the values of positions of game $\mathcal{D}'_C$ are related as follows:

$$c_t(u) = p(Nt + u, Nt - u) \quad \text{for } |u| \leqslant Nt. \tag{4}$$

By the choice of the initial configuration, the following implication holds: if $|u| > Nt > rt$ then $c_t(u) = 1$. To extend the relation to this area, we extend the value function $p(x_1, x_2)$ by setting $p(x_1, x_2) = 1$ if either $x_1 < 0$ or $x_2 < 0$. In other words, we introduce dummy positions with negative values of coordinates; we regard them as terminals of value 1. Note that for the game evaluation functions $[\ldots]$ the equality $[p_1, \ldots, p_k, 1, \ldots, 1] = [p_1, \ldots, p_k]$ holds, that is, extra arguments with the value 1 do not affect the function value. So, the dummy positions do not change the values of real positions of a game.

The initial configuration $c_0 = (\ldots 11011 \ldots)$ satisfies this relation for any game: position $(0, 0)$ is a P-position.

To maintain the relation (4), we should choose an appropriate modulus and difference sets.

Note that the Boolean functions $[p_1, \ldots, p_n]$ defined by Eq. (2) form a complete basis: any Boolean function is represented by a circuit with gates $[\ldots]$.

Thus, it remains to check that the functions from the standard complete basis can be expressed in the basis [...]:

$$\neg x = [x], \quad x \vee y = [[x],[y]], \quad x \wedge y = [[x,y]].$$

Now take a circuit in the basis [...] computing the transition function of 2CA C. The circuit is a sequence of assignments $s_1, \ldots, s_N$ of the form

$$s_j := [\text{list of arguments}],$$

where arguments of the jth assignment may be the input variables or the values of previous assignments s_i, $i < j$. The value of the last assignment s_N is equal to the value of the transition function $\delta(u_{-r}, \ldots, u_{-1}, u_0, u_1, \ldots, u_r)$.

For technical reasons we require that the last assignment s_N does not contain the input variables u_i. This is easy to satisfy: just start a circuit with assignments in the form $s_{i+r+1} = [u_i]$; $s_{i+3r+2} = [s_{i+r+1}]$, where $-r \leqslant i \leqslant r$, and replace variable u_i in the following assignments by s_{i+3r+2}. The size of the modified circuit is obviously greater than r.

We extend the relation (4) to intermediate positions as follows:

$$
\begin{aligned}
p(Nt + i, Nt - i) &= c_t(i), \\
p(Nt + i + j, Nt - i + j) &= s_j, \quad 1 \leqslant j < N,
\end{aligned}
\tag{5}
$$

where s_j is the value of jth assignment of the circuit for values $c_t(i-r), \ldots, c_t(i)$, $\ldots, c_t(i+r)$ of the input variables.

Proposition 2. *There exist sets $\mathcal{D}_j$ such that relation (5) holds for the values of modular game $\mathcal{D}'_C$ with the difference sets $\mathcal{D}_j$.*

The choice of $\mathcal{D}_j$ depends on the arguments of assignments s_j. If an input variable u_k is an argument of s_j then we include in $\mathcal{D}_{2j}$ vector $(j-k, j+k)$. If the value of an intermediate assignment s_k is an argument of s_j then we include in $\mathcal{D}_{2j}$ vector $(j-k, j-k)$. Sets $\mathcal{D}_{2j+1}$ are irrelevant and may be arbitrary.

Relation (5) can be verified for $\mathcal{D}_j$ by induction on the sum of coordinates.

Note that game $\mathcal{D}'_C$ has the following property: if there is a legal move from (x_1, x_2) to (y_1, y_2) then either $x_1 + x_2 \equiv 0 \pmod{2N}$ or the residue of $y_1 + y_2$ modulo $2N$ is less than the residue of $x_1 + x_2$. (Standardly, we assume that the residues take values $0, 1, \ldots, 2N-1$). Note also that $x_1 + x_2 \not\equiv y_1 + y_2 \pmod{2N}$, since the input variables are not arguments of the final assignment.

4.2   Second Step: Simulation of a 2CA by a $(2N + 2)$-Dimensional Subtraction Game

To exclude modular conditions we use the trick suggested in [20].

Using the 2-dimensional modular game $\mathcal{D}'_C$ defined above we construct a $(2N+2)$-dimensional FDG $\mathcal{D}_C$ with the difference set

$$\mathcal{D} = \left\{ (a_1, a_2, 0^{2N}) + e^{(j)} - e^{(k)} : (a_1, a_2) \in D_j, \ k = j - a_1 - a_2 \pmod{2N} \right\}.$$

Here $e^{(i)}$ is the $(i+2)$th coordinate vector: $e^{(i)}_{i+2} = 1$, $e^{(i)}_s = 0$ for $s \neq i + 2$.

Proposition 3. *The value of a position* $(x_1, x_2, 0^{2N}) + e^{(2r)}$ *of the game* $\mathcal{D}_C$ *is equal to the value of a position* (x_1, x_2) *of the modular game* $\mathcal{D}'_C$ *if* $2r \equiv x_1 + x_2$ *(mod* $2N$*).*

Proof. We proceed by induction on $t = x_1 + x_2$. The base case $t = 0$ holds, by the convention on the values of dummy positions (with negative coordinates).

The induction step. A legal move at a position $(Nt+i+j, Nt-i+j, 0^{2N}) + e^{(2j)}$ is to a position $(Nt + i + j, Nt - i + j, 0^{2N}) - (a_1, a_2, 0^{2N}) + e^{(2s)}$, where $2s \equiv 2j - a_1 - a_2$ (mod $2N$) and $(a_1, a_2) \in \mathcal{D}_{2j}$. It corresponds to a move from $(Nt + i + j, Nt - i + j)$ to $(Nt + i + j - a_1, Nt - i + j - a_2)$ in the modular game. $\qquad\square$

From Propositions 2 and 3 we derive:

Corollary 1. *For any 2CA C there exist an integer N and a $(2+2N)$-dimensional FDG $\mathcal{D}_C$ such that the relation*

$$c_t(u) = p(Nt + u, Nt - u, 0, 0, \ldots, 0, 1) \quad \text{holds for } |u| \leqslant Nt.$$

5   From Turing Machines to Cellular Automata

In this section we outline a way to simulate a Turing machine by a binary cellular automaton. It is standard, except some specific requirements.

Let $M = (Q, \{0, 1\}, \Gamma, \Lambda, \delta_M, 1, 2, 3)$ be a Turing machine, where the input alphabet is binary, $Q = \{1, \ldots, q\}$, $q \geqslant 3$ is the set of states, $\Gamma = \{0, 1, \ldots, \ell\}$ is the tape alphabet, $\ell > 1$ is the blank symbol, $\delta_M \colon Q \times \Gamma \to Q \times \Gamma \times \{+1, -1\}$ is the transition function, and $1, 2, 3$ are the start, accept, and reject states, respectively.

We encode a configuration of M by a doubly infinite string $c \colon \mathbb{Z} \to A$, where $A = \{0, \ldots, q\} \times \{0, \ldots, \ell\}$. The head position is indicated by a pair (q, a), $q > 0$, $a \in \Gamma$; the content of any other cell is encoded as $(0, a)$, $a \in \Gamma$.

Let $c_0, \ldots, c_t, \ldots$ be a sequence of encoded configurations produced by M from the start configuration c_0. It is easy to see that $c_{t+1}(u)$ is determined by $c_t(u - 1)$, $c_t(u)$, $c_t(u + 1)$. In this way we obtain CA $C_M = (A, \delta_C)$ over the alphabet A with the transition function $\delta_C \colon A^3 \to A$ thus simulating operation of M in encoded configurations. It is easy to see that $\Lambda_C = (0, \ell)$ is the blank symbol: $\delta_C(\Lambda_C, \Lambda_C, \Lambda_C) = \Lambda_C$.

The next step is to simulate C_M by a 2CA $C_M^{(2)}$. To do so, we use an automaton $C'_M = (A', \delta'_C)$ isomorphic to C_M, where $A' = \{0, \ldots, L - 1\}$ and $L = (|Q| + 1) \cdot |\Gamma|$. The transition function δ'_C is defined as follows

$$\delta'_C(i, j, k) = \pi(\delta_C(\pi^{-1}(i), \pi^{-1}(j), \pi^{-1}(k))),$$

where $\pi \colon A \to A'$ is a bijection. To keep the relation between the start configurations, we require that $\pi(\Lambda_C) = 0$, $\pi((1, \ell)) = 1$. Recall that 1 is the start state of M and ℓ is the blank symbol of M.

To construct the transition function of $C_M^{(2)}$, we encode symbols of A' by binary words of length $L + 2$ as follows

$$\varphi(a) = 1^{1+L-a}0^a1.$$

In particular, $\varphi(0) = \varphi(\pi(\Lambda_C)) = 1^{L+2}$ and $\varphi(1) = \varphi(\pi(1, \ell)) = 1^L01$. The encoding φ is naturally extended to words in the alphabet A' (finite or infinite).

Thus, the start configuration of M with the empty tape corresponds to the configuration $\ldots 1110111 \ldots$ of $C_M^{(2)}$. Recall that 1 is the blank symbol of $C_M^{(2)}$.

Slightly abusing notation, we will denote by φ the extended encoding of configurations in alphabet A' by doubly infinite binary words.

We align configurations in the following way: if $i = q(L+2)+k$, $0 \leqslant k < L+2$ then $\varphi(c)(i)$ is the kth bit of $\varphi(c(q))$.

The size of a neighborhood of $C_M^{(2)}$ is $r = 2(L + 2)$. To define the transition function $\delta_M^{(2)}$, we use a local inversion property of the encoding φ: looking at the r-neighborhood of an ith bit of $\varphi(c)$, where $i = q(L+2) + k$, $0 \leqslant k < L+2$, one can restore symbols $c(q-1)$, $c(q)$, $c(q+1)$, and position k of the bit provided the neighborhood contains zeroes (0 is the non-blank symbol of $C_M^{(2)}$). Note that if the neighborhood of a bit does not contain zeroes then the bit is a part of encoding of the blank symbol 0 of C_M' and, moreover, $c(q-1) = c(q) = c(q+1) = 0$.

Lemma 1. *There exists a function* $\delta_C^{(2)} : \{0,1\}^{2r+1} \to \{0,1\}$ *such that a 2CA* $C_M^{(2)} = (\{0,1\}, \delta_C^{(2)})$ *simulates* C_M': *starting from* $b_0 = \ldots 1110111 \ldots$, *it produces the sequence of configurations* $b_0, b_1, \ldots$ *such that* $b_t = \varphi(c_t)$ *for any* t, *where* (c_t) *is the sequence of configurations produced by* C_M' *starting from the configuration* $c_0 = \ldots 0001000 \ldots$

We skip the proof because of the space limitation.

6   Parallel Execution of a Turing Machine

The last construction needed in the main proof is a Turing machine U simulating an operation of a Turing machine M on all inputs. The idea of simulation is well-known, but again we need to specify some details of the construction.

We assume that on each input of length n the machine M makes at most $T(n) > n$ steps. The alphabet of U includes the set $A = \{0, \ldots, q\} \times \{0, \ldots, \ell\}$ (we use notation from the previous section) and additional symbols.

Machine U operates in *stages* while its tape is divided into *zones*. The zones are surrounded by the delimiters, say, $\triangleleft$ and $\triangleright$. We assume that $\triangleleft$ is placed in the cell 0. Also the zones are separated by a delimiter, say, $\diamond$. An operation of M on a particular input w is simulated inside a separate zone.

Each zone consists of three blocks. The first one is of size 1. It carries $(0, 1)$ iff M accepts the input written in the second block. Otherwise it carries $(0, 0)$. (Note that it does not distinguish an unfinished computation and the negative result.) The last block contains a configuration of M represented by a word over the alphabet A, as described in Sect. 5. The blocks in a zone are separated by a delimiter, say #.

At the start of a stage k there are $k - 1$ zones corresponding to the inputs $w_1, w_2, \ldots, w_{k-1}$ of M. We order binary words by their lengths and words of equal length are ordered lexicographically. The last block of a zone i contains the configuration of M after running $k - 1 - i$ steps on input w_i.

During the stage k, machine U moves along the tape from $\triangleleft$ to $\triangleright$ and in each zone simulates the next step of operation of M. At the end of the stage machine U writes a fresh zone with the input w_k and the initial configuration of M on this input. The initial configuration is extended in both directions by white space of size $T(n)$.

When the operation of M on an input w_k is finished, machine U updates the resulting block if necessary and it does not change the zone on the subsequent stages.

In the arguments below we need U to satisfy specific properties.

Proposition 4. *If $T(n) = C \cdot n^2 \cdot 2^n$ for some integer constant $C \geqslant 1$ then there exists U operating as it described above such that*

1. *For all sufficiently large n machine U produces the result of operation of M on the input w of length n in time $< 2^{4n}$.*
2. *The head of U visits the first blocks of zones only on steps t that are divisible by 3.*

The proof of this proposition is quite technical but not difficult.

7   Proof of the Main Theorem

As a hard language, we use the language of the bounded halting problem

$$L = \{(\langle M \rangle, x, 0^k) : M \text{ accepts } x \text{ after at most } 2^k \text{ steps}\},$$

which is obviously **EXP**-complete. The following proposition is an easy corollary of the time hierarchy theorem [24].

Proposition 5. $L \notin \mathbf{DTIME}(2^{n/2})$, *where n is the input size.*

On the other hand, by standard techniques (universal Turing machine and counter) we achieve the following upper bound.

Proposition 6. $L \in \mathbf{DTIME}(n^2 \cdot 2^n)$, *where n is the input size.*

Thus, for some constant C there exists a Turing machine M recognizing L such that M makes at most $T(n) = C \cdot n^2 \cdot 2^n$ steps on inputs of size n.

For machine M apply the construction from Sect. 6 and Proposition 4 to construct the machine U. Then convert U into 2CA $C_U^{(2)}$ as described in Sect. 5. We put an additional requirement on bijection π, namely, $\pi(0, (0, 1)) = L - 1$. It locates the result of computation of M in the third bit of the encoding of the resulting block. Finally, construct $O(1)$-dimensional FDG $\mathcal{D}_C$ as it described in Sect. 4. The dimension $2N + 2$ of the game is determined by machine M. By Corollary 1, the symbol $c(t, u)$ on the tape of $C_U^{(2)}$ equals the value of position $(Nt + u, Nt - u, 0, 0, \ldots, 0, 1)$ of the game.

Define the function $\rho \colon w \mapsto (Nt + u, Nt - u, 0, 0, \ldots, 0, 1)$ as follows. Set $t = 2^{4n}$, where n is the size of w. Set u be the position of the bit carrying the result of computation of M on input w in the image of the resulting block of the zone k corresponding to the input w.

Proposition 7. (i) There exists a polynomial reduction of the language L to the language $\mathcal{P}(\mathcal{D})$ based on the function ρ; (ii) $u = O(2^{3n})$.

The reduction by the function ρ is correct for sufficiently large n, by the previous constructions and by Proposition 4. (The remaining finite set of inputs causes no problem.) The property 2 of Proposition 4 guarantees that at the moment $t = 2^{4n}$ the head of U is not on the resulting block. Thus the third bit of the encoding of the block is 1 iff M accepts w.

Part (i) of Proposition 7 implies the first part of Theorem 1. To prove the second part, we use the part (ii) of Proposition 7. It guarantees that the reduction transforms a word of size n to a vector of size at most $5n$. So if an algorithm $\mathcal{A}$ solves the game $\mathcal{D}_C$ in time $T_\mathcal{A}(m)$, then the composition of the reduction and $\mathcal{A}$ recognizes L in time $\mathrm{poly}(n) + T_\mathcal{A}(5n)$. By Proposition 5 this value is $\Omega(2^{n/2})$. We conclude that $T_\mathcal{A}(m) = \Omega(2^{m/11})$.

Acknowledgment. The authors are grateful to the anonymous referee for several helpful remarks improving both, the results and their presentation.

References

1. Albert, M., Nowakowski, R., Wolfe, D.: Lessons in Play: An Introduction to Combinatorial Game Theory. Taylor & Francis, Abington (2007)
2. Berlekamp, E.R., Conway, J.H., Guy, R.K.: Winning Ways for Your Mathematical Plays, vol. 1–4. A.K. Peters, Natick (2001–2004)
3. Boros, E., Gurvich, V., Oudalov, V.: A polynomial algorithm for a two parameter extension of Wythoff NIM based on the Perron-Frobenius theory. Int J. Game Theory **42**(4), 891–915 (2013). https://doi.org/10.1007/s00182-012-0338-6
4. Boros, E., Gurvich, V., Ho, N.B., Makino, K., Mursic, P.: Tetris hypergraphs and combinations of impartial games. CoRR abs/1701.02819 (2017). http://arxiv.org/abs/1701.02819
5. Boros, E., Gurvich, V., Ho, N.B., Makino, K., Mursic, P.: On the Sprague-Grundy function of exact k-nim. Discrete Appl. Math. **239**, 1–14 (2018)
6. Boros, E., Gurvich, V., Ho, N.B., Makino, K., Mursic, P.: Sprague-Grundy function of matroids and related hypergraphs. Theor. Comput. Sci. **799**, 40–58 (2019)
7. Boros, E., Gurvich, V., Ho, N.B., Makino, K., Mursic, P.: Sprague-Grundy function of symmetric hypergraphs. J. Comb. Theory Ser. A **165**, 176–186 (2019)
8. Bouton, C.L.: Nim, a game with a complete mathematical theory. Ann. Math. 2nd Ser. **3**, 35–39 (1901–1902)

9. Conway, J.H.: On Numbers and Games. Academic Press, London, New York, San Francisco (1976)
10. Demaine, E.D., Hearn, R.A.: Playing games with algorithms: algorithmic combinatorial game theory. CoRR abs/cs/0106019v2 (2008). http://arxiv.org/abs/cs/0106019v2
11. Duchêne, E., Rigo, M.: Invariant games. Theor. Comput. Sci. **411**, 3169–3180 (2010)
12. Fraenkel, A.: How to beat your Wythoff games' opponent on three fronts. Am. Math. Mon. **89**, 353–361 (1982)
13. Fraenkel, A.: Wythoff games, continued fractions, cedar trees and Fibonacci searches. Theor. Comput. Sci. **29**, 49–73 (1984)
14. Golomb, S.W.: A mathematical investigation of games of "take-away". J. Comb. Theory **1**(4), 443–458 (1966)
15. Grundy, P.M., Smith, C.: Disjunctive games with the last player losing. Proc. Camb. Philos. Soc. **52**, 527–533 (1956)
16. Gurvich, V., Heubach, S., Ho, N.B., Chikin, N.: Slow k-nim integers. Electron. J. Comb. Number Theory **20**, 1–19 (2020)
17. Gurvich, V., Ho, N.B.: Slow k-nim. RUTCOR Research Report RRR-03-2015 (2015)
18. Hopcroft, J.E., Motwani, R., Ullman, J.D.: Introduction to Automata Theory, Languages, and Computation, 3rd edn. Addison-Wesley Longman Publishing Co., Inc., Boston (2006)
19. Jenkyns, T.A., Mayberry, J.P.: The skeleton of an impartial game and the Nim-function of Moore's Nim_k. Int J. Game Theory **9**, 51–63 (1980). https://doi.org/10.1007/BF01784796
20. Larsson, U., Wästlund, W.: From heaps of matches to the limits of computability. Electron. J. Comb. **20**(3), #P41 (2013)
21. Moore, E.: A generalization of the game called Nim. Ann. Math. Second Ser. **11**(3), 93–94 (1910)
22. von Neumann, J., Morgenstern, O.: Theory of Games and Economic Behaviour. Princeton University Press, Princeton (1944)
23. Shaefer, T.J.: On the complexity of some two-person perfect-information games. J. Comput. Syst. Sci. **16**, 185–225 (1978)
24. Sipser, M.: Introduction to the Theory of Computation. Cengage Learning, Boston (2013)
25. Wythoff, W.: A modification of the game of Nim. Nieuw Archief voor Wiskunde **7**, 199–202 (1907)

Parameterized Complexity of Fair Feedback Vertex Set Problem

Lawqueen Kanesh[2(✉)], Soumen Maity[1], Komal Muluk[1(✉)],
and Saket Saurabh[2,3]

[1] Indian Institute of Science Education and Research, Pune, India
soumen@iiserpune.ac.in, komalmuluk15@gmail.com
[2] The Institute of Mathematical Sciences, Chennai, India
lawqueenkanesh091@gmail.com, saket@imsc.res.in
[3] University of Bergen, Bergen, Norway

Abstract. Given a graph $G = (V, E)$, a subset $S \subseteq V(G)$ is said to be a feedback vertex set of G if $G - S$ is a forest. In the FEEDBACK VERTEX SET (FVS) problem, we are given an undirected graph G, and a positive integer k, the question is whether there exists a feedback vertex set of size at most k. This problem is extremely well studied in the realm of parameterized complexity. In this paper, we study three variants of the FVS problem: UNRESTRICTED FAIR FVS, RESTRICTED FAIR FVS, and RELAX FAIR FVS. In UNRESTRICTED FAIR FVS problem, we are given a graph G and a positive integer ℓ, the question is does there exists a feedback vertex set $S \subseteq V(G)$ (of any size) such that for every vertex $v \in V(G)$, v has at most ℓ neighbours in S. First, we study UNRESTRICTED FAIR FVS from different parameterizations such as treewidth, treedepth and neighbourhood diversity and obtain several results (both tractability and intractability). Next, we study RESTRICTED FAIR FVS problem, where we are also given an integer k in the input and we demand the size of S to be at most k. This problem is trivially NP-complete; we show that RESTRICTED FAIR FVS problem when parameterized by the solution size k and the maximum degree Δ of the graph G, admits a kernel of size $\mathcal{O}((k+\Delta)^2)$. Finally, we study RELAX FAIR FVS problem, where we want that the size of S is at most k and for every vertex outside S, that is, for all $v \in V(G) \setminus S$, v has at most ℓ neighbours in S. We give an FPT algorithm for RELAX FAIR FVS problem running in time $c^k n^{\mathcal{O}(1)}$, for a fixed constant c.

Keywords: Feedback vertex set · Parameterized complexity · FPT · W[1]-hard

This project has received funding from the European Research Council (ERC) under the European Union's Horizon 2020 research and innovation programme (grant no. 819416), and the Swarnajayanti Fellowship grant DST/SJF/MSA-01/2017-18.

S. Maity—The author's research was supported in part by the Science and Engineering Research Board (SERB), Govt. of India, under Sanction Order No. MTR/2018/001025.

H. Fernau (Ed.): CSR 2020, LNCS 12159, pp. 250–262, 2020.
https://doi.org/10.1007/978-3-030-50026-9_18

1   Introduction

FEEDBACK VERTEX SET (FVS) problem is one of Karp's 21 NP-complete problems [8]. This problem has been studied extensively in computational complexity theory, as it is one of the fundamental problems in the theory. The FVS problem has applications in operating systems, database systems, and VLSI design. It is used for resource allocation in operating systems. In the system resource allocation graph, the FVS problem is used for deadlock recovery [14].

FVS is a vertex deletion problem which demands to find a set $S \subseteq V(G)$ of size at most k such that the remaining graph $G - S$ is a forest. The set $S \subseteq V(G)$ is called a feedback vertex set of G. For a vertex deletion problem, given a graph $G = (V, E)$ and a property Π; we are asked to find a set $S \subseteq V(G)$ such that the subgraph obtained after the deletion of vertices in S, that is, the graph $G - S$ satisfies the desired property Π. In the optimization version of deletion problems (vertex deletion/edge deletion), we focus on minimizing the total number of elements required to remove to achieve the desired property on the remaining graph. Modified versions of deletion problems, called *fair deletion* problems were introduced by Lin and Sahni in 1989 [12]. Unlike usual deletion problems, fair deletion problems aim to minimize the maximum number of neighbours contributed to the solution set by a single vertex in the graph. Given a set $S \subseteq V(G)$, the fair cost of S is $\max_{v \in V}\{|N(v) \cap S|\}$.

In this paper, we study the parameterized complexity of three variants of FVS, namely, UNRESTRICTED FAIR FEEDBACK VERTEX SET (Unrestricted Fair FVS), RESTRICTED FAIR FEEDBACK VERTEX SET (Restricted Fair FVS), and RELAX FAIR FEEDBACK VERTEX SET (Relax Fair FVS). We define these variants as follows:

UNRESTRICTED FAIR FEEDBACK VERTEX SET (UNRESTRICTED FAIR FVS)
Input: An undirected graph G and a positive integer ℓ.
Question: Does there exist a feedback vertex set $S \subseteq V(G)$ such that $|N(v) \cap S| \le \ell$ for all $v \in V(G)$?

RESTRICTED FAIR FEEDBACK VERTEX SET (RESTRICTED FAIR FVS)
Input: An undirected graph G and positive integers k and ℓ.
Question: Does there exist a feedback vertex set $S \subseteq V(G)$ with $|S| \le k$ such that $|N(v) \cap S| \le \ell$ for all $v \in V(G)$?

RELAX FAIR FEEDBACK VERTEX SET (RELAX FAIR FVS)
Input: An undirected graph G and positive integers k and ℓ.
Question: Does there exist a feedback vertex set $S \subseteq V(G)$ with $|S| \le k$ such that $|N(v) \cap S| \le \ell$ for all $v \in V(G) \setminus S$?

A problem with input size n and parameter k is said to be 'fixed-parameter tractable (FPT)' if it has an algorithm that runs in time $\mathcal{O}(f(k)n^c)$, where

f is some (usually computable) function, and c is a constant that does not depend on k or n. What makes the theory more interesting is a hierarchy of intractable parameterized problem classes above FPT which helps in distinguishing those problems that are not fixed-parameter tractable. Closely related to fixed-parameter tractability is the notion of preprocessing. A reduction to a problem kernel, or equivalently, problem kernelization means to apply a data reduction process in polynomial time to an instance (x, k) such that for the reduced instance (x', k') it holds that (x', k') is equivalent to (x, k), $|x'| \leq g(k)$ and $k' \leq g(k)$ for some function g only depending on k. Such a reduced instance is called a problem kernel. We refer to [4] for further details on parameterized complexity.

In this paper, we consider UNRESTRICTED FAIR FVS, RESTRICTED FAIR FVS and Relax Fair FVS problems under structural parameters and solution size. Our results are as follows:

- UNRESTRICTED FAIR FVS problem is W[1]-hard when parameterized by the treewidth or treedepth of the input graph.
- UNRESTRICTED FAIR FVS admits an algorithm with running time $\mathcal{O}^*(3^k)$,[1] where k is the neighbourhood diversity of the input graph.
- RESTRICTED FAIR FVS admits a kernel of size $\mathcal{O}((\Delta + k)^2)$. Here, Δ is the maximum degree of the graph, and k is the solution size.
- RELAX FAIR FVS admits an algorithm with running time $\mathcal{O}^*(82^k)$, where k is the solution size. This is our main technical result.

Related Work. FVS is extremely well studied in the realms of parameterized complexity. Downey and Fellows [5], Bodlaender [2] proposed the first FPT algorithm with the running time $\mathcal{O}^*(17(k^4)!)$. After a series of improvements, in 2010, Cao et al. gave an FPT algorithm running in time $\mathcal{O}^*(3.83^k)$ [3]. The fastest known randomized algorithm is given by Li and Nederlof [11] and runs in time $\mathcal{O}^*(2.7^k)$. Recently, Iwata and Kobayashi [6] announced the fastest known deterministic algorithm for FVS running in time $\mathcal{O}^*(3.460^k)$.

The study of fair deletion problems in the realm of parameterized complexity was initiated by Masařík and Toufar [13] and Knop et al. [9]. Among several results, they showed that UNRESTRICTED FAIR VERTEX COVER problem is W[1]-hard when parameterized by a combined parameter of the treedepth and the feedback vertex set number of the graph. Jacob et al. [7] studied RESTRICTED d-HITTING SET and RESTRICTED FAIR FVS parameterized by solution size. For UNRESTRICTED FAIR FVS they designed an FPT algorithm parameterized by solution size and the treewidth of the input graph and using this designed an FPT algorithm for RESTRICTED FAIR FVS running in time $\mathcal{O}^*(k^{\mathcal{O}(k)})$.

2   Preliminaries

Throughout the paper, we adopt the following notations. Let G be a graph, $V(G)$ and $E(G)$ denote the vertex set and the edge set of graph G, respectively. Let n

---

[1] We use the $\mathcal{O}^*$ notation to hide polynomial factors (in the input size) in the running time.

and m denote the number of vertices and the number of edges of G, respectively. For a graph G and a set $X \subseteq V(G)$, by $G - X$ we denote the graph G induced on $V(G) \setminus X$. By $N_G(v)$, we denote the neighbourhood of v in G and by $N_G[v]$ we denote the closed neighbourhood of v in G. Degree of a vertex v in graph G is denoted by $d_G(v)$. A *path* $P = \{v_1, \ldots, v_n\}$ is an ordered collection of vertices such that there is an edge between every consecutive vertices in P. A *cycle* $C = \{v_1, \ldots, v_n\}$ is a path $P = \{v_1, \ldots, v_n\}$ with an extra edge $v_1 v_n$. The subscript in the notations will be omitted if it is clear from the context.

For standard notations and definitions in graph theory and parameterized complexity, we refer to West [15] and Cygan et al. [4], respectively. We now define several graph parameters being used throughout the paper.

Definition 1. For a graph $G = (V, E)$, the parameter *feedback vertex set* is the cardinality of the smallest set $S \subseteq V(G)$ such that the graph $G - S$ is a forest and it is denoted by $fvs(G)$.

Treewidth is a well-known graph parameter introduced by Bertelè and Brioshi [1].

Definition 2. A *tree decomposition* of a graph G is a pair (T, X), where T is a rooted tree and $X = \{X_t \mid t \in V(T)\}$. Every node t of T is assigned a subset $X_t \subseteq V(G)$, called a *bag*, such that following properties are satisfied:

- $\bigcup_{t \in V(T)} X_t = V(G)$, that is, each vertex in G is in at least one bag;
- For every edge $uv \in E(G)$, there is $t \in V(T)$ such that $u, v \in X_t$;
- For every vertex $v \in V(G)$ the graph $T[\{t \in V(T) \mid v \in X_t\}]$ is a connected subtree of T.

The *width* of a tree decomposition is the size of its largest bag X_i minus one. The treewidth $tw(G)$ of a graph G is the smallest width of a tree decomposition among all possible tree decompositions of G.

Given a rooted forest F, its *transitive closure* is a graph H in which $V(H)$ contains all the nodes of the rooted forest and $E(H)$ contain an edge between two vertices only if those two vertices form an ancestor-descendant pair in the forest F.

Definition 3. The *treedepth* of a graph G is the minimum height of a rooted forest F whose transitive closure contains the graph G. It is denoted by $td(G)$.

Neighbourhood diversity is another graph parameter introduced by Lampis [10].

Definition 4. The neighbourhood diversity of a graph, denoted by $nd(G)$, is the least integer k for which we can partition the set of vertices of the graph into k classes, where two vertices u and v belong to the same class if and only if $N(u) \setminus \{v\} = N(v) \setminus \{u\}$. For a vertex v, $N(v) = \{u \ : \ (u, v) \in E\}$.

Two vertices $u, v \in V(G)$ are said to be *twin* vertices if they satisfy the criteria $N(u) \setminus \{v\} = N(v) \setminus \{u\}$.

3 UNRESTRICTED FAIR FEEDBACK VERTEX SET

It is clear that if we do not put any restriction on the size of feedback vertex set S of a graph, then any graph has a trivial feedback vertex set $S = V(G)$. However, observe that, in UNRESTRICTED FAIR FVS problem, though we dropped the constraint on the size of S, the problem does not become easy. In this section, we present a parameterized reduction from FAIR VERTEX COVER (FAIR VC) problem to the UNRESTRICTED FAIR FVS problem. In the FAIR VC problem, we are given a graph $G = (V, E)$ and a positive integer ℓ. The objective is to decide whether there exists a set $S \subseteq V(G)$ such that S covers all edges in the graph G and $\max_{v \in V(G)}\{|N(v) \cap S|\} \leq \ell$. It is known that the FAIR VC problem is $W[1]$-hard when parameterized by $fvs(G)$ and $td(G)$ [9]. We study UNRESTRICTED FAIR FVS problem with respect to parameters such as treewidth, treedepth, and neighbourhood diversity. And we obtain the following hardness results.

Theorem 1 ($\star^2$). UNRESTRICTED FAIR FVS *is $W[1]$-hard when parameterized by the treedepth of the input graph.*

Theorem 2 ($\star$). UNRESTRICTED FAIR FVS *is $W[1]$-hard when parameterized by treewidth of the input graph.*

Next, we give an FPT algorithm for UNRESTRICTED FAIR FVS problem with respect to neighbourhood diversity, denoted by $nd(G)$. The idea behind the algorithm is based on the following observations:

Observation 1. *Consider a graph G and a feedback vertex set X of G. If u and v are twin vertices, where $u \in X$ and $v \notin X$, then we can replace u by v in X and the resultant set $X' = (X \setminus \{u\}) \cup \{v\}$ is another feedback vertex set of G.*

Two feedback vertex sets X and X' are said to be of the 'same type' if one set can be obtained from other by replacing twin vertices, otherwise they are said to be of the 'different type'.

Observation 2. *If X and X' are two feedback vertex sets of the same type of a graph G, then*

$$\max_{v \in V(G)} \{|N(v) \cap X|\} = \max_{v \in V(G)} \{|N(v) \cap X'|\}.$$

If any two feedback vertex sets X and X' are of the same type, they have the same fair cost.

Theorem 3. *There exists an FPT algorithm running in time $3^k k^2 n^{\mathcal{O}(1)}$ for* UNRESTRICTED FAIR FVS *problem, where k is the neighbourhood diversity of the graph.*

---

[2] Due to paucity of space, the proofs of statements marked with a $\star$ have been omitted.

Proof. If the neighbourhood diversity of a graph is bounded by an integer k, that is, $nd(G) \leq k$, then there exists a partition $P = \{V_1, V_2, V_3, ..., V_k\}$ of $V(G)$ into k classes, such that all vertices in one class have the same neighbourhood, that is, $N(u) \setminus \{v\} = N(v) \setminus \{u\}$, if $u, v \in V_i$.

We observe a few facts about the partition P. Each class V_i could either be a clique or an independent set. Given a partition P of $V(G)$, the partition graph Q of G is defined as follows. The vertex set $V(Q)$ is $\{V_1, V_2, V_3, ..., V_k\}$. There is an edge $(V_i, V_j) \in E(Q)$ if $(u, v) \in E(G)$ for each $u \in V_i$ and $v \in V_j$. Let A be the $k \times k$ adjacency matrix of the partition graph Q, where

$$A(i, j) = \begin{cases} 1, & \text{if } V_i V_j \in E(Q) \\ 0 & \text{otherwise} \end{cases}$$

From observations 1 and 2, since two feedback vertex sets of the same type have the same fair cost, it suffices to check the fair cost of feedback vertex sets which are of different types. Given a partition $P = \{V_1, V_2, V_3, ..., V_k\}$ of $V(G)$, where k is neighbourhood diversity of G, we construct feedback vertex sets of different types as follows:

- (a) If class V_i is a clique and X is a feedback vertex set of graph G, then X contains either
  1) all vertices of class V_i,
  2) any $|V_i| - 1$ vertices from class V_i, or
  3) any $|V_i| - 2$ vertices from class V_i.

Clearly, there is no other case possible for cliques apart from those mentioned in (a). If we consider only $|V_i| - 3$ vertices from the clique V_i in X, then the remaining 3 vertices will form a triangle in $G - X$.

- (b) If class V_i is an independent set and X is a feedback vertex set of graph G, then X contains either
  1) all vertices of class V_i,
  2) any $|V_i| - 1$ vertices from class V_i, or
  3) no vertex from class V_i.

Now, when V_i is an independent set, we will show that it is sufficient to consider the three cases given in (b). Apart from the cases mentioned in (b), suppose X is a feedback vertex set such that $X \neq \emptyset$ and X contains at most $|V_i| - 2$ vertices of V_i. Then let $\{u, v\} \subseteq V_i \setminus X$. Since X is a feedback vertex set, u and v have at most one common neighbour in $G - X$. If they have two common neighbours then they form a cycle. Since u, v have at most one common neighbour, all the vertices of V_i have the same common neighbour. Therefore, even if X does not contain any vertex from V_i, $G - X$ will not contain any cycle. Thus, in this case, if there exists a solution X containing at most $|V_i| - 2$ vertices from V_i, then there also exists a solution containing no vertex from V_i, and this case is covered in $(b) - 3$.

Given a partition of the vertex set $V(G)$ into k classes according to the neighbourhood diversity, 3^k possible subsets $X \subseteq V(G)$ can be built as per the criteria mentioned in (a) and (b) above. According to Observation 2, it does not matter which vertices are being chosen from a class as all vertices in a class are twin vertices. Note that, not each of 3^k choices produces a feedback vertex set. If X is a feedback vertex set, we check whether X satisfies the cost criteria, that is, $\max_{v \in V(G)}\{|N(v) \cap X|\} \leq \ell$. Let n_i be the number of vertices selected from class V_i in X. For a vertex $v \in V_i$, where V_i is an independent set, verify if

$$|N(v) \cap X| = \sum_{j=1, j \neq i}^{k} A(i,j)n_j \leq \ell.$$

For a vertex $v \in V_i$, where V_i is a clique, verify if

$$|N(v) \cap X| = n_i - 1 + \sum_{j=1, j \neq i}^{k} A(i,j)n_j \leq \ell.$$

If for at least one X, the above conditions are true for all k classes, then the given instance is a yes-instance of UNRESTRICTED FAIR FVS problem. Otherwise, it is a no-instance.

Since $|V(Q)| = k$, computation of the adjacency matrix A requires $O(k^2)$ time. We have total 3^k possible subsets X to verify. We can check if a subset $X \subseteq V(G)$ is a feedback vertex set of graph G in time $\mathcal{O}(n^2)$. For a given X, computation of $(n_1, n_2, \ldots, n_k)$ requires $\mathcal{O}(k)$ time. Verification of the cost conditions corresponding to k classes takes $\mathcal{O}(k)$ time. Hence, the time complexity of this algorithm is $3^k k^2 n^{\mathcal{O}(1)}$.   □

4   RESTRICTED FAIR FEEDBACK VERTEX SET

In this section, we give a kernel for RESTRICTED FAIR FVS with respect to parameter $\Delta + k$, where Δ is the maximum degree in the graph and k is the size of restricted fair feedback vertex set. We first give a couple of reductions to reduce an instance (G, k, ℓ) of RESTRICTED FAIR FVS problem.

Reduction Rule 1. *If G contains a vertex v of degree at most 1, delete v. The new instance is $(G - \{v\}, k, \ell)$.*

Reduction Rule 1 is safe because for a given instance (G, k, ℓ) of RESTRICTED FAIR FVS, if the graph G has a vertex of degree at most one, then this vertex is not part of any cycle. Thus, its removal does not change the solution.

Reduction Rule 2. *If there exists a sequence $\{v_1, v_2, \ldots, v_p\}$ with $p > 3$, $v_i \neq v_j$, $d(v_i) = 2$ and $(v_i, v_{i+1}) \in E$ for all $i = 1, 2, \ldots, p-1$, then delete vertices $v_3, v_4, \ldots, v_{p-1}$ and introduce an edge between v_2 and v_p. In other words, we replace the sequence $\{v_1, \ldots, v_p\}$ by $\{v_1, v_2, v_p\}$. The reduced instance is $(G - \{v_3, v_4, \ldots, v_{p-1}\}, k, \ell)$.*

Reduction Rule 2 is safe because every minimum FVS of a graph contains at most one vertex from a sequence of degree-2 vertices. If (G, k, ℓ) is a yes-instance, there is a fair FVS S of size at most k. If S contains a vertex from a long degree-2 sequence, then we replace it by the middle vertex v_2 from reduced degree-2 sequence of length three in the graph $G - \{v_3, v_4, \ldots, v_{p-1}\}$ and obtain a solution of reduced instance. Observe that, for every vertex in the reduced instance, the number of neighbours in the solution does not exceed ℓ. Conversely, if $(G - \{v_3, v_4, \ldots, v_{p-1}\}, k, \ell)$ is a yes-instance of RESTRICTED FAIR FVS problem, then the solution S' of $(G - \{v_3, v_4, \ldots, v_{p-1}\}, k, \ell)$ also satisfies the instance (G, k, ℓ).

Lemma 1 ($\star$). *If (G, k, ℓ) is a yes-instance of* RESTRICTED FAIR FVS *problem and none of the above reductions is applicable, then $|V(G)| \leq k + 8\Delta k - 3$.*

Reduction Rule 3. *If the total number of vertices in the reduced graph is more than $(k + 8\Delta k - 3)$, then conclude that we are dealing with a no-instance.*

Reduction 3 is safe because of Lemma 1. This gives us the following theorem.

Theorem 4. RESTRICTED FAIR FVS *problem admits a kernel with $\mathcal{O}((\Delta + k)^2)$ vertices when parameterized by $\Delta + k$.*

5   RELAX FAIR FEEDBACK VERTEX SET

In this section, we give an FPT algorithm for RELAX FAIR FVS problem with respect to the solution size. An instance (G, k) of FVS can be trivially reduced to an instance (G, k, k) of RELAX FAIR FVS problem. It is easy to see that (G, k) is a yes-instance of FVS if and only if (G, k, k) is a yes-instance of RELAX FAIR FVS problem. This shows that RELAX FAIR FVS problem is NP-hard. We give an FPT algorithm for RELAX FAIR FVS problem using solution size as a parameter. Towards designing an algorithm for RELAX FAIR FVS, we define another problem, which we call as DISJOINT-RELAX FAIR FEEDBACK VERTEX SET (DISJOINT-RELAX FFVS) (to be defined shortly). Firstly, we design an FPT algorithm for RELAX FAIR FVS using assumed FPT algorithm for DISJOINT-RELAX FFVS. Then, we design an FPT algorithm for DISJOINT-RELAX FFVS. DISJOINT-RELAX FFVS is defined as follows.

DISJOINT-RELAX FAIR FEEDBACK VERTEX SET (DISJOINT-RELAX FFVS)
Input: An undirected graph G, a feedback vertex set Y of G, two vectors $\mathbf{w}, \mathbf{n} \in \mathbb{N}^n$, two integers k, ℓ.
Question: Does there exist a feedback vertex set X of G with $\sum_{v \in X} w(v) \leq k$ such that $X \cap Y = \emptyset$, and $|N(v) \cap X| \leq \ell - n(v)$ for all $v \in V(G) \setminus X$?

We note that in the definition of DISJOINT-RELAX FFVS, there are two vectors $\mathbf{w}, \mathbf{n}$ in the input, which we use to store the cost (weight) of taking a vertex into a solution and the number of neighbours of a vertex which have been already taken in a partial solution, respectively. This will become clear when we describe the algorithm. In Sect. 5.1 we obtain the following result.

Theorem 5. DISJOINT-RELAX FFVS *problem is solvable in time* $\mathcal{O}^*(81^k)$, *where k is the solution size.*

Next, we give FPT algorithm for RELAX FAIR FVS assuming an FPT algorithm for DISJOINT-RELAX FFVS. Before we proceed with the algorithm, notice that, if a graph does not have a feedback vertex set of size at most k, it does not have a relax fair feedback vertex set (a solution to RELAX FAIR FVS problem) of size at most k. We first find a feedback vertex set S' of G of size at most k and then we use S' to obtain a relax fair feedback vertex set S of G. We try all possible ways an optimal solution S can intersect S'. Let $X' = S' \cap S$ be one such guess. For each guess of X', we set $Y = S' \setminus X'$, $n(v) = |N(v) \cap X'|$ and $w(v) = 1$ for all $v \in G - X'$, and solve the DISJOINT-RELAX FFVS problem on the instance $(G - X', Y, \mathbf{w}, \mathbf{n}, k - |X'|, \ell)$. Here, $\mathbf{n}$ denotes the vector of $n(v)$ values and $\mathbf{w}$ denotes the vector of $w(v)$ values; $w(v)$ is the weight of vertex v. If for some set X', we find a relax fair feedback vertex set X of $G - X'$ of size $k - |X'|$ that is disjoint from Y, we can output $S = X \cup X'$. Otherwise, we conclude that the given instance is no-instance. The number of all such guesses is $\sum_{i=0}^{k} \binom{k}{i}$. Thus, to obtain an FPT algorithm for the RELAX FAIR FVS problem, it is sufficient to solve the DISJOINT-RELAX FFVS problem in FPT time.

Theorem 6 ($\star$). RELAX FAIR FVS *problem is solvable in time* $\mathcal{O}^*(82^k)$, *where k is the solution size.*

5.1 Algorithm for DISJOINT-RELAX FFVS Problem

Let $(G, Y, \mathbf{w}, \mathbf{n}, k, \ell)$ be an instance of DISJOINT-RELAX FFVS problem and let $H = G - Y$. We proceed only if $G[Y]$ is acyclic. If $G[Y]$ contains a cycle then we conclude that it is a no-instance. We define an update operation of vector $\mathbf{n}$ for all the vertices in the neighbourhood of v as follows:

$$\texttt{update}(\mathbf{n}, v) = \begin{cases} n(x) + 1 & \text{for } x \in N(v) \\ n(x) & \text{otherwise} \end{cases}$$

We give a few reduction rules that simplify the input instance.

Reduction Relax-FFVS 1. If there is a vertex $v \in V(H)$ such that $n(v) > \ell$, that is, v has more than ℓ neighbours in the solution, then delete v, decrease k by $w(v)$, and update the $\mathbf{n}$ vector for all the vertices in $N(v)$. The new instance is $(G - v, Y, \mathbf{w}, \texttt{update}(\mathbf{n}, v), k - w(v), \ell)$.

Reduction Relax-FFVS 2. If there is a vertex $v \in V(H)$ such that $G[Y \cup \{v\}]$ contains a cycle, then delete v, decrease k by $w(v)$, and update $\mathbf{n}$ vector for all the vertices in $N(v)$. The new instance is $(G - v, Y, \mathbf{w}, \texttt{update}(\mathbf{n}, v), k - w(v), \ell)$.

Reduction Relax-FFVS 3. If there exists a vertex $v \in V(H)$ such that $d_G(v) = 0$, then delete v. The new instance is $(G - v, Y, \mathbf{w}, \mathbf{n}, k, \ell)$.

Reduction Relax-FFVS 4. If there exists a vertex $v \in V(H)$ such that $d_G(v) = 1$ and $n(v) < \ell$, then delete v. The new instance is $(G - v, Y, \mathbf{w}, \mathbf{n}, k, \ell)$.

It is easy to see that Reduction Relax-FFVS 1, 2, 3 and 4 are safe.

Reduction Relax-FFVS 5. If there exists a vertex $v \in V(H)$ such that $d_G(v) = 1$ and $n(v) = \ell$, then delete v and update $w(u) = w(u) + w(v)$ where u is the neighbour of v in G; the weights remain the same for other vertices. The new instance is $(G - v, Y, \mathbf{w}', \mathbf{n}, k, \ell)$ where $\mathbf{w}'$ is the updated $\mathbf{w}$ vector.

It can be proved that Reduction Relax-FFVS 5 is safe. Although due to page constraint we omit the safeness proof here. All the Reduction rules Relax-FFVS 1, 2, 3, 4 and 5 can be applied in polynomial time.

Let $(G, Y, \mathbf{w}, \mathbf{n}, k, \ell)$ be an instance of DISJOINT-RELAX FFVS problem. The algorithm first applies Reduction Relax-FFVS 1, 2, 3, 4 and 5 exhaustively. We denote the reduced instance by $(G, Y, \mathbf{w}, \mathbf{n}, k, \ell)$. If $G[Y]$ is cyclic or if there exists a vertex $v \in Y$ with $n(v) > \ell$, then it is a no-instance. So from now onwards we assume that $G[Y]$ is indeed a forest and $n(v) \leq \ell$ for all $v \in Y$. For the reduced instance if $k < 0$, then return that it is a no-instance. Thus, from now onwards we assume that $k \geq 0$. Below we give a branching strategy with a non-trivial measure function.

Branching Relax-FFVS 1. If there is a vertex $v \in V(H)$ such that it has at least two neighbours in Y, then either v goes to the solution or v goes to Y. That is, we call the algorithm on instances $(G - v, Y, \mathbf{w}, \texttt{update}(\mathbf{n}, v), k - w(v), \ell)$ and $(G, Y \cup \{v\}, \mathbf{w}, \mathbf{n}, k, \ell)$.

After the application of Branching Relax-FFVS 1 exhaustively, every vertex in $V(H)$ has at most one neighbour in Y. When Reduction Relax-FFVS 3, 4, 5 and Branching Relax-FFVS 1 cannot be applied further, we get that every leaf node in H has exactly one neighbour in Y.

Branching Relax-FFVS 2. Consider a leaf node u and a vertex v from the same connected component of H such that both of them have only one neighbour in $G[Y]$ and the neighbours belong to different components of $G[Y]$. Let $P = ux_1x_2\ldots x_m v$ be the path in H which joins vertices u and v. If all of the vertices $x_1 x_2 \ldots x_m$ are of degree two in G, that is, they do not have any neighbour in Y, then we branch by moving vertex u to the solution in the first branch; vertex v to the solution in the second branch; vertex x_1 to the solution in the third branch; vertex x_m to the solution in the fourth branch; by moving the path P to Y in the fifth branch; by taking a set S in the solution in the sixth branch where $S = \{x_i, x_{i+1}, \ldots, x_{j-1}, x_j\}$ is a set of consecutive vertices in P with the minimum value of $\sum_{x \in S} w(x)$ such that $1 < i, j < m$, $n(x_{i-1}) < \ell$, $n(x_{j+1}) < \ell$.

Thus we call the algorithm on the instances $(G - u, Y, \mathbf{w}, \texttt{update}(\mathbf{n}, u), k - w(u), \ell)$ in first branch; $(G - v, Y, \mathbf{w}, \texttt{update}(\mathbf{n}, v), k - w(v), \ell)$ in second branch; $(G - x_1, Y, \mathbf{w}, \texttt{update}(\mathbf{n}, x_1), k - w(x_1), \ell)$ in third branch; $(G - x_m, Y, \mathbf{w}, \texttt{update}(\mathbf{n}, x_m), k - w(x_m), \ell)$ in fourth branch; $(G, Y \cup V(P), \mathbf{w}, \mathbf{n}, k, \ell)$ in fifth branch; $(G-S, Y, \mathbf{w}, \mathtt{new}(\mathbf{n}, S), k - \sum_{x \in S} w(x), \ell)$ in sixth branch where $\mathtt{new}$ operation updates vector $\mathbf{n}$ for all vertices in $N(S)$ as follows:

$$\mathtt{new}(\mathbf{n}, S) = \begin{cases} n(x) + |N(x) \cap S| & \text{for } x \in N(S) \\ n(x) & \text{otherwise} \end{cases}$$

Branching Relax-FFVS 3. Consider a leaf node u and a vertex v from the same connected component of H such that both of them have exactly one neighbour in $G[Y]$ and the neighbours belong to the same connected component of $G[Y]$. Let $P = ux_1x_2 \ldots x_m v$ be the path in H which joins vertices u and v. If all of the vertices $x_1x_2 \ldots x_m$ are of degree two in G, then we branch by considering all branches of Branching Relax-FFVS 2, except the fifth branch.

Branching Relax-FFVS 4. Consider two leaf nodes u and v from the same connected component of H; both of them have exactly one neighbour in Y. We perform Branching Relax-FFVS 4 if the neighbours of u and v are in different components of $G[Y]$. Let P be the path in H joining vertices u and v. If no intermediate vertex of P has a neighbour in Y and there is exactly one vertex z with $d_H(z) \geq 3$ on path $P = ux_1x_2 \ldots x_m zy_1y_2 \ldots y_r v$, then we branch by moving vertex u to the solution in the first branch; vertex v to the solution in the second branch; vertex z to the solution in the third branch; vertex x_1 to the solution in the fourth branch; vertex x_m to the solution in the fifth branch; vertex y_1 to the solution in the sixth branch; vertex y_r to the solution in the seventh branch; by moving the path P in Y in the eighth branch; by taking a set of consecutive vertices S from P in the solution in the ninth branch where $S = \{x_i, x_{i+1}, \ldots, x_{j-1}, x_j\}$ or $\{y_{i'}, y_{i'+1}, \ldots, y_{j'-1}, y_{j'}\}$ with the minimum value of $\sum_{x \in S} w(x)$ such that $1 < i, j < m$ and $n(x_{i-1}) < \ell, n(x_{j+1}) < \ell$, or $1 < i', j' < r$ and $n(y_{i'-1}) < \ell, n(y_{j'+1}) < \ell$.

Thus we call the algorithm on the instances $(G - u, Y, \mathbf{w}, \mathtt{update}(\mathbf{n}, u), k - w(u), \ell)$ in the first branch; $(G - v, Y, \mathbf{w}, \mathtt{update}(\mathbf{n}, v), k - w(v), \ell)$ in the second branch; $(G - z, Y, \mathbf{w}, \mathtt{update}(\mathbf{n}, z), k - w(z), \ell)$ in the third branch; $(G - x_1, Y, \mathbf{w}, \mathtt{update}(\mathbf{n}, x_1), k - w(x_1), \ell)$ in the fourth branch; $(G - x_m, Y, \mathbf{w}, \mathtt{update}(\mathbf{n}, x_m), k - w(x_m), \ell)$ in the fifth branch; $(G - y_1, Y, \mathbf{w}, \mathtt{update}(\mathbf{n}, y_1), k - w(y_1), \ell)$ in the sixth branch; $(G - y_r, Y, \mathbf{w}, \mathtt{update}(\mathbf{n}, y_r), k - w(y_r), \ell)$ in the seventh branch; $(G, Y \cup V(P), \mathbf{w}, \mathbf{n}, k, \ell)$ in the eighth branch; and $(G - S, Y, \mathbf{w}, \mathtt{new}(\mathbf{n}, S), k - \sum_{x \in S} w(x), \ell)$ in the ninth branch.

Branching Relax-FFVS 5. Consider two leaf nodes u and v from the same connected component of H; both of them have exactly one neighbour in Y. Let P be the path in H joining vertices u and v. If no intermediate vertex of P has a neighbour in Y and there is exactly one vertex z with $d_H(z) \geq 3$ on path $P = ux_1x_2 \ldots x_m zy_1y_2 \ldots y_r v$, then we branch by considering all branches of Branching Relax-FFVS 4, except the eighth branch.

Analysis of the Branching Algorithm: We apply all the Reduction Relax-FFVS 1, 2, 3, 4, and 5 in the order in which they are stated. When none of the Reduction Relax-FFVS 1, 2, 3, 4, and 5 is applicable, we apply Branching Relax-FFVS 1. After the application of Branching Relax-FFVS 1 exhaustively, we apply Branching Relax-FFVS 2 or 3 depending on the neighbours of u and v in $G[Y]$. When Branching Relax-FFVS 1, 2, and 3 are not applicable, we seek two leaf nodes u and v in H and apply Branching Relax-FFVS 4 or 5 depending on the position of neighbours of u and v in $G[Y]$. Observe that, there always exist two leaf nodes in a forest such that we can apply Branching Relax-FFVS 2, 3, 4, or 5 unless the forest is an empty graph. Thus we can branch until H becomes an empty graph.

The correctness of this algorithm follows from the safeness of our reduction rules and the fact that branching rules Branching Relax-FFVS 1, 2, 3, 4, and 5 are exhaustive. Thus if $I = (G, Y, \mathbf{w}, \mathbf{n}, k, \ell)$ is a yes-instance of DISJOINT-RELAX FFVS problem, we get a yes-instance at one of the leaf nodes of the branching tree. If all the leaves of the branching tree contain no-instances, we conclude that the instance I is a no-instance of DISJOINT-RELAX FFVS problem.

To estimate the running time of the algorithm for an instance $I = (G, Y, \mathbf{w}, \mathbf{n}, k, \ell)$, we define its measure $\mu(I) = k + \gamma(I)$, where $\gamma(I)$ is the number of connected components of $G[Y]$. The maximum value that $\mu(I)$ can take is $2k$. Observe that Reduction Relax-FFVS 1 to 5 do not increase the measure. In Branching Relax-FFVS 1, when v goes to the solution, k decreases by $w(v)$ and $\gamma(I)$ remains the same. Thus $\mu(I)$ decreases by at least 1. In the other branch, v goes into Y, then k remains the same and $\gamma(I)$ decreases by at least 1. Thus $\mu(I)$ decreases by at least 1. Thus we have a branching vector $(1,1)$ for Branching Relax-FFVS 1. For Branching Relax-FFVS 2, clearly $\mu(I)$ decreases by at least 1 in the first, second, third, fourth and sixth branch as k value decreases by at least 1. In the fifth branch, when we include $V(P)$ in Y, $\gamma(I)$ drops by 1 and k remains the same, therefore $\mu(I)$ decreases by 1. Thus, we have a branching vector $(1,1,1,1,1,1)$. Similarly, we have a branching vector $(1, 1, 1, 1, 1)$ for Branching Relax-FFVS 3. In Branching Relax-FFVS 4, clearly, $\mu(I)$ decreases by at least one in the first, second, third, fourth, fifth, sixth, seventh and ninth branch as k decreases by at least 1. In the eighth branch, $\mu(I)$ drops by 1 as $\gamma(I)$ drops by 1. Thus the branching vector is $(1,1,1,1,1,1,1,1,1)$. Similarly, we have a branching vector $(1, 1, 1, 1, 1, 1, 1, 1)$ for Branching Relax-FFVS 5. As the maximum number of branches is 9, the running time of our algorithm is $9^{\mu(I)} n^{\mathcal{O}(1)}$. Since, we have $\mu(I) \leq 2k$, the running time of our algorithm is $81^k n^{\mathcal{O}(1)}$.

References

1. Bertelè, U., Brioschi, F.: Nonserial Dynamic Programming. Academic Press Inc., New York (1972)
2. Bodlaender, H.L.: On disjoint cycles. Int. J. Found. Comput. Sci. **5**, 59–68 (1994)
3. Cao, Y., Chen, J., Liu, Y.: On feedback vertex set: new measure and new structures. Algorithmica **73**(1), 63–86 (2015)
4. Cygan, M., et al.: Lower bounds for kernelization. Parameterized Algorithms, pp. 523–555. Springer, Cham (2015). https://doi.org/10.1007/978-3-319-21275-3_15

5. Downey, R.G., Fellows, M.R.: Fixed-parameter tractability and completeness I: basic results. SIAM J. Comput. **24**, 873–921 (1995)
6. Iwata, Y., Kobayashi, Y.: Improved analysis of highest-degree branching for feedback vertex set. In: 14th International Symposium on Parameterized and Exact Computation, IPEC 2019, 11–13 September, 2019, Munich, Germany, LIPIcs, vol. 148, pp. 22:1–22:11 (2019)
7. Jacob, A., Raman, V., Sahlot, V.: Deconstructing parameterized hardness of fair vertex deletion problems. In: Du, D.-Z., Duan, Z., Tian, C. (eds.) COCOON 2019. LNCS, vol. 11653, pp. 325–337. Springer, Cham (2019). https://doi.org/10.1007/978-3-030-26176-4_27
8. Karp, R.M.: Reducibility among combinatorial problems. In: Complexity of Computer Computations, pp. 85–103. Plenum Press (1972)
9. Knop, D., Masařík, T., Toufar, T.: Parameterized complexity of fair vertex evaluation problems. In: 44th International Symposium on Mathematical Foundations of Computer Science, MFCS 2019, 26–30 August 2019, Aachen, Germany, LIPIcs, vol. 138, pp. 33:1–33:16 (2019)
10. Lampis, M.: Algorithmic meta-theorems for restrictions of treewidth. Algorithmica **64**(1), 19–37 (2012)
11. Li, J., Nederlof, J.: Detecting feedback vertex sets of size k in $O^*(2.7^k)$ time. In: Proceedings of the Thirty First Annual ACM-SIAM Symposium on Discrete Algorithms, SODA, pp. 971–981 (2020)
12. Lin, L., Sahni, S.: Fair edge deletion problems. IEEE Trans. Comput. **38**(5), 756–761 (1989)
13. Masařík, T., Toufar, T.: Parameterized complexity of fair deletion problems. Discrete Appl. Math. **278**, 51–61 (2020)
14. Silberschatz, A., Galvin, P.B., Gagne, G.: Operating System Concepts. Wiley Publishing, New York (2008)
15. West, D.B.: Introduction to Graph Theory. Prentice Hall, Upper Saddle River (2000)

The Power of Leibniz-Like Functions
as Oracles

Jaeyoon Kim, Ilya Volkovich[(✉)] [iD], and Nelson Xuzhi Zhang

CSE Division, University of Michigan, Ann Arbor, MI, USA
{jaeykim,ilyavol,xuzhizh}@umich.edu

Abstract. A *Leibniz-like* function χ is an arithmetic function (i.e., $\chi : \mathbb{N} \to \mathbb{N}$) satisfying the product rule (which is also known as "Leibniz's rule"): $\chi(MN) = \chi(M) \cdot N + M \cdot \chi(N)$. In this paper we study the computational power of efficient algorithms that are given oracle access to such functions. Among the results, we show that certain families of Leibniz-like functions can be use to factor integers, while many other families can used to compute the radicals of integers and other number-theoretic functions which are *believed* to be as hard as integer factorization [1,2].

Keywords: Integer factorization · Number-theoretic functions · Square-free integers · Möbius function · Oracles

1 Introduction

The domains of univariate polynomials and integer numbers share many properties. Nevertheless, the corresponding factorization problems differ significantly in their computational complexity. While numerous efficient algorithms for polynomial factorization have been devised (see, e.g., the surveys [6,7]), integer factorization keeps resisting nearly four decades of attempts. Indeed, the security of the RSA cryptosystem is based on this presumed hardness.

A key difference between the domains is the absence of a standard notion of a derivative in the context of integers. Observe that if a_i is a root of a polynomial $f(x)$ of multiplicity at least 2 then $x - a_i$ is a factor of $f'(x)$ (the derivative of f). More generally, if $f(x) = \prod_{i=1}^{k} (x - a_i)^{e_i}$ then[1] $\frac{f(x)}{\gcd(f,f')} = \prod_{i=1}^{k} (x - a_i)$. This observation allows us to compute the so-called "square-free" part of f, efficiently. Indeed, this procedure is carried out as the first step in many polynomial factorization algorithms. By contrast, there is no known efficient algorithm for computing the "square-free" part of an integer[2]. Indeed, this problem is *believed* to be as hard as (complete) factorization (see, e.g., [1,2,13]). The above motivates an introduction of a derivative-like notion to the domain of integers.

---

[1] For sufficiently large fields.

[2] If $N = \prod_{i=1}^{k} p_i^{\alpha_i}$ then its "square-free" part (or "radical") is defined as $\prod_{i=1}^{k} p_i$.

© Springer Nature Switzerland AG 2020
H. Fernau (Ed.): CSR 2020, LNCS 12159, pp. 263–275, 2020.
https://doi.org/10.1007/978-3-030-50026-9_19

A *Leibniz-like* function χ is an arithmetic function (i.e., $\chi : \mathbb{N} \to \mathbb{N}$) satisfying the product rule (which is also known as "Leibniz's rule"): $\chi(MN) = \chi(M) \cdot N + M \cdot \chi(N)$. Following the standard approach, to get a clear view on the role of such functions, we will assume that they are given via oracle access. That is, we can efficiently query χ on any input $N \in \mathbb{N}$.

The introduction of oracles allows us to separate the "hard" elements of the problem from the "easy" ones. In addition, in the context of integer factorization, oracles can model extra information on N obtained by means of side-channel attacks. Indeed, in the past decades we have seen several relations between the complexity of various number-theoretic problems, including integer factorization [4,8,10–13]. These relations were typically expressed in terms of oracle access (i.e., via Turing reductions). In particular, Miller [10] and some subsequent works have shown that using randomness one can efficiently factor integers, given Euler's Totient function Φ as an oracle. Woll [13] and Landau [8], independently, have shown that under the same premises one can compute the square-free part of an integer, deterministically. It is, though, remains an open question whether one can extend these results to obtain a complete factorization. A progress in this direction was recently made in [11].

1.1   Results

We outline our main results that identify properties of families of Leibniz-like functions. Our first observation is that a Leibniz-like function χ is uniquely determined by its value on the set of primes $\mathcal{P}$ (for more details see Lemma 8). Therefore, we identify the families in terms of their restrictions to $\mathcal{P}$. More technical versions are available in further sections. Our first result affirms our initial motivation given rise to efficient algorithms for computing the square-free part of an integer. We are also able to extend this to compute the Möbius function under additional technical condition. This function play an important role in analytic number theory. See Sect. 2.1 for the exact definition.

Theorem 1. *Let* $\chi : \mathbb{N} \to \mathbb{N} \cup \{0\}$ *be a Leibniz-like function such that for every prime* $p \in \mathcal{P}$ *it holds that* $\gcd(p, \chi(p)) = 1$. *Then there exists an algorithm that given* $N \in \mathbb{N}$, *as an input, and any* χ *as above via oracle access, runs in time* $\mathrm{polylog}(N)$ *and outputs the radical of* N. *If, in addition, for every prime* p, $\chi(p)$ *is an odd number, then the algorithm can be extended to also compute* $\mu(N)$, *where* $\mu(N)$ *is the Möbius function.*

This theorem yields to the following:

Corollary 1. *There exists an algorithm that, given* $N \in \mathbb{N}$ *as an input, and* χ *satisfying one of the following conditions for all primes* $p \in \mathcal{P}$, *via oracle access, runs in time* $\mathrm{polylog}(N)$ *and outputs the radical of* N.

1. $\chi_{q-1}(p) \stackrel{\Delta}{=} p - 1$. *Motivation - resembles Euler's Totient function.*

2. $\chi_\pi(p) \stackrel{\Delta}{=} \pi(p)$. *Here $\pi(x)$ is the prime counting function. That is, $\pi(p)$ returns the rank of the prime p in $\mathcal{P}$.*

3. $\chi_1(p) \stackrel{\Delta}{=} 1$.

In addition, the function that satisfies $\chi(p) = 1$ can also be used to compute $\mu(N)$.

We remark (again) that the problems of computing $\mu(N)$ or merely deciding if a number is square-free are *believed* to be as hard as (complete) integer factorization [1,2]. Next, show that if a family of Leibniz-like functions can be "parametrized" efficiently, then it can be used to obtain complete factorization, under some technical conditions.

Definition 1. *A family of Leibniz-like functions $\{\chi_a(\cdot)\}_{a \in I}$ is called an* efficient filter *if it there exists an algorithm $\mathcal{A}$ that given $N \in \mathbb{N}$, as an input, runs in time $\mathrm{polylog}(N)$ and outputs a set S such that for any distinct primes $p, q \leq \sqrt{N}$ there exists $a \in S$ such that: $p \mid \chi_a(p)$ or $q \mid \chi_a(q)$ but* **not** *both.*

Theorem 2. *There exists an algorithm that, given an integer N and oracle access to an efficient filter, runs in time $\mathrm{polylog}(N)$ and outputs the complete factorization of N.*

From this theorem, we can draw the following conclusion:

Corollary 2. *There exists an algorithm that, given $N \in \mathbb{N}$ as an input, and χ satisfying one of the following conditions for all $p \in \mathcal{P}$, via oracle access, runs in time $\mathrm{polylog}(N)$ and outputs the factorization of N.*

1. $\chi_T(p) = \begin{cases} 1 & p \leq T \\ 0 & \text{otherwise} \end{cases}$

2. $\chi_{(a,q)}(p) = \begin{cases} 1 & p \equiv a \pmod{q} \\ 0 & \text{otherwise} \end{cases}$

Remark: A more general definition of an efficient filter is given in Definition 11.

Finally, we exhibit a different family of Leibniz-like functions that can be used to compute $\Omega(N)$ - the number of prime factors of N with multiplicity as well as to completely factor N.

Theorem 3. *For $\ell \in \mathbb{N}$, let χ_{q^ℓ} denote the Leibniz-like function satisfying*

$$\chi_{q^\ell}(p) = p^\ell, \text{ for all } p \in \mathcal{P}.$$

Then there exists an algorithm that given $N \in \mathbb{N}$, as an input, and the function χ_q via oracle access, runs in time $\mathrm{polylog}(N)$ and outputs $\Omega(N)$. In addition, there exists an algorithm that given the functions $\{\chi_{q^\ell}\}_{\ell=2}^{\log N + 1}$ via oracle access, runs in time $\mathrm{polylog}(N)$ and outputs the prime factorization of N.

2   Preliminaries

Let $\mathcal{P}$ denote the set of all primes and $\mathcal{P}^*$ the set of prime powers. By the fundamental theorem of arithmetic, each integer $N \in \mathbb{N}$ has a unique prime factorization. That is, N can be uniquely written as $N = \prod_{i=1}^{k} p_i^{\alpha_i}$ such that for all i: $p_i \in \mathcal{P}$ and $\alpha_i \in \mathbb{N}$. We require the following technical result. Similar ideas and techniques were previously used in [12].

Lemma 1. *Let $N_1 > N_2 \in \mathbb{N}$. Then there exists a prime $q \leq 1.2 \ln(N_1 - N_2)$ such that $N_1 \not\equiv N_2 \pmod{q}$.*

The following result is inspired by Newton's Identities.

Definition 2. *For $i \geq 0$, we denote by $P_i(N_1, \ldots, N_k) \triangleq \sum_{j=1}^{k} N_j^i$ - the sum of i-th powers of integers and by $Q_i(N_1, \ldots, N_k)$ the i-th elementary symmetric polynomial. That is, the sum of all distinct products of i distinct variables amongst $N_1, \ldots, N_k$.*

Newton's Identities establish a relation between the two stating that

$$kQ_k(N_1, \ldots, N_k) = \sum_{i=1}^{k} Q_{k-i}(N_1, \ldots, N_k) P_i(N_1, \ldots, N_k)$$

The above gives rise to the following:

Observation 1. *There exists a polynomial-time algorithm* NEWTON *that, given $\{P_i(N_1, \ldots, N_k)\}_{i \in [k]}$ as inputs, outputs $\{Q_i(N_1, \ldots, N_k)\}_{i \in [k]}$.*

Finally, we recall Vieta's Theorem.

Lemma 2. *Let $Q_i(N_1, \ldots, N_k) = d_i$. Then $\sum_i (-1)^i d_{j-i} x^i = \prod_i^k (x - N_1)$.*

Putting all together, we obtain the following:

Lemma 3. *Let $N_1, \ldots, N_k \in \mathbb{N}$ (not necessarily distinct integers). There exists an algorithm that, given the values $\{P_i(N_1, \ldots, N_k)\}_{i \in [k]}$ as inputs, runs in time* polylog$(N_1 + N_2 + \ldots + N_k)$ *and outputs $N_1, \ldots, N_k$.*

Let us first sketch out the idea of the proof. Given $\{P_i(N_1, \ldots, N_k)\}_{i \in [k]}$, we first apply the Algorithm of Observation 1 to obtain $\{Q_i(N_1, \ldots, N_k)\}_{i \in [k]}$. We then use the univariate polynomial factorization algorithm of [9] to obtain $N_1, \ldots, N_k$. The correctness follows from Lemma 2. Due to space constraints, the complete proof is given in the full version of the paper.

2.1   Arithmetic Functions

In this paper we study the computational power of certain kind of arithmetic functions given as oracles.

Definition 3. *We call a function* $f : \mathbb{N} \to \mathbb{C}$ *an arithmetic function. Unless explicitly stated otherwise, we will assume that the range of our arithmetic functions is* $\mathbb{N}$ *or* $\mathbb{Z}$, *and* $|f(N)| \leq 2^{\mathrm{polylog}(N)}$.

Example 1. Below are examples of common arithmetic functions.

1. $\Omega(N) \overset{\Delta}{=} \sum_{i=1}^{k} \alpha_i$ - the number of prime factors of N with multiplicity.
2. $\omega(N) \overset{\Delta}{=} k$ - the number of prime factors of N **without** multiplicity.
3. $\Phi(N)$ - Euler's totient function
4. $\mu : \mathbb{N} \to \{-1, 0, 1\}$ - Möbius Function:

$$\mu(N) = \begin{cases} 0 & \text{if } N \text{ is not square-free} \\ (-1)^k & \text{if } N \text{ is a product of } k \text{ (distinct) primes} \end{cases}$$

5. $\nu_p(N)$ - p-adic valuation of an integer N: for a prime $p \in \mathcal{P}$, $\nu_p(N)$ is the largest integer m such that $p^m \mid N$ while $p^{m+1} \nmid N$.

While several relations between these (and other) function have been established [4,8,10–13], there is no known polynomial-time algorithm for any of above functions, expect $\nu_p(N)$. Indeed, these function are *believed* to be as hard as (complete) integer factorization [1,2]. We will use the following notation throughout the paper.

Definition 4. $\mathrm{D}_f(N) \overset{\Delta}{=} \gcd(N, f(N))$. *We say that* $\mathrm{D}_f(N)$ *is* non-trivial *if* $1 < \mathrm{D}_f(N) < N$.

2.2   Square-Free Numbers

Definition 5. *Consider* $N \in \mathbb{N}$ *and its unique factorization* $N = \prod_{i=1}^{k} p_i^{\alpha_i}$.

We define the radical *of* N *as* $rad(N) \overset{\Delta}{=} \prod_{i=1}^{k} p_i$. N *is called* square-free *or* radical *iff* $\forall i : \alpha_i = 1$. *In other words,* N *is radical iff* $N = rad(N)$.
We define the set $\mathbb{SQF} \subseteq \mathbb{N}$ *as the set of all square-free integers.*
We define the set $\mathbb{SQF}^* \subseteq \mathbb{N}$ *as set of powers of all square-free integers.*

The following is immediate from the definition:

Observation 2. *For all* $M, N \in \mathbb{N} : rad(N \cdot M) = lcm\,(rad(N), rad(M))$.

While computing the radical (i.e., square-free part) of a polynomial can be carried out easily (see, e.g., [14]). There is no known polynomial-time algorithm for computing the radical of a given integer number N. Moreover, even testing if a given N is square-free (i.e., membership in $\mathbb{SQF}$) is considered a hard computational problem (see [1,2]). Indeed, this is equivalent to computing $|\mu(N)|$. We complete this section with the following result.

Lemma 4 ([5]). *There exists an algorithm that given $N = m^k$ for some square-free integer m as an input, runs in time $\log(N)$ and outputs m and k.*

2.3   Rough Numbers

Definition 6 (Roughness). *Let $r \in \mathbb{N}$. A number $N \in \mathbb{N}$ is called r-rough if all its prime factors are greater than or equal to r. In other words, for all i: $p_i \geq r$. We will call a number N rough (without the parameter r) if it is $(\log N + 1)$-rough.*

The following statements are immediate from the definition and the fact that every positive integer is 2-rough.

Lemma 5. *For all $N \in \mathbb{N} : 1 \leq \Omega(N) \leq \log N$.*

Corollary 3. *Let N be a rough number. Then for all primes $p \in \mathcal{P} : \nu_p(N) < p$.*

For the purposes of factorization and related tasks, we will assume w.l.o.g that each N given as an input is rough, since we can simply brute-force all the primes smaller than $\log N + 1$.

2.4   Partial Factorization Algorithms

Definition 7 (The set of irreducibles). *The set of irreducibles of an arithmetic function $f : \mathbb{N} \to \mathbb{N}$ is defined as follows:*

$$\mathrm{IS}(f) = \{N \in \mathbb{N} \mid \mathrm{D}_f(N) = N \text{ or } \mathrm{D}_f(N) = 1\}.$$

In other words, the set of irreducibles of an arithmetic function f corresponds to set of integers for which the function f "fails" to produce a non-trivial factor. This could be thought of as an extension of the set of primes $\mathcal{P}$ (which constitutes the set of the irreducible elements of $\mathbb{N}$) to a specific function. Indeed, for every such f we have that $\mathcal{P} \subseteq \mathrm{IS}(f)$. This allows us to relax the notion of factorization.

Definition 8 (f-factorization). *Let $f : \mathbb{N} \to \mathbb{N}$ be an arithmetic function and let $N \in \mathbb{N}$. A product $N = \prod_{i=1}^{k} N_i$ is called an f-factorization of N if for each $i \in [k] : N_i \in \mathrm{IS}(f)$.*

We remark that unlike the standard factorization (into prime factors) an f-factorization may be not unique. Furthermore, in [3,5] efficient algorithms have been shown to "refine" a given computation into a "gcd-free" one.

Given $N \in \mathbb{N}$ as an input, many factorization algorithms proceed by first finding a non-trivial factor M of N and then recursing on both M and N/M, until these become irreducible. Given an oracle access to an arithmetic function f, we can extend this idea by using $f(N)$ as a candidate for a non-trivial factor of N and then recurse until $f(N)$ fails to produce such a factor. Observe that this procedure will result in an f-factorization. The following lemma summarizes this discussion.

Lemma 6. *There exists an algorithm that given $N \in \mathbb{N}$, as an input, and an arithmetic function $f : \mathbb{N} \to \mathbb{N}$ as an oracle, runs in time $\mathrm{polylog}(N)$ and outputs an f-factorization of N.*

The above framework can used to compute the radical of an integer and even complete factorization, given the "right" f. In particular, recall that $\mathcal{P}^*$ and $\mathbb{SQF}^*$ denote the sets of all natural powers of all the primes and square-free numbers, respectively. Then the following is immediate given Lemmas 4 and 6.

Corollary 4. *Suppose that $f : \mathbb{N} \to \mathbb{N}$ is an arithmetic function such that $\mathrm{IS}(f) \subseteq \mathcal{P}^*$. Then there exists an algorithm that given $N \in \mathbb{N}$, as an input, and f as an oracle, runs in time $\mathrm{polylog}(N)$ and outputs the factorization of N.*

Proof. Apply Lemma 6 to obtain an f-factorization $N_1, \ldots, N_k$ of N. By assumption N_i, for each $i \in [k] : N_i \in \mathrm{IS}(f) \subseteq \mathcal{P}^*$. The primes and their powers can be then recovered by Lemma 4.   □

Similarly, by further incorporating Observation 2, we obtain the following corollary.

Corollary 5. *Suppose that $f : \mathbb{N} \to \mathbb{N}$ is an arithmetic function such that $\mathrm{IS}(f) \subseteq \mathbb{SQF}^*$. Then there exists an algorithm that given $N \in \mathbb{N}$, as an input, and f as an oracle, runs in time $\mathrm{polylog}(N)$ and outputs $rad(N)$.*

Indeed, we can now easily re-establish the results of Woll [13] and Landau [8] (computing $rad(N)$, given $\Phi(N)$) by observing that $\mathrm{IS}(\Phi) \subseteq \mathbb{SQF}$.

2.5   Leibniz-Like Functions and Their Properties

In this paper we will consider the computational power of a special type of arithmetic functions. This is our main conceptual contribution. We begin with a formal definition.

Definition 9 (Leibniz-like Function). *An arithmetic function $\chi : \mathbb{N} \to \mathbb{C}$ is said to be Leibniz-like if for all $M, N \in \mathbb{N}$ it holds that $\chi(M \cdot N) = M \cdot \chi(N) + \chi(M) \cdot N$.*

The following is immediate from the definition.

Lemma 7. *Let χ be a Leibniz-like function and let $N = \prod_{i=1}^{k} N_i^{e_i}$ (not necessarily irreducible). Then $\chi(N) = \sum_{i=1}^{k} \frac{e_i \cdot \chi(N_i) \cdot N}{N_i}$.*

Given the above, we observe that a Leibniz-like function is uniquely determined by its values on the set of all primes $\mathcal{P}$.

Lemma 8 (Uniqueness of Extension). *Let $f : \mathcal{P} \to \mathbb{C}$ be an arbitrary function. Then there exists a unique Leibniz-like function $\chi : \mathbb{N} \to \mathbb{C}$ such that $\chi(p) = f(p)$ for all $p \in \mathcal{P}$.*

Proof. Let $N \in \mathbb{N}$ and let $N = \prod_{i=1}^{k} p_i^{\alpha_i}$ be N's (unique) prime factorization. We define $\chi(N) \stackrel{\Delta}{=} \sum_{i=1}^{k} \frac{\alpha_i \cdot f(p_i) \cdot N}{p_i}$. One could verify that $\chi(\cdot)$ is indeed a Leibniz-like function consistent with $f(\cdot)$ on $\mathcal{P}$. The uniqueness follows from Lemma 7. $\square$

Based on the above, given $f : \mathcal{P} \to \mathbb{C}$ we will denote by χ_f its unique extension. Furthermore, we observe that a sum of two Leibniz-like functions is itself a Leibniz-like function. This implies that the set of all Leibniz-like functions forms a linear space.

Lemma 9. *Let χ_f and $\chi_{f'}$ be two Leibniz-like functions that are the extensions of f and f', respectively. Then $(\chi_f + \chi_{f'})(N) \stackrel{\Delta}{=} \chi_f(N) + \chi_{f'}(N)$ is a Leibniz-like function and constitutes the extension of $f + f'$.*

For the rest of the paper we will focus on functions $\chi : \mathbb{N} \to \mathbb{N} \cup 0$. We will now explore some useful properties of $D_\chi(N)$ (recall Definition 4).

Lemma 10. *Let χ be a differential multiplicative function and let $N = \prod_{s \in S} p_s^{\alpha_s} \cdot \prod_{t \in T} p_t^{\alpha_t}$ be N's (unique) prime factorization where:*

- *for each $s \in S : p_s \nmid \alpha_s \cdot \chi(p_s)$.*
- *for each $t \in T : p_t \mid \alpha_t \cdot \chi(p_t)$.*

Then $D_\chi(N) = \prod_{s \in S} p_s^{\alpha_s - 1} \cdot \prod_{t \in T} p_t^{\alpha_t}$ and, in particular, $N/D_\chi(N)$ is square-free.

Proof. For a prime $p \in \mathcal{P}$, let us denote by $\nu_p(N)$ the largest integer m such that $p^m \mid N$ while $p^{m+1} \nmid N$. Fix $s \in S$. By definition, $\nu_{p_s}(N) = \alpha_s$. Now let us consider $\nu_{p_s}(\chi(N))$. By Lemma 7, $\chi(N) = \sum_{i=1}^{k} \frac{\alpha_i \cdot \chi(p_i) \cdot N}{p_i}$. Therefore, $\nu_{p_s}(\chi(N)) \geq \alpha_s - 1$ since $p_s^{\alpha_s - 1}$ divides every term in the summation. Yet, $p_s^{\alpha_s} \nmid \chi(N)$ since $p_s^{\alpha_s}$ divides every term in the summation **except** $\frac{\alpha_s \chi(p_s) N}{p_s}$ as $p_s \nmid \alpha_s \cdot \chi(p_s)$. This implies that $\nu_{p_s}(D_\chi(N)) = \nu_{p_s}(\chi(N)) = \alpha_s - 1$. Now let us fix $t \in T$. As $p_t \mid \alpha_t \cdot \chi(p_t)$, we have that $p_t^{\alpha_t}$ divides every term in the summation above. Therefore, $\nu_{p_t}(D_\chi(N)) = \nu_{p_t}(\chi(N)) = \alpha_t$. $\square$

Corollary 6. *Suppose that χ is a Leibniz-like function and let N be rough then:*

- *If there exists p_i such that $p_i \nmid \chi(p_i)$ then $D_\chi(N) < N$.*
- *If there exists p_j such that $p_j \mid \chi(p_j)$ then $D_\chi(N) > 1$.*
- *If $N \notin \mathbb{SQF}$ then $D_\chi(N) > 1$.*

Proof. By Corollary 3, each $\alpha_i < p_i$. Consequently, $\alpha_i \nmid p_i$. Therefore, $p_i \mid \alpha_i \cdot \chi(p_i)$ iff $p_i \mid \chi(p_i)$ and the claim follows from Lemma 10. $\square$

As a subsequent corollary we obtain the following statements, which imply Theorem 1.

Corollary 7. *Let χ be such that $\forall p \in \mathcal{P} : D_\chi(p) = 1$. Then for all rough N we have $N/D_\chi(N) = rad(N)$.*

Observation 3. *Let $N \in \mathbb{N}$ be a product of k distinct odd primes and let χ be such that $\forall p \in \mathcal{P} : \chi(p)$ is odd. Then $\chi(N) \equiv k \pmod 2$.*

A particular instance satisfying the premises of the observation is the function χ_1. That is, a Leibniz-like function satisfying $\chi_1(p) = 1$ for all $p \in \mathcal{P}$. We conclude this section by observing that when N is a product of two prime powers, the algorithm can be extended to obtain a complete factorization of N.

Lemma 11. *Let $p, q \in \mathcal{P}$ and $\alpha, \beta \in \mathbb{N}$. Then there exists an algorithm that given N of the form $N = p^\alpha q^\beta$, as an input, and χ_1 via oracle access, runs in time $\mathrm{polylog}(N)$ and outputs the factorization of N. Namely, outputs p, q and α, β.*

Proof. By Corollary 7, $N/D_\chi(N) = pq$. On the other hand, $\chi_1(pq) = p + q$. Consequently, p and q are the solutions of the quadratic equation $x^2 - \chi_1(pq)x + pq = 0$, which could be found efficiently. Given p and q, we can recover α and β by repeated division, since $\alpha, \beta < \log N$. $\square$

3 Families of Oracles

In this section we prove a relaxed version of Theorem 1 as a motivation for the proof of Theorem 2. Rather than considering a single oracle function we will assume that we have an access to a family of functions that could be "parametrized" efficiently.

Definition 10. *A family of Leibniz-like functions $\{\chi_a(\cdot)\}_{a \in I}$ is called an efficient cover if it there exists an algorithm $\mathcal{A}$ that given $N \in \mathbb{N}$, as an input, runs in time $\mathrm{polylog}(N)$ and outputs a set S such that there exists $a \in S$ and a prime $p \mid N$ satisfying: $p \nmid \chi_a(p)$.*

We now show that an efficient cover can be used to construct an arithmetic function with $\mathrm{IS}(f) \subseteq \mathbb{SQF}$. This, in turn, can be combined with Corollary 5 to obtain an efficient algorithm that computes $rad(N)$ of a given $N \in \mathbb{N}$.

Lemma 12. *There exists an arithmetic function $f : \mathbb{N} \to \mathbb{N}$ with $\mathrm{IS}(f) \subseteq \mathbb{SQF}$ that is efficiently computable given an efficient cover $\{\chi_a(\cdot)\}_{a \in I}$ via oracle access. Algorithm 1 provides the outline.*

Input: $N \in \mathbb{N}$, $\mathcal{A}$, $\{\chi_a(\cdot)\}_{a \in I}$ via oracle access
Output: $f(N)$

1: **for** $M = 2 \ldots \log N$ **do**
2:     **if** $M \mid N$ **then**
3:         **return** M
       // N is rough
4: **for** $a \in \mathcal{A}(N)$ **do**
5:     **if** $D_{\chi_a}(N) < N$ **then**
6:         **return** $D_{\chi_a}(N)$

Algorithm 1: Computing $f(N)$

Proof. Let $N \notin \mathbb{SQF}$. If there exist $M \leq \log N$ such that $M \mid N$ then we are done. Otherwise, N is rough. Let $a \in \mathcal{A}(N)$ be as guaranteed by the definition of a cover. Then by Corollary 6, $D_{\chi_a}(N)$ is non-trivial. ☐

In order to extend the algorithm to yield a complete factorization, we require a family that satisfies additional technical conditions.

Definition 11. *A family of Leibniz-like functions $\{\chi_a(\cdot)\}_{a \in I}$ is an* efficient filter *if there exists an algorithm $\mathcal{A}$ that given $N \in \mathbb{N} \setminus \mathcal{P}^*$, as an input, runs in time $\mathrm{polylog}(N)$ and outputs a set S such that there exists $a \in S$ and two distinct primes $p, q \mid N$ satisfying: $p \mid \chi_a(p)$ and $q \nmid \chi_a(q)$.*

Lemma 13. *There exists an arithmetic function $f : \mathbb{N} \to \mathbb{N}$ with $\mathrm{IS}(f) \subseteq \mathcal{P}^*$ that is efficiently computable given an efficient filter $\{\chi_a(\cdot)\}_{a \in I}$ via oracle access. Algorithm 1 provides the outline.*

Theorem 2 following by combining Lemma 13 with Corollary 4. In the following sections we give two examples of families of Leibniz-like functions that constitute efficient filters.

3.1 Special Case: χ_T

Let $\{\chi_T\}_{T \in \mathbb{N}}$ be a family of "threshold" Leibniz-like functions that is defined as follows for all $p \in \mathcal{P}$:

$$\chi_T(p) = \begin{cases} 1 & p \leq T \\ 0 & \text{otherwise} \end{cases}$$

Let us now describe $\mathcal{A}$:
On input $N \in \mathbb{N}$, $\mathcal{A}$ outputs the set $\left\{ N, \lfloor N^{1/2} \rfloor, \lfloor N^{1/3} \rfloor, \ldots, \lfloor N^{1/\log N} \rfloor \right\}$.

Analysis: Let $N = \prod_{i=1}^{k} p_i^{\alpha_i}$ with $k \geq 2$. Let $M = N^{1/\Omega(N)}$ (see Example 1 for reference). By the properties of the geometric mean, there exists i, j such that $p_i < M < p_j$. Therefore, $p_j \mid \chi_M(p_j) = 0$ and $p_i \nmid \chi_M(p_i) = 1$. Lemma 5 completes the analysis. ☐

3.2 Special Case: $\chi_{(a,Q)}$

Let $\{\chi_{(a,q)}\}_{q\in\mathcal{P}\,,\,0\le a<q}$ be a family of Leibniz-like functions s.t. for all $p \in \mathcal{P}$:

$$\chi_{(a,q)}(p) = \begin{cases} 1 & p \equiv a \pmod{q} \\ 0 & \text{otherwise} \end{cases}$$

The function $\chi_{(a,q)}$ is motivated by the following intuition: $\chi_{(a,q)}$ extracts information about prime factors of congruent to $a \pmod q$. Then using Chinese Remainder Theorem to combine all information from small prime moduli, we should be able to reconstruct the prime factorization.

As a first step for intuition purposes, we observe the following, which in particular implies that $\chi_{(a,q)}$ constitutes an efficient cover.

Observation 4. *For all $p, q \in \mathcal{P} : \sum\limits_{a=0}^{q} \chi_{(a,q)}(p) = 1$.*

Next, we show that $\chi_{(a,q)}$ is also an efficient filter by describing and analyzing an appropriate $\mathcal{A}$. On input $N \in \mathbb{N}$, set $y = 1.2 \ln(N_1 - N_2)$ and output the set $\{(a,q) \mid q \in \mathcal{P} \le y\,,\,0 \le a < q\}$. Since y is "small", we can simply brute-force all the primes.

Analysis: Let $N = \prod_{i=1}^{k} p_i^{\alpha_i}$ with $k \ge 2$. Suppose $p_1 > p_k$. Then $p_1 - p_k < N$. By Lemma 1, there exists $q \in \mathcal{P} \le y$ such that $p_1 \not\equiv p_k \pmod{q}$. Let $a \overset{\Delta}{=} p_k \bmod q$. Then $p_1 \mid \chi_{(a,q)}(p_1) = 0$ and $p_k \nmid \chi_{(a,q)}(p_k) = 1$. $\qquad\square$

4 · Non-coprime Case

So far we have studied several families of Leibniz-like function χ where $D_\chi(p) = 1$ for every $p \in \mathcal{P}$. In this section, we are going to change the gear a little and explore families of oracles that do not satisfy this property. First of all, observe that by definition if $D_\chi(p) \neq 1$ for a prime $p \in \mathcal{P}$, it must be the case that $D_\chi(p) = p$. Applying Lemma 7 provides the following (natural) extension.

Observation 5. *Suppose that χ is a Leibniz-like function such that for all $p \in \mathcal{P} : D_\chi(p) \neq 1$. Then for all $N \in \mathbb{N} : D_\chi(N) = N$.*

Given the above, it is natural to consider the "normalized" function.

Definition 12. *For χ as above, we consider $\bar\chi(N) \overset{\Delta}{=} \chi(N)/N$.*

Our first observation is that the normalized function exhibits a "log-like" behavior. Indeed, for any $x, y : \log(x \cdot y) = \log(x) + \log(y)$.

Observation 6. *For any $M, N \in \mathbb{N} : \bar\chi(M \cdot N) = \bar\chi(M) + \bar\chi(N)$.*

For the rest of the section, we will focus on a particular family of functions.

Definition 13. *For $\ell \in \mathbb{N}$, let χ_{q^ℓ} denote the Leibniz-like function satisfying $\forall p \in \mathcal{P} : \chi_{q^\ell}(p) = p^\ell$.*

The next main property of this family of function follows from Lemma 7.

Lemma 14. *Let $N = \prod_{i=1}^{k} p_i^{\alpha_i}$ and $\ell \geq 1$. Then $\bar{\chi}_{q^\ell}(N) = \sum_{i=1}^{k} \alpha_i p_i^{\ell-1}$.*

In particular, given χ_q via oracle access, one can efficiently compute $\Omega(N)$.

Corollary 8. $\bar{\chi}_p(N) = \Omega(N)$.

We remind that while $1 \leq \Omega(N) \leq \log N$ (see Lemma 5), there is no known polynomial-time algorithm for computing the exact value of $\Omega(N)$. Indeed, computing $\Omega(N)$ is *believed* to be as hard computing the complete factorization (see, e.g., [1]). Likewise, it unlikely to have a polynomial-time algorithm that computes the factorization of N even if given in addition the value of $\chi_q(N)$, since we could just "guess" that value. In order to elaborate on the discussion and provide more intuition, we provide a "baby step" towards the proof of Theorem 3, by showing that one can efficiently factor N of the form $N = p^\alpha r^\beta$, given in addition $\chi_q(N), \chi_{q^2}(N), \chi_{q^3}(N)$.

Observation 7. *Let $N = p^\alpha r^\beta$. Then $\bar{\chi}_q(N) \cdot \bar{\chi}_{q^3}(N) - (\bar{\chi}_{q^3}(N))^2 = \alpha\beta \cdot (p-r)^2$.*

In order to obtain p and r, the algorithm will "guess" α, β such that $\alpha + \beta \leq \log N$ and then mimic the argument of Lemma 11. We leave the details as an exercise for the reader. The next result follows from Lemma 3. The proof is given in the full version of the paper. Theorem 3 follows from Lemma 15 and Corollary 8.

Lemma 15. *There exists an algorithm that, given $N \in \mathbb{N}$ as an input, computes the functions $\{\chi_{q^\ell}\}_{\ell=2}^{\log N + 1}$ via oracle access, runs in time $\mathrm{polylog}(N)$ and outputs the complete factorization of N.*

5   Discussion and Open Questions

In this paper we have introduced the concept of Leibniz-like functions and discussed its computational power when given via an oracle access. In particular, we have shown that certain families of Leibniz-like functions can be used to compute number-theoretic functions for which no polynomial-time algorithms are known. We conclude with a few open questions.

1. Can one extend Theorem 1 to obtain a complete factorization?
2. A simpler version: Can we prove Theorem 2 using a single oracle?
3. One important conclusion of the results in this paper pertain to the complexity of computing an extension of a given Leibniz-like function χ from $\mathcal{P}$ to $\mathbb{N}$. In particular, χ_1 is trivial to compute on $\mathcal{P}$ whereas Corollary 1 and Lemma 11 suggest that there is no polynomial-time algorithm for computing its extension to $\mathbb{N}$. Is there any "interesting" Leibniz-like function χ for which computing the extension to $\mathbb{N}$ is "easy"? A trivial example of such a function would be χ that satisfies $\chi(p) = 0$ for all $p \in \mathcal{P}$. We note that this example can be easily generalized to the case when $\chi(p) = 0$ for all, but finitely many $p \in \mathcal{P}$.

4. Finally, can we compute other "interesting" number-theoretic function given oracle access to another (family of) Leibniz-like function(s)?

Acknowledgements. The authors would also like to thank the anonymous referees for their detailed comments and suggestions.

References

1. Adleman, L.M., McCurley, K.S.: Open problems in number theoretic complexity, II. In: Adleman, L.M., Huang, M.-D. (eds.) ANTS 1994. LNCS, vol. 877, pp. 291–322. Springer, Heidelberg (1994). https://doi.org/10.1007/3-540-58691-1_70
2. Bach, E.: Intractable problems in number theory. In: Goldwasser, S. (ed.) CRYPTO 1988. LNCS, vol. 403, pp. 77–93. Springer, New York (1990). https://doi.org/10.1007/0-387-34799-2_7
3. Bach, E., Driscoll, J.R., Shallit, J.: Factor refinement. J. Algorithms **15**(2), 199–222 (1993). https://doi.org/10.1006/jagm.1993.1038
4. Bach, E., Miller, G.L., Shallit, J.: Sums of divisors, perfect numbers and factoring. SIAM J. Comput. **15**(4), 1143–1154 (1986). https://doi.org/10.1137/0215083
5. Bernstein, D.: Factoring into coprimes in essentially linear time. J. Algorithms **54**(1), 1–30 (2005). https://doi.org/10.1016/j.jalgor.2004.04.009
6. Gathen, J.V.Z.: Who was who in polynomial factorization. In: Trager, B.M. (ed.) Symbolic and Algebraic Computation, International Symposium, ISSAC. p. 2. ACM (2006). https://doi.org/10.1145/1145768.1145770
7. Kaltofen, E.: Polynomial factorization: a success story. In: Sendra, J.R. (ed.) Symbolic and Algebraic Computation, International Symposium, ISSAC. pp. 3–4. ACM (2003). https://doi.org/10.1145/860854.860857
8. Landau, S.: Some remarks on computing the square parts of integers. Inf. Comput. **78**(3), 246–253 (1988). https://doi.org/10.1016/0890-5401(88)90028-4
9. Lenstra, A.K., Lenstra, H.W., Lovász, L.: Factoring polynomials with rational coefficients. Mathematische Annalen **261**(4), 515–534 (1982)
10. Miller, G.L.: Riemann's hypothesis and tests for primality. J. Comput. Syst. Sci. **13**(3), 300–317 (1976). https://doi.org/10.1016/S0022-0000(76)80043-8
11. Morain, F., Renault, G., Smith, B.: Deterministic factoring with oracles. CoRR abs/1802.08444 (2018). http://arxiv.org/abs/1802.08444
12. Shallit, J., Shamir, A.: Number-theoretic functions which are equivalent to number of divisors. Inf. Process. Lett. **20**(3), 151–153 (1985). https://doi.org/10.1016/0020-0190(85)90084-5
13. Woll, H.: Reductions among number theoretic problems. Inf. Comput. **72**(3), 167–179 (1987). https://doi.org/10.1016/0890-5401(87)90030-7
14. Yun, D.Y.Y.: On square-free decomposition algorithms. In: SYMSAC 1976, Proceedings of the third ACM Symposium on Symbolic and Algebraic Manipulation, Yorktown Heights, New York, USA, 10–12 August 1976, pp. 26–35. ACM (1976). https://doi.org/10.1145/800205.806320

Optimal Skeleton Huffman Trees Revisited

Dmitry Kosolobov$^{(\boxtimes)}$ and Oleg Merkurev

Ural Federal University, Ekaterinburg, Russia
dkosolobov@mail.ru, o.merkuryev@gmail.com

Abstract. A skeleton Huffman tree is a Huffman tree in which all disjoint maximal perfect subtrees are shrunk into leaves. Skeleton Huffman trees, besides saving storage space, are also used for faster decoding and for speeding up Huffman-shaped wavelet trees. In 2017 Klein et al. introduced an optimal skeleton tree: for given symbol frequencies, it has the least number of nodes among all optimal prefix-free code trees (not necessarily Huffman's) with shrunk perfect subtrees. Klein et al. described a simple algorithm that, for fixed codeword lengths, finds a skeleton tree with the least number of nodes; with this algorithm one can process each set of optimal codeword lengths to find an optimal skeleton tree. However, there are exponentially many such sets in the worst case. We describe an $\mathcal{O}(n^2 \log n)$-time algorithm that, given n symbol frequencies, constructs an optimal skeleton tree and its corresponding optimal code.

Keywords: Huffman tree · Skeleton tree · Dynamic programming

1 Introduction

The Huffman code [10] is one of the most fundamental primitives of data compression. Numerous papers are devoted to Huffman codes and their variations; see the surveys [1] and [16]. In this paper we investigate the skeleton Huffman trees, introduced by Klein [11], which are code trees with all (disjoint) maximal perfect subtrees shrunk into leaves (precise definitions follow). We describe the first polynomial algorithm that, for code weights $w_1, w_2, \ldots, w_n$, constructs the smallest in the number of nodes skeleton tree among all trees formed by optimal prefix-free codes. Klein et al. [12] called such trees *optimal skeleton trees*.

The idea of the skeleton tree is simple: to determine the length of a given codeword in the input bit stream, the standard decoding algorithm for Huffman codes descends in the tree from the root to a leaf; instead, one can descend to a leaf of the skeleton tree, where the remaining length is uniquely determined by the height of the corresponding shrunk perfect subtree. While presently the decoding is performed by faster table methods [16] and, in general, the entropy encoding is implemented using superior methods like ANS [3], there are still important

Supported by the Russian Science Foundation (RSF), project 18-71-00002.

H. Fernau (Ed.): CSR 2020, LNCS 12159, pp. 276–288, 2020.
https://doi.org/10.1007/978-3-030-50026-9_20

applications for the trees of optimal codes in compressed data structures, where one has to perform the tree descending. For instance, two such applications are in compressed pattern matching [17] and in the so-called Huffman-shaped wavelet trees [9,15], the basis of FM-indexes [5]: the access time to the FM-index might be decreased by skeleton trees in exchange to slower search operations [2].

The skeleton trees were initially introduced only for canonical trees (in which the depths of leaves do not decrease when listed from left to right), as they showed good performance in practice [11]. However, as it was noticed in [12], the smallest skeleton trees might be induced by neither Huffman nor canonical trees. In order to find optimal skeleton trees, Klein et al. [12] described a simple algorithm that, for fixed codeword lengths, builds a skeleton tree with the least number of nodes. As a consequence, to find an optimal skeleton tree, one can process each set of optimal codeword lengths using this algorithm. However, there are exponentially many such sets in the worst case [8] and, hence, such algorithm is not polynomial.

To develop a polynomial algorithm, we first prove that it suffices to consider only Huffman codes, not all optimal codes. Then, we investigate properties of Huffman trees resembling the known sibling property [7]. It turns out that all Huffman trees, for fixed code weights $w_1, w_2, \ldots, w_n$, share a similar layered structure and our dynamic programming $\mathcal{O}(n^2 \log n)$-time algorithm is based on it. Since normally in practice there are few choices of optimal codeword lengths for given weights, our result is mostly of theoretical value but the found properties, we believe, might be interesting by themselves.

The paper is organized as follows. In Sect. 2 we define all basic concepts and overview some known results. In Sect. 3 we explore the layered structure of Huffman trees that underlies our construction. Section 4 begins with a simpler cubic algorithm and, then, proceeds to improve its time to $\mathcal{O}(n^2 \log n)$.

2   Huffman and Skeleton Trees

Throughout the paper, all trees are rooted and binary. A tree is *full* if all its nodes have either two or zero children. The *depth* of a node is the length of the path from the root to the node. A *subtree* rooted at a node is the tree consisting of the node and all its descendants. A *perfect tree* is a full tree in which all leaves have the same depth. A perfect subtree is *maximal* if it is not a subtree of another perfect subtree. Two subtrees are *disjoint* if they do not share common nodes. To *shrink a subtree* is to remove all nodes of the subtree except its root.

Fix n symbols with positive weights $w_1, w_2, \ldots, w_n$ (typically, symbol frequencies). Their *prefix-free code* is a sequence of n binary *codewords* such that no codeword is a prefix of another codeword. Denote by $\ell_1, \ell_2, \ldots, \ell_n$ the codeword lengths. The *quantized source* or *q-source*, as defined in [4], is a sequence $q_1, q_2, \ldots, q_m$ such that, for each ℓ, q_ℓ is the number of codewords of length ℓ and all lengths are at most m. Note that $\sum_{\ell=1}^{m} q_\ell = n$. It is well known that the lengths and the q-source of any prefix-free code satisfy Kraft's inequality [13]:

$$\sum_{i=1}^{n} \frac{1}{2^{\ell_i}} = \sum_{\ell=1}^{m} \frac{q_\ell}{2^\ell} \le 1. \tag{1}$$

Conversely, every set of lengths and every q-source satisfying (1) correspond to a (non-unique) prefix-free code. The code is *optimal* if the sum $\sum_{i=1}^{n} \ell_i w_i$ is minimal among all possible prefix-free codes. In particular, when $w_1, w_2, \ldots, w_n$ are symbol frequencies in a message, optimal prefix-free codes minimize the total length of the encoded message obtained by the codeword substitution.

We assume that the reader is familiar with Huffman's algorithm [10] (either its version with heap or queue [14]): given positive weights $w_1, w_2, \ldots, w_n$, it builds a full tree whose n leaves are labelled with $w_1, w_2, \ldots, w_n$ and each internal node is labelled with the sum of weights of its children. This tree represents an optimal prefix-free code for the weights: all left edges of the tree are marked with zero, all right edges with one, and each w_i is associated with the codeword written on the corresponding root-leaf path. The obtained tree is not necessarily unique: swapping siblings and arbitrarily breaking ties during the construction (some weights or sums of weights might be equal), one can build many different trees, even exponentially many in some cases as shown by Golomb [8]; see Fig. 1. We call all such trees for $w_1, w_2, \ldots, w_n$ *Huffman trees* and they represent *Huffman codes*. The order of siblings and the choice among equal-weighted nodes are the only ties emerging in Huffman's algorithm and leading to different trees. In view of this, the following lemma is straightforward.

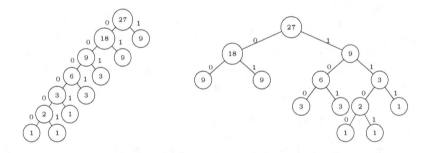

Fig. 1. Two Huffman trees for the weights $1, 1, 1, 3, 3, 9, 9$.

Lemma 1. *By swapping subtrees rooted at equal-weighted nodes or swapping siblings in a Huffman tree for $w_1, w_2, \ldots, w_n$, one obtains another Huffman tree and all Huffman trees for $w_1, w_2, \ldots, w_n$ can be reached by these operations.*

By analogy to Huffman trees, each prefix-free code forms a tree with n leaves labelled with $w_1, w_2, \ldots, w_n$ (and internal nodes labelled with sums of weights of their children). For a given tree, its *skeleton tree* [11] is obtained by choosing all (disjoint) maximal perfect subtrees and then shrinking them. An *optimal skeleton tree* [12] is a skeleton tree with the least number of nodes among all skeleton trees for the trees formed by optimal prefix-free codes. Figure 2 gives an example from [12] showing that optimal skeleton trees are not necessarily obtained from Huffman trees: only the left tree is Huffman's, the skeleton trees are drawn in gray, and both codes are optimal (note also that not every optimal code can be obtained by Huffman's algorithm).

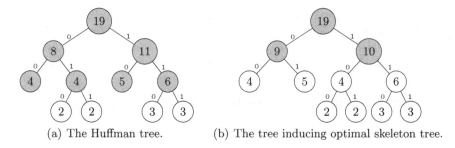

(a) The Huffman tree.   (b) The tree inducing optimal skeleton tree.

Fig. 2. Trees for the weights $2, 2, 3, 3, 4, 5$; the gray nodes form skeleton trees.

To find an optimal skeleton tree for $w_1, w_2, \ldots, w_n$, Klein et al. [12] addressed the following problem: given a q-source $q_1, q_2, \ldots, q_n$ of an optimal prefix-free code of size n, find a prefix-free code having this q-source whose tree induces a skeleton tree with the least number of nodes. Since q_ℓ in optimal codes can be non-zero only for $\ell < n$, we ignore $q_{n+1}, q_{n+2}, \ldots$ in the q-source specifying only $q_1, q_2, \ldots, q_n$ (though q_n can be omitted too). Let us sketch the solution of [12].

Denote by $\mathsf{pop}(x)$ the number of ones in the binary representation of integer x. Since $q_1, q_2, \ldots, q_n$ is the q-source of an optimal code, (1) is an equality rather than inequality: $\sum_{\ell=1}^{n} \frac{q_\ell}{2^\ell} = 1$. Hence, any resulting code tree is full and so is its skeleton tree. Consider such skeleton tree. As it is full, the smallest such tree has the least number of leaves. Each of its leaves is a shrunk perfect subtree (possibly, one-node subtree). Split all q_ℓ depth-ℓ leaves of the corresponding code tree into r subsets, each of which is a power of two in size and consists of all leaves of a shrunk subtree. Then, $r \geq k$ if $q_\ell = \sum_{i=1}^{k} 2^{m_i}$ for some $m_1 > m_2 > \cdots > m_k$, i.e., $r \geq \mathsf{pop}(q_\ell)$. Therefore, the skeleton tree has at least $\sum_{\ell=1}^{n} \mathsf{pop}(q_\ell)$ leaves.

The bound is attainable. Shrinking a perfect subtree with 2^m depth-ℓ leaves, we decrease q_ℓ by 2^m and increment $q_{\ell-m}$ by 1, which does not affect the sum (1) since $\frac{1}{2^{\ell-m}} = \frac{2^m}{2^\ell}$. Based on this, we initialize with zeros some $q'_1, q'_2, \ldots, q'_n$ and, for $\ell \in \{1, 2, \ldots, n\}$, increment $q'_{\ell-m_1}, \ldots, q'_{\ell-m_k}$, where $q_\ell = \sum_{i=1}^{k} 2^{m_i}$ is the binary representation of q_ℓ. In the end, $q'_1, q'_2, \ldots, q'_n$ satisfy (1) and $\sum_{\ell=1}^{n} q'_\ell = \sum_{\ell=1}^{n} \mathsf{pop}(q_\ell)$. By a standard method, we build a full tree having, for each ℓ, q'_ℓ depth-ℓ leaves and it is precisely the sought skeleton tree. By appropriately "expanding" its leaves into perfect subtrees, one can construct the corresponding code tree (see details in [12]). Thus, Klein et al. proved the following lemma.

Lemma 2. *Let $q_1, q_2, \ldots, q_n$ be a q-source of a size-n code such that $\sum_{\ell=1}^{n} \frac{q_\ell}{2^\ell} = 1$. The smallest skeleton tree in the number of nodes for a tree of a prefix-free code having this q-source has $\sum_{\ell=1}^{n} \mathsf{pop}(q_\ell)$ leaves and, thus, $2 \sum_{\ell=1}^{n} \mathsf{pop}(q_\ell) - 1$ nodes.*

Lemma 2 implies that, for weights admitting a Huffman tree of height h, any optimal skeleton tree has at most $2h \log_2 n$ nodes. In particular, as noted in [11], the skeleton tree has $\mathcal{O}(\log^2 n)$ nodes if a Huffman tree is of height $\mathcal{O}(\log n)$.

By Lemma 2, one can find an optimal skeleton tree for weights $w_1, w_2, \ldots, w_n$ by searching the minimum of $\sum_{\ell=1}^{n} \mathsf{pop}(q_\ell)$ among all q-sources $q_1, q_2, \ldots, q_n$ yielding optimal codes. Such algorithm is exponential in the worst case as it follows from [8]. We are to develop a polynomial-time algorithm for this problem.

3   Layered Structure of Huffman Trees

The following property of monotonicity in trees of optimal codes is well known.

Lemma 3. *If the weights of nodes u and u' in the tree of an optimal prefix-free code are w and w', and $w > w'$, then the depth of u is at most that of u'.*

Due to Lemma 2, our goal is to find a q-source $q_1, q_2, \ldots, q_n$ yielding an optimal prefix-free code for $w_1, w_2, \ldots, w_n$ that minimizes $\sum_{\ell=1}^{n} \mathsf{pop}(q_\ell)$. Huffman trees represent optimal codes; are there other optimal codes whose q-sources we have to consider? The following lemma answers this question in negative.

Lemma 4. *Let $q_1, q_2, \ldots, q_n$ be a q-source of an optimal prefix-free code for weights $w_1, w_2, \ldots, w_n$. Then, there is a Huffman code having the same q-source.*

Proof. The proof is by induction on n. The case $n \le 2$ is trivial, so assume $n > 2$. Let $w_1 \ge w_2 \ge \cdots \ge w_n$. Consider an optimal prefix-free code for $w_1, w_2, \ldots, w_n$ with the q-source $q_1, q_2, \ldots, q_n$ and lengths $\ell_1, \ell_2, \ldots, \ell_n$. Let $\ell = \max\{\ell_1, \ell_2, \ldots, \ell_n\}$. The tree for the code is full and, due to Lemma 3, has depth-ℓ leaves with weights w_{n-1} and w_n. By swapping depth-ℓ leaves, we can make w_{n-1} and w_n siblings. Shrinking these siblings into one leaf of weight $w' = w_{n-1} + w_n$, we obtain the tree for a code with the q-source $q_1, q_2, \ldots, q_{\ell-2}, q_{\ell-1}+1, q_\ell-2$. The tree represents an optimal prefix-free code for the weights $w_1, w_2, \ldots, w_{n-2}, w'$: its total cost is $\sum_{i=1}^{n} \ell_i w_i - w'$ and any prefix-free code with a smaller cost would induce a code of cost smaller than $\sum_{i=1}^{n} \ell_i w_i$ for $w_1, w_2, \ldots, w_n$ (which is impossible) by expanding a leaf of weight w' in its tree into two leaves of weights w_{n-1} and w_n. Since the code is optimal, by Lemma 3, the smallest $q_\ell - 2$ weights in the set $\{w_1, w_2, \ldots, w_{n-2}, w'\}$ mark all depth-ℓ leaves in the tree and the next smallest $q_{\ell-1} + 1$ weights mark all depth-$(\ell-1)$ leaves; the weight w' is in depth $\ell - 1$ and, so, is in the second group. By the inductive hypothesis, there is a Huffman tree for the weights $w_1, w_2, \ldots, w_{n-2}, w'$ with the q-source $q_1, q_2, \ldots, q_{\ell-2}, q_{\ell-1} + 1, q_\ell - 2$. Again by Lemma 3, the smallest $q_\ell - 2$ weights in the set $\{w_1, w_2, \ldots, w_{n-2}, w'\}$ mark all depth-ℓ leaves in the Huffman tree and the next smallest $q_{\ell-1} + 1$ (w' among them) mark depth $\ell - 1$. The leaf of weight w' can be expanded into two leaves of weights w_{n-1} and w_n, thus producing a Huffman tree for the weights $w_1, w_2, \ldots, w_n$ (since Huffman's algorithm by its first step unites w_{n-1} and w_n into w') with the q-source $q_1, q_2, \ldots, q_n$. □

By Lemma 4, it suffices to consider only q-sources of Huffman codes. Instead of processing them all, we develop a different approach. Consider a Huffman tree and all its nodes with a given depth ℓ. Denote by $v_1, v_2, \ldots, v_i$ all distinct weights of these nodes such that $v_1 > v_2 > \cdots > v_i$. For $1 \leq j \leq i$, let h_j be the number of depth-ℓ nodes of weight v_j, and k_j be the number of depth-ℓ leaves of weight v_j (so that $k_j \leq h_j$). Lemma 3 implies that all nodes of weight v such that $v_1 > v > v_i$ (if any) have depth ℓ, i.e., are entirely inside the depth-ℓ "layer". Based on this observation, one can try to tackle the problem using the dynamic programming that, given parameters $\ell, v_1, h_1, k_1, v_i, h_i, k_i$, computes the minimal sum $\sum_{\ell' \geq \ell} \mathsf{pop}(q_{\ell'})$ among all q-sources induced by Huffman trees in which the depth-ℓ "layer" is compatible with $v_1, h_1, k_1, v_i, h_i, k_i$. The main challenge in this approach is to figure out somehow all possible configurations of the depth-$(\ell+1)$ layer compatible with the depth-ℓ parameters. In what follows, we prove a number of structural lemmas that resolve such issues and simplify the method; in particular, it turns out, the parameters ℓ, v_i, h_i, k_i can be omitted.

Lemma 5. *The depths of equal-weight nodes in Huffman tree differ by at most* 1.

Proof. Since all weights are positive and the tree is full, the parent of any node of weight w has weight larger than w. Then, by Lemma 3, the depths of all nodes of weight w are at least the depth of that parent. Hence, the result follows. □

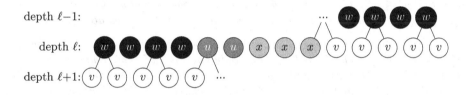

Fig. 3. A "layer": all nodes whose weights appear on depth ℓ ($w > u > x > v$).

Due to Lemmas 3 and 5, for each ℓ, the depth-ℓ nodes with the largest and smallest weights (among all depth-ℓ nodes) can have equal-weighted nodes with depths $\ell - 1$ and $\ell + 1$, respectively, but all other depth-ℓ nodes cannot have equal-weighted nodes of different depths. See Fig. 3. We are to show that this layout is quite stable: in all Huffman trees the nodes of a given weight are the same in number and have the same children, and the heaviest nodes in each layer, in a sense, determine the lightest nodes.

Since all node weights appear in the queue of Huffman's algorithm in non-decreasing order, the following claim is immediate.

Lemma 6. *In a Huffman tree, for any $v \leq v'$ and $w \leq w'$, if a pair of siblings has weights v and v', and a different pair of siblings has weights w and w', then we have either $w' \leq v$ or $v' \leq w$.*

Lemma 7. *One of the following (mutually exclusive) alternatives holds for the set of all nodes of a given weight w in any Huffman tree for $w_1, w_2, \ldots, w_n$:*

1. all the nodes are leaves;

2. *all the nodes except one are leaves and this one node has two children with weights w' and w'' such that $w' < w''$ and $w' + w'' = w$;*
3. *all the nodes except m of them, for some $m > 0$, are leaves and each of these m nodes has two children with weights $w/2$.*

In different Huffman trees the sets of all nodes of weight w have the same size and the same alternative holds for them with the same w', w'', and m, respectively.

Proof. It suffices to prove that if two nodes of weight w are not leaves, then their children have weights $w/2$. Suppose, to the contrary, that the children have weights, respectively, w', w'', and v', v'' such that $w' < w''$ and $v' \le v''$. But then $v' < w''$ and $w' < v''$, which is impossible by Lemma 6.

As the swapping operations of Lemma 1 do not change the number of nodes of weight w and the weights of their children, the last part of the lemma (the preservation of the alternative with given w', w'' or m) is straightforward. □

The heaviest layer nodes determine the lightest ones as follows.

Lemma 8. *In a Huffman tree, for $\ell \ge 0$, choose (if any) the largest w among all weights of depth-ℓ nodes such that there is a non-leaf node of weight w (maybe not of depth ℓ). Let t be a node of weight w with children of weights w' and w'' such that $w' \le w''$. Then, the weights of depth-ℓ nodes are at least w', the weights of depth-$(\ell+1)$ nodes are at most w'', and the next alternatives are possible:*

1. *$w' < w''$, the depth of t is ℓ, and only one node of depth $\ell + 1$ has weight w'';*
2. *$w' < w''$, the depth of t is $\ell - 1$, and only one node of depth ℓ has weight w';*
3. *$w' = w'' = w/2$, for some m, exactly m non-leaf nodes of depth ℓ have weight w, and exactly $2m + \delta$ nodes of depth $\ell + 1$ have weight $w/2$, where $\delta = 1$ if some node of weight $w/2$ has a sibling of smaller weight, and $\delta = 0$ otherwise.*

Proof. Due to Lemma 6, any node of depth $\ell+1$ with weight larger than w'' has a sibling of weight at least w''. Hence, their parent, whose depth is ℓ, has weight larger than $2w'' \ge w$, which contradicts the choice of w. Analogously, any node of depth ℓ with weight less than w' has a sibling of weight at most w' and, hence, their parent, whose depth is $\ell - 1$, has weight smaller than $2w' \le w$, which is impossible by Lemma 3 since there is a node of depth ℓ with weight w. Thus, the first part of the lemma is proved. Let us consider t and its alternatives.

Let $w' < w''$. By Lemma 5, the depth of t differs from ℓ by at most one. Since the weight of t is larger than w'', t cannot have depth $\ell + 1$. Suppose the depth of t is ℓ. By Lemma 6, any node v of depth $\ell + 1$ and weight w'' whose parent is not t has a sibling of weight at least w''. Hence, its parent, whose depth is ℓ, is of weight at least $2w'' > w$, which contradicts the choice of w. Thus, there is no such v. Suppose the depth of t is $\ell - 1$. By Lemma 6, any node v of depth ℓ and weight w' whose parent is not t has a sibling of weight at most w' and their parent, whose depth is $\ell - 1$, has weight at most $2w' < w$, which contradicts Lemma 3 since we have a depth-ℓ node of weight w. Thus, there is no such v.

Let $w' = w'' = w/2$. Lemma 6 implies that at most one node of weight $w/2$ can have a sibling of smaller weight. Suppose such node v exists. The weight of its parent is less than w but larger than $w/2$. Since all nodes of depth ℓ have weights at most $w/2$, the depth of the parent is at most ℓ and, by Lemma 3, at least ℓ. Hence, v has depth $\ell + 1$ and, therefore, the total number of nodes of depth $\ell + 1$ with weights $w/2$ is $2m + 1$. The remaining details are obvious. □

The layout described in Lemmas 7 and 8 can be seen as a more detailed view on the well-known sibling property [7]: listing all node weights layer by layer upwards, one can obtain a non-decreasing list in which all siblings are adjacent.

4   Algorithm

Our scheme is as follows. Knowing the heaviest nodes in a given depth-ℓ layer of a Huffman tree and which of them are leaves, one can deduce from Lemma 8 the lightest layer nodes. Then, by Lemma 3, the heaviest nodes of the depth-$(\ell+1)$ layer either immediately precede the lightest nodes of depth ℓ in the weight-sorted order or have the same weight; in any case they can be determined and the depth-$(\ell+1)$ layer can be recursively processed. Let us elaborate details.

In the beginning, the algorithm constructs a Huffman tree for $w_1, w_2, \ldots, w_n$ and splits all nodes into classes $c_1, c_2, \ldots, c_r$ according to their weights: all nodes of class c_u are of the same weight denoted $\hat{w}_u$ and $\hat{w}_1 > \hat{w}_2 > \cdots > \hat{w}_r$. In addition to $\hat{w}_u$, the algorithm calculates for each class c_u the following attributes:

- the total number of nodes in c_u, denoted $|c_u|$;
- the number of leaves in c_u, denoted λ_u;
- provided $\lambda_u \neq |c_u|$ (i.e., not all class nodes are leaves), the weights of two children of an internal node with weight $\hat{w}_u$ (by Lemma 7, the choice of the node is not important), denoted w'_u and w''_u and such that $w'_u \leq w''_u$;
- the number δ_u such that $\delta_u = 1$ if $\lambda_u \neq |c_u|$ and a node of weight w'_u has a sibling of weight $<w'_u$, and $\delta_u = 0$ otherwise (δ_u serves as δ from Lemma 8);
- the number t_u such that $t_u = 0$ if $\lambda_t = |c_t|$ for all $t \geq u$, and $t_u = \min\{t \geq u : \lambda_t \neq |c_t|\}$ otherwise (t_u is used to identify a node t as in Lemma 8).

It is straightforward that the attributes, for all classes, can be computed in $\mathcal{O}(n \log n)$ overall time. By Lemma 7, the class information (i.e., class weights and attributes) is the same in all possible Huffman trees for $w_1, w_2, \ldots, w_n$.

We use a dynamic programming with parameters c_u, h, k: c_u is the class to which the heaviest nodes in a depth-ℓ layer of a Huffman tree belong, $h \leq |c_u|$ is the number of nodes from c_u with depth ℓ, and $k \leq h$ is the number of leaves from c_u with depth ℓ (note that ℓ is not known). Informally, the algorithm recursively computes from c_u, h, k the minimal sum $\sum_{\ell' \geq \ell} \mathsf{pop}(q_{\ell'})$, where $q_{\ell'}$ is the number of leaves with depth ℓ', among all Huffman trees that, for some ℓ, have a depth-ℓ layer compatible with the parameters c_u, h, k. Using the class information, one can deduce from c_u, h, k the weight $\hat{w}_{u'}$ of the heaviest nodes of the lower depth-$(\ell+1)$ layer, and the number h' of the nodes of weight $\hat{w}_{u'}$ with depth $\ell + 1$: if $t_u = 0$, our layer is lowest and we simply return $\mathsf{pop}(k + \sum_{i=u+1}^{r} \lambda_i)$; otherwise, we denote $t = t_u$ and consider several cases according to Lemma 8:

1. $w'_t < w''_t$ and either $t \neq u$ or $k \neq h$; then $\hat{w}_{u'} = w''_t$ and $h' = 1$;
2. $w'_t < w''_t$, $t = u$, and $k = h$ (i.e., as Lemmas 7 and 8 imply, the only non-leaf node of weight $\hat{w}_t$ has depth $\ell - 1$); then either $\hat{w}_{u'} = w'_t$ and $h' = |c_{u'}| - 1$ if there are at least two nodes of weight w'_t, or, otherwise, $\hat{w}_{u'}$ is the weight immediately preceding w'_t in the sorted set of all weights and $h' = |c_{u'}|$;

3.1 $w'_t = w''_t$, $t = u$, and $2(h - k) + \delta_u \neq 0$; then $\hat{w}_{u'} = \hat{w}_u/2$ and $h' = 2(h - k) + \delta_u$;

3.2 $w'_t = w''_t$, $t = u$, and $2(h - k) + \delta_u = 0$; then $\hat{w}_{u'}$ is the weight immediately preceding $\hat{w}_u/2$ in the sorted set of all weights and $h' = |c_{u'}|$;

3.3 $w'_t = w''_t$ and $t \neq u$; then $\hat{w}_{u'} = \hat{w}_t/2 \; (= w'_t)$ and $h' = 2(|c_t| - \lambda_t) + \delta_t$.

Thus, the parameters c_u, h, k uniquely determine the weight $\hat{w}_{u'}$ (and, hence, the class $c_{u'}$) of the heaviest nodes of the subsequent depth-$(\ell+1)$ layer and the number h' of nodes of weight $\hat{w}_{u'}$ with depth $\ell+1$ (recall that ℓ denotes the depth of the "(c_u, h, k)-layer" and is unknown). The number of leaves of weight $\hat{w}_{u'}$ with depth $\ell + 1$, denoted k', is a "free" parameter of the recursion. We loop through all possible k' and, thus, compute the minimum of the sums $\sum_{\ell' \geq \ell} q_{\ell'}$.

The parameter k' is not arbitrary: it cannot exceed neither h' nor $\lambda_{u'}$, and cannot make $\lambda_{u'} - k'$ (the number of leaves from $c_{u'}$ with depth ℓ) larger than $|c_{u'}| - h'$ (the total number of nodes from $c_{u'}$ with depth ℓ). Processing all k' subject to these restrictions, we compute the answer, $f(c_u, h, k)$, as follows:

$$f(c_u, h, k) = \min_{\substack{0 \leq k' \leq \min\{h', \lambda_{u'}\} \\ \text{and} \\ \lambda_{u'} - |c_{u'}| + h' \leq k'}} \left\{ f(c_{u'}, h', k') + \mathsf{pop}\left(k + \lambda_{u'} - k' + \sum_{i=u+1}^{u'-1} \lambda_i \right) \right\}. \quad (2)$$

Thus, if (2) is indeed a solution, $f(c_1, 1, 0)$ gives the answer for the problem: the minimal $\sum_{\ell=1}^{n} \mathsf{pop}(q_\ell)$ for all q-sources $q_1, q_2, \ldots, q_n$ of optimal prefix-free codes; the restriction to Huffman trees in the design of the recursion is justified by Lemma 4. Using memoization instead of the mere recursion in (2), one obtains a polynomial time algorithm. The optimal skeleton tree itself is constructed by applying the algorithm of Klein et al. [12] to a q-source $q_1, q_2, \ldots, q_n$ achieving the minimum, which can be found by the standard backtracking technique.

It is not immediate, however, that every choice of the parameter k' in (2) yields a "layered structure" (designated by recursive calls) corresponding to a valid Huffman tree for $w_1, w_2, \ldots, w_n$; in principle, one could imagine a situation when the assignments of $c_{u'}, h', k'$ leading to the minimum in (2) do not correspond to any Huffman tree at all. Fortunately, this is not the case. Let us prove that whenever a triple (c_u, h, k) is reached on an $(\ell+1)$st level of the recursion (2) starting from $f(c_1, 1, 0)$ (so that the triple $(c_1, 1, 0)$ is on the first level), it is guaranteed that there exists a Huffman tree for $w_1, w_2, \ldots, w_n$ in which the heaviest nodes of the depth-ℓ layer are from the class c_u, exactly h nodes of weight $\hat{w}_u$ have depth ℓ, exactly k nodes of these h nodes are leaves, and every (higher) depth-ℓ' layer, for $\ell' < \ell$, in the tree is analogously compatible with a corresponding triple $(c_{u'}, h', k')$ on level ℓ' leading to (c_u, h, k) in the recursion.

Obviously, the only node of depth zero (the root) is from the class c_1 with $|c_1| = 1$ in all possible Huffman trees. Suppose that a triple (c_u, h, k) is reached in (2) on level $\ell + 1$ starting from $f(c_1, 1, 0)$ and there is a Huffman tree $\mathcal{T}$ in which the heaviest nodes with depth ℓ are from c_u, exactly h of the class c_u nodes have depth ℓ, k of these h nodes are leaves, and all (higher) depth-ℓ' layers, for $\ell' < \ell$, are compatible with their corresponding triples from the call stack of the recursion. The argument above shows that the triple (c_u, h, k) uniquely determines analogous parameters $c_{u'}$ and h' (but not k') for the layer of depth $\ell + 1$ in $\mathcal{T}$. Fix an arbitrary integer k' satisfying the conditions imposed by the minimum in (2). It is straightforward that by swapping subtrees rooted at weight-$\hat{w}_{u'}$ nodes of depths ℓ and $\ell + 1$, one can transform $\mathcal{T}$ into a tree having exactly k' leaves from $c_{u'}$ of depth $\ell + 1$. By Lemma 1, the obtained tree is a Huffman tree too and the transformations do not affect the nodes of class c_u and nodes with depths less than ℓ. Therefore, the claim is proved and the correctness of the algorithm immediately follows from it.

Let us estimate the running time. The sum $\sum_{i=u+1}^{u'-1} \lambda_i$ in (2) can be calculated in $\mathcal{O}(1)$ time if one has precomputed sums $\sum_{i=1}^{v} \lambda_i$, for all $v \leq r$. All other parameters in (2), except k', are precomputed and accessible in $\mathcal{O}(1)$ time. Since $h \leq |c_u|$ and $\sum_{i=1}^{r} |c_i| = 2n - 1$ (the number of nodes in any full tree with n leaves is $2n - 1$), the number of different pairs (c_u, h) is $\mathcal{O}(n)$. Hence, the number of triples (c_u, h, k) is $\mathcal{O}(n^2)$. For each of the triples, the algorithm runs through at most $\mathcal{O}(n)$ appropriate k's in (2). Therefore, the overall running time is $\mathcal{O}(n^3)$ and we have proved the following theorem.

Theorem 1. *An optimal skeleton tree for positive weights* $w_1, w_2, \ldots, w_n$ *can be constructed in* $\mathcal{O}(n^3)$ *time.*

The table used by the algorithm of Theorem 1 is of size $\mathcal{O}(n^2)$. The main slowdown making the running time cubic is in the processing of $\mathcal{O}(n)$ possible parameters k' in (2). We are to optimize this aspect of the algorithm.

For $i, j \in \mathbb{Z}$, denote $[i..j] = \{i, i+1, \ldots, j\}$. Given parameters c_u, h, k, the conditions under the minimum in (2) determine a range $[k'_{min}..k'_{max}]$ of k' that should be processed. As a first attempt for optimization, one might try to pass in the function f not one k' but the whole range $[k'_{min}..k'_{max}]$, as if the function looked like $f(c_{u'}, h', [k'_{min}..k'_{max}])$. However, in this case it is not clear how to take into account the effects of the choice of $k' \in [k'_{min}..k'_{max}]$ on higher (with smaller depth) layers, namely, on the following summand from (2):

$$\mathsf{pop}\left(k + \lambda_{u'} - k' + \sum_{i=u+1}^{u'-1} \lambda_i\right). \tag{3}$$

It turns out that the range $[k'_{min}..k'_{max}]$ can be split into $\mathcal{O}(\log n)$ disjoint subranges so that, for each subrange, the effect of the choice of k' on higher layers is well determined. The splitting is based on the following lemma.

Lemma 9. *In* $\mathcal{O}(\log b)$ *time one can split any range* $[a..b-1]$ *into subranges* $[i_0..i_1-1], [i_1..i_2-1], \ldots, [i_{m-1}..i_m-1]$ *such that* $a = i_0 < i_1 < \cdots < i_m = b$, $m \leq 2\log_2 b$, *and for each* $j \in [0..m-1]$, *the length of the subrange* $[i_j..i_{j+1}-1]$ *is a power of two and* $\mathsf{pop}(i_j + x) = \mathsf{pop}(i_j) + \mathsf{pop}(x)$ *whenever* $0 \leq x < i_{j+1} - i_j$.

Proof. We use the operation $\mathsf{ctz}(x)$ counting trailing zeros in the binary representation of x.[1] The subranges are generated iteratively: given i_j (initially $i_0 = a$), we compute $d_j = \max\{d \leq \mathsf{ctz}(i_j): i_j + 2^d \leq b\}$ decrementing $d = \mathsf{ctz}(i_j)$ until $i_j + 2^d \leq b$, and put $i_{j+1} = i_j + 2^{d_j}$; we stop when $i_{j+1} = b$. Since $\mathsf{ctz}(x + 2^d) > d$ for $d = \mathsf{ctz}(x)$, the numbers d_j first increase on each iteration ($d_0 < d_1 < \cdots$), then decrease ($\cdots > d_{m-2} > d_{m-1}$), but always $d_j \leq \log_2 b$. Therefore, at most $2\log_2 b$ subranges are created. Finally, since each i_j, by construction, has at least d_j trailing zeros, we have $\mathsf{pop}(i_j + x) = \mathsf{pop}(i_j) + \mathsf{pop}(x)$ for $0 \leq x < 2^{d_j}$. □

Suppose that we compute $f(c_u, h, k)$ using the recursion (2). The range $[k'_{min}..k'_{max}]$ for k' determined by the conditions under the minimum in (2) is bijectively mapped onto a contiguous range of arguments for pop in (3), namely, onto $[\Delta - k'_{max}..\Delta - k'_{min}]$, where $\Delta = k + \lambda_{u'} + \sum_{i=u+1}^{u'-1} \lambda_i$. According to Lemma 9, in $\mathcal{O}(\log n)$ time one can split the range $[k'_{min}..k'_{max}]$ into $\mathcal{O}(\log n)$ disjoint subranges $[i_0+1..i_1], [i_1+1..i_2], \ldots, [i_{m-1}+1..i_m]$, for some $i_1 < i_2 < \cdots < i_m$, such that, for any $j \in [1..m]$, the length of $[i_{j-1}+1..i_j]$ is a power of two, denoted by 2^{d_j}, and one has $\mathsf{pop}(\Delta - k') = \mathsf{pop}(\Delta - i_j) + \mathsf{pop}(i_j - k')$ whenever $i_j - 2^{d_j} < k' \leq i_j$. Therefore, the minimum in (2) restricted to any such subrange $[i_{j-1}+1..i_j]$ can be transformed as follows:

$$\min\{f(c_{u'}, h', k') + \mathsf{pop}(\Delta - k'): i_{j-1} < k' \leq i_j\} =$$
$$\mathsf{pop}(\Delta - i_j) + \min\{f(c_{u'}, h', i_j - s) + \mathsf{pop}(s): 0 \leq s < 2^{d_j}\}. \quad (4)$$

In order to compute the minimum of the values $f(c_{u'}, h', i_j - s) + \mathsf{pop}(s)$, we devise a new function $g(c_u, h, k_{max}, d)$ that is similar to f but, instead of simply computing $f(c_u, h, k_{max})$, it finds the minimum of $f(c_u, h, k_{max} - s) + \mathsf{pop}(s)$ for all $s \in [0..2^d-1]$. In other words, the function g computes the minimum of $f(c_u, h, k)$ in the range $k_{max} - 2^d < k \leq k_{max}$ taking into account the effect on the higher layer (something that f does not care). Using (4), one can calculate (2) by finding the following minimum: $\min\{\mathsf{pop}(\Delta - i_j) + g(c_{u'}, h', i_j, d_j): j \in [1..m]\}$, which contains $\mathcal{O}(\log n)$ invocations of g since $m = \mathcal{O}(\log n)$ due to Lemma 9. The implementation of the function g itself is rather straightforward:

$$g(c_u, h, k_{max}, 0) = f(c_u, h, k_{max}),$$
$$g(c_u, h, k_{max}, d) = \min \begin{cases} g(c_u, h, k_{max}, d-1), \\ g(c_u, h, k_{max} - 2^{d-1}, d-1) + 1. \end{cases} \quad (5)$$

While computing $g(c_u, h, k_{max}, d)$ with $d > 0$, we choose whether to set the dth bit of s in order to minimize $f(c_u, h, k_{max} - s) + \mathsf{pop}(s)$: if set, it affects $\mathsf{pop}(s)$ increasing it by one and we still can choose $d - 1$ lower bits, hence, the answer is the same as for $g(c_u, h, k_{max} - 2^{d-1}, d-1) + 1$; if not set, the answer is the same as for $g(c_u, h, k_{max}, d-1)$. Thus, $g(c_u, h, k_{max}, d)$ implemented as in (5) indeed computes the required $\min\{f(c_u, h, k_{max} - s) + \mathsf{pop}(s): 0 \leq s < 2^d\}$.

The correctness of the algorithm should be clear by this point. Let us estimate the running time. The splitting process in the described implementation

---
[1] If not supported, it can be implemented using a precomputed table of size $\mathcal{O}(n)$ [6].

of $f(c_u, h, k)$ takes $\mathcal{O}(\log n)$ time, by Lemma 9, which is a significant improvement over the previous $\mathcal{O}(n)$ time. Since the number of tripes (c_u, h, k) is $\mathcal{O}(n^2)$ in total, there are at most $\mathcal{O}(n^2 \log n)$ quadruples (c_u, h, k_{max}, d) that can serve as parameters for g (note that $0 \leq d \leq \log_2 n$). Every such quadruple is processed in constant time according to (5). Therefore, the overall running time of such algorithm with the use of memoization is $\mathcal{O}(n^2 \log n)$.

Theorem 2. *An optimal skeleton tree for positive weights* $w_1, w_2, \ldots, w_n$ *can be constructed in* $\mathcal{O}(n^2 \log n)$ *time.*

References

1. Abrahams, J.: Code and parse trees for lossless source encoding. In: Proceedings of the Compression and Complexity of Sequences, pp. 145–171. IEEE (1997). https://doi.org/10.1109/SEQUEN.1997.666911
2. Baruch, G., Klein, S.T., Shapira, D.: A space efficient direct access data structure. J. Discrete Algorithms **43**, 26–37 (2017). https://doi.org/10.1016/j.jda.2016.12.001
3. Duda, J.: Asymmetric numeral systems. arXiv preprint arXiv:0902.0271 (2009)
4. Ferguson, T.J., Rabinowitz, J.H.: Self-synchronizing Huffman codes. IEEE Trans. Inf. Theory **30**(4), 687–693 (1984). https://doi.org/10.1109/TIT.1984.1056931
5. Ferragina, P., Manzini, G.: Opportunistic data structures with applications. In: Proceedings of the Symposium on Foundations of Computer Science (FOCS), pp. 390–398. IEEE (2000). https://doi.org/10.1109/SFCS.2000.892127
6. Fredman, M.L., Willard, D.E.: Surpassing the information theoretic bound with fusion trees. J. Comput. Syst. Sci. **47**(3), 424–436 (1993). https://doi.org/10.1016/0022-0000(93)90040-4
7. Gallager, R.G.: Variations on a theme by Huffman. IEEE Trans. Inf. Theory **24**(6), 168–174 (1978). https://doi.org/10.1109/TIT.1978.1055959
8. Golomb, S.W.: Sources which maximize the choice of a Huffman coding tree. Inf. Control **45**(3), 263–272 (1980). https://doi.org/10.1016/S0019-9958(80)90648-8
9. Grossi, R., Gupta, A., Vitter, J.S.: High-order entropy-compressed text indexes. In: Proceedings of the Symposium on Discrete Algorithms (SODA), pp. 841–850. SIAM (2003)
10. Huffman, D.A.: A method for the construction of minimum-redundancy codes. Proc. Inst. Radio Eng. (IRE) **40**(9), 1098–1101 (1952). https://doi.org/10.1109/JRPROC.1952.273898
11. Klein, S.T.: Skeleton trees for the efficient decoding of Huffman encoded texts. Inf. Retrieval **3**(1), 7–23 (2000). https://doi.org/10.1023/A:1009910017828
12. Klein, S.T., Serebro, T.C., Shapira, D.: Optimal skeleton Huffman trees. In: Fici, G., Sciortino, M., Venturini, R. (eds.) SPIRE 2017. LNCS, vol. 10508, pp. 241–253. Springer, Cham (2017). https://doi.org/10.1007/978-3-319-67428-5_21
13. Kraft, L.G.: A device for quantizing, grouping, and coding amplitude modulated pulses. Master's thesis, MIT, Cambridge, Massachusetts (1949)
14. van Leeuwen, J.: On the construction of Huffman trees. In: Proceedings of the International Colloquium on Automata, Languages and Programming (ICALP), pp. 382–410. Edinburgh University Press (1976)
15. Mäkinen, V., Navarro, G.: New search algorithms and time/space tradeoffs for succinct suffix arrays. Technical report C-2004-20, University of Helsinki, Finland, April 2004

16. Moffat, A.: Huffman coding. ACM Comput. Surv. **52**(4), 1–35 (2019). https://doi.org/10.1145/3342555
17. Shapira, D., Daptardar, A.: Adapting the Knuth-Morris-Pratt algorithm for pattern matching in Huffman encoded texts. Inf. Process. Manag. **42**(2), 429–439 (2006). https://doi.org/10.1016/j.ipm.2005.02.003

The Subtrace Order and Counting First-Order Logic

Dietrich Kuske[(✉)]

Technische Universität Ilmenau, Ilmenau, Germany
dietrich.kuske@tu-ilmenau.de

Abstract. We study the subtrace relation among Mazurkiewicz traces which generalizes the much-studied subword order. Here, we consider the 2-variable fragment of a counting extension of first-order logic with regular predicates. It is shown that all definable trace languages are effectively recognizable implying that validity of a sentence of this logic is decidable (this problem is known to be undecidable for virtually all stronger logics already for the subword relation).

Keywords: Mazurkiewicz traces · Counting logic · Subword relation

1 Introduction

The subword relation is one of the simplest nontrivial examples of a well-quasi ordering [7] and can be used in the verification of infinite state systems [4]. It can be understood as embeddability of one word into another. This embeddability relation has been considered for other classes of structures like trees, posets, semilattices, lattices, graphs etc. [8,11,19,21]; this paper initiates its consideration for the class of Mazurkiewicz traces. (The prefix order on the set of traces has been studied extensively before, both order-theoretically (cf. [5]) and under logical aspects (e.g., [15]).)

These traces were first investigated by Cartier and Foata [2] to study the combinatorics of free partially commutative or, equivalently, trace monoids. Later, Mazurkiewicz [16] used them to relate the interleaving and the partial-order semantics of a distributed system (see [3] for surveys on the many results on trace monoids).

Many of the above mentioned papers on the embeddability relation study its logical aspects. Regarding the subword relation, they provide a rather sharp description of the border between decidable and undecidable fragments of first-order logic: For the subword order alone, the $\exists^*$-theory is decidable [12] and the $\exists^*\forall^*$-theory is undecidable [9]. For the subword order together with regular predicates, the two-variable theory is decidable [9] (this holds even for the two-variable fragment of a counting extension of first-order logic [14]) and the three-variable theory [9] as well as the $\exists^*$-theory are undecidable [6] (even if we only consider singleton predicates, i.e., constants). If one restricts the universe from all words to a particular language, an even more diverse picture appears [14].

© Springer Nature Switzerland AG 2020
H. Fernau (Ed.): CSR 2020, LNCS 12159, pp. 289–302, 2020.
https://doi.org/10.1007/978-3-030-50026-9_21

All the undecidability results hold for the subtrace relation since it generalizes the subword relation. The strongest decidability result for the subword relation is the decidability of the 2-variable fragment of a counting extension of first-order logic [14]. The proof shows that every definable unary relation is an effectively regular language. It proceeds by quantifier elimination and relies crucially on the fact that the downwards closure, the upwards closure, and the "incomparability language" (i.e., the set of words that are incomparable to some element of the language) of a regular language are effectively regular. These three preservation results hold since the subword relation and the incomparability relation are unambiguous rational transductions [9].

Considering the subtrace relation, the main result of this paper shows the decidability of the 2-variable fragment of the extension of first-order logic by threshold-counting quantifiers. This extends results by Karandikar and Schnoebelen [9] and by Kuske and Zetzsche [14] from words to traces. As their proofs for words, we proceed by quantifier elimination and rely on the preservation properties mentioned above, but this time for trace languages. Differently from the study of subwords, here we cannot use rational relations for traces since they do not preserve recognizability (and are not available for other classes of structures at all).

To substitute the use of rational relations, we consider the internal structure of a trace, i.e., we consider a trace not as an element of a monoid, but as a labeled directed graph. Now monadic second order (abbreviated MSO) logic can be used to make statements about such a graph. Generalizing Büchi's result, Thomas [20] showed that a set of traces is recognizable if, and only if, it is the set of models of some MSO-sentence. With this shift of view, we have to prove the preservation results not for recognizable, but for MSO-definable sets of traces. This is rather straightforward for the upwards closure since a trace has a subtrace satisfying some MSO-sentence σ if, and only if, some induced subgraph satisfies σ which is easily expressible in MSO logic. Since we consider also threshold counting quantifiers, we have to express, e.g., that there are two non-isomorphic induced subgraphs satisfying σ. Since isomorphism is not expressible in MSO logic, the solution relies on "leftmost" or "canonical" subgraphs. When talking about the incomparability relation, we are interested in traces (i.e., graphs) that are neither a sub- nor a supergraph. We base the solution on the largest prefix of one trace that is a subtrace of the other trace as well as on the combinatorics of traces and, in particular, on MSO logic.

Methodwise, we derive the decidability without the use of rational relations. Instead, our arguments are based on the rich theory of traces and in particular on the relation between recognizability and MSO-definability in this setting. It remains to be explored whether these ideas can be transfered to other settings where rational relations are not available.

2   Definitions and Main Result

2.1   Traces and Subtraces

A *dependence alphabet* is a pair (Σ, D) where Σ is a finite alphabet and the *dependence relation* $D \subseteq \Sigma^2$ is symmetric and reflexive.

A *trace over* (Σ, D) is (an isomorphism class of) a directed acyclic graph $t = (V, E, \lambda)$ with node-labels from Σ (i.e., $\lambda \colon V \to \Sigma$) such that, for all $x, y \in V$,

- $(x, y) \in E \implies (\lambda(x), \lambda(y)) \in D$ and
- $(\lambda(x), \lambda(y)) \in D \implies (x, y) \in E$ or $x = y$ or $(y, x) \in E$.

The set of all traces is denoted $\mathbb{M}(\Sigma, D)$, 1 is the unique trace with empty set of nodes. For two traces $s = (V_s, E_s, \lambda_s)$ and $t = (V_t, E_t, \lambda_t)$, we define their product $s \cdot t = u = (V_u, E_u, \lambda_u)$ setting $V_u = V_s \uplus V_t$, $\lambda_u = \lambda_s \cup \lambda_t$, and $E_u = E_s \cup E_t \cup \{(x, y) \in V_s \times V_t \mid (\lambda_s(x), \lambda_t(y)) \in D\}$.

This operation is easily seen to be associative with neutral element 1, i.e., $\mathbb{M}(\Sigma, D)$ forms a monoid that we call *trace monoid (induced by (Σ, D))*.

Let $a \in \Sigma$. Abusing notation, we denote the singleton trace $(\{x\}, \emptyset, \{(x, a)\})$ by a. Then the monoid $\mathbb{M}(\Sigma, D)$ is generated by the set Σ of singleton traces.

Note that $\mathbb{M}(\Sigma, \{(a, a) \mid a \in \Sigma\}) \cong (\mathbb{N}, +)^{|\Sigma|}$ and $\mathbb{M}(\Sigma, \Sigma \times \Sigma) \cong \Sigma^*$. Further, the direct and the free product of two trace monoids is a trace monoid, again. But there are also trace monoids not arising by free and direct products from free monoids (consider, e.g., the dependence alphabet with $\Sigma = \{a_1, a_2, a_3, a_4\}$ and $(a_i, a_j) \in D \iff |i - j| \leq 1$). See [3] for a collection of surveys on the many results known for traces.

Let $t = (V, E, \lambda)$ be a trace. To simplify notation, we write $X \subseteq t$ for "X is a set of nodes of t", i.e., for $X \subseteq V$.

Now let $X \subseteq t$. Then $t{\restriction}_X$ denotes the subgraph of t induced by X, i.e., $(X, E \cap X^2, \lambda{\restriction}_X)$. Note that $s = t{\restriction}_X$ is a trace that we call *subtrace of t (induced by X)*. We denote this fact by $s \sqsubseteq_{\mathrm{sub}} t$ and call t a *supertrace* of s.

It can be observed that $s \sqsubseteq_{\mathrm{sub}} t$ if, and only if, there are a natural number $n \geq 0$ and traces $s_1, s_2, \ldots, s_n$ and $t_0, t_1, \ldots, t_n$ such that $s = s_1 s_2 \cdots s_n$ and $t = t_0 s_1 t_1 s_2 t_2 \cdots s_n t_n$.

2.2   Recognizable Sets

Let $(M, \cdot, 1)$ be some monoid. A set $S \subseteq M$ is *recognizable* if there exists a monoid homomorphism $\eta \colon (M, \cdot, 1) \to M'$ into some finite monoid M' such that $\eta(s) = \eta(t)$ and $s \in S$ imply $t \in S$ for all $s, t \in M$. We call the triple $(M', \eta, \eta(S))$ an *automaton accepting S*.

2.3   The Logic $\mathbf{C^2}$ and the Main Result

Let (Σ, D) be some dependence alphabet and let $\mathcal{R}$ denote the class of recognizable subsets of $\mathbb{M}(\Sigma, D)$. We consider the structure

$$\mathcal{S} = (\mathbb{M}(\Sigma, D), \sqsubseteq_{\mathrm{sub}}, \mathcal{R})$$

whose universe is the set of traces, whose only binary relation is the subtrace relation and that has a unary relation for each recognizable subset of $M(\Sigma, D)$. We will make statements about this structure using some variant of classical first-order logic. More precisely, the formulas of C^2 are defined by the following syntax:

$$\varphi := x_1 \sqsubseteq_{\mathrm{sub}} x_2 \mid x_1 = x_2 \mid x_1 \in S \mid \varphi \vee \varphi \mid \neg\varphi \mid \exists^{\geq k} x_1\, \varphi$$

where x_1, x_2 are variables from $\{x, y\}$, $S \in \mathcal{R}$ is some recognizable set, and $k \in \mathbb{N}$. Note that we allow only two variables, namely x and y. The semantics of these formulas is as expected with the understanding that $\exists^{\geq k} x_1\, \varphi$ holds if there are at least k mutually distinct traces $t_1, t_2, \cdots, t_k \in M(\Sigma, D)$ that all make the formula φ true. Note that $\exists^{\geq 1}$ is the usual existential quantifier and that $\exists^{\geq 0} x\, \varphi$ is always true. Now we can formulate the main result of this paper and sketch its proof from results to be demonstrated in later sections:

Theorem 2.1. *If $\varphi(x)$ is a formula from C^2 with a single free variable, then the set of traces $S(\varphi) = \{t \in M(\Sigma, D) \mid S \models \varphi(t)\}$ is recognizable.*

Even more, from the dependence alphabet (Σ, D) and the formula φ, one can compute an automaton accepting this set. Consequently, the C^2-theory of $(M(\Sigma, D), \sqsubseteq_{\mathrm{sub}}, \mathcal{R})$ is decidable uniformly in (Σ, D).

Proof. The proof proceeds by induction on the construction of the formula φ, the most interesting case being $\varphi = \exists^{\geq k} x\, \psi(x, y)$. Using arguments like de Morgan's laws and basic arithmetic, one can reduce this to the case that $\psi(x, y)$ is a conjunction of possibly negated formulas of the following form:

(a) $x \sqsubseteq_{\mathrm{sub}} y$, $x \sqsubseteq_{\mathrm{sub}} x$, $y \sqsubseteq_{\mathrm{sub}} x$, $y \sqsubseteq_{\mathrm{sub}} y$
(b) $x \in S$ and $y \in S$ for $S \in \mathcal{R}$
(c) $\exists^{\geq \ell} x\colon \alpha(x, y)$ and $\exists^{\geq \ell} y\colon \alpha(x, y)$

Since formulas of the form (c) have at most one free variable, we can apply the induction hypothesis, i.e., replace them by formulas of the form (b). Since $\mathcal{R}$ is closed under Boolean operations, there are $S_i, T_i \in \mathcal{R}$ such that the formula ψ is equivalent to the formula

$$y \in T_1 \vee (x \sqsubseteq_{\mathrm{sub}} y \wedge y \not\sqsubseteq_{\mathrm{sub}} x \wedge x \in S_2 \wedge y \in T_2)$$
$$\vee (x \not\sqsubseteq_{\mathrm{sub}} y \wedge y \sqsubseteq_{\mathrm{sub}} x \wedge x \in S_3 \wedge y \in T_3)$$
$$\vee (x \not\sqsubseteq_{\mathrm{sub}} y \wedge y \not\sqsubseteq_{\mathrm{sub}} x \wedge x \in S_4 \wedge y \in T_4).$$

Since the order relations between x and y in this formula are mutually exclusive, the formule φ is equivalent to a Boolean combination of formulas of the form $y \in T$ and

$$\exists^{\geq \ell} x\colon (x\, \theta_1\, y \wedge y\, \theta_2\, x \wedge x \in S \wedge y \in T)$$

with $\theta_1, \theta_2 \in \{\sqsubseteq_{\mathrm{sub}}, \not\sqsubseteq_{\mathrm{sub}}\}$, $\ell \leq k$ and $S, T \in \mathcal{R}$. Depending on θ_1 and θ_2, this last formula defines a Boolean combination of T and sets of traces t satisfying

S contains $\geq \ell$ traces s that are a proper subtrace of (a proper supertrace of, are incomparable with, resp.) t.

Theorems 3.4, 4.5, and 5.13 demonstrate that these sets are effectively recognizable which completes this proof.   □

The proofs of the three results on recognizable trace languages (Theorems 3.4, 4.5, and 5.13) are the content of the remaining paper. But before, we formulate a simple consequence that describes the expressive power of the logic C^2.

Corollary 2.2. *Let $R \subseteq \mathbb{M}(\Sigma, D)^2$. Then the following are equivalent:*

1. *There is some $\varphi(x, y) \in C^2$ such that $R = \{(s, t) \in \mathbb{M}(\Sigma, D)^2 \mid S \models \varphi(s, t)\}$.*
2. *R is a finite union of relations of the form $\{(s, t) \in S \times T \mid s\, \theta_1\, t\, \theta_2\, s\}$ where S and T are recognizable subsets of $\mathbb{M}(\Sigma, D)$ and $\theta_1, \theta_2 \in \{\sqsubseteq_{\mathrm{sub}}, \not\sqsubseteq_{\mathrm{sub}}\}$.*

By Mezei's theorem (cf. [1]), this can be reformulated as "R is a Boolean combination of recognizable subsets of the monoid $\mathbb{M}(\Sigma, D)^2$ and the subtrace relation."

2.4 Auxiliary Definitions

Let $E \subseteq V^2$ be a binary relation (e.g., a partial order or an acyclic relation). Then $vE = \{w \in V \mid (v, w) \in E\}$ and $Ev = \{w \in V \mid (w, v) \in E\}$ for $v \in V$. A set $X \subseteq V$ is *downwards closed wrt. E* if $Ex \subseteq X$ for all $x \in X$. By $X{\downarrow}_E$, we denote the least downwards closed subset of V containing X. A node $v \in V$ is *maximal in V* if $vE = \emptyset$, $\max(V, E)$ denotes the set of maximal elements of V. Dually, we define upwards closed sets, $X{\uparrow}_E$, and minimal elments of V.

Let $t = (V, E, \lambda) \in \mathbb{M}(\Sigma, D)$ be a trace. Then $|t| = |V|$ denotes the size of t, i.e., its number of nodes. We write $|t|_a$ for the number of nodes of t that are labeled by a (for $a \in \Sigma$). By $\mathrm{alphmin}(t)$, we denote the set of letters $\lambda(v)$ for $v \in \min(t)$.

Let $s, t \in \mathbb{M}(\Sigma, D)$ be traces. We call s a *prefix* of t (denoted $s \sqsubseteq_{\mathrm{pref}} t$) if there exists a trace s' with $s \cdot s' = t$. The set of all prefixes of t forms a finite lattice under the relation $\sqsubseteq_{\mathrm{pref}}$. Even more, any set L of traces that all are prefixes of some trace t have a least upper bound that we denote $\sup(L)$ and call the *supremum of L*.

Let, again, $t = (V, E, \lambda) \in \mathbb{M}(\Sigma, D)$ be a trace and $A \subseteq \Sigma$. Then $X = \lambda^{-1}(A){\downarrow}_E \subseteq V$ is the set of nodes of t that are dominated by some node whose label belongs to A. We denote $t{\restriction}_X$ by $\partial_A(t)$. This is the smallest prefix s of t such that $|s|_a = |t|_a$ for all letters $a \in A$. We write $\partial_b(t)$ for $\partial_{\{b\}}(t)$ for $b \in \Sigma$. In this context, we also need the definition $D(B) = \bigcup_{b \in B} Db$ for $B \subseteq \Sigma$ of letters that are dependent from some letter in B.

3 Downward Closure

Definition 3.1. *Let S be a set of traces and $k \in \mathbb{N}$. Then $S{\downarrow}^{\geq k}$ is the set of traces t such that there are $\geq k$ traces $s \in S$ with $t \sqsubseteq_{\mathrm{sub}} s$.*

Note that $S{\downarrow}^{\geq 1}$ is the usual downward closure $S{\downarrow}_{\sqsubseteq_{sub}}$ of S as defined above. It is our aim to prove that $S{\downarrow}^{\geq k}$ is effectively recognizable if S is recognizable.

Lemma 3.2. *Let $S \subseteq \mathrm{M}(\Sigma, D)$ be a recognizable trace language. Then the trace language $S{\downarrow}^{\geq 1}$ is effectively recognizable.*

Proof. The set S is effectively rational [17, Theorem 2]. By induction on the rational expression denoting S, one can construct a starfree expression denoting $S{\downarrow}^{\geq 1}$. Since $\mathcal{R}$ is effectively closed under Boolean operations and concatenation (cf. [3]), the result follows.   □

Lemma 3.3. *Let $S \subseteq \mathrm{M}(\Sigma, D)$ be a recognizable set of traces and $k \geq 1$. Then the trace language $S{\downarrow}^{\geq 1} \setminus S{\downarrow}^{\geq k}$ is effectively recognizable.*

Proof. Let n be the size of some automaton accepting S. A pumping argument shows that all traces from $S{\downarrow}^{\geq 1}$ of length $\geq n$ also belong to $S{\downarrow}^{\geq k}$. Consequently, the difference of these two sets is finite and therefore recognizable.   □

Now $S{\downarrow}^{\geq k}$ is effectively recognizable since it is the difference of the two sets from the two lemmas above. Note that a trace t has $\geq k$ *proper* supertraces in S if, and only if, it belongs to $(S \cap S{\downarrow}^{\geq k+1}) \cup (S{\downarrow}^{\geq k} \setminus S)$. Thus, we showed the following result:

Theorem 3.4. *Let $S \subseteq \mathrm{M}(\Sigma, D)$ be recognizable and $k \in \mathbb{N}$ with $k \geq 1$. Then the set of traces t with at least k distinct proper supertraces from S is effectively recognizable.*

4   Upward Closure

Definition 4.1. *Let S be a set of traces and $k \in \mathbb{N}$. Then $S{\uparrow}_{\geq k}$ is the set of traces t such that there are $\geq k$ traces $s \in S$ with $s \sqsubseteq_{sub} t$.*

It is our aim to prove that $S{\uparrow}_{\geq k}$ is effectively recognizable if S is recognizable. The main tool in this section (and also in the following one) is a logic that talks about the internal structure of a trace $t = (V, E, \lambda)$.

The logic C^2 considers traces as elements of the structure $(\mathrm{M}(\Sigma, D), \sqsubseteq_{sub}, \mathcal{R})$ such that it allows to describe "external" properties of traces (e.g., the existence of at least two subtraces in a recognizable set S). We now shift our point of view and look at traces as relational structures. Then logical formulas describe their "internal" properties (e.g., the existence of two a-labeled nodes).

To define the set of MSO-formulas, we fix a set of first-order and a (disjoint) set of monadic second-order variables (the former are usually denoted by small letters, the latter by capital letters). Then *MSO-formulas* are defined by the following syntax (where x and y are first-order variables, X is a second-order variable, and $a \in \Sigma$):

$$\varphi := (x = y) \mid \lambda(x) = a \mid (x, y) \in E \mid x \in X \mid \varphi \vee \varphi \mid \neg \varphi \mid \exists x\, \varphi \mid \exists X\, \varphi.$$

Henceforth, we will speak of "formulas" when we actually mean "MSO-formulas".

The satisfaction relation $\models$ between a trace $t = (V, E, \lambda)$ and a formula φ is defined in the obvious way with the understanding that first-order variables denote single nodes and second-order variables denote sets of nodes of the trace.

Definition 4.2. *Let S be a set of traces. Then S is* definable *if there exists a sentence φ with $S = \{s \in \mathbb{M}(\Sigma, D) \mid s \models \varphi\}$.*

Since the notions "definable" and "recognizable" are effectively equivalent for sets of traces [20], we can reformulate the aim of this section as "if $S \subseteq \mathbb{M}(\Sigma, D)$ is definable, then so is $S{\uparrow}_{\geq k}$".

Consequently, we have to write down a formula that holds in a trace t if, and only if, it has at least k subtraces from S. The idea is to express that there are k distinct subsets of t that all induce traces from S. The problem we face here is that distinct subsets can induce the same subtrace. This problem is solved by choosing the "minimal", "leftmost" or, as we call it, "canonical" set X.

Definition 4.3. *Let t be some trace and $Z \subseteq t$. Then Z is* canonical *in t if $t \models \mathrm{canon}(Z)$, where $\mathrm{canon}(Z)$ is the formula*

$$\forall x, z \colon \left(\begin{array}{c} (\lambda(x) = \lambda(z) \wedge x \notin Z \wedge z \in Z \wedge (x, z) \in E) \\ \rightarrow \exists y \in Z \colon \big((x, y) \in E \wedge (y, z) \in E\big) \end{array} \right).$$

Then we can show that every subtrace of t is induced by precisely one set canonical in t:

Theorem 4.4. *Let $s \sqsubseteq_{\mathrm{sub}} t$ be traces. Then there is a unique canonical set $X \subseteq t$ with $s \cong t{\restriction}_X$.*

Theorem 4.4 allows us to obtain the main result of this section:

Theorem 4.5. *Let $S \subseteq \mathbb{M}(\Sigma, D)$ be definable and $k \in \mathbb{N}$ with $k \geq 1$. Then the set $S{\uparrow}_{\geq k}$ is effectively definable. Similarly, the set of traces with $\geq k$ proper subtraces from S is effectively definable.*

Proof. Let σ be a sentence defining S and consider the sentence

$$\exists X_1, X_2, \ldots, X_k \left(\bigwedge_{1 \leq i \leq k} \big(\sigma{\restriction}_{X_i} \wedge \mathrm{canon}(X_i)\big) \wedge \bigwedge_{1 \leq i < j \leq k} X_i \neq X_j \right)$$

where $\sigma{\restriction}_X$ arises from σ by restricting all quantifications to elements and subsets of X. By Theorem 4.4, it defines the set $S(\sigma){\downarrow}^{\geq k}$. To show the claim about proper subtraces, we require the sets X_i to be different from the set of all nodes.   □

5   Incomparable Traces

For two traces s and t, we write $s \parallel t$ as abbreviation for $t \not\sqsubseteq_{\mathrm{sub}} s \not\sqsubseteq_{\mathrm{sub}} t$.

Definition 5.1. *Let S be a set of traces and $k \in \mathbb{N}$. Then $S^{\parallel}_{\geq k}$ is the set of traces t such that there are $\geq k$ traces $s \in S$ satisfying $t \parallel s$.*

It is our aim to prove that $S^{\parallel}_{\geq k}$ is effectively definable if S is definable.

Two traces s and t are incomparable if, and only if, either $|s| \leq |t|$ and $s \not\sqsubseteq_{\mathrm{sub}} t$, or $|s| > |t|$ and $t \not\sqsubseteq_{\mathrm{sub}} s$. In the following two subsections, we will consider these two cases separately.

5.1   Short Non-subtraces

Definition 5.2. *Let S be a set of traces and $k \in \mathbb{N}$. Then $S^{\mathrm{short}}_{\geq k}$ is the set of traces t such that there are $\geq k$ traces $s \in S$ with $|s| \leq |t|$ and $s \not\sqsubseteq_{\mathrm{sub}} t$.*

Let S be defined by the sentence σ. We have to formulate, as a property of the labeled directed graph $t = (V, E, \lambda)$, the existence of k models s of σ that all are incomparable with t and have length at most $|t|$. The idea is to split a trace s into its largest prefix s_1 that is a subtrace of t and the complementary suffix (using Theorem 4.4, one first shows that s_1 is uniquely defined for any pair of traces (s, t)). Since s_1 is a subtrace of t, Theorem 4.4 ensures that $t \in S^{\mathrm{short}}_{\geq k}$ if, and only if, there are k pairs (X, s_2) such that

(1) $X \subseteq t$ is canonical and $s_2 \in \mathbb{M}(\Sigma, D)$,
(2) $(t{\restriction}_X) \cdot s_2 \models \sigma$,
(3) $t{\restriction}_X = \sup\{s_1 \sqsubseteq_{\mathrm{pref}} (t{\restriction}_X) \cdot s_2 \mid s_1 \sqsubseteq_{\mathrm{sub}} t\}$, and
(4) $1 \leq |s_2| \leq |t| - |X|$.

From Shelah's decomposition theorem [18, Theorem 2.4], we obtain a finite family $(\tau_j, \nu_j)_{j \in J}$ of sentences such that Condition (2) is equivalent to

(2') there exists $j \in J$ with $t{\restriction}_X \models \tau_j$ (equivalently, $(t, X) \models \tau_j{\restriction}_X$) and $s_2 \models \nu_j$.

Thus, we express Condition (2) as a Boolean combination of properties of (t, X) and of s_2. Our next aim is to also express Condition (3) in such a manner.

To this end, let $t, s_2 \in \mathbb{M}(\Sigma, D)$ and $X \subseteq t$. One first shows that Condition (3) holds if, and only if, for all $a \in \mathrm{alphmin}(s_2)$, the trace $(t{\restriction}_X) \cdot a$ is not a subtrace ot t. Let $\mathbb{U}(t, X)$ denote the set of letters $a \in \Sigma$ that violate this last condition, i.e., that satisfy $(t{\restriction}_X) \cdot a \sqsubseteq_{\mathrm{sub}} t$. If X is canonical, we can express the statement $a \in \mathbb{U}(t, X)$ by a formula:

Lemma 5.3. *Let $t = (V, E, \lambda) \in \mathbb{M}(\Sigma, D)$ be a trace, $X \subseteq t$ be canonical, and $a \in \Sigma$. Then $a \in \mathbb{U}(t, X)$ if, and only if, there exists $y \in V$ with $\lambda(y) = a$ and $yE \cap X = \emptyset$.*

Hence, for $A \subseteq \Sigma$, there are formulas $\alpha_A(X)$ and sentences β_A such that

– $(t, X) \models \alpha_A$ if, and only if, $A \cap \mathbb{U}(t, X) = \emptyset$ for all $t \in \mathbb{M}(\Sigma, D)$ and $X \subseteq t$ canonical and
– $s_2 \models \beta_A$ if, and only if, $A = \mathrm{alphmin}(s_2)$.

In summary, we found a family of formulas $(\alpha_A(X), \beta_A)_{A \subseteq \Sigma}$ such that Condition (3) is equivalent to

(3') there exists $A \subseteq \Sigma$ with $(t, X) \models \alpha_A$ and $s_2 \models \beta_A$.

Thus, $t \in S_{\geq k}^{\mathrm{short}}$ if, and only if, there are k pairs (X, s_2) all satisfying the conditions (1), (2'), (3'), and (4). We group these pairs according to their first component. Then $t \in S_{\geq k}^{\mathrm{short}}$ if, and only if, there exist $\ell \leq k$, a function $f \colon \{1, \ldots, \ell\} \to \{0, 1, \ldots, k\}$ with $\sum_{1 \leq i \leq \ell} f(i) = k$, and sets $A_1, A_2, \ldots, A_\ell \subseteq \Sigma$ such that there are mutually distinct canonical sets $X_i \subseteq t$ satisfying, for all $i \in [\ell]$, the existence of some $j \in J$ with

– $(t, X_i) \models \tau_j \wedge \alpha_{A_i}$,
– there are $f(i)$ many traces s_2 of length $1 \leq |s_2| \leq |t| - |X_i|$ satisfying $s_2 \models \nu_j \wedge \beta_{A_i}$.
  From $\nu_j \wedge \beta_{A_i}$, one can compute a number N such that this holds if, and only if, $|t| - |X_i| \geq N$. Hence this is a property of (t, X_i) that can be expressed by a formula.

All this can be translated into a sentence that only talks about the trace t. Consequently, we obtain

Proposition 5.4. *Let $S \subseteq \mathbb{M}(\Sigma, D)$ be definable and $k \in \mathbb{N}$ with $k \geq 1$. Then $S_{\geq k}^{\mathrm{short}}$ is effectively definable.*

5.2   Long Non-supertraces

Definition 5.5. *Let T be a set of traces and $k \in \mathbb{N}$. Then $T_{\geq k}^{\mathrm{long}}$ is the set of traces s such that there are $\geq k$ traces $t \in T$ with $|s| < |t|$ and $s \not\sqsubseteq_{\mathrm{sub}} t$.*

We have to formulate, as a property of the labeled directed graph $s = (V, E, \lambda)$, the existence of k traces $t \in T$ that all are incomparable with s and have length at least $|s| + 1$. The first idea is, again, to split the trace s into its largest prefix s_1 that is a subtrace of t and the complementary suffix. Since this time, we have to formulate properties of s, we would then have to "fill" the prefix s_1 with arbitrarily many nodes to obtain the trace t (more precisely: the minimal prefix of t that contains s_1 as a subtrace). Since this cannot be done with logical formulas, we have to bound this number of "missing pieces". The central notion here is the following:

Definition 5.6. *Let $t = (V, E, \lambda)$ be a trace and $X \subseteq t$. The number of holes of X in t equals $\mathrm{nh}(X, t) = |X{\downarrow}_E \setminus X|$.*
  Now let s be a trace. If $s \sqsubseteq_{\mathrm{sub}} t$, then $\mathrm{nh}(s, t) = \mathrm{nh}(X, t)$ where $X \subseteq t$ is canonical with $s = t{\restriction}_X$. If s is not a subtrace of t, then $\mathrm{nh}(s, t) = \infty$.

The following lemma describes, in terms of the number of holes and the length difference, when a trace is a subtrace of a longer trace:

Lemma 5.7. *Let s, t be traces with $|s| < |t|$. Then $s \parallel t$ if, and only if, $s \neq \sup\{s' \sqsubseteq_{\text{pref}} s \mid \text{nh}(s', t) \leq |t| - |s|\}$.*

Recall that we have to express, as a property of the labeled directed graph $s = (V, E, \lambda)$, the existence of k properly longer traces $t \in T$ with $s \not\sqsubseteq_{\text{sub}} t$. In doing so, the previous characterisation is particularly useful if the length difference of t and s is fixed. The following lemma, whose proof uses a straightforward pumping argument, allows to do precisely this:

Lemma 5.8. *One can compute a number $n \in \mathbb{N}$ such that the following holds for all $k \in \mathbb{N}$ and $s \in T_{\geq k}^{\text{long}}$: There exist k traces $t \in T$ such that $|s| < |t|$, $s \not\sqsubseteq_{\text{sub}} t$, and $|t| \leq |s| + k \cdot (n + 1)$.*

Thus, it suffices to characterize, for all $k \geq 0$ and all length differences $N > 0$, those traces s that allow $\geq k$ traces $t \in T$ with $|t| = |s| + N$ and $s \neq \sup\{s' \sqsubseteq_{\text{pref}} s \mid \text{nh}(s', t) \leq N\}$.

Grouping these traces t according to $\sup\{s' \sqsubseteq_{\text{pref}} s \mid \text{nh}(s', t) \leq N\}$, it suffices to characterize those pairs (s_1, s_2) with $s_2 \neq 1$ (where we think of $s_1 s_2$ as a factorisation of s) that allow $\geq k$ pairs (t_1, t_2) of traces such that

(a) $t_1 t_2 \in T$ and t_1 is the minimal prefix of $t_1 t_2$ with $s_1 \sqsubseteq_{\text{sub}} t_1$,
(b) $|s_1 s_2| + N = |t_1 t_2|$, and
(c) $s_1 = \sup\{s' \sqsubseteq_{\text{pref}} s_1 s_2 \mid \text{nh}(s', t_1 t_2) \leq N\}$.

Note that t_1 is the minimal prefix of $t_1 t_2$ with $s_1 \sqsubseteq_{\text{sub}} t_1$ if, and only if, $s_1 \sqsubseteq_{\text{sub}} t_1$ and, for all prefixes $t' \sqsubseteq_{\text{pref}} t_1$ with $s_1 \sqsubseteq_{\text{sub}} t'$, we have $t' = t_1$. This allows to reformulate the second half of Condition (a) as a condition on the pair (s_1, t_1), only. Since T is definable, Shelah's decomposition theorem allows us to compute a finite family $(\mu_j, \nu_j)_{j \in J}$ of pairs of sentences such that $t_1 t_2 \in T$ if, and only if, there exists $j \in J$ with $t_1 \models \mu_j$ and $t_2 \models \nu_j$.

Consequently, for $k \geq 0$, $N > 1$, and a fixed index $j \in J$, it suffices to characterize those pairs (s_1, s_2) with $s_2 \neq 1$ that allow $\geq k$ pairs (t_1, t_2) of traces such that, besides Conditions (b) and (c), also the following holds:

(a_j) $t_1 \models \mu_j$, $s_1 \sqsubseteq_{\text{sub}} t_1$, and $s_1 \sqsubseteq_{\text{sub}} t' \Rightarrow t' = t_1$ for all $t' \sqsubseteq_{\text{pref}} t_1$ and $s_2 \models \nu_j$.

Let (t_1, t_2) be a pair of traces with these properties. At this point, it comes in handy that $\text{nh}(s_1, t_1 t_2) \leq N \cdot |\Sigma|$ (this holds for any traces s_1, t_1, and t_2). Further, since t_1 is the smallest prefix of $t_1 t_2$ with $s_1 \sqsubseteq_{\text{sub}} t_1$, we get $\text{nh}(s_1, t_1 t_2) = \text{nh}(s_1, t_1) = |t_1| - |s_1|$.

Consequently, we can group these pairs (t_1, t_2) according to the length difference $|t_1| - |s_1|$ (which can be bounded by $N \cdot |\Sigma|$ by the above). Hence, it suffices to characterize, for $k \geq 0$, $N > 0$, $j \in J$ and for a fixed length difference ℓ, those pairs (s_1, s_2) of traces with $s_2 \neq 1$ that allow $\geq k$ pairs (t_1, t_2) of traces such that, besides (a_j) and (c), the following holds:

(b$_\ell$) $|s_1| + \ell = |t_1|$ and $|t_2| = |s_2| + N - \ell$.

Note that Conditions (a$_j$) and (b$_\ell$) form a Boolean combination of properties of the pairs (s_1, t_1) and (s_2, t_2), respectively. Our next aim is to ensure that this also holds for Condition (c) which forms the main work in this section.

Lemma 5.9. *Let s_1, s_2, t_1, and t_2 be traces such that t_1 is the minimal prefix of $t_1 t_2$ with $s_1 \sqsubseteq_{\mathrm{sub}} t_1$. Then Condition (c) is equivalent to*

(c$_1$) For all $a \in \Sigma$, there exists a trace s' with $\partial_a(s_1) \sqsubseteq_{\mathrm{pref}} s' s_1$ and $\mathrm{nh}(s', t_1) \le N$ and

(c$_2$) For all $b \in \mathrm{alphmin}(s_2)$ and all s' with $\partial_{D(b)}(s_1) \sqsubseteq_{\mathrm{pref}} s' s_1$, we have $\mathrm{nh}(s'b, t_1 t_2) > N$.

Condition (c$_1$) only depends on the pair (s_1, t_1). Since Σ is finite, Condition (c$_2$) is a Boolean combination of properties of s_2 and of properties of the triple (s_1, t_1, t_2). We now reformulate this last condition using the following lemma.

Lemma 5.10. *Let $b \in \Sigma$, let s_1, t_1, t_2, and s' be traces such that t_1 is the minimal prefix of $t_1 t_2$ with $s_1 \sqsubseteq_{\mathrm{sub}} t_1$ and $\partial_b(s_1) \sqsubseteq_{\mathrm{pref}} s' s_1$. Then $\mathrm{nh}(s'b, t_1 t_2) > N$ if, and only if, one of the following holds:*

- *$N < \mathrm{nh}(s'b, t_1) < \infty$ or*
- *$\mathrm{nh}(s'b, t_1) = \mathrm{nh}(b, t_2) = \infty$ or*
- *$\mathrm{nh}(s'b, t_1) = \infty$, $b \sqsubseteq_{\mathrm{sub}} t_2$, and $\mathrm{nh}\big(\sup(s', \partial_{D(B)}(s_1)), t_1)\big) + \mathrm{nh}(b, t_2) > N$ where $B \subseteq \Sigma$ is the set of letters appearing before the first b in t_2.*

Replacing $\mathrm{nh}(s'b, t_1 t_2) > N$ in Condition (c$_2$) by the properties from the above lemma, it turns into a Boolean combination (c$_2'$) of statements

(i) $\mathrm{nh}(b, t_2) < h$ for $h \in \mathbb{N} \cup \{\infty\}$,
(ii) "A is the set of letters appearing in t_2 before the first b" for $A \subseteq \Sigma$,
(iii) "for all traces s' with $\partial_b(s_1) \sqsubseteq_{\mathrm{pref}} s' s_1$" followed by a Boolean combination of statements of the form
  - $N \ge \mathrm{nh}(s'b, t_1)$ and $\mathrm{nh}(s'b, t_1) = \infty$ for $b \in \Sigma$, and
  - $\mathrm{nh}(\sup\big(\sup(s', \partial_{D(B)}(s_1)), t_1)\big) < N_1$ for $B \subseteq \Sigma$ and $N_1 \in \mathbb{N}$.

To finish, let $k \in \mathbb{N}$, $N > 0$, $j \in J$, and $\ell \in \mathbb{N}$. Further, let H denote the set of pairs (s_1, s_2) of traces such that conditions (a$_j$), (b$_\ell$), (c$_1$), and (c$_2'$) hold for at least k pairs of traces (t_1, t_2). The conjunction of these four properties, that talks about the quadruple (s_1, s_2, t_1, t_2), forms a Boolean combination of properties that talk about the pairs (s_1, t_1) and (s_2, t_2), respectively.

Now we can transform the statement "$(s_1, s_2) \in H$" into a Boolean combination of statements of the following form:

(A) there are $\ge k_1$ traces t_1 satisfying a Boolean combination of statements of the form
  - $t_1 \models \mu_j$,

- $s_1 \sqsubseteq_{\mathrm{sub}} t_1$ and $s_1 \sqsubseteq_{\mathrm{sub}} t' \Rightarrow t' = t_1$ for all $t' \sqsubseteq_{\mathrm{pref}} t_1$,
- $|s_1| + \ell = |t_1|$,
- $\exists s': \partial_a(s_1) \sqsubseteq_{\mathrm{pref}} s's_1 \wedge \mathrm{nh}(s', t_1) \leq N$, and
- statements of the form (iii).

(B) there are $\geq k_2$ traces t_2 satisfying a Boolean combination of statements of the following forms:

- $t_2 \models \nu_j$,   – $b \sqsubseteq_{\mathrm{sub}} t_2$,
- $|t_2| = |s_2| + N - \ell$,   – $\mathrm{nh}(b, t_2) < h$,
- $b \in \mathrm{alphmin}(s_2)$,   – "$A$ is the set of letters in t_2 before the first b"

Regarding (B), we can formulate all the properties that do not mention s_2 as a formula. This turns all of (B) into a condition on the minimal letters in s_2 and the length of s_2 which both can be expressed by a formula.

It remains to also express (A) as a property of the trace s_1. As a first step, we replace, in the Boolean combination, all references to s_1 by $t \restriction_X$ and add that X is canonical. This gives a formula $\varphi(X)$ talking about (t_1, X) and we have to express that there are $\geq k_1$ models (t_1, X) of $\varphi(X)$ all satisfying $s_1 = t_1 \restriction_X$. This is achieved by the following result:

Lemma 5.11. *Let $\varphi(X)$ be a formula and $k_1, n \in \mathbb{N}$ such that $(t_1, X) \models \varphi$ implies $|t| - |X| = n$. Then one can construct (from φ, k_1, and n) a sentence ψ such that, for all traces s_1 we have $s_1 \models \psi$ if, and only if, there are $\geq k_1$ pairs (t_1, X) with $(t_1, X) \models \varphi$ and $s_1 = t_1 \restriction_X$.*

In summary, we obtained the following:

Proposition 5.12. *Let $T \subseteq \mathbb{M}(\Sigma, D)$ be definable and $k \in \mathbb{N}$ with $k \geq 1$. Then the set $T_{\geq k}^{\mathrm{long}}$ is effectively definable.*

Now the following result follows easily from Propositions 5.4 and 5.12:

Theorem 5.13. *Let $S \subseteq \mathbb{M}(\Sigma, D)$ be definable and $k \in \mathbb{N}$ with $k \geq 1$. Then the set $S_{\geq k}^{\parallel}$ of traces t with at least k distinct traces $s \in S$ with $s \not\sqsubseteq_{\mathrm{sub}} t$ and $t \not\sqsubseteq_{\mathrm{sub}} s$ is effectively definable.*

Thus, we demonstrated how to prove Theorems 3.4, 4.5, and 5.13. This closes the gaps left open in our proof of the main result (Theorem 2.1).

6   Concluding Remarks

The C+MOD^2-theory of $(\Sigma^*, \sqsubseteq_{\mathrm{sub}}, \mathcal{R})$ is decidable [14]. This logic has, in addition to the logic C^2, modulo-counting quantifiers $\exists^{q,r}$. It seems that the only obstacle in proving the analogous result for the subtrace order is the use of Lemma 5.8 in the proof of Proposition 5.12. Whether this lemma has an analogue in the modulo-counting setting is not clear.

The decision algorithms in this paper (as well as those in [9,14] for the subword order) are nonelementary. Karandikar and Schnoebelen [10] prove that the FO^2-theory of the subword order can be decided in triply exponential space if we only allow unary languages (instead of all languages from $\mathcal{R}$), current research improves the upper bound to doubly exponential space and extends the result to the C^2-theory [13]. It is not clear whether such an elementary upper bound also holds for the subtrace relation.

Finally, it remains to be explored whether the methods developed in this paper can be applied in other settings where rational relations are not available.

References

1. Berstel, J.: Transductions and Context-Free Languages. Teubner Studienbücher, Stuttgart (1979)
2. Cartier, P., Foata, D.: Problemes combinatoires de commutation et rearrangements. Lecture Notes in Mathematics, vol. 85. Springer, Heidelberg (1969). https://doi.org/10.1007/BFb0079468
3. Diekert, V., Rozenberg, G.: The Book of Traces. World Scientific Publ. Co., London (1995)
4. Finkel, A., Schnoebelen, P.: Well-structured transition systems everywhere!. Theoret. Comput. Sci. **256**, 63–92 (2001)
5. Gastin, P., Petit, A.: Infinite traces. In: [3], pp. 393–486 (1995)
6. Halfon, S., Schnoebelen, Ph., Zetzsche, G.: Decidability, complexity, and expressiveness of first-order logic over the subword ordering. In: LICS 2017, pp. 1–12. IEEE Computer Society (2017)
7. Higman, G.: Ordering by divisibility in abstract algebras. Proc. London Math. Soc. **2**, 326–336 (1952)
8. Ježek, J., McKenzie, R.: Definability in substructure orderings. I: finite semilattices. Algebra Univers. **61**(1), 59–75 (2009)
9. Karandikar, P., Schnoebelen, Ph.: Decidability in the logic of subsequences and supersequences. In: FSTTCS 2015, Leibniz International Proceedings in Informatics, vol. 45, pp. 84–97. Leibniz-Zentrum für Informatik (2015)
10. Karandikar, P., Schnoebelen, P.: The height of piecewise-testable languages and the complexity of the logic of subwords. Log. Methods Comput. Sci. **15**(2), 6:1–6:27 (2019)
11. Kudinov, O.V., Selivanov, V.L., Yartseva, L.V.: Definability in the subword order. In: Ferreira, F., Löwe, B., Mayordomo, E., Mendes Gomes, L. (eds.) CiE 2010. LNCS, vol. 6158, pp. 246–255. Springer, Heidelberg (2010). https://doi.org/10.1007/978-3-642-13962-8_28
12. Kuske, D.: Theories of orders on the set of words. Theoret. Inf. Appl. **40**, 53–74 (2006)
13. Kuske, D., Schwarz, Ch.: Complexity of counting first-order logic for the subword order (2020, in preparation)
14. Kuske, D., Zetzsche, G.: Languages ordered by the subword order. In: Bojańczyk, M., Simpson, A. (eds.) FoSSaCS 2019. LNCS, vol. 11425, pp. 348–364. Springer, Cham (2019). https://doi.org/10.1007/978-3-030-17127-8_20
15. Madhusudan, P.: Model-checking trace event structures. In: LICS 2003, pp. 371–380. IEEE Computer Society Press (2003)
16. Mazurkiewicz, A.: Concurrent program schemes and their interpretation. Technical report, DAIMI Report PB-78, Aarhus University (1977)

17. McKnight, J.: Kleene's quotient theorems. Pac. J. Math. **XIV**, 1343–1352 (1964)
18. Shelah, S.: The monadic theory of order. Ann. Math. **102**, 379–419 (1975)
19. Thinniyam, R.S.: Defining recursive predicates in graph orders. Log. Methods Comput. Sci. **14**(3), 1–38 (2018)
20. Thomas, W.: On logical definability of trace languages. In: Diekert, V. (ed.) Proceedings of a workshop of the ESPRIT BRA No 3166: Algebraic and Syntactic Methods in Computer Science (ASMICS) 1989, Report TUM-I9002, Technical University of Munich, pp. 172–182 (1990)
21. Wires, A.: Definability in the substructure ordering of simple graphs. Ann. Comb. **20**(1), 139–176 (2016). https://doi.org/10.1007/s00026-015-0295-4

Speedable Left-c.e. Numbers

Wolfgang Merkle[(✉)] and Ivan Titov

Institute of Computer Science, Heidelberg University, Heidelberg, Germany
merkle@math.uni-heidelberg.de, titov@stud.uni-heidelberg.de

Abstract. A left-c.e. real number α is ρ-speedable if there is a computable left approximation $a_0, a_1, \ldots$ of α and a nondecreasing computable function f such that we have $f(n) \geq n$ and

$$\liminf_{n \to \infty} \frac{\alpha - a_{f(n)}}{\alpha - a_n} \leq \rho,$$

and α is speedable if it is ρ-speedable for some $\rho < 1$. Barmpalias and Lewis-Pye [JCSS 89:349–360, 2016] have implicitly shown that Martin-Löf random left-c.e. real numbers are never speedable. We give a straightforward direct proof of this fact and state as open problem whether this implication can be reversed, i.e., whether all nonspeedable left c.e. real numbers are Martin-Löf random. In direction of solving the latter problem, we demonstrate that speedability is a degree property for Solovay degrees in the sense that either all or no real numbers in such a degree are speedable, and that left-c.e. real numbers of nonhigh Turing degree are always speedable. In this connection, we observe that every c.e. Turing degree contains a speedable left-c.e. real number. Furthermore, we obtain a dichotomy result: by definition, left-approximations of nonspeedable real numbers are never speedable, while for any speedable real number all of its left approximations are ρ-speedable for all $\rho > 0$.

Keywords: Left-c.e. real numbers · Speedable ·
Martin-Löf-randomness · Solovay reducibility

1 Speedable Left-c.e Numbers

In the field of algorithmic randomness, left-c.e. real numbers and the structure induced on them by Solovay reducibility have been intensively studied. In what follows, we investigate into the question to what extent computable left-approximations of left-c.e. real numbers can be accelerated. Here Omega numbers are of particular interest, i.e., left-c.e. real numbers that are Martin-Löf random. By a well-known result due to various groups of authors, Omega numbers can be equivalently characterised as the halting probabilities of universal prefix-free Turing machines and by being Solovay complete [2, Section 9.2].

The second author was supported by Landesgraduiertenförderung Baden-Württemberg.

H. Fernau (Ed.): CSR 2020, LNCS 12159, pp. 303–313, 2020.
https://doi.org/10.1007/978-3-030-50026-9_22

Definition 1. *A* LEFT APPROXIMATION *is a converging sequence* $a_0, a_1, \ldots$ *of dyadic rational numbers such that* $a_i < a_{i+1}$. *A real number* α *is* LEFT-C.E. *if there is a computable left approximation with limit* α *[2].*

As usual and if not explicitly specified otherwise, we restrict attention to left-c.e. numbers in the unit interval and identify such numbers with infinite binary sequences where a sequence $A(0)A(1)\ldots$ corresponds to the real number with binary expansion $0.A(0)A(1)\ldots$.

Definition 2. *A function* $f \colon \mathbb{N} \to \mathbb{N}$ *is a* SPEED-UP FUNCTION *if it is non-decreasing and* $n \le f(n)$ *holds for all* n. *A left approximation* $a_0, a_1, \ldots$ *with limit* α *is* ρ-SPEEDABLE *for some real number* ρ *if there is a computable speed-up function* f *such that we have*

$$\liminf_{n \to \infty} \frac{\alpha - a_{f(n)}}{\alpha - a_n} \le \rho, \tag{1}$$

and the left approximation is SPEEDABLE *if it is* ρ-*speedable for some* $\rho < 1$.

A real number α *is* ρ-SPEEDABLE WITH RESPECT TO A GIVEN LEFT APPROXIMATION *if the left approximation converges to* α *and is computable and* ρ-*speedable. A real number is* ρ-SPEEDABLE *if it is* ρ-*speedable with respect to some left-approximation. A real number is* SPEEDABLE *if it is* ρ-*speedable for some* $\rho < 1$, *and the notion of* SPEEDABLE WITH RESPECT TO A GIVEN LEFT APPROXIMATION *is defined in the same manner. A left approximation is* NON-SPEEDABLE *if it is not speedable, and* NONSPEEDABLE *real numbers are defined likewise.*

Apparently, the notions of speedable just introduced have not been considered in the literature before and, in particular, have no obvious connections to a notion by the same name introduced by Soare [4]. Barmpalias and Lewis have been shown implicitly that Martin-Löf random real numbers are nonspeedable [1, Theorem 1.7]. In what follows, we will give a straightforward direct proof of their result. In general, we investigate into the question which left-c.e. real numbers are speedable. For any left approximation $a_0, a_1, \ldots$ of some real number α and any speed-up function f, the values of the fraction in (1) and thus also their limit inferior and limit superior are confined to the half-open real interval $(0, 1]$.

Remark 3. By definition, every left-c.e real is ρ-speedable for every $\rho \ge 1$ via any computable speed-up function. Consequently, a left-c.e. real number is ρ-speedable for all ρ in $(0, 1]$ if and only if it is ρ-speedable for every $\rho > 0$. In what follows, we will use the latter condition in order to express that a left-c.e. real is ρ-speedable for arbitrarily small strictly positive values of ρ.

Remark 4. By definition, for a nonspeedable left-c.e. real number α all limits inferior as in (1) are equal to 1, hence coincide with the corresponding limits superior, which can be at most 1 by the discussion in Remark 3. I.e., for nonspeedable left-c.e. α, the values of the fractions in (1) always converge to 1.

By the next proposition, the limit superior of the fractions in (1) is always equal to 1 unless the real number α is computable.

Proposition 5. *Let $a_0, a_1, \ldots$ be a computable left approximation with limit α. Then α is computable if and only if there is a computable speed-up function f and some $\rho < 1$ where*

$$\limsup_{n \to \infty} \frac{\alpha - a_{f(n)}}{\alpha - a_n} \le \rho. \tag{2}$$

Furthermore, in case α is computable, for every $\rho > 0$ there is a computable speed-up function f that satisfies (2).

Proof. First assume that (2) holds for some computable speed-up function f. Choose some dyadic rational number $\gamma < 1$ and some large enough m such that $\alpha - a_m$ is at most 1 and for all $n > m$ it holds that

$$\frac{\alpha - a_{f(n)}}{\alpha - a_n} \le \gamma, \quad \text{where then} \quad \alpha - a_{f^k(m)} \le \gamma^k$$

follows for all natural numbers k by a straightforward induction argument. Consequently, the real number α is computable.

Next assume that α is computable and for given $\rho > 0$, fix some natural number k where $1/k < \rho$. Then the required speed-up function f is obtained as follows. For given argument n, compute a dyadic rational number a' such that

$$|\alpha - a'| \le \frac{\delta}{3} \quad \text{where} \quad \delta = \frac{a_{n+1} - a_n}{k}, \quad \text{and let} \quad a = a' - \frac{\delta}{2}.$$

Then we have $\alpha - \delta < a < \alpha$, hence if we let $f(n)$ be equal to the minimum $m \ge n$ such that $a < a_m$, it holds that

$$\frac{\alpha - a_{f(n)}}{\alpha - a_n} < \frac{\alpha - a}{a_{n+1} - a_n} < \frac{\delta}{k\delta} = \frac{1}{k} < \rho.$$

$\square$

2   A Dichotomy for Speedability

For the notion of a speedable left-c.e. real number, we obtain the following dichotomy. By definition, no left approximation of a nonspeedable left-c.e. real number can be speedable. On the other hand, all left approximations of speedable left-c.e. real numbers are ρ-speedable for all $\rho > 0$.

Theorem 6. *Every speedable left-c.e. real number is ρ-speedable for any $\rho > 0$ with respect to any of its left approximations.*

Theorem 6 is immediate from Lemmas 7 and 8.

Lemma 7. *Let $a_0, a_1, \ldots$ be a computable left approximation that is ρ-speedable for some $\rho < 1$. Then all computable left approximations with the same limit are also ρ-speedable.*

Proof. Let $b_0, b_1, \ldots$ be any computable left-approximation that has the same limit α as $a_0, a_1, \ldots$. We show that the former left approximation is ρ-speedable via some computable speed-up function g. In fact, it suffices to show the latter for some computable subsequence of $b_0, b_1, \ldots$. Thus we can assume that the sets

$$D_i = \{n \colon b_i \le a_n < b_{i+1}\}$$

are all nonempty, hence the functions defined by

$$m(i) = \max\{f(n) \colon n \in D_i\} \quad \text{and} \quad g(i) = \min\{j \colon i \le j \text{ and } a_{m(i)} < b_j\}$$

are total and computable. By construction, for every n there is a unique i such that n is in D_i, and for such n and i we have

$$\frac{\alpha - b_{g(i)}}{\alpha - b_i} \le \frac{\alpha - a_{f(n)}}{\alpha - a_n}.$$

Since the limit inferior of the terms on the right-hand side is at most ρ, a similar remark holds for the terms on the left-hand side, thus the left approximation $b_0, b_1, \ldots$ is ρ-speedable via g. $\square$

Lemma 8. *Let α be a speedable left-c.e. real number and let $\rho > 0$ be a real number. Then α is ρ-speedable.*

Proof. Let $\rho_{\inf}$ be the infimum of the set of all real numbers γ such that α is γ-speedable. By definition, the latter set is closed upwards, hence contains all $\gamma > \rho_{\inf}$ but no $\gamma < \rho_{\inf}$. Furthermore, we have $\rho_{\inf} < 1$ since α is speedable, and we can assume $\rho < \rho_{\inf}$ because otherwise, we are done. By choosing ρ_1 and ρ_2 close enough together, we can fix dyadic rational numbers ρ_1, ρ_2, and ρ_3 that satisfy

$$0 < \rho_1 < \rho_{\inf} < \rho_3 < \rho_2 < 1 \quad \text{and} \quad \frac{\rho_2}{1 - \rho_2} - \frac{\rho_1}{1 - \rho_1} < \frac{\rho}{1 - \rho}. \tag{3}$$

Then α is ρ_3-speedable, so we can fix a left approximation $a_0, a_1, \ldots$ with limit α that is ρ_3-speedable via some computable speed-up function f. Note that for any real number $x \ne 1$, we have

$$\frac{\alpha - a_{f(n)}}{\alpha - a_n} < x \quad \text{if and only if} \quad \alpha - a_{f(n)} < \frac{x}{1 - x}(a_{f(n)} - a_n), \tag{4}$$

which is immediate by writing $\alpha - a_n$ as the sum of $\alpha - a_{f(n)}$ and $a_{f(n)} - a_n$. Note further, that the equivalence (4) does not depend on the choice of the speed-up function f and remains valid if we replace both occurrences of $<$ by $>$. As a consequence we have

$$\rho_1 < \frac{\alpha - a_{f(n)}}{\alpha - a_n} < \rho_2 \quad \text{if and only if} \quad \frac{\rho_1(a_{f(n)} - a_n)}{1 - \rho_1} < \alpha - a_{f(n)} < \frac{\rho_2(a_{f(n)} - a_n)}{1 - \rho_2}. \tag{5}$$

By the choice of ρ_1, ρ_2, and f, in (5) the first and thus also the third inequality, from left to right, hold for almost all n, while the second and fourth one hold for infinitely many n. Fix n_0 such that the third strict inequality in (5) holds for all $n \geq n_0$. Let $g(n) = f(n)$ for all $n < n_0$ and for all other n choose $g(n)$ to be minimum such that we have

$$g(n) > \max\{g(n-1), f(n)\} \quad \text{and} \quad a_{g(n)} > a_{f(n)} + \frac{\rho_1}{1-\rho_1}\left(a_{f(n)} - a_n\right).$$

Then g is total by the choice of n_0 and is actually a computable speed-up function. Furthermore, for the infinitely many n for which the third strict inequality in (5) holds, we have

$$\alpha - a_{g(n)} < \alpha - a_{f(n)} - \frac{\rho_1}{1-\rho_1}\left(a_{f(n)} - a_n\right)$$

$$< \frac{\rho_2}{1-\rho_2}\left(a_{f(n)} - a_n\right) - \frac{\rho_1}{1-\rho_1}\left(a_{f(n)} - a_n\right)$$

$$< \frac{\rho}{1-\rho}\left(a_{f(n)} - a_n\right) < \frac{\rho}{1-\rho}\left(a_{g(n)} - a_n\right).$$

The relations hold for these n, from left to right and top to bottom, by definition of g, by (5), by choice of ρ_1 and ρ_2, and finally by definition of g. For all such n, by (4) we have

$$\frac{\alpha - a_{g(n)}}{\alpha - a_n} < \rho,$$

hence α is ρ-speedable. □

Lemma 9. *Let the left-c.e. real number α be speedable and let $\rho > 0$. Then α is ρ-speedable with respect to some computable left approximation via the speed-up function $n \mapsto n + 1$.*

Proof. By Theorem 6, we can fix a computable left-approximation $a_0, a_1, \ldots$ of α that is ρ-speedable via some computable speed-up function f that can be assumed to be strictly monotonic. Inductively, let $g(0) = 0$ and let $g(i + 1) = f(g(i))$, which can also be written as $g(i) = f^{(i)}(0)$. Let $i(n)$ be the maximum index i such that $g(i) \leq n$. By choice of f and by definition of the functions g and i, we then have for all n and for $i = i(n)$

$$g(i) \leq n < g(i+1) = f(g(i)) \leq f(n) < f(g(i+1)) = g(i+2).$$

In particular, for all n the interval $[n, f(n)]$ is contained in the "double interval" $[g(n(i)), g(n(i) + 2)]$, hence

$$\frac{\alpha - a_{g(i(n)+2)}}{\alpha - a_{g(i(n))}} \leq \frac{\alpha - a_{f(n)}}{\alpha - a_n}.$$

This inequality remains valid if we apply the lim inf operator on both sides, hence the limit inferior of the terms on the left-hand side is at most ρ. As a consequence, at least one of the left approximations $a_{g(0)}, a_{g(2)}, \ldots$ and $a_{g(1)}, a_{g(3)}, \ldots$ witnesses the conclusion of the lemma. □

3   Martin-Löf Random and Nonhigh Left-c.e. Real Numbers

Recall that a SOLOVAY TEST is a computable sequence $I_0, I_1, \ldots$ of subintervals of the unit interval that are bounded by dyadic rational numbers such that the sum of the interval lengths is finite. Recall further that a real number is not Martin-Löf random if and only if it is covered by such a Solovay test in the sense of being contained in infinitely many of the intervals I_n.

Theorem 10 (Barmpalias, Lewis-Pye [1]). *Martin-Löf random left-c.e. real numbers are never speedable*

Proof. We show that speedable left-c.e. real numbers are never Martin-Löf random. Fix some speedable real number α and by Lemma 9 fix a computable left approximation $a_0, a_1, \ldots$ of α such that for some dyadic rational number $\gamma < 1$ and for infinitely many n it holds that

$$\frac{\alpha - a_{n+1}}{\alpha - a_n} < \gamma, \quad \text{hence} \quad \alpha < a_n + \frac{a_{n+1} - a_n}{1 - \gamma}.$$

by (4). Consequently, because of $a_n < a_{n+1} < \alpha$, for all such n the real number α is contained in the interval

$$I_n = [a_{n+1}, a_{n+1} + \frac{1}{1-\gamma}(a_{n+1} - a_n)], \quad \text{where} \quad \sum_{n=0}^{\infty} |I_n| = \frac{1}{1-\gamma}(\alpha - a_0).$$

Since the a_n are computable, the intervals I_n form a Solovay test, which by construction covers α. As a consequence, α is not Martin-Löf random.   □

Recall that a left-c.e. real number α is HIGH if the jump of α and of the halting problem are Turing equivalent, and that the latter holds if and only if there is a function computable with oracle α that dominates all computable functions.

Theorem 11. *All nonhigh left-c.e. real numbers are speedable.*

Proof. Let α be a nonhigh left-c.e. real number and let $a_0, a_1, \ldots$ be any computable left approximation with limit α. Similar to the case of computable α, there is a function f computable in α such that

$$\liminf_{n \to \infty} \frac{\alpha - a_{f(n)}}{\alpha - a_n} < 1, \quad \text{hence} \quad \frac{\alpha - a_{f(n)}}{\alpha - a_n} \leq \gamma \quad \text{for some } \gamma < 1 \qquad (6)$$

and almost all n. Since α is nonhigh, there is a computable function g that is not dominated by f where we can assume that g is strictly monotonic and thus is a speed-up function. So $f(n) < g(n)$ for infinitely many n. Since (6) holds for almost all of these n, the left approximation $a_0, a_1, \ldots$ and thus also the real number α is speedable via g.   □

Theorems 10 and 11 leave open whether high left-c.e. real numbers that are not Martin-Löf random are speedable or nonspeedable. In particular, we do not know whether all nonspeedable left-c.e. real numbers are Martin-Löf random. As a partial result in the direction of theses questions, by the following remark every computably enumerable Turing degree contains a speedable left-c.e. real number.

Remark 12. Let A be an infinite recursively enumerable set with computable enumeration $z_0, z_1, \ldots$ and let $\alpha = 0.A(0)A(1)\ldots$. If we let

$$a_i = \sum_{n \in \{z_0, \ldots, z_i\}} 2^{-(n+1)},$$

then $a_0, a_1, \ldots$ is a computable left-approximation of α which is $1/2$-speedable via the speed-up function $n \mapsto n+1$. For a proof, let $i+1$ be any of the infinitely many true stage of the given enumeration, i.e., $z = z_{i+1}$ differs from z_0 through z_i and all numbers in the intersection of A with $\{0, \ldots, z\}$ occur already among the numbers $z_0, \ldots, z_{i+1}$. Then we have

$$\alpha - a_{n+1} \leq 2^{-z}, \quad \text{hence} \quad \frac{\alpha - a_{n+1}}{\alpha - a_n} = \frac{\alpha - a_{n+1}}{\alpha - a_{n+1} + 2^{-z}} \leq \frac{1}{2}.$$

Next we derive further examples of speedable left-c.e. real numbers that are non-high where in addition these numbers have initial segments of high Kolmogorov complexity. Recall that all left-c.e. Martin-Löf random real numbers are equal to the halting probability of an additively optimal prefix-free machine and that Chaitin's Ω is such a real number. We write **0** and **1** for the infinite sequence of zeroes and ones, respectively. Furthermore, for a set of natural numbers H and two infinite binary sequences $A = A(0)A(1)\ldots$ and $B = B(0)B(1)\ldots$, we write $A \oplus_H B$ for the infinite binary sequence where

$$A \oplus_H B(n) = \begin{cases} A(i) & \text{in case } n \in H \text{ and } |\{j < n : j \in H\}| = i, \\ B(i) & \text{in case } n \in \overline{H} \text{ and } |\{j < n : j \in \overline{H}\}| = i. \end{cases}$$

E.g., if $H = \{0, 3, 6, \ldots\}$, then $\mathbf{0} \oplus_H \mathbf{1}$ is equal to the sequence $011011\ldots$.

Proposition 13. *For every infinite and co-infinite c.e. set H, the real number $\mathbf{1} \oplus_H \Omega$ is a speedable left-c.e. real number*

Proof. First we fix $(\omega_i)_{i \in \omega}$ as the canonical approximation of Ω from below and $h_0, h_1, \ldots$ as an enumeration of H and define a computable sequence $(a_n)_{n \in \omega}$ in the following way:

$$a_0 := \omega_0,$$

$$a_1 := \omega_1,$$

$$a_{2i} := \mathbf{1} \oplus_{\{h_0, \ldots, h_{i-1}\}} \omega_i,$$

$$a_{2i+1} := \mathbf{1} \oplus_{\{h_0, \ldots, h_{i-1}\}} \omega_{i+1}.$$

For $\alpha := 1 \oplus_H \Omega$, we obtain $\lim_{n \to \infty} a_n = \alpha$.

To show the monotony of $(a_n)_{n \in \omega}$, we need some additional preparation steps.

Given a real number $a \in [0,1)$, define for every i a rational number $x_{a,i} := a \upharpoonright h_i$, and binary sequence $y_{a,i}$ via $y_{a,i}(j) := a(h_i + j) \; \forall j \in \omega$.

Now we can obtain the monotony of $(a_i)_{i \in \omega}$ immediately from the following observations:

for every i, $a_{2i} < a_{2i+1}$ since $\omega_i < \omega_{i+1}$, and $a_{2i+1} < a_{2i+2}$ since

$$a_{2i+1} = x_{a_{2i+1},i} + \frac{1}{2^{h_i}} \cdot 0, y_{a_{2i+1},i} < x_{a_{2i+1},i} + \frac{1}{2^{h_i+1}} + \frac{1}{2^{h_i+1}} \cdot 0, y_{a_{2i+1},i} = a_{2i+2}.$$

Thus, the sequence $(a_n)_{n \in \omega}$ is a computable left approximation of α.

Now we consider the set I of true stages of H

$$I := \{i \in \omega : \forall j > i (h_j > h_i)\}.$$

In our case, I contains the indexes of all h_i corresponding to the last changes in the 1-part of prefixes of $1 \oplus_H \Omega$.

I is infinite, since H is, thus, for every subset $I' \subseteq I$, at least one of the sets I' and $I \backslash I'$ should be infinite.

We define I' as a subset of I containing all the final stages of H which are at the same time the final stages of α

$$I' := \{i \in I : (a_{2i+2} \upharpoonright h_i) = (\alpha \upharpoonright h_i)\}.$$

Note that $(a_{2i+2} \upharpoonright h_i) = (a_{2i+1} \upharpoonright h_i)$ for every $i \in \omega$. As we know from the previous observation, at least one of sets I' and $I' \backslash I$ is infinite. It remains to consider both cases separately and obtain in each case the speedability of α.

<u>Case 1</u>: I' is infinite.

If I' is still infinite, then for every $i \in I'$ we consider the value $\frac{\alpha - a_{2i+1}}{\alpha - a_{2i+2}}$:

$a_{2i+1} = (\alpha \upharpoonright h_i) + \frac{1}{2^{h_i}} y_{a_{2i+1},i}$,

$a_{2i+2} = (\alpha \upharpoonright h_i) + \frac{1}{2^{h_i}} y_{a_{2i+2},i}$.

These two observations easily imply

$$\frac{\alpha - a_{2i+2}}{\alpha - a_{2i+1}} = \frac{0, y_{\alpha,i} - 0, y_{a_{2i+2},i}}{0, y_{\alpha,i} - 0, y_{a_{2i+1},i}}.$$

We can represent $y_{a_{2i+1},i}$ in the following way:

$$y_{a_{2i+1},i} = \underbrace{1 \ldots 1}_{k_i} 0 \underbrace{1 \ldots 1}_{l_i} \ldots$$

where the block $\underbrace{1 \ldots 1}_{l_i}$ is defined as the (possibly empty) block of ones already enumerated into the 1-part of α after the first "0" appearing in a_{2i+1} after $\alpha \upharpoonright h_i$ on the position $h_i + k_i + 1$ for $k_i \geq 0$.

Noting that "0" on the position $k_i + 1$ in $y_{a_{2i+1},i}$ is, in fact, "0" on the position $h_i + k_i + 1$ in a_{2i+1}, which belongs obviously to the Ω-part of a_{2i+1}, we can easily see that $y_{a_{2i+2},i}$ has the following form:

$$y_{a_{2i+2},i} = 1\underbrace{1\ldots1}_{k_i}\underbrace{1\ldots1}_{l_i}0\ldots$$

The appearance of the first "1" above is implied by $a_{2i+2}(h_i) = a_{2(i+1)}(h_i) = 1$. One remains to consider two possibilities for the next bit of $y_{a_{2i+1},i}$:

1. If $y_{a_{2i+1},i} = 0,\underbrace{1\ldots1}_{k_i}0\underbrace{1\ldots1}_{l_i}0\ldots$, then:

$$0, y_{a_{2i+1},i} < 0,\underbrace{1\ldots1}_{k_i}0\underbrace{1\ldots1}_{l_i}1,$$

$$0, y_{a_{2i+2},i} \geq 0,1\underbrace{1\ldots1}_{k_i}\underbrace{1\ldots1}_{l_i},$$

$$0, y_{\alpha,i} < 1.$$

Replacing left sides by the right ones from the inequalities above, we can get an upper bound for $\frac{\alpha - a_{2i+2}}{\alpha - a_{2i+1}}$:

$$\frac{0, y_{\alpha,i} - 0, y_{a_{2i+2},i}}{0, y_{\alpha,i} - 0, y_{a_{2i+1},i}} < \frac{2^{-(k_i+l_i+1)}}{2^{-(k_i+1)} + 2^{-(k_i+1+l_i+1)}} \leq \frac{2}{3}.$$

2. If $y_{a_{2i+1},i} = \underbrace{1\ldots1}_{k_i}0\underbrace{1\ldots1}_{l_i}1\ldots$, then $y_{a_{2i+2},i} = 1\underbrace{1\ldots1}_{k_i}\underbrace{1\ldots1}_{l_i}01\ldots$

The last "1" appears since $y_{a_{2i+2},i}(k_i + l_i + 1) = a_{2i+2}(h_i + 1 + k_i + l_i + 1)$ and either $a_{2i+2}(h_i + 1 + k_i + l_i + 1) = a_{2i+1}(h_i + 1 + k_i + l_i) = 1$, if it belongs to the Ω-part of a_{2i+2}, or $a_{2i+2}(h_i + 1 + k_i + l_i + 1) = 1$ directly, if it belongs to the 1-part of a_{2i+2}. Thus:

$$0, y_{a_{2i+1},i} < 0,\underbrace{1\ldots1}_{k_i}1,$$

$$0, y_{a_{2i+2},i} \geq 0,1\underbrace{1\ldots1}_{k_i}\underbrace{1\ldots1}_{l_i}01,$$

$$0, y_{\alpha,i} < 1.$$

In the similar way as in the first case, we get an upper bound for $\frac{\alpha - a_{2i+2}}{\alpha - a_{2i+1}}$:

$$\frac{0, y_{\alpha,i} - 0, y_{a_{2i+2},i}}{0, y_{\alpha,i} - 0, y_{a_{2i+1},i}} < \frac{2^{-(k_i+l_i+2)} + 2^{-(k_i+l_i+3)}}{2^{-(k_i+1)}} \leq \frac{3}{4}.$$

Thus, assuming I' to be infinite, α is at least $\frac{3}{4}$-speedable via the speed-up function $f(n) := n + 1$.

Case 2: $I \backslash I'$ is infinite.

In this case, there exists infinitely many true stages i of H such that at least one change of the Ω-part of α_{h_i+1} will happen after the stage i.

Now we can show the $\frac{1}{2}$-speedability of α via the same speed-up function as in the first case.

For every $i \in I \backslash I'$, we consider an index of the last change in the Ω-part of $\alpha \restriction (h_i + 1)$, that is, such $j > 2i + 2$ that

$$a_{j-1} \restriction (h_i + 1) \neq a_j \restriction (h_i + 1) = \Omega \restriction (h_i + 1).$$

From the definition of I', we know that j is also the index of the last change happened in a whole $\alpha \restriction (h_i + 1)$, so

$$a_{j-1} \restriction (h_i + 1) \neq a_j \restriction (h_i + 1) = \alpha \restriction (h_i + 1).$$

Due to $a_{j-1}(h_i) = a_j(h_i) = \alpha(h_i) = 1$, since $i \in I$ and $j > 2i + 2$, one obtains the following relations

$$\alpha - a_{j-1} < 2^{-(h_i+1)} \quad \text{and} \quad a_j - a_{j-1} > 2^{-(h_i+1)},$$

which obviously imply the sought-for estimation

$$\frac{\alpha - a_j}{\alpha - a_{j-1}} < \frac{1}{2}.$$

In both cases α is at least $\frac{3}{4}$-speedable, thus, due to Theorem 6, speedable. □

Note that for any H as in the proposition, it holds that $\Omega \oplus_H 1$ is a left-c.e. real number, which in the case that H is computable is not Martin-Löf random and has the same Turing degree as the halting problem, in particular, is high.

4 Solovay Degrees

A real number α is Solovay reducible to a real number β, for short $\alpha \leq_S \beta$, if there is a constant c and a partial computable function g from dyadic rational numbers to dyadic rational numbers such that for every dyadic rational number $q < \beta$, we have $g(q) < \alpha$ and $\alpha - g(q) < c(\beta - q)$. As usual, we restrict Solovay reducibility to the set of left-c.e. real numbers and consider the degree structure induced by Solovay reducibility on this set. It is known that the Martin-Löf random left-c.e. real numbers form a Solovay degree and, since Solovay reducibility implies Turing reducibility, that the set of nonhigh left-c.e. real numbers is equal to a union of Solovay degrees. By Theorems 11 and 10, all degrees just mentioned contain either only nonspeedable or only speedable left-c.e. real numbers. By the next theorem, this is no coincidence.

Theorem 14. *Speedability is a degree property with respect to Solovay reducibility in the sense that either every or no left-c.e. real number in a Solovay degree is speedable.*

Proof. Fix any pair of left-c.e. real numbers α and β such that $\alpha \equiv_S \beta$ and β is speedable. Let $a_0, a_1, ..$ and $b_0, b_1, ..$ be computable left approximations of α and β, respectively. It can be shown [2, Proposition 9.1.2] that $\beta \leq_S \alpha$ implies the existence of a constant c and a computable function g such that for all n we have

$$\beta - b_{g(n)} < c(\alpha - a_n).$$

Similarly, $\beta \leq_S \alpha$ implies the existence of a constant d and a strictly monotonic computable function h such that for all n we have

$$\alpha - b_{h(n)} < d(\beta - b_n).$$

The function g can be assumed to be strictly monotonic, hence $b_{g(0)}, b_{g(1)}, \ldots$ is a computable left-approximation of β. Since the real number β is speedable, its left approximation $b_{g(0)}, b_{g(1)}, \ldots$ is ρ-speedable for every $\rho > 0$ according to Theorem 6. As a consequence, there is a computable function f_β such that

$$\liminf_{n \to \infty} \frac{\beta - b_{f_\beta(g(n))}}{\beta - b_{g(n)}} \leq \frac{1}{2cd + 1}, \quad \text{hence} \quad \frac{\beta - b_{f_\beta(g(n))}}{\beta - b_{g(n)}} < \frac{1}{2cd} \quad (7)$$

holds for infinitely many n. Define the function f_α inductively by $f_\alpha(0) = 0$ and for all $n > 0$ by

$$f_\alpha(n) := 1 + \max\{n, f_\alpha(n-1), h(f_\beta(g(n)))\}.$$

Then for each of the infinitely many n for which (7) holds, we have

$$\alpha - a_{f_\alpha(n)} < d(\beta - b_{f_\beta(g(n))}) < \frac{1}{2c}(\beta - b_{g(n)}) < \frac{1}{2}(\alpha - a_n).$$

So, the real number α is $\frac{1}{2}$-speedable via the computable speed-up function f_α, hence is speedable. $\square$

References

1. Barmpalias, G., Lewis-Pye, A.: Differences of halting probabilities. J. Comput. Syst. Sci. **89**, 349–360 (2017)
2. Downey, R., Hirschfeldt, D.: Algorithmic Randomness and Complexity. Springer, Berlin (2010). https://doi.org/10.1007/978-0-387-68441-3
3. Nies, A.: Computability and Randomness. Oxford University Press, Oxford (2009)
4. Soare, R.I.: Computational complexity, speedable and levelable sets. J. Symb. Log. **42**, 545–563 (1977)

The Complexity of Controlling Condorcet, Fallback, and k-Veto Elections by Replacing Candidates or Voters

Marc Neveling, Jörg Rothe[(✉)], and Roman Zorn

Institut für Informatik, Heinrich-Heine-Universität Düsseldorf, Düsseldorf, Germany
{marc.neveling,rothe,roman.zorn}@hhu.de

Abstract. Electoral control models malicious ways of tampering with the outcome of elections via structural changes and has turned out to be one of the central themes in computational social choice. While the standard control types—adding/deleting/partitioning either voters or candidates—have been studied quite comprehensively, much less is known for the control actions of replacing voters or candidates. Continuing the work of Loreggia et al. [18, 19] and Erdélyi, Reger, and Yang [10], we study the computational complexity of control by replacing candidates or voters in Condorcet, fallback, and k-veto elections.

1 Introduction

Bartholdi, Tovey, and Trick [1] were the first to propose control of elections as a malicious way of tampering with their outcome via changing their structure, e.g., by adding or deleting voters or candidates. They introduced the constructive variant where the goal of an election chair is to make a favorite candidate win. Focusing on plurality and Condorcet elections, they studied the complexity of the associated control problems, showing either resistance (NP-hardness) or vulnerability (membership in P). Complementing their work, Hemaspaandra, Hemaspaandra, and Rothe [15] introduced the destructive variant of control where the chair's goal is to prevent a despised candidate's victory. Pinpointing the complexity of destructive control in plurality and Condorcet, they also studied the constructive and destructive control complexity of approval voting. Since then, plenty of voting rules have been analyzed in terms of their control complexity, as surveyed by Faliszewski and Rothe [13] and Baumeister and Rothe [3].

The computational complexity of *replacing* voters or candidates was first studied by Loreggia et al. [18, 19] and later on by Erdélyi, Reger, and Yang [10]. Replacement control models voting situations in which the number of candidates or voters are predefined and cannot be changed by the chair. For instance a parliament often consists of a fixed number of seats whose occupants must be replaced if they are removed from their seat. From another viewpoint, the chair might try to veil its election tampering by replacement control actions such that

© Springer Nature Switzerland AG 2020
H. Fernau (Ed.): CSR 2020, LNCS 12159, pp. 314–327, 2020.
https://doi.org/10.1007/978-3-030-50026-9_23

the number of participating candidates and voters is the same as before. Then, the election might seem to be unchanged at first glance.

Compared with the standard control types (adding/deleting/partitioning voters or candidates), much less is known for the control action of replacing voters or candidates. It can be seen as a combination of adding and deleting them, with the additional constraint that the same number of voters/candidates must be added as have been deleted. Other types of combining control attacks, namely *multimode control*, have been investigated by Faliszewski, Hemaspaandra, and Hemaspaandra [12]. Although some types of multimode control seem to be similar to replacement control the key difference lies in the tightly coupled control types of replacement control while in multimode control the combined types of standard election control can often times be handled separately. This leads to the interesting situation that resistances of voting rules against certain types of standard control do not transfer trivially to related types of replacement control, whereas this indeed can happen for multimode control.

Our contribution is to study the complexity of control by replacing either voters or candidates in Condorcet, fallback, and k-veto elections. The complexity of control under the *standard types* has been studied and completely settled for Condorcet voting, as pointed out above, by Bartholdi, Tovey, and Trick [1] and Hemaspaandra, Hemaspaandra, and Rothe [15]; for fallback voting by Erdélyi et al. [7–9,11]; and for veto (i.e., 1-veto) elections by Lin [16,17] (who also settled some cases of standard control in k-veto for $k \geq 2$), Chen et al. [6], and Maushagen and Rothe [20–22]. Among these rules, fallback voting (a hybrid system due to Brams and Sanver [5] that combines Bucklin with approval voting) is special in that it is one of the two natural voting rules with a polynomial-time winner problem that are currently known to have the most resistances against standard control attacks, the other one being normalized range voting [23].

In the related area of judgment aggregation, control by replacing judges has been introduced by Baumeister et al. [2] and further studied by Baumeister, Rothe, and Selker [4].

2   Preliminaries

An election is a pair (C, V) with C being a set of m candidates and V a set of n voters. Voters express their preferences over the candidates by, e.g., linear orders over C, such as $c\ b\ a\ d$ for $C = \{a, b, c, d\}$, where the leftmost candidate is the most preferred one by this voter and preference (strictly) decreases from left to right. A voting rule $\mathcal{R}$ then maps each election (C, V) to a subset $W \subseteq C$ of the candidates, called the $\mathcal{R}$ *winners* (or simply the *winners* if $\mathcal{R}$ is clear from the context) of election (C, V). For candidates $a, b \in C$, denote the number of votes in (C, V) preferring a to b by $N_{(C,V)}(a, b)$. We will study the following voting rules:

- k-veto: A candidate gains a point from each vote in which she is ranked higher than in the last k positions (i.e., the candidates in the last k positions are vetoed). The candidate(s) with the most points (i.e., with the fewest vetoes) win(s) the election.

- *Condorcet*: A Condorcet winner is a candidate a who beats all other candidates in pairwise contests, i.e., for each other candidate b, it holds that $N_{(C,V)}(a, b) > N_{(C,V)}(b, a)$. Note that a Condorcet winner does not always exist, but if there is one, he or she is unique.
- *Fallback*: In a fallback election (C, V), each voter v submits her preferences as a subset of candidates $S_v \subseteq C$ that she approves of and, in addition, a strict linear ordering of those candidates (e.g., if a voter v approves of the candidates $S_v = \{c_1, ..., c_k\}$ and orders them lexicographically, her vote would be denoted as $c_1 \cdots c_k \mid C \setminus S_v$). Let $score_{(C,V)}(c) = |\{v \in V \mid c \in S_v\}|$ be the *number of approvals of* c and $score^i_{(C,V)}(c)$ be the *number of level i approvals of* c (i.e., the number of voters who approve of c and rank c in their top i positions). The fallback winner(s) will then be determined as follows: (1) A candidate c is a *level ℓ winner* if $score^\ell_{(C,V)}(c) > |V|/2$. Letting i be the smallest integer such that there is a level i winner, the candidate(s) with the most level i approvals win(s). (2) If there is no fallback winner on any level, the candidate(s) with the most approvals win(s).

Unlike the original papers on electoral control that in particular investigated the control actions of adding and deleting either candidates or voters [1,15], we will consider control by *replacing* either candidates or voters, which combines adding and deleting them and was introduced by Loreggia et al. [18,19] and later on also studied by Erdélyi, Reger, and Yang [10].

For a given voting rule $\mathcal{R}$, define the following problems:

$\mathcal{R}$-CONSTRUCTIVE-CONTROL-BY-REPLACING-CANDIDATES

Given: An election $(C \cup D, V)$, where D with $C \cap D = \emptyset$ is a set of spoiler candidates, a distinguished candidate $c \in C$, and an integer $r \in \mathbb{N}$.

Question: Are there subsets $C' \subseteq C \setminus \{c\}$ and $D' \subseteq D$ of equal size (at most r) such that c is an $\mathcal{R}$ winner of the election $((C \setminus C') \cup D', V)$?

We are given an election $(C \cup D, V)$ in this problem, i.e., all votes in V express preferences over all the candidates in $C \cup D$. But only the candidates from C are taken into account before the control action, and some of those have then been replaced by the same number of candidates from D. In any case, we implicitly assume that missing candidates do not show up in the votes, i.e., all votes from V are restricted to those candidates actually occurring in the election at hand.

$\mathcal{R}$-CONSTRUCTIVE-CONTROL-BY-REPLACING-VOTERS

Given: An election $(C, V \cup W)$ with registered voters V, as yet unregistered voters W, a distinguished candidate $c \in C$, and an integer $r \in \mathbb{N}$.

Question: Are there subsets $V' \subseteq V$ and $W' \subseteq W$ of equal size (at most r) such that c is an $\mathcal{R}$ winner of the election $(C, (V \setminus V') \cup W')$?

In short, we denote the former problem as $\mathcal{R}$-CCRC and the latter as $\mathcal{R}$-CCRV. We will also consider the *destructive* variants of these problems, denoted by $\mathcal{R}$-DCRC and $\mathcal{R}$-DCRV, in which the goal is to prevent the distinguished candidate from being an $\mathcal{R}$ winner. We focus on the so-called *nonunique-winner model* in which we do not care if the distinguished candidate is the *only* winner as long as he or she is *a* winner (respectively, *not even a winner* in the destructive variants). By contrast, in the *unique-winner model* a control action is considered successful only if the distinguished candidate is the *unique* winner (respectively, *not a unique winner*). We note in passing that, with slight modifications, our proofs work for the *unique-winner model* as well.

We assume the reader to be familiar with the basic notions from complexity theory; in particular, with the complexity classes P and NP and the notions of NP-hardness and NP-completeness. For our proofs, we define the following well-known NP-complete problems [14]:

EXACT-COVER-BY-THREE-SETS (X3C)	
Given:	A set $B = \{b_1, b_2, ..., b_{3s}\}$ with $s \geq 1$ and a family $\mathcal{S} = \{S_1, S_2, ..., S_t\}$ of subsets $S_i \subseteq B$ with $|S_i| = 3$ for each i, $1 \leq i \leq t$.
Question:	Is there a subfamily $\mathcal{S}' \subseteq \mathcal{S}$ such that every element of B appears in exactly one subset of $\mathcal{S}'$?

HITTING-SET	
Given:	A set $B = \{b_1, b_2, ..., b_s\}$ with $s \geq 1$, a family $\mathcal{S} = \{S_1, S_2, ..., S_t\}$ of subsets $S_i \subseteq B$, and an integer q with $1 \leq q \leq s$.
Question:	Is there a subset $B' \subseteq B$, $|B'| \leq q$, such that each $S_i \in \mathcal{S}$ is *hit* by B' (i.e., $S_i \cap B' \neq \emptyset$ for all $S_i \in \mathcal{S}$)?

We call a voting rule *immune* to a type of control if it is never possible for the chair to reach her goal by this control action; otherwise, the voting rule is said to be *susceptible* to this control type. A susceptible voting rule is said to be *vulnerable* to this control type if the associated control problem is in P, and it is said to be *resistant* to it if the associated control problem is NP-hard. Note that all considered control problems are in NP, so resistance implies NP-completeness.

3   Overview of Results

Table 1 gives an overview of our complexity results for control by replacing candidates or voters in Condorcet, fallback, and k-veto. Let us compare them with the results for control by adding/deleting candidates or voters in the same voting systems. For fallback voting (which is vulnerable only to destructive control by replacing, adding, and deleting voters and is resistant in all other cases), we have the same results for replacement control as Erdélyi et al. [7] obtained for control by adding/deleting candidates or voters. Also, for k-veto, the results for replacement control are very similar to those by Lin [16,17] for control by adding/deleting candidates or voters.[1] For Condorcet voting, however, while we again have the same results for replacing voters as Bartholdi et al. [1] and Hemaspaandra et al. [15] obtained for control by adding/deleting voters, our results differ from theirs for candidate control: We show vulnerability for both the constructive and destructive cases of replacement control, whereas Bartholdi et al. [1] show that Condorcet voting is vulnerable to constructive control by deleting candidates and Hemaspaandra et al. [15] show it is vulnerable to destructive control by adding candidates, yet they show immunity for constructive control by adding candidates [1] and destructive control by deleting candidates [15].

Table 1. Overview of complexity results. "R" stands for *resistant* and "V" for *vulnerable*. Results marked by "†" are due to Erdélyi et al. [10] and "‡" means that the case of $k = 1$ is due to Loreggia et al. [19] while we show resistance for the cases $k > 1$.

Problem	Condorcet	Fallback	k-Veto
CCRV	R	R	V $(k \leq 2)^\dagger / R(k \geq 3)^\dagger$
DCRV	V	V	V $(k \geq 1)^\dagger$
CCRC	V	R	R $(k \geq 1)^\ddagger$
DCRC	V	R	R $(k \geq 1)^\ddagger$

4 Condorcet Voting

We will start with Condorcet and show that it is vulnerable to three types of control, yet resistant to the fourth one, starting with the resistance proof.

Theorem 1. *Condorcet is resistant to constructive control by replacing voters.*

Proof. We prove NP-hardness by reducing X3C to Condorcet-CCRV. A similar reduction was used by Bartholdi, Tovey, and Trick [1] to prove that Condorcet-CCAV (where CCAV stands for "constructive control by adding voters") is NP-hard.

Let $(B, \mathcal{S})$ be an X3C instance with $B = \{b_1, \ldots, b_{3s}\}$, $s \geq 2$ (which may be assumed, as X3C is trivially solvable when $s = 1$), and $\mathcal{S} = \{S_1, \ldots, S_t\}$, $t \geq 1$. The set of candidates is $C = B \cup \{c\}$ with c being the distinguished candidate. The list V of votes is constructed as follows:

- There are $2s - 3$ registered votes of the form $b_1 \cdots b_{3s} \ c$ in V and
- for each j, $1 \leq j \leq t$, there is one unregistered vote of the form $S_j \ c \ B \setminus S_j$ in W.

The ordering of candidates in S_j and $B \setminus S_j$ does not matter in any of those votes. Finally, set $r = s$.

---

[1] One minor difference is that while Erdélyi et al. [10] show that k-veto is resistant to constructive control by replacing voters for $k \geq 3$ (with vulnerability holding for $k \leq 2$), Lin [16,17] shows that it is resistant to constructive control by deleting voters for $k \geq 4$ (with vulnerability holding for $k \leq 3$).

Analyzing the election (C, V), b_1 is the Condorcet winner; in particular, c loses against every $b_i \in B$ with a deficit of $2s - 3$ votes, i.e.,

$$N_{(C,V)}(b_i, c) - N_{(C,V)}(c, b_i) = 2s - 3.$$

We will now show that $(B, \mathcal{S})$ is a yes-instance of X3C if and only if c can be made the Condorcet winner of the election by replacing s votes from V with votes from W.

From left to right, assume there is an exact cover $\mathcal{S}' \subseteq \mathcal{S}$ of B. We remove s votes of the form $b_1 \cdots b_{3s} \ c$ from the election and replace them by the votes of the form $S_j \ c \ B \setminus S_j$ for all $S_j \in \mathcal{S}'$. Let (C, V') be the resulting election. Since $\mathcal{S}'$ is an exact cover of B, for each $b_i \in B$,

$$N_{(C,V')}(b_i, c) - N_{(C,V')}(c, b_i) = (2s - 3 - s + 1) - (s - 1) = -1 < 0.$$

Thus c now defeats each $b_i \in B$ in pairwise comparison and, therefore, has been made the Condorcet winner of (C, V').

From right to left, assume that c can be made a Condorcet winner of the election by replacing at most s votes. Recall that c has a deficit of

$$N_{(C,V)}(b_i, c) - N_{(C,V)}(c, b_i) = 2s - 3$$

to every $b_i \in B$ in the original election. Thus *exactly* s votes need to be removed from the election, for otherwise c's deficit of at least $s - 2$ to every other candidate cannot be caught up on, since at least one other candidate is in front of c in every unregistered vote. With s removed votes, c's deficit to every other candidate is now decreased to $s - 3$. However, none of the s votes from W replacing the removed votes can rank some $b_i \in B$ in front of c more than once, as otherwise we would have

$$N_{(C,V')}(b_i, c) \geq s - 1 \quad \text{and} \quad N_{(C,V')}(c, b_i) \leq s - 2$$

for at least one $b_i \in B$ in the resulting election (C, V'), and c would not win. Let $\mathcal{S}' \subseteq \mathcal{S}$ be the set such that each $S_j \in \mathcal{S}'$ corresponds to the vote $S_j \ c \ B \setminus S_j$ from W that is added to the election to replace a removed vote. Every unregistered voter ranks three candidates of B in front of c. By the pigeonhole principle, in order for the s new votes to rank each of the $3s$ candidates of B in front of c only once, $\mathcal{S}'$ needs to be an exact cover of B. $\qed$

By contrast, we show vulnerability to destructive control by replacing voters for Condorcet via a simple algorithm.

Theorem 2. *Condorcet is vulnerable to destructive control by replacing voters.*

Proof. To prove membership in P, we will provide an algorithm that solves the problem in polynomial time and outputs, if possible, which of the registered voters must be replaced by which unregistered voters for c to not win.

The input to our algorithm is an election $(C, V \cup W)$, the distinguished candidate $c \in C$, and an integer r. The algorithm will output either a pair

(V', W') with $V' \subseteq V$, $W' \subseteq W$ and $|V'| = |W'| \leq r$ (i.e., in V' are voters that must be removed and in W' are voters that must be added to the election instead for c to not win), or that control is impossible.

First, the algorithm checks whether c is already not winning the election (C, V) and outputs $(\emptyset, \emptyset)$ if this is the case, and we are done.

Otherwise, c currently wins, and the algorithm iterates over all candidates $d \in C \setminus \{c\}$ and first checks whether $N_{(C,V)}(c, d) - N_{(C,V)}(d, c) + 1 \leq 2r$ (if this is not the case, d loses to c in any case and we can skip this candidate.)

Let $V' \subseteq V$ contain at most r votes from V preferring c to d and let $W' \subseteq W$ contain at most r votes from W preferring d to c. If one of them is smaller than the other, remove votes from the larger one until they are equal in size.

Then we check whether

$$N_E(C, (V \cup W') \setminus V')(c, d) \leq N_E(d, c)$$

in the election $E = (C, (V \cup W') \setminus V')$. If this is the case, c does not beat d in direct comparison, so c cannot win the election. The algorithm then outputs (V', W').

Otherwise, d cannot beat c and the algorithm proceeds to the next candidate. If, after all iterations, no candidate was found that beats or ties c, the algorithm outputs "control impossible." Obviously, this algorithm runs in polynomial-time and solves the problem.   □

Bartholdi, Tovey, and Trick [1] observed that, due to the Weak Axiom of Revealed Preference, Condorcet voting is immune to constructive control by adding candidates, and Hemaspaandra, Hemaspaandra, and Rothe [15] made the same observation regarding destructive control by deleting candidates. For control by *replacing* candidates, however, Condorcet is susceptible both in the constructive and in the destructive case.

In the constructive case, for instance, if $C = \{b, c\}$ and there is one spoiler candidate in $D = \{d\}$ and only one vote $b\ c\ d$ over $C \cup D$, we can turn c (who does not win according to $b\ c$) into a Condorcet winner by replacing b with d (so we now have $c\ d$).

For susceptibility in the destructive case, just consider $C' = \{c, d\}$ and $D' = \{b\}$, and replace d with b, all else being equal.

Moreover, since in Condorcet elections the direct comparison between two candidates cannot be influenced by deleting or adding other candidates to the election, Condorcet-CCRC and Condorcet-DCRC are both easy to solve.

Theorem 3. *Condorcet is vulnerable to constructive control by replacing candidates.*

Proof. To prove membership in P, we will provide an algorithm that solves the problem in polynomial time and outputs, if possible, which of the original candidates must be replaced by which spoiler candidates for c to win.

The input to our algorithm is an election $(C \cup D, V)$, the distinguished candidate $c \in C$, and a positive integer r. The algorithm will output either a pair

(C', D') with $C' \subseteq C \setminus \{c\}$, $D' \subseteq D$ and $|C'| = |D'| \le r$ (i.e., in C' are candidates that must be removed and in D' are candidates that must be added to the election for c to win), or that control is impossible.

First, we check whether c already wins the election (C, V) and output $(\emptyset, \emptyset)$ if this is the case, and we are done.

Otherwise, let $C' \subseteq C \setminus \{c\}$ be the set of candidates from $C \setminus \{c\}$ that beat or tie c in direct comparison and let $D' \subseteq D$ be a set of at most $|C'|$ candidates from D that c beats in direct comparison.

If $|C'| \le r$ and $|C'| = |D'|$, we output (C', D'), and otherwise we output "control impossible."

Obviously, the algorithm solves the problem and runs in polynomial time. $\square$

Theorem 4. *Condorcet is vulnerable to destructive control by replacing candidates.*

Proof. An algorithm that solves the problem works as follows: Given an election $(C \cup D, V)$, a distinguished candidate $c \in C$, and an integer r, it checks whether c is not winning the election (C, V) and outputs $(\emptyset, \emptyset)$ if this is the case.

Otherwise, it checks whether there is a candidate $d \in D$ who beats or ties c in direct comparison, whether there is another candidate $b \in C$ with $b \ne c$ and whether $r \ge 1$. If these conditions are satisfied, it outputs $(\{b\}, \{d\})$, and otherwise "control impossible".

This algorithm outputs either a successful pair (C', D') with $C' \subseteq C \setminus \{c\}$, $D' \in D$, and $|C'| = |D'| \le r$ if c can be prevented from winning by replacing at most r candidates, or else "control impossible." Obviously, the algorithm is correct and runs in polynomial time. $\square$

5   Fallback Voting

We will now consider fallback voting and show that it is vulnerable to one type of control and resistant to the others.

Theorem 5. *Fallback is resistant to constructive control by replacing voters.*

Proof. To prove NP-hardness, we will modify the reduction from X3C that Erdélyi and Rothe [11] (and Erdélyi et al. [7]) used to show NP-hardness of fallback-CCAV.

Let (B, S) be an X3C instance with $B = \{b_1, \ldots, b_{3s}\}$, $s \ge 2$, and $S = \{S_1, \ldots, S_t\}$, $t \ge 1$. The set of candidates is $C = B \cup D \cup \{c\}$ with c being the distinguished candidate and $D = \{d_1, \ldots, d_{t(3s-4)}\}$ a set of $t(3s - 4)$ dummy candidates. In V (corresponding to the registered voters), there are the $3s - 1$ votes:

- $2s - 1$ votes of the form $B \mid D \cup \{c\}$ and
- for each i, $1 \le i \le s$, one vote $d_i \mid B \cup (D \setminus \{d_i\}) \cup \{c\}$.

In W (corresponding to the unregistered voters), there are the following t votes:

– For each j, $1 \leq j \leq t$, let

$$D_j = \{d_{(j-1)(3s-4)+1}, \ldots, d_{j(3s-4)}\}$$

and include in W the vote

$$D_j \; S_j \; c \mid (B \setminus S_j) \cup (D \setminus D_j).$$

Finally, set $r = s$.

Having no approvals in (C, V), c does not win. We will show that $(B, \mathcal{S})$ is a yes-instance of X3C if and only if c can be made a fallback winner of the constructed election by replacing at most s votes from V with as many votes from W.

From left to right, suppose there is an exact cover $\mathcal{S}' \subseteq \mathcal{S}$ of B. Remove s votes $B \mid D \cup \{c\}$ from the election and add, for each $S_j \in \mathcal{S}'$, the vote $D_j \; S_j \; c \mid (B \setminus S_j) \cup (D \setminus D_j)$ instead. Let $(C, \widehat{V})$ be the resulting election. It follows that

– $score_{(C,\widehat{V})}(d_i) \leq 2$ for every $d_i \in D$,
– $score_{(C,\widehat{V})}(b_i) = s$ for every $b_i \in B$ ($s - 1$ approvals from the remaining registered voters and one approval from the added voters since $\mathcal{S}'$ is an exact cover of B), and
– $score_{(C,\widehat{V})}(c) = s$.

Thus no candidate has a majority on any level and c is one of the winners since she ties all candidates of B for the most approvals overall.

From right to left, suppose c can be made a fallback winner of the election by replacing at most s votes from V with as many votes from W. Since c has no approvals in (C, V) and we can only add at most s approvals for c, the only chance for c to win is to have the most approvals in the last stage of the election. Regardless of which votes we remove or add to the election, every dummy candidate can have at most two approvals, which will at least be tied by c if we add $s \geq 2$ unregistered votes to the election. We need to remove s votes $B \mid D \cup \{c\}$ from the election; otherwise, some $b_i \in B$ would have at least s approvals, whereas c could gain no more than $s - 1$ approvals from adding unregistered votes. Each $b_i \in B$ receives $s - 1$ approvals from the remaining registered votes of the original election and c reveices s approvals from the added votes. Additionally, every added voter approves of three candidates from B. Hence, in order for c to at least tie every candidate from B, each $b_i \in B$ can only be approved by at most one of the added votes. Since there are s added votes, there must be an exact cover of B. $\square$

By contrast, we establish vulnerability of the destructive case of control by replacing voters for fallback voting. The proof employs a rather involved polynomial-time algorithm solving this problem and is omitted here due to space limitations.

Theorem 6. *Fallback is vulnerable to destructive control by replacing voters.*

Turning to control by replacing candidates, fallback is resistant in both the constructive and the destructive case.

Theorem 7. *Fallback is resistant to constructive and destructive control by replacing candidates.*

Proof. Erdélyi and Rothe [11] (see also the subsequent journal version by Erdélyi et al. [7]) showed that fallback is resistant to constructive and destructive control by deleting candidates. In the former problem (denoted by fallback-CCDC), we are given a fallback election (C, V), a distinguished candidate $c \in C$, and an integer r, and we ask whether c can be made a fallback winner by deleting at most r votes. In the destructive variant (denoted by fallback-DCDC), for the same input we ask whether we can prevent c from winning by deleting at most r votes. To prove the theorem, we will reduce

- fallback-CCDC to fallback-CCRC and
- fallback-DCDC to fallback-DCRC, respectively.

Let $((C, V), c, r)$ be an instance of fallback-CCDC (or fallback-DCDC). We construct from (C, V) a fallback election $(C \cup D, V')$ with (dummy) spoiler candidates $D = \{d_1, \ldots, d_r\}$, $D \cap C = \emptyset$, where we extend the votes of V to the set of candidates $C \cup D$ by letting all voters disapprove of all candidates in D, thus obtaining V'. Our distinguished candidate remains c, and r remains the limit on the number of candidates that may be replaced.

Since all candidates from D are irrelevant to the election and can be added to the election without changing the winner(s), it is clear that c can be made a fallback winner of (C, V) by deleting up to r candidates from C if and only if c can be made a fallback winner of $(C \cup D, V')$ by deleting up to r candidates from C and adding the same number of dummy spoiler candidates from D. This gives the desired reduction in both the constructive and the destructive case. $\square$

6 k-Veto

Erdélyi, Reger, and Yang [10] solved the two cases of control by replacing voters for k-veto (recall Table 1 in Sect. 3), while Loreggia et al. [19] solved the two cases of control by replacing candidates for veto only (i.e., for k-veto with $k = 1$). We solve these cases for k-veto with $k \geq 2$.

Theorem 8. *For $k \geq 2$, k-veto is resistant to constructive control by replacing candidates.*

Proof. To prove NP-hardness of k-veto-CCRC for $k \geq 2$, we will modify the reduction provided by Lin [16] to prove that k-veto-CCAC and k-veto-CCDC are NP-hard. Since his reduction was designed so as to prove both cases at once but we only need the "adding candidates" part, we will simplify the reduction.

Let $(B, \mathcal{S}, q)$ be an instance of HITTING-SET with $B = \{b_1, \ldots, b_s\}$, $s \geq 1$, $\mathcal{S} = \{S_1, \ldots, S_t\}$, $t \geq 1$, and integer q, $1 \leq q < s$ (without loss of generality, we may assume that $q < s$ since $(B, \mathcal{S}, q)$ is trivially a yes-instance if $q \geq s$).

We construct an instance $((C \cup B, V), c, q)$ of k-veto-CCRC with candidates $C = \{c, d\} \cup C' \cup X \cup Y$, where

$$C' = \{c'_1, \ldots, c'_{k-1}\},$$
$$X = \{x_1, \ldots, x_{k-1}\}, \text{ and}$$
$$Y = \{y_1, \ldots, y_q\},$$

and spoiler candidates B. Let V contain the following votes:

- $(t + 2s)(s - q + 1)$ votes $Y \cdots c\, C'$;
- $(t + 2s)(s - q + 1) - s + q$ votes $Y \cdots d\, X$;
- for each i, $1 \le i \le t$, one vote $Y \cdots c\, X\, S_i$;
- for each i, $1 \le i \le s$, one vote $Y \cdots d\, X\, b_i$; and
- for each i, $1 \le i \le s$, $(t + 2s)(s - q + 1) + q$ votes $Y \cdots c\, B \setminus \{b_i\}\, X\, b_i$.

Let $M = (t + 2s)(s - q + 1)$. Without the spoiler candidates, vetoes are assigned to the other candidates as follows:

c	d	$c' \in C'$	$y \in Y$	$x \in X$
$M(s + 1) + sq + t$	$M + q$	M	0	$M(s + 1) + q(s + 1) + t$

We show that $(B, \mathcal{S}, q)$ is a yes-instance of HITTING-SET if and only if c can be made a k-veto winner of the election by replacing q candidates from C with candidates from B.

From left to right, assume there is a hitting set $B' \subseteq B$ of $\mathcal{S}$ of size q (since $q < s$, if B' is a hitting set of size less than q, we fill B' up by adding arbitrary candidates from $B \setminus B'$ to B' until $|B'| = q$). We then replace the candidates from Y with the candidates from B'. Since c, d, and candidates from C' have $(t + 2s)(s - q + 1)$ vetoes and candidates from X and B' have at least $(t + 2s)(s - q + 1) + q$ vetoes, c is a k-veto winner.

From right to left, assume c can be made a k-veto winner of the election by replacing q candidates. Since the q candidates from Y have zero vetoes but c has at least one veto, we need to remove each candidate of Y (and no other candidate), and in turn we need to add q candidates from B. Note that c cannot have more than $(t + 2s)(s - q + 1)$ vetoes, for otherwise c would lose to the candidates from C'. Let $B' \subseteq B$ be the set of q candidates from B that are added to the election. Since $|B'| = q > 0$, c will lose all $s((t + 2s)(s - q + 1) + q)$ vetoes from the last group of voters. Furthermore, in order to tie the candidates in C', c cannot gain any vetoes from the third group of voters. Thus the q added candidates from B need to be a hitting set of $\mathcal{S}$. Also note that with the q added candidates from B, c also ties d (who lost q vetoes from the fourth group of voters) and beats the candidates from X and the added candidates from B. $\square$

The same result can be shown for destructive control by replacing candidates in k-veto elections via a similar proof, which is again omitted here due to space limitations.

Theorem 9. *For $k \geq 2$, k-veto is resistant to destructive control by replacing candidates.*

7   Conclusions and Open Problems

We have extended to Condorcet, fallback, and k-veto elections the study of control by replacing voters or candidates initiated by Loreggia et al. [18,19] and pursued later on by Erdélyi, Reger, and Yang [10]. Our complexity results for the associated control problems are summarized in Table 1.

We can observe that our results follow the results for the standard control types, i.e., if one of these voting rules is vulnerable to control by adding or to control by deleting voters or candidates then this voting rule is vulnerable to control by replacing candidates or voters as well (respectively, if it is resistant to a type of standard control then it is also resistant to the corresponding type of replacement control). As Loreggia et al. [18,19] have shown, this is not necessarily the case and there even exist voting rules (albeit artificial ones) that are resistant (respectively, vulnerable) to a type of replacement control, yet vulnerable (respectively, resistant) to both of the corresponding types of standard control. We therefore propose to continue the study of electoral control by replacing voters or candidates for other natural voting rules and it would be especially interesting to find a natural voting rule for which the complexity of the standard controls types—in particular, control by adding or deleting voters or candidates—differs from the complexity of control by replacing them.

Another interesting and suprisingly still open problem is the complexity of CCRV for 2-approval (in which the voters assign one point each to the two top-ranked candidates in their preferences). Pinpointing the complexity of this problem would complete the dichotomy of k-approval-CCRV with regards to k, since 1-approval-CCRV (i.e., plurality-CCRV) is polynomial-time solvable and k-approval-CCRV is NP-hard for $k \geq 3$, as shown by Loreggia et al. [19].

Admittedly, resistance in terms of NP-hardness—being a *worst-case* measure of complexity only—may not be the last word in wisdom. Indeed, Walsh [25,26] and Rothe and Schend [24] address this issue in electoral control and other manipulative attacks and survey approaches of how to circumvent it. As an ambitious long-term goal, we therefore propose to complement our worst-case complexity analysis by a typical-case analysis of the problems considered here.

Acknowledgments. We thank the anonymous reviewers for their helpful comments. This work was supported in part by DFG grants RO-1202/14-2 and RO-1202/21-1.

References

1. Bartholdi III, J., Tovey, C., Trick, M.: How hard is it to control an election? Math. Comput. Modell. **16**(8/9), 27–40 (1992)

2. Baumeister, D., Erdélyi, G., Erdélyi, O., Rothe, J.: Control in judgment aggregation. In: Proceedings of the 6th European Starting AI Researcher Symposium, pp. 23–34. IOS Press, August 2012

3. Baumeister, D., Rothe, J.: Preference aggregation by voting. In: Rothe, J. (ed.) Economics and Computation. STBE, pp. 197–325. Springer, Heidelberg (2016). https://doi.org/10.1007/978-3-662-47904-9_4

4. Baumeister, D., Rothe, J., Selker, A.-K.: Complexity of bribery and control for uniform premise-based quota rules under various preference types. In: Walsh, T. (ed.) ADT 2015. LNCS (LNAI), vol. 9346, pp. 432–448. Springer, Cham (2015). https://doi.org/10.1007/978-3-319-23114-3_26

5. Brams, S., Sanver, R.: Voting systems that combine approval and preference. In: Brams, S., Gehrlein, W., Roberts, F. (eds.) The Mathematics of Preference, Choice, and Order: Essays in Honor of Peter C. Fishburn. Studies in Choice and Welfare, pp. 215–237. Springer, Heidelberg (2009). https://doi.org/10.1007/978-3-540-79128-7_12

6. Chen, J., Faliszewski, P., Niedermeier, R., Talmon, N.: Elections with few voters: candidate control can be easy. In: Proceedings of the 29th AAAI Conference on Artificial Intelligence, pp. 2045–2051. AAAI Press (Jan 2015)

7. Erdélyi, G., Fellows, M., Rothe, J., Schend, L.: Control complexity in Bucklin and fallback voting: a theoretical analysis. J. Comput. Syst. Sci. $81(4)$, 632–660 (2015)

8. Erdélyi, G., Fellows, M., Rothe, J., Schend, L.: Control complexity in Bucklin and fallback voting: an experimental analysis. J. Comput. Syst. Sci. $81(4)$, 661–670 (2015)

9. Erdélyi, G., Piras, L., Rothe, J.: The complexity of voter partition in Bucklin and fallback voting: solving three open problems. In: Proceedings of the 10th International Conference on Autonomous Agents and Multiagent Systems, IFAAMAS, pp. 837–844, May 2011

10. Erdélyi, G., Reger, C., Yang, Y.: Towards completing the puzzle: solving open problems for control in elections. In: Proceedings of the 18th International Conference on Autonomous Agents and Multiagent Systems, IFAAMAS, pp. 846–854, May 2019

11. Erdélyi, G., Rothe, J.: Control complexity in fallback voting. In: Proceedings of Computing: the 16th Australasian Theory Symposium, Australian Computer Society Conferences in Research and Practice in Information Technology Series, vol. 32, no. 8, pp. 39–48, January 2010

12. Faliszewski, P., Hemaspaandra, E., Hemaspaandra, L.: Multimode control attacks on elections. J. Artif. Intell. Res. 40, 305–351 (2011)

13. Faliszewski, P., Rothe, J.: Control and bribery in voting. In: Brandt, F., Conitzer, V., Endriss, U., Lang, J., Procaccia, A. (eds.) Handbook of Computational Social Choice, chap. 7, pp. 146–168. Cambridge University Press (2016)

14. Garey, M., Johnson, D.: Computers and Intractability: A Guide to the Theory of NP-Completeness. W. H Freeman and Company, New York (1979)

15. Hemaspaandra, E., Hemaspaandra, L., Rothe, J.: Anyone but him: the complexity of precluding an alternative. Artif. Intell. $171(5–6)$, 255–285 (2007)

16. Lin, A.: The complexity of manipulating k-approval elections. In: Proceedings of the 3rd International Conference on Agents and Artificial Intelligence, pp. 212–218. SciTePress, January 2011

17. Lin, A.: Solving hard problems in election systems. Ph.D. thesis, Rochester Institute of Technology, Rochester, NY, USA, March 2012

18. Loreggia, A.: Iterative voting and multi-mode control in preference aggregation. Intelligenza Artificiale $8(1)$, 39–51 (2014)

19. Loreggia, A., Narodytska, N., Rossi, F., Venable, B., Walsh, T.: Controlling elections by replacing candidates or votes (extended abstract). In: Proceedings of the 14th International Conference on Autonomous Agents and Multiagent Systems, IFAAMAS, pp. 1737–1738, May 2015

20. Maushagen, C., Rothe, J.: Complexity of control by partitioning veto and maximin elections and of control by adding candidates to plurality elections. In: Proceedings of the 22nd European Conference on Artificial Intelligence, pp. 277–285. IOS Press, August/September 2016

21. Maushagen, C., Rothe, J.: Complexity of control by partition of voters and of voter groups in veto and other scoring protocols. In: Proceedings of the 16th International Conference on Autonomous Agents and Multiagent Systems, IFAAMAS, pp. 615–623, May 2017

22. Maushagen, C., Rothe, J.: Complexity of control by partitioning veto elections and of control by adding candidates to plurality elections. Ann. Math. Artif. Intell. **82**(4), 219–244 (2017). https://doi.org/10.1007/s10472-017-9565-7

23. Menton, C.: Normalized range voting broadly resists control. Theory Comput. Syst. **53**(4), 507–531 (2013). https://doi.org/10.1007/s00224-012-9441-0

24. Rothe, J., Schend, L.: Challenges to complexity shields that are supposed to protect elections against manipulation and control: a survey. Ann. Math. Artif. Intell. **68**(1–3), 161–193 (2013). https://doi.org/10.1007/s10472-013-9359-5

25. Walsh, T.: Is computational complexity a barrier to manipulation? Ann. Math. Artif. Intell. **62**(1–2), 7–26 (2011). https://doi.org/10.1007/s10472-011-9255-9

26. Walsh, T.: Where are the hard manipulation problems? J. Artif. Intell. Res. **42**, 1–29 (2011)

On the Transformation of LL(k)-linear Grammars to LL(1)-linear

Alexander Okhotin🄳 and Ilya Olkhovsky[(✉)]

St. Petersburg State University, 7/9 Universitetskaya nab.,
Saint Petersburg 199034, Russia
alexander.okhotin@spbu.ru, ilianolhin@gmail.com

Abstract. It is proved that every LL(k)-linear grammar can be transformed to an equivalent LL(1)-linear grammar. The transformation incurs a blow-up in the number of nonterminal symbols by a factor of $m^{2k-O(1)}$, where m is the size of the alphabet. A close lower bound is established: for certain LL(k)-linear grammars with n nonterminal symbols, every equivalent LL(1)-linear grammar must have at least $n \cdot (m-1)^{2k-O(\log k)}$ nonterminal symbols.

1 Introduction

The LL(k) parsing is one of the most well-known linear-time parsing techniques. In this method, a parse tree of an input string is constructed top-down, along with reading the string from left to right. A parser selects each rule by looking ahead by at most k symbols. The family of *LL(k) grammars*, to which this algorithm is applicable, was introduced and systematically studied in the papers by Knuth [5], Lewis and Stearns [7] and Rozenkrantz and Stearns [10]. In particular, Kurki-Suonio [6] and, independently, Rozenkrantz and Stearns [10] proved that LL($k+1$) grammars are more powerful than LL(k) grammars, and thus there is a strict hierarchy of languages defined by LL(k) grammars, for different k.

An important subclass of LL(k) grammars, the *LL(k)-linear grammars*, was first studied by Ibarra et al. [3] and by Holzer and Lange [2], who proved that all languages defined by these grammars belong to the complexity class NC^1. Learning algorithms for LL(1)-linear grammars and related subclasses were studied by de la Higuera and Oncina [1], and language-theoretic properties of these grammars have recently been investigated by Jirásková and Klíma [4]. LL(k)-linear grammars are weaker in power than both LL(k) non-linear grammars and linear non-LL grammars, since the language $\{\, a^n b^n c \mid n \geqslant 0 \,\} \cdot \{a, b\}$ can be defined by a linear grammar and by an LL(1) grammar, but not by any LL(k)-linear grammar [8].

Whether LL(k)-linear grammars form a hierarchy with respect to the length of the look-ahead k, remains unexplored. The first contribution of this paper is a proof that every language defined by an LL(k)-linear grammar, for some

Research supported by RFBR grant 18-31-00118.

H. Fernau (Ed.): CSR 2020, LNCS 12159, pp. 328–340, 2020.
https://doi.org/10.1007/978-3-030-50026-9_24

k, is defined by an LL(1)-linear grammar; therefore, in the case of LL(k)-linear grammars, the hierarchy with respect to k collapses. The proof is constructive: it is shown how to transform any given LL(k)-linear grammar to an LL(1)-linear grammar that defines the same language.

Next, it is shown that the proposed tranformation is close to being optimal in terms of the number of nonterminal symbols. The transformation of an LL(k)-linear grammar to an LL(1)-linear grammar increases the number of nonterminal symbols by a factor of $m^{2k-O(1)}$, where m is the size of the alphabet. A lower bound of $(m-1)^{2k-O(\log k)}$ on this factor is established.

2   Definitions

Definition 1. *A (formal) grammar is a quadruple $G = (\Sigma, N, R, S)$, where Σ is the* alphabet *of the language being defined, N is the set of* nonterminal symbols *(that is, syntactic categories defined in the grammar), R is a finite set of* rules, *each of the form $A \to \alpha$, with $A \in N$ and $\alpha \in (\Sigma \cup N)^*$, and $S \in N$ is a nonterminal symbol representing all well-formed sentences in the language, known as the* initial symbol.

Each rule $A \to X_1 \ldots X_\ell$ in R states that every string representable as a concatenation of ℓ substrings of the form $X_1, \ldots, X_\ell$, has the property A. This is formalized as follows.

Definition 2. *Let $G = (\Sigma, N, R, S)$ be a grammar. A* parse tree *is a rooted tree with leaves labelled with symbols from Σ, and with internal nodes labelled with nonterminal symbols from N. For each node labelled with $A \in N$, with its successors labelled with $X_1, \ldots, X_\ell$, there must be a rule $A \to X_1 \ldots X_\ell$ in the grammar. All successors are ordered, and if w is the string of symbols in the leaves, and A is the nonterminal symbol in the root, this is said to be a* parse tree *of w from A.*

The language defined by a nonterminal symbol A, denoted by $L_G(A)$, is the set of all strings $w \in \Sigma^*$, for which there exists a parse tree from A. The language defined by the grammar is $L(G) = L_G(S)$.

A grammar is called *linear*, if each rule in R is of the form $A \to uBv$, with $u, v \in \Sigma^*$ and $B \in N$, or of the form $A \to w$, with $w \in \Sigma^*$. A parse tree for a linear grammar consists of a path labelled with nonterminal symbols, with each rule $A \to uBv$ spawning off the leaves u to the left and v to the right.

A *top-down parser* attempts to construct a parse tree of an input string, while reading it from left to right. At every point of its computation, the parser's memory configuration is a pair $(X_1 \ldots X_\ell, v)$, with $\ell \geqslant 0$ and $X_1, \ldots, X_\ell \in \Sigma \cup N$, where v is the unread portion of the input string uv. The parser tries to parse v as a concatenation $X_1 \cdot \ldots \cdot X_\ell$. This sequence of symbols is stored in a stack, with X_1 as the top of the stack.

The initial configuration is (S, w), where w is the entire input string. At each point of the computation, the parser sees the top symbol of the stack and the

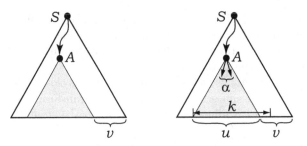

Fig. 1. (left) v follows A; (right) $T_k(A, First_k(uv)) = A \to \alpha$.

first k symbols of the unread input—the *look-ahead string*—where $k \geqslant 1$ is a constant. If there is a nonterminal symbol $A \in N$ at the top of the stack, the parser determines a rule $A \to \alpha$ for this symbol, pops this symbol, and pushes the right-hand side of the rule onto the stack.

$$(A\beta, v) \xrightarrow{A \to \alpha} (\alpha\beta, v)$$

If the top symbol of the stack is a symbol $a \in \Sigma$, the parser checks that the unread portion of the input begins with the same symbol, and then pops this symbol from the stack and reads it from the input.

$$(a\beta, av) \xrightarrow{\text{READ } a} (\beta, v)$$

The parser accepts in a configuration $(\varepsilon, \varepsilon)$, which indicates that the whole input string has been parsed as S.

It remains to explain how the parser chooses a rule to apply to a nonterminal symbol A. Denote by $\Sigma^{\leqslant k}$ the set of strings of length at most k. Let $x \in \Sigma^{\leqslant k}$ be the first k unread input symbols; if this string is shorter than k, this indicates the end of the input approaching. The rule is chosen by accessing a look-up table $T_k \colon N \times \Sigma^{\leqslant k} \to R \cup \{-\}$, which contains either a rule to apply, or a marker indicating a syntax error.

For a string $w \in \Sigma^*$, denote its first k symbols, by $First_k(w)$; if the length of w is less than k, then $First_k(w) = w$. This definition is extended to languages as $First_k(L) = \{ First_k(w) \mid w \in L \}$.

Definition 3. *Let $G = (\Sigma, N, R, S)$ be a grammar. A string $v \in \Sigma^*$ is said to follow $A \in N$, if there exists a parse tree containing a subtree with a root A, so that the leaves to the right of this subtree form the string v, as in Fig. 1 (left). Denote:*

$$\text{Follow}(A) = \{ v \mid v \text{ follows } A \}$$
$$\text{Follow}_k(A) = First_k(\{ v \mid v \text{ follows } A \})$$

The LL(k) parsing algorithm requires the grammar to satisfy the condition in the following definition.

Definition 4. *Let $k \geqslant 1$ and let $G = (\Sigma, N, R, S)$ be a grammar. An LL(k) table for G is a partial function $T_k \colon N \times \Sigma^{\leqslant k} \to R$ that satisfies the following condition, illusrated in Fig. 1 (right): $T_k(A, x) = A \to \alpha$ if and only if there exists a parse tree containing a subtree with a root A and with leaves u, with further leaves v to the right of this subtree, where $\mathrm{First}_k(uv) = x$ and the rule applied to A is $A \to \alpha$.*

If such a function exists, then G is said to be LL(k).

This property ensures that the parser can always determine the rule to apply *uniquely*, by looking at the next k symbols of the input string.

3 General Plan of the Transformation

The goal is to transform an arbitrary LL(k)-linear grammar G to an LL(1)-linear grammar G' that defines the same language. Choosing a rule for a nonterminal symbol A in the original grammar requires the next k symbols of the input. In the new grammar, the general plan is to use a *buffer* for up to $k - 1$ next input symbols, so that the parser reads them before having to choose a rule for A. In the new grammar, this buffer shall be attached to every nonterminal symbol, so that they are of the form $_u A$, with $A \in N$ and $u \in \Sigma^{\leqslant k-1}$. The goal is to have $L_{G'}(_u A) = \{\, w \mid uw \in L_G(A) \,\}$.

Upon a closer inspection, there is a certain problem with this plan. If there is a rule $A \to s$ in the original grammar, with $s \in \Sigma^*$ and $|s| < k - 1$, then, in order to choose a rule for A, an LL(1) parser needs to know *more symbols than there are in s and in its own 1-symbol lookahead*. Assuming that all these symbols are buffered, if the parser has to apply a "short" rule $A \to s$, it would not be able to match the entire buffer against input symbols, and the remaining buffered symbols could not be placed anywhere.

Example 1. The following grammar is linear LL(3).

$$S \to aabSaa$$
$$S \to a \qquad\qquad \text{(short rule)}$$

In order to distinguish between these two rules, a hypothetical LL(1) parser buffers up to two first symbols using the following rules.

$$_\varepsilon S \to a\,_a S$$
$$_a S \to a\,_{aa} S$$

Once the parser has aa in the buffer and sees that the next symbol is b, it knows that the correct rule is $S \to aabSaa$, and simulates it by the following rule.

$$_{aa} S \to b\,_\varepsilon Saa$$

However, the rule $S \to a$ in the original grammar cannot be similarly adapted for the nonterminal $_{aa} S$ in the new grammar, and the construction fails.

The cause of this problem is a *short rule* that defines a substring of length less than $k - 1$ in the middle of the input. Accordingly, the first step of the proposed transformation is to eliminate such rules.

4   Elimination of "Short" Rules

The first step in the transformation of an LL(k)-linear grammar to an LL(1)-linear grammar is the elimination of the so-called *short rules*, that is, rules of the form $A \to w$, with $w \in \Sigma^*$, $|w| < k - 1$ and Follow$(A) \neq \{\varepsilon\}$.

Lemma 1. *For every LL(k)-linear grammar $G = (\Sigma, N, R, S)$ there exists an LL(k)-linear grammar G' without short rules that defines the same language. The number of nonterminal symbols in G' is at most $|\Sigma^{\leqslant k-1}| \cdot |N|$.*

Proof. In the new grammar $G' = (\Sigma, N', R', S_\varepsilon)$, nonterminals are of the form A_u, with $A \in N$ and $u \in$ Follow$_{k-1}(A)$. The goal is that every nonterminal A_u defines all strings defined by A in G, with a suffix u appended: $L_{G'}(A_u) = \{ wu \mid w \in L_G(A) \}$.

For every nonterminal symbol A_u and for every rule $A \to w_1 B w_2 \in R$, the new grammar has a rule defined as follows. Let s denote the first $k - 1$ symbols of $w_2 u$ and t denote the rest of $w_2 u$, so that $st = w_2 u$ and $|s| = \min(|w_2 u|, k - 1)$.

The corresponding rule in G' defers the string s to the nonterminal B, and appends the rest of the symbols in the end; these include all the remaining symbols of u.

$$A_u \to w_1 B_s t$$

This is illustrated in Fig. 2.

For rules of the form $A \to s$ with $s \in \Sigma^*$, the corresponding rule in the new grammar appends the suffix to s.

$$A_u \to su$$

The correctness proof is comprised of several assertions: namely, that G' defines the desired language, has no short rules and is LL(k).

Claim. If a string w is defined by A_u in the new grammar, then $w = xu$ and A defines x in the original grammar.

Claim. If a string x is defined by A in the original grammar, then A_u defines xu in the new grammar.

Both claims are established by induction on the height of the respective parse trees. Together, the above two claims establish that $L_{G'}(A_u) = L_G(A)u$ for each nonterminal symbol A_u.

Separate proofs are needed to show that the new grammar has no short rules and the construction preserves the LL(k) property. All proofs are omitted due to space constraints.   □

Example 2 (continued from Example 1). The linear LL(3) grammar in Example 1 is transformed as follows.

$$S_\varepsilon \to a \mid aabS_{aa}$$
$$S_{aa} \to aaa \mid aabS_{aa}aa$$

The rule $S_\varepsilon \to a$ is not short, because Follow$(S_\varepsilon) = \{\varepsilon\}$.

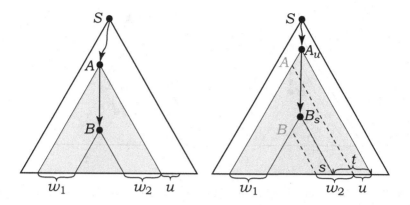

Fig. 2. Simulating a rule $A \to w_1 B w_2$ in G by the rule $A_u \to w_1 B_s t$ in G'.

5   Reduction to One-Symbol Look-Ahead

Once all short rules are eliminated, the following second construction is applied to reduce the length of the look-ahead strings to 1.

Lemma 2. *For every LL(k)-linear grammar $G = (\Sigma, N, R, S)$ without short rules, there exists an LL(1)-linear grammar G' that describes the same language.*

Proof. In the new grammar G', nonterminal symbols are of the form $_uA$, with $A \in N$ and $u \in \Sigma^{\leqslant k}$. The left subscript u of a nonterminal $_uA$ is a buffer for up to $k - 1$ last symbols read by a parser. The goal is to have $L_{G'}(_uA) = \{\, w \mid uw \in L_G(A) \,\}$.

While the buffer is underfull, the parser reads extra symbols one by one and appends them to the buffer. As soon as the buffer is filled, the parser sees a nonterminal symbol $_uA$ with $u \in \Sigma^{k-1}$, as well as a one-symbol look-ahead a. Altogether, the parser has all k symbols needed to determine a rule to apply to A, which is given in the entry $T(A, ua)$ in the LL(*k*) table for G.

The initial symbol of the new grammar, $_\varepsilon S$, is S with an empty buffer.

There are three types of rules in the grammar G'. First, there are rules for filling the buffer. For each nonterminal $_uA$ with $|u| < k - 1$, and for each symbol $a \in \Sigma$, there is a rule that appends this symbol to the buffer.

$$_uA \to a\,_{ua}A$$

Second, there are rules obtained from the corresponding rules in G. For each $_uA \in N'$ and $a \in \Sigma$ with $|u| = k - 1$ and with $T(A, ua)$ defined, the new grammar contains one rule defined as follows. If $T(A, ua) = A \to sBt$, then one of u, s is a prefix of the other; there are two cases, depending on which string is longer. If s is not shorter than u, then $s = us'$ for some $s' \in \Sigma^*$, and the new rule is obtained by removing the prefix u from s.

$$_uA \to s'\,_\varepsilon Bt$$

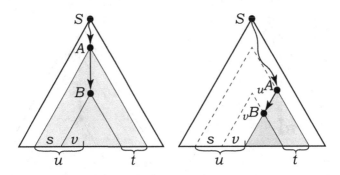

Fig. 3. Simulating a rule $A \rightarrow sBt$ in G by the rule $_uA \rightarrow _vBt$ in G', where $u = sv$.

If u is longer than s then $u = sv$ for some $v \in \Sigma^+$ and the new rule is obtained by removing s and passing the remaining buffer contents v to B, as illustrated in Fig. 3.

$$_uA \rightarrow _vBt$$

If $T(A, ua) = A \rightarrow s$, then $s = ux$ because this is not a short rule, and a is the first symbol of x if $x \neq \varepsilon$. Then the new grammar contains the following rule.

$$_uA \rightarrow x$$

At last, there are rules for the case when the end of the string has been reached. Namely, for each $_uA \in N'$ with $|u| \leqslant k-1$ and with $T(A, u)$ defined, since there are no short rules in G, it is known that $u \in L_G(A)$. Then the grammar G' contains a null rule.

$$_uA \rightarrow \varepsilon$$

The resulting grammar G' is linear. The proof that G' is $LL(1)$ and defines the same language as G is given in a series of claims.

Claim. $x \in L_{G'}(_uA)$ if and only if $ux \in L_G(A)$.

In each direction, the proof is by induction on the height of the respective parse trees.

Claim. For each $u \in \Sigma^{\leqslant k-1}$, if $y \in \text{Follow}(_uA)$, then $y \in \text{Follow}(A)$.

Here the proof considers a parse tree and a subtree followed by y, and is carried out by induction on the depth of that subtree in the tree.

Claim. The grammar G' is LL(1).

The proof naturally relies on the LL(k) property of G.   □

Example 3 (continued from Example 2). The linear LL(3) without short rules constructed in Example 2, is transformed to LL(1) as follows.

$$S_\varepsilon \to a \mid aabS_{aa}$$
$$S_{aa} \to aaa \mid aabS_{aa}aa$$

The rule $S_\varepsilon \to a$ is not short, because $\mathrm{Follow}(S_\varepsilon) = \{\varepsilon\}$.

Both constructions in Lemma 1 and Lemma 2 entail a blow-up in the number of nonterminal symbols by the factor of $|\Sigma^{\leqslant k-1}|$ each. Hence, combining these Lemmata, the desired result can be obtained.

Theorem 1. *For every LL(k)-linear grammar $G = (\Sigma, N, R, S)$ there exists an LL(1)-linear grammar with $|N| \cdot |\Sigma^{\leqslant k-1}|^2$ nonterminal symbols that describes the same language.*

6 Lower Bound

The above construction, applied to an LL(k)-linear grammar with a set of nonterminal symbols N, produces a new LL(1)-linear grammar with as many as $|N| \cdot |\Sigma^{\leqslant k-1}|^2$ nonterminal symbols. The next result is that almost as many nonterminal symbols are in the worst case necessary.

Theorem 2. *For every $m \geqslant 3$, $k \geqslant 4$ and $n \geqslant 1$, there exists a language described by an LL(k)-linear grammar G over an m-symbol alphabet, with n nonterminal symbols, so that every LL(1)-linear grammar for the same language has at least $n \cdot (m-1)^{2k-3-\lceil \log_{m-1} k \rceil}$ nonterminal symbols.*

The grammar witnessing the lower bound is defined over an m-symbol alphabet $\Sigma \cup \{\#\}$, where $\#$ is a special symbol not in Σ.

This grammar shall have rules of the form $A \to xAf(x)$, with $x \in \Sigma^{k-1}\#$, for some function $f \colon \Sigma^{k-1}\# \to \Sigma$, as well as a rule $A \to \varepsilon$. The function shall be defined in a way that in order to detect where the rule $A \to \varepsilon$ should be applied, an LL(1)-linear parser would have to buffer almost $2k$ symbols. Since the arguments to f are strings of a fixed length k, this function can be equally regarded as a function of k arguments.

For every integer $C \geqslant 1$, consider the last C arguments of f. Once the parser reads the first $k - C$ symbols $c_1 \ldots c_{k-C}$ of a presumed block $x \in \Sigma^{k-1}\#$, and has C further symbols to read, $d_{k-C+1} \ldots d_{k-1}\#$, in order to compute f on this block, it needs to remember its projection to the last $C - 1$ arguments, $g \colon \Sigma^{C-1} \to \Sigma$, defined by $g(d_{k-C+1} \ldots d_{k-1}) = f(c_1 \ldots c_{k-C}d_{k-C+1} \ldots d_{k-1}\#)$. The goal is to choose C and f, so that the number of these functions g is as large as possible.

Lemma 3. *For $C = \lceil \log_{|\Sigma|} k \rceil + 1$, there exists a surjective function $f \colon \Sigma^{k-1}\# \to \Sigma$, for which all projections g_w obtained by substituting the first $k - C$ arguments of f with w ($g_w(w') = f(ww')$), for $w \in \Sigma^{k-C}$, are pairwise distinct.*

Proof. There are $|\Sigma|^{|\Sigma|^{C-1}}$ possible projections $g \colon \Sigma^{C-1}\# \to \Sigma$. In order to map distinct strings $w \in |\Sigma|^{k-C}$ to distinct functions g_w, the number of strings should not exceed the number of possible projections.

$$|\Sigma|^{k-C} \leqslant |\Sigma|^{|\Sigma|^{C-1}}$$

This inequality holds true, because $C = \log_{|\Sigma|} k + 1$ implies $k - C \leqslant |\Sigma|^{C-1}$.

Then, strings $w \in \Sigma^{k-C}$ can be injectively mapped to functions $g_w \colon \Sigma^{C-1}\# \to \Sigma$, so that the desired function f is defined as $f(ww') = g_w(w')$, for all $w \in \Sigma^{k-C}$ and $w' \in \Sigma^{C-1}\#$. In order to ensure that f is surjective, it is sufficient to choose the functions g_w so that every symbol $a \in \Sigma$ is in the image of some function g_w, which does not cause any problems. $\square$

With f fixed, the grammar $G = (\Sigma \cup \{\#\}, N, R, S)$ is defined to have $N = \{A_1, \ldots, A_n\}$ and $S = A_1$, with the following rules. Each nonterminal symbol has the rules for reproducing itself surrounded by x on the left, and $f(x)$ on the right, for all possible strings $x \in \Sigma^{k-1}\#$.

$$A_i \to xA_i f(x) \qquad\qquad (1 \leqslant i \leqslant n,\ x \in \Sigma^{k-1}\#)$$

In the middle of the string, there are two possibilities. First, there can be nothing, which shall be detected only k symbols later, when the marker ($\#$) is not found at the end of a k-symbol block.

$$A_i \to \varepsilon \qquad\qquad (1 \leqslant i \leqslant n)$$

The other possibility is that the number of the nonterminal symbol is explicitly written in the middle. This is detected immediately; the reason for having this case is that it forces a parser to remember this number.

$$A_i \to b\#^i \qquad\qquad (1 \leqslant i \leqslant n)$$

Finally, the number of the nonterminal symbol can be changed by applying a rule

$$A_i \to \#A_{i+1} \qquad\qquad (1 \leqslant i < n)$$

This is the LL(k)-linear grammar promised in Theorem 2. It remains to prove that every LL(1)-linear grammar $G' = (\Sigma \cup \{\#\}, N', R', S')$ for the same language must have at least as many nonterminal symbols as stated in the theorem.

The proof uses the following strings in $L(G)$ (the grammar also defines other strings besides these).

$$\#^{i-1}x_1 \ldots x_\ell f(x_\ell) \ldots f(x_1), \qquad \text{with } i, \ell \geqslant 1,\ x_1, \ldots, x_\ell \in \Sigma^{k-1}\#$$
$$\#^{i-1}x_1 \ldots x_\ell b\#^i f(x_\ell) \ldots f(x_1), \qquad \text{with } i, \ell \geqslant 1,\ x_1, \ldots, x_\ell \in \Sigma^{k-1}\#$$

Each string is obtained by first using the rules $A_1 \to \#A_2$, ..., $A_{i-1} \to \#A_i$, and then applying rules of the form $A_i \to xA_i f(x)$. For the first string, this is followed by the rule $A_i \to \varepsilon$, whereas the second string uses the rule $A_i \to b\#^i$ instead.

Consider the stack contents of an LL(1) parser for G' after reading a string of the following form, for some $i \in \{1, \ldots, n\}$, $x_1, \ldots, x_k \in \Sigma^{k-1}\#$, and $a_1, \ldots, a_{k-C} \in \Sigma$.

$$u = \#^{i-1} x_1 \ldots x_k a_1 \ldots a_{k-C}$$

Since there exist strings in $L(G)$ that begin with u followed by any symbol from Σ, the stack contents begin with a nonterminal symbol. Since the grammar is linear, the stack contents are Av, where $A \in N'$ and $v \in (\Sigma \cup \{\#\})^*$. The general plan of the proof is to show that, for a certain large set of strings u of the above form, the corresponding nonterminal symbols A must be pairwise distinct.

The first claim is that most of the information the parser stores at this point is encoded in the nonterminal symbol.

Lemma 4. *Let $G' = (\Sigma \cup \{\#\}, N', R', S')$ be any LL(1)-linear grammar that describes the language $L(G)$, and let Av be the stack contents after reading a string of the form $u = \#^{i-1} x_1 \ldots x_k a_1 \ldots a_{k-C}$, with $i \in \{1, \ldots, n\}$, $x_1, \ldots, x_k \in \Sigma^{k-1}\#$, and $a_1, \ldots, a_{k-C} \in \Sigma$. Then, $|v| \leqslant 2$.*

Proof. Let $a_{k-C+1}, \ldots, a_{k-1} \in \Sigma$ be any symbols following u, and consider the following two blocks from $\Sigma^{k-1}\#$.

$$x_{k+1} = a_1 \ldots a_{k-1}\#$$
$$x_{k+2} = f(x_{k+1}) f(x_k) \ldots f(x_3)\#$$

Consider the following two strings, one obtained by extending u with the block x_{k+1}, and the other by extending u with both blocks, along with filling in the matching symbols in the rest of the string.

$$w_1 = \#^{i-1} x_1 \ldots x_k \boldsymbol{x_{k+1}} f(x_{k+1}) f(x_k) \ldots f(x_3) f(x_2) f(x_1)$$
$$w_2 = \#^{i-1} x_1 \ldots x_k \boldsymbol{x_{k+1} x_{k+2}} f(x_{k+2}) f(x_{k+1}) f(x_k) \ldots f(x_3) f(x_2) f(x_1)$$

Both strings begin with u and are in $L(G)$, and have the following prefix.

$$w = \#^{i-1} x_1 \ldots x_k x_{k+1} f(x_{k+1}) f(x_k) \ldots f(x_4)$$

The longest common prefix of w_1 and w_2 is $w f(x_3)$. There exist strings in $L(G)$ with prefix w followed by any symbol, so the stack contents after reading w again begin with a nonterminal symbol. Let $A'v'$ be the contents of the stack after w is read. At this point, the parser sees the symbol $f(x_3)$. There are at least two strings in $L(G)$ that begin with $w f(x_3)$—namely, w_1 and w_2—and therefore the right-hand side of the rule in the LL(1)-table for $(A', f(x_3))$ contains a nonterminal symbol. For the same reason, there must be a nonterminal symbol in the stack at least until $f(x_3)$ is finally read. Until that moment, the parser can push further symbols onto the stack, whereas the string v lies at the bottom of the stack and remains untouched. The resulting stack contents after reading $wf(x_3)$ are $\gamma v''$, where $\gamma \in (\Sigma \cup \{\#\})^* N$, $v'' \in (\Sigma \cup \{\#\})^*$, and v is a suffix of v''.

To accept the string $w_1 = wf(x_3)f(x_2)f(x_1)$, the parser, after reading $wf(x_3)$, must have at most two symbols on the stack: that is, $|v''| \leqslant 2$. Since v is a suffix of v'', this proves the lemma. $\square$

Now consider any *two* distinct strings of the same general form as in Lemma 4. It is claimed that the parser must have different nonterminal symbols in the stack after reading these strings.

Lemma 5. *Let $G' = (\Sigma \cup \{\#\}, N', R', S')$ be any LL(1)-linear grammar for the language $L(G)$, and consider two strings of the following form.*

$$u_1 = \#^{i-1}x_1 \ldots x_k a_1 \ldots a_{k-C} \ (A_i \in N, \ x_1, \ldots, x_k \in \Sigma^{k-1}\#, \ a_1, \ldots, a_{k-C} \in \Sigma)$$
$$u_2 = \#^{j-1}y_1 \ldots y_k b_1 \ldots b_{k-C} \ (A_j \in N, \ y_1, \ldots, y_k \in \Sigma^{k-1}\#, \ b_1, \ldots, b_{k-C} \in \Sigma)$$

Assume that either $i \neq j$, or $f(x_3) \ldots f(x_k) \neq f(y_3) \ldots f(y_k)$, or $a_1 \ldots a_{k-C} \neq b_1 \ldots b_{k-C}$. Let Av and Bv' be the parser's stack contents after reading these strings. Then, $A \neq B$.

Proof. Suppose, for the sake of a contradiction, that $A = B$, and accordingly the stack contents after reading u_1 and u_2 are Av and Av', respectively.

Denote $\tilde{a} = a_1 \ldots a_{k-C}$ and $\tilde{b} = b_1 \ldots b_{k-C}$. If $\tilde{a} \neq \tilde{b}$, then, by the construction of f, there is a string $z = d_{k-C+1} \ldots d_{k-1}\#$ of length C, with $d_{k-C+1}, \ldots, d_{k-1} \in \Sigma$, that satisfies $f(\tilde{a}z) \neq f(\tilde{b}z)$; if $\tilde{a} = \tilde{b}$, then let $z = d_{k-C+1} \ldots d_{k-1}\#$ be any string with $d_{k-C+1}, \ldots, d_{k-1} \in \Sigma$.

The following two strings are in $L(G)$.

$$u_1 z f(\tilde{a}z)f(x_k) \ldots f(x_1)$$
$$u_1 z b \#^i f(\tilde{a}z)f(x_k) \ldots f(x_1)$$

Since the stack contains Av after reading their common prefix u_1, both remaining suffixes must be in $L_{G'}(Av)$. The prefixes of these strings containing all but the last $|v|$ symbols are in $L_{G'}(A)$. Denote them by w_1 and w_2, respectively.

$$w_1 = zf(\tilde{a}z)f(x_k) \ldots f(x_{|v|+1})$$
$$w_2 = zb\#^i f(\tilde{a}z)f(x_k) \ldots f(x_{|v|+1})$$

(note that $|v| \leqslant 2$ by Lemma 4)

By the assumption, u_1 and u_2 must differ either in the number of sharp signs in the beginning ($i \neq j$), or in one of the images of the last $k - 2$ complete blocks ($f(x_3) \ldots f(x_k) \neq f(y_3) \ldots f(y_k)$), or in one of the symbols in the last incomplete block ($\tilde{a} \neq \tilde{b}$). There are accordingly three cases to consider.

- Let $i \neq j$. The parser's stack contents after reading u_2 is Av', and therefore the string $u_2 w_2 v'$ must be accepted.

$$u_2 w_2 v' = \#^{j-1}y_1 \ldots y_k b_1 \ldots b_{k-C}zb\#^i f(\tilde{a}z)f(x_k) \ldots f(x_{|v|+1})v'$$

However, the mismatch between the prefix $\#^{j-1}$ and the substring $b\#^i$ in the middle means that the string is not in $L(G)$. This is a contradiction.

- In the case when $f(x_3)\ldots f(x_k) \neq f(y_3)\ldots f(y_k)$, after reading u_2, the parser has Av' in its stack, and thus must accept u_2w_1v'.

$$u_2w_1v' = \#^{j-1}y_1\ldots y_kb_1\ldots b_{k-C}zf(\widetilde{a}z)f(x_k)\ldots f(x_{|v|+1})v'$$

However, since $f(x_3)\ldots f(x_k) \neq f(y_3)\ldots f(y_k)$, this string is not in L(G) and should not be accepted, contradiction.
- Assume that $\widetilde{a} \neq \widetilde{b}$. Then, $f(\widetilde{a}z) \neq f(\widetilde{b}z)$ by the choice of z. As in the previous case, the parser accepts u_2w_1v', which is not in $L(G)$. $\square$

Proof (of Theorem 2). For all $i \in \{1,\ldots,n\}$, $d_3,\ldots,d_k \in \Sigma$ and $a_1,\ldots,a_{k-C} \in \Sigma$, let $x_1,\ldots,x_k \in \Sigma^{k-1}\#$, with $f(x_j) = d_j$ for all $j \in \{3,\ldots,n\}$. Then the corresponding string $u_{i;d_3,\ldots,d_k;a_1,\ldots,a_{k-C}}$ is defined as follows.

$$u_{i;d_3,\ldots,d_k;a_1,\ldots,a_{k-C}} = \#^{i-1}x_1\ldots x_ka_1\ldots a_{k-C}$$

By Lemma 5, upon reading different strings of this form, the LL(1)-linear parser must have pairwise distinct nonterminal symbols in its stack. Therefore, there must be at least as many nonterminal symbols as there are such strings, that is, $n \cdot (m-1)^{2k-C-2}$, as claimed. $\square$

7 Conclusion

The collapse of the hierarchy of LL(k)-linear languages establishes the *LL-linear languages* as a robust language family that deserves future investigation.

In particular, the succinctness tradeoff between LL(k)-linear grammars with different k has been determined only with respect to the number of nonterminal symbols. It would be interesting to know whether the elimination of look-ahead similarly affects the total length of description (the number of symbols needed to describe the grammar). The witness languages constructed in this paper do not establish any lower bounds on that, and more research is accordingly needed to settle this question.

Another suggested line of research concerns LL subfamilies of other families of grammars [9]. For instance, is there a similar lookahead hierarchy for LL(k)-linear conjunctive grammars and LL(k)-linear Boolean grammars? [8]

References

1. de la Higuera, C., Oncina, J.: Inferring deterministic linear languages. In: Kivinen, J., Sloan, R.H. (eds.) COLT 2002. LNCS (LNAI), vol. 2375, pp. 185–200. Springer, Heidelberg (2002). https://doi.org/10.1007/3-540-45435-7_13
2. Holzer, M., Lange, K.-J.: On the complexities of linear LL(1) and LR(1) grammars. In: Ésik, Z. (ed.) FCT 1993. LNCS, vol. 710, pp. 299–308. Springer, Heidelberg (1993). https://doi.org/10.1007/3-540-57163-9_25
3. Ibarra, O.H., Jiang, T., Ravikumar, B.: Some subclasses of context-free languages in NC^1. Inf. Process. Lett. **29**(3), 111–117 (1988). https://doi.org/10.1016/0020-0190(88)90047-6

4. Jirásková, G., Klíma, O.: Deterministic biautomata and subclasses of deterministic linear languages. In: Martín-Vide, C., Okhotin, A., Shapira, D. (eds.) LATA 2019. LNCS, vol. 11417, pp. 315–327. Springer, Cham (2019). https://doi.org/10.1007/978-3-030-13435-8_23

5. Knuth, D.E.: Top-down syntax analysis. Acta Informatica **1**, 79–110 (1971). https://doi.org/10.1007/BF00289517

6. Kurki-Suonio, R.: Notes on top-down languages. BIT Numer. Math. **9**(3), 225–238 (1969). https://doi.org/10.1007/BF01946814

7. Lewis II, P.M., Stearns, R.E.: Syntax-directed transduction. J. ACM **15**(3), 465–488 (1968). https://doi.org/10.1145/321466.321477

8. Okhotin, A.: Expressive power of LL(k) Boolean grammars. Theor. Comput. Sci. **412**(39), 5132–5155 (2011). https://doi.org/10.1016/j.tcs.2011.05.013

9. Okhotin, A.: Underlying principles and recurring ideas of formal grammars. In: Klein, S.T., Martín-Vide, C., Shapira, D. (eds.) LATA 2018. LNCS, vol. 10792, pp. 36–59. Springer, Cham (2018). https://doi.org/10.1007/978-3-319-77313-1_3

10. Rosenkrantz, D.J., Stearns, R.E.: Properties of deterministic top-down grammars. Inf. Control **17**, 226–256 (1970). https://doi.org/10.1016/S0019-9958(70)90446-8

On Computing the Hamiltonian Index
of Graphs

Geevarghese Philip[1,3]([✉]) (iD), M. R. Rani[2](iD), and R. Subashini[2](iD)

[1] Chennai Mathematical Institute, Chennai, India
gphilip@cmi.ac.in
[2] National Institute of Technology Calicut, Calicut, India
{rani_p150067cs,suba}@nitc.ac.in
[3] UMI ReLaX, Chennai, India

Abstract. For an integer $r \geq 0$ the *r-th iterated line graph* $L^r(G)$ of a graph G is defined by: (i) $L^0(G) = G$ and (ii) $L^r(G) = L(L^{(r-1)}(G))$ for $r > 0$, where $L(G)$ denotes the line graph of G. The *Hamiltonian Index* $h(G)$ of G is the smallest r such that $L^r(G)$ has a Hamiltonian cycle [Chartrand, 1968]. Checking if $h(G) = k$ is NP-hard for any fixed integer $k \geq 0$ even for subcubic graphs G [Ryjáček et al., 2011]. We study the parameterized complexity of this problem with the parameter treewidth, $tw(G)$, and show that we can find $h(G)$ in time $\mathcal{O}^\star((1 + 2^{(\omega+3)})^{tw(G)})$ where ω is the matrix multiplication exponent. This generalizes various prior results on computing $h(G)$ including an $\mathcal{O}^\star((1 + 2^{(\omega+3)})^{tw(G)})$-time algorithm for checking if $h(G) = 1$ holds [Misra et al., CSR 2019].

The NP-hard EULERIAN STEINER SUBGRAPH problem takes as input a graph G and a specified subset K of *terminal* vertices of G and asks if G has an Eulerian subgraph H containing all the terminals. A key ingredient of our algorithm for finding $h(G)$ is an algorithm which solves EULERIAN STEINER SUBGRAPH in $\mathcal{O}^\star((1+2^{(\omega+3)})^{tw(G)})$ time. To the best of our knowledge this is the first FPT algorithm for EULERIAN STEINER SUBGRAPH, and generalizes previous results on various special cases.

1   Introduction

All graphs in this article are finite and undirected, and are without self-loops or multiple edges unless explicitly stated. We use $\mathbb{N}$ to denote the set of non-negative integers, and $V(G), E(G)$, respectively, to denote the vertex and edge sets of graph G. A graph is *Eulerian* if it has a closed Eulerian trail, and *Hamiltonian* if it has a Hamiltonian cycle. The vertex set of the *line graph* of a graph G—denoted $L(G)$—is the edge set $E(G)$ of G, and two vertices e, f are adjacent in $L(G)$ if and only if the edges e and f share a vertex in G. Let r be a non-negative integer. The *r-th iterated line graph* $L^r(G)$ of G is defined by: (i) $L^0(G) = G$, and (ii) $L^r(G) = L(L^{(r-1)}(G))$ for $r > 0$. If $G = P_\ell$ for a non-negative integer ℓ then $L^\ell(G)$ is K_1, the graph with one vertex and no edges, and $L^r(G)$ is the empty graph for all $r > \ell$. If G is a connected graph which is *not* a path then $L^r(G)$ is nonempty for *all* $r \geq 0$ [5].

Full version on arXiv: https://arxiv.org/abs/1912.01990.

© Springer Nature Switzerland AG 2020
H. Fernau (Ed.): CSR 2020, LNCS 12159, pp. 341–353, 2020.
https://doi.org/10.1007/978-3-030-50026-9_25

An *edge Hamiltonian path* of a graph G is any permutation Π of the edge set $E(G)$ of G such that every pair of consecutive edges in Π has a vertex in common, and an *edge Hamiltonian cycle* of G is an edge Hamiltonian path of G in which the first and last edges also have a vertex in common.

Theorem 1. *The following are equivalent for a graph G:*

- *Its line graph $L(G)$ is Hamiltonian*
- *G has an edge Hamiltonian cycle* [6]
- *G contains a closed trail T such that every edge in G has at least one end-point in T* [13]

Chartrand showed that for essentially all connected graphs G, $L^r(G)$ becomes Hamiltonian for *some* integer $r \geq 0$.

Theorem 2. [6] *If G is a connected graph on n vertices which is not a path, then $L^r(G)$ is Hamiltonian for all integers $r \geq (n-3)$.*

This led Chartrand to define the *Hamiltonian Index $h(G)$* of a connected graph G which is not a path, to be the *smallest* non-negative integer r such that $L^r(G)$ is Hamiltonian [6]. The main focus of the current work is the algorithmic question of *computing $h(G)$*. Checking if $h(G) = 0$ holds is the same as checking if graph G is Hamiltonian. This is long known to be NP-complete, even when the input graph is planar and *subcubic* (has maximum degree at most 3) [12]. Checking if $h(G) = 1$ holds is the same as checking if (i) G is *not* Hamiltonian, and (ii) the line graph $L(G)$ is Hamiltonian. Bertossi [1] showed that the latter problem is NP-complete, and Ryjáček et al. proved that this holds even if graph G is subcubic [24]. Indeed they showed that checking whether $h(G) = t$ is NP-complete for *any* fixed integer $t \geq 0$, even when the input graph G is subcubic.

Our Problems and Results. In this work we take up the *parameterized complexity analysis* of the problem of computing the Hamiltonian Index. Briefly put, an instance of a *parameterized problem* is a pair (x, k) where x is an instance of a classical problem and k is a (usually numerical) *parameter* which captures some aspect of x. A primary goal is to find a *fixed-parameter tractable* (or FPT) algorithm for the problem, one which solves the instance in time $\mathcal{O}(f(k) \cdot |x|^c)$ where $f()$ is a function of the parameter k alone, and c is a constant independent of x and k; this running time is abbreviated as $\mathcal{O}^\star(f(k))$. The design of FPT algorithms is a vibrant field of research; we refer the interested reader to standard textbooks [7,9].

Since checking whether $h(G) = t$ is NP-complete for *any* fixed $t \geq 0$ for subcubic graphs, neither the value $h(G)$ nor the maximum or average degree of G is a sensible parameter for this problem. We choose the *treewidth* of the input graph G as our parameter. This is motivated by prior related work as well, as we describe below. Thus the main problem which we take up in this work is

---

HAMILTONIAN INDEX (HI)   **Parameter:** tw

Input: A connected undirected graph $G = (V, E)$ which is not a path, a tree decomposition $\mathcal{T} = (T, \{X_t\}_{t \in V(T)})$ of G of width tw, and $r \in \mathbb{N}$.

Question: Is $h(G) \leq r$?

Our first main result is that this problem is fixed-parameter tractable. ω denotes the matrix multiplication exponent; it is known that $\omega < 2.3729$ holds [26].

Theorem 3. *There is an algorithm which solves an instance $(G, \mathcal{T}, tw, r)$ of* HAMILTONIAN INDEX *in* $\mathcal{O}^*((1 + 2^{(\omega+3)})^{tw})$ *time.*

From this and from Theorem 2 we get the next result.

Corollary 4. *There is an algorithm which takes as input a graph G and a tree decomposition $\mathcal{T}$ of width tw of G as input, and outputs the Hamiltonian Index $h(G)$ of G in* $\mathcal{O}^*((1 + 2^{(\omega+3)})^{tw})$ *time.*

The input to a *Steiner subgraph problem* consists of a graph G and a specified set K of *terminal vertices* of G, and the objective is to find a subgraph of G which (i) contains all the terminals, and (ii) satisfies some other specified set of constraints, usually including connectivity constraints on the set K. The archetypal example is the STEINER TREE problem where the goal is to find a *connected* subgraph of G of the smallest size (number of edges) which contains all the terminals. STEINER TREE and a number of its variants have been the subject of extensive research [14]. A key part of our algorithm for computing $h(G)$ consists of solving:

EULERIAN STEINER SUBGRAPH (ESS)	**Parameter:** tw
Input: An undirected graph $G = (V, E)$, a set of "terminal" vertices $K \subseteq V$, and a tree decomposition $\mathcal{T} = (T, \{X_t\}_{t \in V(T)})$ of G, of width tw.	
Question: Does there exist an Eulerian subgraph $G' = (V', E')$ of G such that $K \subseteq V'$?	

G' is an *Eulerian Steiner subgraph* of G *for the terminal set K*. The second main result of this work is

Theorem 5. *There is an algorithm which solves an instance $(G, K, \mathcal{T}, tw)$ of* EULERIAN STEINER SUBGRAPH *in* $\mathcal{O}^*((1 + 2^{(\omega+3)})^{tw})$ *time.*

Related Work. The parameterized complexity of computing $h(G)$ *per se* has not, to the best of our knowledge, been previously explored. The two special cases of checking if $h(G) \in \{0, 1\}$ *have* been studied with the treewidth tw of the input graph G as the parameter; we now summarize the main existing results. Checking whether $h(G) = 0$ holds—that is, whether G is Hamiltonian—was long known to be solvable in $\mathcal{O}^*(tw^{\mathcal{O}(tw)})$ time (See, e.g., [10, Exercise 11.36]). This was suspected to be the best possible till, in a breakthrough result in 2011, Cygan et al. [8] showed that this can be done in randomized $\mathcal{O}^*(4^{tw})$ time. More recently, Bodlaender et al. [3] and Fomin et al. [11] showed, independently and using different techniques, that this can be done in *deterministic* $\mathcal{O}^*(2^{\mathcal{O}(tw)})$ time.

Recall that a *vertex cover* of graph G is any subset $S \subseteq V(G)$ such that every edge in $E(G)$ has at least one of its two endpoints in the set S. A subgraph G' of a graph G is said to be a *dominating Eulerian subgraph* of G if (i) G' is Eulerian, and (ii) $V(G')$ is a vertex cover of G. Note that—in conformance with the literature (e.g. [19]) on this subject—the word "dominating" here denotes the existence of a *vertex cover*, and *not* of a *dominating set*. The input to the DOMINATING EULERIAN SUBGRAPH (DES) problem consists of a graph G and a tree decomposition $\mathcal{T}$ of G of width tw, and the question is whether G has a dominating Eulerian subgraph; the parameter is tw. The input to the EDGE HAMILTONIAN PATH (EHP) (respectively, EDGE HAMILTONIAN CYCLE (EHC)) problem consists of a graph G and a tree decomposition $\mathcal{T}$ of G of width tw, and the question is whether G has an edge Hamiltonian path (resp. cycle); the parameter is tw. Observe that a closed trail in graph G is an Eulerian subgraph of G. So Theorem 1 tells us that EHC is equivalent to DES.

The parameterized complexity of checking whether $h(G) = 1$ holds was first taken up by Lampis et al. in 2014 [18,19], albeit indirectly: they addressed EHC and EHP. They showed that EHP is in FPT if and only if EHC is in FPT, and that these problems (and hence DES) can be solved in $\mathcal{O}^\star(tw^{\mathcal{O}(tw)})$ time. Very recently Misra et al. [20] investigated an optimization variant of EDGE HAMILTONIAN PATH which they called LONGEST EDGE-LINKED PATH (LELP). An *edge-linked path* is a sequence of edges in which every consecutive pair has a vertex in common. Given a graph G, $k \in \mathbb{N}$, and a tree decomposition $\mathcal{T}$ of G of width tw as input the LELP problem asks whether G has an edge-linked path of length at least k. Note that setting $k = |E(G)|$ yields EHP as a special case. Misra et al. [20] gave an algorithm which solves LELP (and hence, EHP, EHC and DES) in $\mathcal{O}^\star((1 + 2^{(\omega+3)})^{tw})$ time. This gives the current best algorithm for checking if $h(G) = 1$ holds. In the full version of this paper [21] we derive an alternative algorithm for these problems, with the same running time.

Theorem 6. [20,21] *There is an algorithm which solves an instance $(G, \mathcal{T}, tw)$ of* EDGE HAMILTONIAN PATH *(respectively,* EDGE HAMILTONIAN CYCLE *or* DOMINATING EULERIAN SUBGRAPH*) in* $\mathcal{O}^\star((1 + 2^{(\omega+3)})^{tw})$ *time.*

To the best of our knowledge, ours is the first FPT algorithm for EULERIAN STEINER SUBGRAPH. A subgraph H of a graph G is a *spanning* subgraph of G if H contains every vertex of G. A graph G is *supereulerian* if it has a spanning subgraph H which is Eulerian. To the best of our knowledge, EULERIAN STEINER SUBGRAPH problem is not mentioned in the literature, but we did find quite a bit of existing work on the special case—obtained by setting $K = V(G)$—of checking if an input graph G is supereulerian [4,17]. Pulleyblank observed already in 1979 that this latter problem is NP-complete even on subcubic graphs [22]. This implies that EULERIAN STEINER SUBGRAPH is NP-complete as well. Richey et al. [23] showed in 1985 that the problem can be solved in polynomial time on *series-parallel* graphs. More recently, Sau and Thilikos showed in 2010 that the problem can be solved in $\mathcal{O}^\star(2^{\mathcal{O}(\sqrt{n})})$ time on *planar graphs* with n vertices [25]. Now consider the following parameterization:

SPANNING EULERIAN SUBGRAPH (SES)   **Parameter:** tw
Input: An undirected graph $G = (V, E)$ and a tree decomposition $\mathcal{T} = (T, \{X_t\}_{t \in V(T)})$ of G, of width tw.
Question: Does G have a spanning Eulerian subgraph?

Setting $K = V(G)$ in Theorem 5 we get the following conclusion.

Corollary 7. *There is an algorithm which solves an instance $(G, \mathcal{T}, tw)$ of* SPANNING EULERIAN SUBGRAPH *in $\mathcal{O}^\star((1 + 2^{(\omega+3)})^{tw})$ time.*

It is known that series-parallel graphs have treewidth at most 2 and are planar, and that planar graphs on n vertices have treewidth $\mathcal{O}(\sqrt{n})$ [2]. Further, given a planar graph G of treewidth t we can, in polynomial time, output a tree decomposition of G of width $\mathcal{O}(t)$ [16]. These facts together with Corollary 7 subsume the results of Richey et al. and Sau and Thilikos, respectively.

Organization of the Rest of the Paper. We outline the proof of Theorem 5 in Sect. 2 and the proof of our main result Theorem 3 in Sect. 3. We conclude in Sect. 4. We have omitted the proofs of most statements (those marked with a (†)) in this Extended Abstract, for want of space. The full version of this Extended Abstract is available online on arXiv [21] and contains all the proofs as well as a more complete discussion of our problems, results, and techniques.

2 An **FPT** Algorithm for EULERIAN STEINER SUBGRAPH

In this section we describe an algorithm which takes an instance $(G, K, \mathcal{T}, tw)$ of EULERIAN STEINER SUBGRAPH as input and tells in $\mathcal{O}^\star((1 + 2^{(\omega+3)})^{tw})$ time whether graph G has a subgraph which is (i) Eulerian, and (ii) contains every vertex in the terminal set K. We assume, without loss of generality, that $\mathcal{T}$ is itself a nice tree decomposition of width tw. The rest of our algorithm for EULERIAN STEINER SUBGRAPH consists of dynamic programming (DP) over the bags of this nice tree decomposition, and is modelled after the algorithm of Bodlaender et al. [3] for STEINER TREE. We pick an arbitrary terminal $v^\star \in K$ and add it to every bag of $\mathcal{T}$; *from now on we use $\mathcal{T}$ to refer to the resulting "nearly-nice" tree decomposition in which the bags at all the leaves and the root are equal to $\{v^\star\}$.* The next definition captures how an Eulerian subgraph $G' = (V', E')$ of G which contains all the terminals K interacts with the structures defined by node t.

Definition 8 (Valid partitions, witness for validity). *For a bag X_t and subsets $X \subseteq X_t$, $O \subseteq X$, we say that a partition $P = \{X^1, X^2, \ldots X^p\}$ of X is valid for the combination (t, X, O) if there exists a subgraph $G'_t = (V'_t, E'_t)$ of G_t such that*

1. $X_t \cap V(G'_t) = X$.
2. *G'_t has exactly p connected components $C_1, C_2, \ldots, C_p$ and for each $i \in \{1, 2, \ldots, p\}$, $X^i \subseteq V(C_i)$. That is, the vertex set of each connected component of G'_t has a non-empty intersection with set X, and P is the partition of X defined by the subgraph G'_t.*
3. *Every terminal vertex from $K \cap V_t$ is in $V(G'_t)$.*
4. *The set of odd-degree vertices in G'_t is exactly the set O.*

Such a subgraph G'_t of G_t is a witness *for partition P being valid for the combination (t, X, O) or, in short: G'_t is a witness for $((t, X, O), P)$.*

Definition 9 (Completion). *For a bag X_t and subsets $X \subseteq X_t$, $O \subseteq X$ let P be a partition of X which is valid for the combination (t, X, O). Let H be a residual subgraph with respect to t such that $V(H) \cap X_t = X$. We say that $((t, X, O), P)$ completes H if there exists a subgraph G'_t of G_t which is a witness for $((t, X, O), P)$, such that the graph $G'_t \cup H$ is an Eulerian Steiner subgraph of G for the terminal set K. We say that G'_t is a* certificate *for $((t, X, O), P)$ completing H.*

Lemma 10 (†). *Let $(G, K, \mathcal{T}, tw)$ be an instance of* Eulerian Steiner Sub-graph, *and let t be an arbitrary node of $\mathcal{T}$. Let $X \subseteq X_t$, $O \subseteq X$, and let $\mathcal{A}$ be a collection of partitions of X, each of which is valid for the combination (t, X, O). Let $\mathcal{B}$ be a representative subset of $\mathcal{A}$, and let H be an arbitrary residual subgraph of G with respect to t such that $V(H) \cap X_t = X$ holds. If there is a partition $P \in \mathcal{A}$ such that $((t, X, O), P)$ completes H then there is a partition $Q \in \mathcal{B}$ such that $((t, X, O), Q)$ completes H.*

Lemma 11 (†). *Let $(G, K, \mathcal{T}, tw)$ be an instance of* Eulerian Steiner Sub-graph, *let r be the root node of $\mathcal{T}$, and let $v^\star$ be the terminal vertex which is present in every bag of $\mathcal{T}$. Then $(G, K, \mathcal{T}, tw)$ is a **yes** instance of* Eulerian Steiner Subgraph *if and only if the partition $P = \{\{v^\star\}\}$ is valid for the combination $(r, X = \{v^\star\}, O = \emptyset)$.*

Lemma 12 (†). *Let $(G, K, \mathcal{T}, tw)$ be an instance of* Eulerian Steiner Sub-graph, *let r be the* root *node of $\mathcal{T}$, and let $v^\star$ be the terminal vertex which is present in every bag of $\mathcal{T}$. Let $H = (\{v^\star\}, \emptyset)$, $X = \{v^\star\}$, $O = \emptyset$, and $P = \{\{v^\star\}\}$. Then $(G, K, \mathcal{T}, tw)$ is a **yes** instance if and only if $((r, X, O), P)$ completes H.*

A naïve implementation of our algorithm would consist of computing, for each node t of the tree decomposition $\mathcal{T}$—starting at the leaves and working up towards the root—and subsets $O \subseteq X \subseteq X_t$, the set of all partitions P which are valid for the combination (t, X, O). At the root node r the algorithm would apply Lemma 11 to decide the instance $(G, K, \mathcal{T}, tw)$. Since a bag X_t can have up to $tw + 2$ elements (including the special terminal $v^\star$) the running time of this algorithm could have a factor of tw^{tw} in it, since X_t can have these many partitions. To avoid this we turn to the completion-based alternate characterization of **yes** instances—Lemma 12—and the fact—Lemma 10—that representative subset computations do not "forget" completion properties. After computing a set $\mathcal{A}$ of valid partitions for each combination (t, X, O) we compute a representative subset $\mathcal{B} \subseteq \mathcal{A}$ and throw away the remaining partitions $\mathcal{A} \setminus \mathcal{B}$. Thus the number of partitions which we need to remember for any combination (t, X, O) never exceeds 2^{tw}. We now describe the steps of the DP algorithm for each type of node in $\mathcal{T}$. We use $VP[t, X, O]$ to denote the set of **valid partitions** for the combination (t, X, O) which we store in the DP table for node t.

Leaf Node t: In this case $X_t = \{v^\star\}$. Set $VP[t, \{v^\star\}, \{v^\star\}] = \emptyset$, $VP[t, \{v^\star\}, \emptyset] = \{\{\{v^\star\}\}\}$, and $VP[t, \emptyset, \emptyset] = \{\emptyset\}$.

Introduce Vertex Node t: Let t' be the child node of t, and let v be the vertex introduced at t. Then $v \notin X_{t'}$ and $X_t = X_{t'} \cup \{v\}$. For each $X \subseteq X_t$ and $O \subseteq X$,

1. If v is a terminal vertex, then
   - if $v \notin X$ or if $v \in O$ then set $VP[t, X, O] = \emptyset$
   - if $v \in (X \setminus O)$ then for each partition P' in $VP[t', X \setminus \{v\}, O]$, add the partition $P = (P' \cup \{\{v\}\})$ to the set $VP[t, X, O]$

2. If v is *not* a terminal vertex, then
   - if $v \in O$ then set $VP[t, X, O] = \emptyset$
   - if $v \in (X \setminus O)$ then for each partition P' in $VP[t', X \setminus \{v\}, O]$, add the partition $P = P' \cup \{\{v\}\}$ to the set $VP[t, X, O]$
   - if $v \notin X$ then set $VP[t, X, O] = VP[t', X, O]$

3. Set $\mathcal{A} = VP[t, X, O]$. Compute a representative subset $\mathcal{B} \subseteq \mathcal{A}$ and set $VP[t, X, O] = \mathcal{B}$.

Introduce Edge Node t: Let t' be the child node of t, and let uv be the edge introduced at t. Then $X_t = X_{t'}$ and $uv \in (E(G_t) \setminus E(G_{t'}))$. For each $X \subseteq X_t$ and $O \subseteq X$,

1. Set $VP[t, X, O] = VP[t', X, O]$.

2. If $\{u, v\} \subseteq X$ then:
   (a) Construct a set of *candidate partitions* $\mathcal{P}$ as follows. Initialize $\mathcal{P} = \emptyset$.
       - if $\{u, v\} \subseteq O$ then add all of $VP[t', X, O \setminus \{u, v\}]$ to $\mathcal{P}$.
       - if $\{u, v\} \cap O = \{u\}$ then add all of $VP[t', X, (O \setminus \{u\}) \cup \{v\}]$ to $\mathcal{P}$.
       - if $\{u, v\} \cap O = \{v\}$ then add all of $VP[t', X, (O \setminus \{v\}) \cup \{u\}]$ to $\mathcal{P}$.
       - if $\{u, v\} \cap O = \emptyset$ then add all of $VP[t', X, O \cup \{u, v\}]$ to $\mathcal{P}$.
   (b) For each candidate partition $P' \in \mathcal{P}$, if vertices u, v are in different blocks of P'—say $u \in P'_u, v \in P'_v$; $P'_u \neq P'_v$—then merge these two blocks of P' to obtain P. That is, set $P = (P' \setminus \{P'_u, P'_v\}) \cup (P'_u \cup P'_v)$. Now set $\mathcal{P} = (\mathcal{P} \setminus \{P'\}) \cup P$.
   (c) Add all of $\mathcal{P}$ to the list $VP[t, X, O]$.

3. Set $\mathcal{A} = VP[t, X, O]$. Compute a representative subset $\mathcal{B} \subseteq \mathcal{A}$ and set $VP[t, X, O] = \mathcal{B}$.

Forget Node t: Let t' be the child node of t, and let v be the vertex forgotten at t. Then $v \in X_{t'}$ and $X_t = X_{t'} \setminus \{v\}$. Recall that $P(v)$ is the block of partition P which contains element v, and that $P - v$ is the partition obtained by eliding v from P. For each $X \subseteq X_t$ and $O \subseteq X$,

1. Set $VP[t, X, O] = \{P' - v \; ; \; P' \in VP[t', X \cup \{v\}, O], |P'(v)| > 1\}$.

2. If v is *not* a terminal vertex then set $VP[t, X, O] = VP[t, X, O] \cup VP[t', X, O]$.

3. Set $\mathcal{A} = VP[t, X, O]$. Compute a representative subset $\mathcal{B} \subseteq \mathcal{A}$ and set $VP[t, X, O] = \mathcal{B}$.

Join Node t: Let t_1, t_2 be the children of t. Then $X_t = X_{t_1} = X_{t_2}$. For each $X \subseteq X_t, O \subseteq X$:

1. Set $VP[t, X, O] = \emptyset$

2. For each $O_1 \subseteq O$ and $\hat{O} \subseteq (X \setminus O)$:

(a) Let $O_2 = O \setminus O_1$.
(b) For each pair of partitions $P_1 \in VP[t_1, X, O_1 \cup \hat{O}]$, $P_2 \in VP[t_2, X, O_2 \cup \hat{O}]$, add their join $P_1 \sqcup P_2$ to the set $VP[t, X, O]$.
3. Set $\mathcal{A} = VP[t, X, O]$. Compute a representative subset $\mathcal{B} \subseteq \mathcal{A}$ and set $VP[t, X, O] = \mathcal{B}$.

The key insight in the proof of correctness of our DP is that the processing at every node in $\mathcal{T}$ preserves the following *Correctness Criteria*. Let t be a node of $\mathcal{T}$, let $X \subseteq X_t, O \subseteq X$, and let $VP[t, X, O]$ be the set of partitions computed by the DP for the combination (t, X, O).

1. **Soundness:** Every partition $P \in VP[t, X, O]$ is valid for the combination (t, X, O).
2. **Completeness:** For any residual subgraph H with respect to t with $V(H) \cap X_t = X$, if there exists a partition P of X such that $((t, X, O), P)$ completes H then the set $VP[t, X, O]$ contains a partition Q of X such that $((t, X, O), Q)$ completes H. Note that
   – the two partitions P, Q must both be valid for the combination (t, X, O); and
   – Q can potentially be the same partition as P.

The processing at each of the non-leaf nodes computes a representative subset as a final step. This step does not negate the correctness criteria.

Observation 13 (†). *Let t be a node of $\mathcal{T}$, let $X \subseteq X_t, O \subseteq X$, and let $\mathcal{A}$ be a set of partitions which satisfies the correctness criteria for the combination (t, X, O). Let $\mathcal{B}$ be a representative subset of $\mathcal{A}$. Then $\mathcal{B}$ satisfies the correctness criteria for the combination (t, X, O).*

Lemma 14 (†). *Let t be a node of the tree decomposition $\mathcal{T}$ and let $X \subseteq X_t, O \subseteq X$ be arbitrary subsets of X_t, X respectively. The collection $\mathcal{A}$ of partitions computed by the DP for the combination (t, X, O) satisfies the correctness criteria.*

Theorem 5. *There is an algorithm which solves an instance $(G, K, \mathcal{T}, tw)$ of* EULERIAN STEINER SUBGRAPH *in $\mathcal{O}^*((1 + 2^{(\omega+3)})^{tw})$ time.*

Proof. We first modify $\mathcal{T}$ to make it a "nearly-nice" tree decomposition rooted at r as described at the start of this section. We then execute the dynamic programming steps described above on $\mathcal{T}$. We return **yes** if the element $\{\{v^*\}\}$ is present in the set $VP[r, X = \{v^*\}, O = \emptyset]$ computed by the DP, and **no** otherwise. From Lemma 12 we know that $(G, K, \mathcal{T}, tw)$ is a **yes** instance of EULERIAN STEINER SUBGRAPH if and only if the combination $((r, X = \{v^*\}, O = \emptyset), P = \{\{v^*\}\})$ completes the residual graph $H = (\{v^*\}, \emptyset)$. By induction on the structure of the tree decomposition $\mathcal{T}$ and using Observation 13 and Lemma 14 we get that the set $VP[r, X = \{v^*\}, O = \emptyset]$ computed by the algorithm satisfies the correctness criteria. And since $\{\{v^*\}\}$ is the unique partition of set $\{v^*\}$ we get that the set $VP[r, X = \{v^*\}, O = \emptyset]$ computed by the algorithm will contain the partition $\{\{v^*\}\}$ if and only if $(G, K, \mathcal{T}, tw)$ is a **yes** instance of EULERIAN STEINER SUBGRAPH.

Note that we compute representative subsets as the last step in the computation at each bag. So we get, while performing computations at an intermediate node t, that the number of partitions in any set $VP[t', X', \cdot]$ for any *child* node t' of t and subset X' of $X_{t'}$ is at most $2^{(|X'|-1)}$.

The computation at each **leaf node** of $\mathcal{T}$ can be done in constant time. For an **introduce vertex node** or an **introduce edge node** or a **forget node** t and a fixed pair of subsets $X \subseteq X_t, O \subseteq X$, the computation of set $\mathcal{A}$ involves— in the worst case—spending polynomial time for each partition P' in some set $VP[t', X' \subseteq X, \cdot]$. Since the number of partitions in this latter set is at most $2^{(|X'|-1)} \leq 2^{(|X|-1)}$ we get that the set $\mathcal{A}$ can be computed in $\mathcal{O}^\star(2^{(|X|-1)})$ time, and that the set $\mathcal{B}$ can be computed in $\mathcal{O}^\star(2^{(|X|-1)} \cdot 2^{(\omega-1)\cdot|X|}) = \mathcal{O}^\star(2^{\omega\cdot|X|})$ time. Since the number of ways of choosing the subset $O \subseteq X$ is $2^{|X|}$ the entire computation at an introduce vertex, introduce edge, or forget node t can be done in time

$$\sum_{|X|=0}^{|X_t|} \binom{|X_t|}{|X|} 2^{|X|} \mathcal{O}^\star(2^{\omega\cdot|X|}) = \mathcal{O}^\star((1 + 2 \cdot 2^\omega)^{tw}).$$

For a **join node** t and a fixed subset $X \subseteq X_t$ we guess three pairwise disjoint subsets $\hat{O}, O_1, O_2$ of X in time $4^{|X|}$. For each guess we go over all partitions $P_1 \in VP[t_1, X, O_1 \cup \hat{O}], P_2 \in VP[t_2, X, O_2 \cup \hat{O}]$ and add their join $P_1 \sqcup P_2$ to the set $\mathcal{A}$. Since the number of partitions in each of the two sets $VP[t_1, X, O_1 \cup \hat{O}], VP[t_2, X, O_2 \cup \hat{O}]$ is at most $2^{(|X|-1)}$, the size of set $\mathcal{A}$ is at most $2^{(2|X|-2)}$. The entire computation at the join node can be done in time

$$\sum_{|X|=0}^{|X_t|} \binom{|X_t|}{|X|} 4^{|X|} (2^{(2|X|-2)} + \mathcal{O}^\star(2^{(2|X|-2)} \cdot 2^{(\omega-1)\cdot|X|})) = \mathcal{O}^\star((1 + 2^{(\omega+3)})^{tw}).$$

The entire DP over $\mathcal{T}$ can thus be done in $\mathcal{O}^\star((1 + 2^{(\omega+3)})^{tw})$ time. $\square$

3   Finding the Hamiltonian Index

In this section we prove Theorem 3: we describe an algorithm which takes an instance $(G, \mathcal{T}, tw, r)$ of HAMILTONIAN INDEX as input and outputs in $\mathcal{O}^\star((1 + 2^{(\omega+3)})^{tw})$ time whether graph G has Hamiltonian Index at most r. If $r \geq (|V(G)| - 3)$ holds, then our algorithm returns **yes**. If $r < (|V(G)| - 3)$ holds, then it checks, for each $i = 0, 1, \ldots, r$ in increasing order, whether $h(G) = i$ holds. From Theorem 2, we know that this procedure correctly solves HAMILTONIAN INDEX. We now describe how we check if $h(G) = i$ holds for increasing values of i. For $i = 0$ we apply an algorithm of Bodlaender et al., and for $i = 1$ we leverage a classical result of Harary and Nash-Williams.

Theorem 15 [3]. *There is an algorithm which takes a graph G and a tree decomposition of G of width tw as input, runs in $\mathcal{O}^\star((5 + 2^{(\omega+2)/2})^{tw})$ time, and tells whether G is Hamiltonian.*

Theorem 16 [13]. *Let G be a connected graph with at least three edges. Then $L(G)$ is Hamiltonian if and only if G has a dominating Eulerian subgraph.*

For checking if $h(G) \in \{2,3\}$ holds we make use of a structural result of Hong et al. [15]. For a connected subgraph H of graph G the *contraction G/H* is the graph obtained from G by replacing all of $V(H)$ with a single vertex v_H and adding edges between v_H and $V(G) \setminus V(H)$ such that the number of edges in G/H between v_H and any vertex $v \in V(G) \setminus V(H)$ is equal to the number of edges in G with one end point at v and the other in $V(H)$. Note that the graph G/H is, in general, a multigraph with multiedges incident on v_H. Let V_2 be the set of all vertices of G of degree two, and let $\hat{V} = V(G) \setminus V_2$. A *lane* of G is either (i) a path whose end-vertices are in $\hat{V}$ and internal vertices (if any) are in V_2, or (ii) a cycle which contains exactly one vertex from $\hat{V}$. The *length* of a lane is the number of edges in the lane. An *end-lane* is a lane which has a degree-one vertex of G as an end-vertex.

For $i \in \{2,3\}$, let U_i be the union of lanes of length *less than* i. Let $C_1^i, C_2^i, \ldots,$[3] $C_{p_i}^i$ be the connected components of $G[\hat{V}] \cup U_i$. Then each C_j^i consists of components of $G[\hat{V}]$ connected by lanes of length less than i. Let $H^{(i)}$ be the graph obtained from G by contracting each of the connected subgraphs $C_1^i, C_2^i, \ldots, C_{p_i}^i$ to a distinct vertex. Let D_j^i denote the vertex of $H^{(i)}$ obtained by contracting subgraph C_j^i of G. Let $\tilde{H}^{(i)}$ be the graph obtained from $H^{(i)}$ by these steps:

1. Delete all lanes beginning and ending at the same vertex D_j^i.
2. If there are two vertices D_j^i, D_k^i in $H^{(i)}$ which are connected by ℓ_1 lanes of length at least $i + 2$ and ℓ_2 lanes of length i or $i + 1$ such that $\ell_1 + \ell_2 \geq 3$ holds, then delete an arbitrary subset of these lanes such that there remain ℓ_3 lanes with length at least $i + 2$ and ℓ_4 lanes of length i or $i + 1$, where

$$
(\ell_3, \ell_4) = \begin{cases} (2,0) & \text{if } \ell_1 \text{ is even and } \ell_2 = 0; \\ (1,0) & \text{if } \ell_1 \text{ is odd and } \ell_2 = 0; \\ (1,1) & \text{if } \ell_2 = 1; \\ (0,2) & \text{if } \ell_2 \geq 2. \end{cases}
$$

3. Delete all end-lanes of length i, and replace each lane of length i or $i + 1$ by a single edge.

Theorem 17 [15, See Theorem 3]. *Let G be a connected graph with $h(G) \geq 2$ and with at least one vertex of degree at least three, and let $\tilde{H}^{(2)}, \tilde{H}^{(3)}$ be graphs constructed from G as described above. Then*

- $h(G) = 2$ *if and only if $\tilde{H}^{(2)}$ has a spanning Eulerian subgraph; and*
- $h(G) = 3$ *if and only if $h(G) \neq 2$ and $\tilde{H}^{(3)}$ has a spanning Eulerian subgraph.*

For checking if $h(G) = i$ holds for $i \in \{4, 5, \ldots\}$ we appeal to a reduction due to Xiong and Liu [27]. Let $\mathcal{L} = \{L_1, L_2, \ldots, L_t\}$ be a set of lanes (called *branches* in [27]) in G, each of length at least 2. A *contraction* of G by $\mathcal{L}$, denoted $G//\mathcal{L}$, is a graph obtained from G by contracting one edge of each lane in $\mathcal{L}$. Note that $G//\mathcal{L}$ is not, in general, unique.

Theorem 18 [27, Theorem 20]. *Let G be a connected graph with $h(G) \geq 4$ and let $\mathcal{L}$ be the set of all lanes of length at least 2 in G. Then $h(G) = h(G//\mathcal{L}) + 1$.*

We can now prove

Theorem 3. *There is an algorithm which solves an instance $(G, \mathcal{T}, tw, r)$ of* HAMILTONIAN INDEX *in $\mathcal{O}^{\star}((1 + 2^{(\omega+3)})^{tw})$ time.*

Proof. We first apply Theorem 15 to check if G is Hamiltonian. If G is Hamiltonian then we return **yes**. If G is not Hamiltonian and $r = 0$ holds then we return **no**. Otherwise we apply Theorem 6 and Theorem 16 to check if $L(G)$ is Hamiltonian. If $L(G)$ is Hamiltonian then we return **yes**. If $L(G)$ is not Hamiltonian and $r = 1$ holds then we return **no**.

At this point we know—since G is connected, is not a path, and is not Hamiltonian—that G has at least one vertex of degree at least three, and that $h(G) \geq 2$ holds. We construct the graph $\tilde{H}^{(2)}$ of Theorem 17 and use Corollary 7 to check if $\tilde{H}^{(2)}$ has a spanning Eulerian subgraph. If it does then we return **yes**. If it does not and $r = 2$ holds then we return **no**. Otherwise we construct the graph $\tilde{H}^{(3)}$ of Theorem 17 and use Corollary 7 to check if $\tilde{H}^{(3)}$ has a spanning Eulerian subgraph. If it does then we return **yes**. If it does not and $r = 3$ holds then we return **no**.

At this point we know that $h(G) \geq 4$ holds. We compute the set $\mathcal{L}$ of all lanes of G of length at least 2, and a contraction $G' = G//\mathcal{L}$. We construct a tree decomposition $\mathcal{T}'$ of G' from $\mathcal{T}$ as follows: For each edge xy of G which is contracted to get G', we introduce a new vertex v_{xy} to each bag of $\mathcal{T}$ which contains at least one of $\{x, y\}$. We now delete vertices x and y from all bags. It is easy to verify that the resulting structure $\mathcal{T}'$ is a tree decomposition of G', of width $tw' \leq tw$. We now recursively invoke the algorithm on the instance $(G', \mathcal{T}', tw', (r-1))$ and return its return value (**yes** or **no**).

The correctness of this algorithm follows from Theorem 15, Theorem 6, Theorem 16, Theorem 17, Corollary 7, and Theorem 18. As for the running time, checking Hamiltonicity takes $\mathcal{O}^{\star}((5 + 2^{(\omega+2)/2})^{tw})$ time (Theorem 15). Checking if $L(G)$ is Hamiltonian takes $\mathcal{O}^{\star}((1 + 2^{(\omega+3)})^{tw})$ time (Theorem 6, Theorem 16). The graphs $\tilde{H}^{(2)}$ and $\tilde{H}^{(3)}$ of Theorem 17 can each be constructed in polynomial time, and checking if each has a spanning Eulerian subgraph takes $\mathcal{O}^{\star}((1 + 2^{(\omega+3)})^{tw})$ time (Corollary 7). The graph G' and its tree decomposition $\mathcal{T}'$ of width tw' can be constructed in polynomial time. Given that $5 + 2^{(\omega+2)/2} < 1 + 2^{(\omega+3)}$ and $tw' \leq tw$ hold, we get that the running time of the algorithm satisfies the recurrence $T(r) = \mathcal{O}^{\star}((1 + 2^{(\omega+3)})^{tw}) + T(r-1)$. Since we recurse only if $r < |V(G)| - 3$ holds we get that the recurrence resolves to $T(r) = \mathcal{O}^{\star}((1 + 2^{(\omega+3)})^{tw})$. □

4   Conclusion

The Hamiltonian Index $h(G)$ of a graph G is a generalization of the notion of Hamiltonicity. It was introduced by Chartrand in 1968, and has received a lot of attention from graph theorists over the years. It is known to be NP-hard to check if $h(G) = t$ holds for any fixed integer $t \geq 0$, even for subcubic graphs G. We initiate the parameterized complexity analysis of the problem of finding $h(G)$ with the treewidth $tw(G)$ of G as the parameter. We show that this problem is in FPT and can be solved in $\mathcal{O}^{\star}((1 + 2^{(\omega+3)})^{tw(G)})$ time. This running time matches that of the current fastest algorithm, due to Misra et al. [20], for checking if $h(G) = 1$ holds. We also derive an algorithm—described in the full version— of our own, with the same running time, for checking if $h(G) = 1$ holds. A key ingredient of our solution for finding $h(G)$ is an algorithm which solves the EULERIAN STEINER SUBGRAPH problem in $\mathcal{O}^{\star}((1 + 2^{(\omega+3)})^{tw(G)})$ time. This is—to the best of our knowledge—the first FPT algorithm for this problem, and it subsumes known algorithms for the special case of SPANNING EULERIAN SUBGRAPH in series-parallel graphs and planar graphs. We note in passing that it is not clear that the algorithm of Misra et al. for solving LELP can be adapted to check for larger values of $h(G)$. We believe that our FPT result on EULERIAN STEINER SUBGRAPH could turn out to be useful for solving other problems as well.

Two different approaches to checking if $h(G) = 1$ holds—Misra et al.'s approach via LELP and our solution using DOMINATING EULERIAN SUBGRAPH— both run in $\mathcal{O}^{\star}((1 + 2^{(\omega+3)})^{tw(G)})$ time. Does this suggest the existence of a matching lower bound, or can this be improved? More generally, can $h(G)$ be found in the same FPT running time as it takes to check if G is Hamiltonian (currently: $\mathcal{O}^{\star}((5 + 2^{(\omega+2)/2})^{tw(G)})$ due to Bodlaender et al.)? Since $tw(G) \leq |V(G)| - 1$, our algorithm implies an $\mathcal{O}^{\star}((1 + 2^{(\omega+3)})^{|V(G)|})$-time exact exponential algorithm for finding $h(G)$. We ask if this can be improved, as a first step, to the classical $\mathcal{O}^{\star}(2^{|V(G)|})$ bound for Hamiltonicity.

References

1. Bertossi, A.A.: The edge Hamiltonian path problem is NP-complete. Inform. Process. Lett. **13**(4–5), 157–159 (1981). https://doi.org/10.1016/0020-0190(81)90048-X
2. Bodlaender, H.L.: A partial k-arboretum of graphs with bounded treewidth. Theor. Comput. Sci. **209**(1), 1–45 (1998)
3. Bodlaender, H.L., Cygan, M., Kratsch, S., Nederlof, J.: Deterministic single exponential time algorithms for connectivity problems parameterized by treewidth. Inform. Comput. **243**, 86–111 (2015)
4. Catlin, P.A.: Supereulerian graphs: a survey. J. Graph Theor. **16**, 177–196 (1992)
5. Catlin, P.A., Janakiraman, I.T.N., Srinivasan, N.: Hamilton cycles and closed trails in iterated line graphs. J. Graph Theor. **14**(3), 347–364 (1990)
6. Chartrand, G.: On Hamiltonian line-graphs. Trans. Am. Math. Soc. **134**(3), 559–566 (1968)
7. Cygan, M., et al.: Parameterized Algorithms. Springer, Heidelberg (2015). https://doi.org/10.1007/978-3-319-21275-3
8. Cygan, M., Nederlof, J., Pilipczuk, M., Pilipczuk, M., van Rooij, J.M., Wojtaszczyk, J.O.: Solving connectivity problems parameterized by treewidth in single exponential time. In: FOCS 2011, pp. 150–159. IEEE (2011)

9. Downey, R.G., Fellows, M.R.: Fundamentals of Parameterized Complexity. Springer, Heidelberg (2013). https://doi.org/10.1007/978-1-4471-5559-1
10. Flum, J., Grohe, M.: Parameterized Complexity Theory. Springer, Heidelberg (2006). https://doi.org/10.1007/3-540-29953-X
11. Fomin, F.V., Lokshtanov, D., Panolan, F., Saurabh, S.: Efficient computation of representative families with applications in parameterized and exact algorithms. J. ACM **63**(4), 29:1–29:60 (2016)
12. Garey, M.R., Johnson, D.S., Tarjan, R.E.: The planar Hamiltonian circuit problem is NP-complete. SIAM J. Comput. **5**(4), 704–714 (1976)
13. Harary, F., Nash-Williams, C.S.J.: On Eulerian and Hamiltonian graphs and line graphs. Can. Math. Bull. **8**(6), 701–709 (1965)
14. Hauptmann, M., Karpiński, M.: A compendium on Steiner tree problems. Inst. für Informatik (2013). http://citeseerx.ist.psu.edu/viewdoc/summary?doi=10.1.1.392.7444
15. Hong, Y., Lin, J.L., Tao, Z.S., Chen, Z.H.: The Hamiltonian index of graphs. Discrete Math. **309**(1), 288–292 (2009)
16. Kammer, F., Tholey, T.: Approximate tree decompositions of planar graphs in linear time. Theor. Comput. Sci. **645**, 60–90 (2016)
17. Lai, H.J., Shao, Y., Yan, H.: An update on supereulerian graphs. WSEAS Trans. Math. **12**(9), 926–940 (2013)
18. Lampis, M., Makino, K., Mitsou, V., Uno, Y.: Parameterized edge hamiltonicity. In: Kratsch, D., Todinca, I. (eds.) WG 2014. LNCS, vol. 8747, pp. 348–359. Springer, Cham (2014). https://doi.org/10.1007/978-3-319-12340-0_29
19. Lampis, M., Makino, K., Mitsou, V., Uno, Y.: Parameterized edge hamiltonicity. Discrete Appl. Math. (2017). https://doi.org/10.1016/j.dam.2017.04.045
20. Misra, N., Panolan, F., Saurabh, S.: On the parameterized complexity of edge-linked paths. In: van Bevern, R., Kucherov, G. (eds.) CSR 2019. LNCS, vol. 11532, pp. 286–298. Springer, Cham (2019). https://doi.org/10.1007/978-3-030-19955-5_25
21. Philip, G., Rani, M.R., Subashini, R.: On computing the Hamiltonian index of graphs. CoRR abs/1912.01990 (2019). http://arxiv.org/abs/1912.01990
22. Pulleyblank, W.R.: A note on graphs spanned by Eulerian graphs. J. Graph Theory **3**(3), 309–310 (1979). https://doi.org/10.1002/jgt.3190030316
23. Richey, M., Parker, R.G., Rardin, R.: On finding spanning Eulerian subgraphs. Naval Res. Logistics Q. **32**(3), 443–455 (1985)
24. Ryjáček, Z., Woeginger, G.J., Xiong, L.: Hamiltonian index is NP-complete. Discrete Appl. Math. **159**(4), 246–250 (2011). https://doi.org/10.1016/j.dam.2010.08.027
25. Sau, I., Thilikos, D.M.: Subexponential parameterized algorithms for degree-constrained subgraph problems on planar graphs. J. Discrete Algorithms **8**(3), 330–338 (2010)
26. Williams, V.V.: Multiplying matrices faster than Coppersmith-Winograd. In: Proceedings of the Forty-Fourth Annual ACM Symposium on Theory of Computing, pp. 887–898. ACM (2012)
27. Xiong, L., Liu, Z.: Hamiltonian iterated line graphs. Discrete Math. **256**(1–2), 407–422 (2002)

A Lower Bound for the Query Phase of Contraction Hierarchies and Hub Labels

Tobias Rupp[(✉)] and Stefan Funke

Institut für Formale Methoden der Informatik, Universität Stuttgart,
Stuttgart, Germany
rupp@fmi.uni-stuttgart.de

Abstract. We prove a $\Omega(\sqrt{n})$ lower bound on the query time for contraction hierarchies (CH) as well as hub labels, two popular speed-up techniques for shortest path routing. Our construction is based on a graph family not too far from subgraphs that occur in real-world road networks, in particular it is planar and has bounded degree. Additionally, we borrow ideas from our lower bound proof to come up with *instance-based* lower bounds for concrete road network instances of moderate size, reaching up to 83% of an upper bound given by a constructed CH.

Keywords: Route planning · Contraction hierarchies · Hub labelling

1 Introduction

While the problem of computing shortest paths in general graphs with non-negative edge weights seems to have been well understood already decades ago, the last 10–15 years have seen tremendous progress when it comes to the specific problem of efficiently computing shortest paths in *real-world road networks*. Here the idea is to spend some time for preprocessing where auxiliary information about the network is computed and stored, such that subsequent queries can be answered much faster than standard Dijkstra's algorithm. One might classify most of the employed techniques into two classes: ones that are based on *pruned graph search* and others that are based on *distance lookups*. Most approaches fall into the former class, e.g., reach-based methods [11,12], highway hierarchies [16], arc-flags-based methods [6], or contraction hierarchies (CH) [10]. Here, Dijkstra's algorithm is given a hand to ignore some vertices or edges during the graph search. The speed-up for road networks compared to plain Dijkstra's algorithm ranges from one up to three orders of magnitudes [10,12]. In practice, this means that a query on a country-sized network like that of Germany (around 20 million nodes) can be answered in less than a *millisecond* compared to few seconds of Dijkstra's algorithm. While these methods directly yield the actual shortest path, the latter class is primarily concerned with the computation of the (exact) distance between given source and target – recovering the actual path often requires some additional effort. Examples of such distance-lookup-based methods

© Springer Nature Switzerland AG 2020
H. Fernau (Ed.): CSR 2020, LNCS 12159, pp. 354–366, 2020.
https://doi.org/10.1007/978-3-030-50026-9_26

are transit nodes [3,4] and hub labels (HL) [1]. They allow for the answering of *distance queries* another one or two orders of magnitudes faster.

There have also been attempts at theoretically explaining the impressive practical performance of these speed-up schemes. These approaches first identify certain properties of a graph, which supposedly characterize 'typical' inputs in the real world and then show that for graphs satisfying these properties, certain speed-up schemes have guaranteed query/construction time or space consumption. Examples of such graph characterizations are given via *highway dimension* [2], *skeleton dimension* [13], or *bounded growth* [7]. It is important to note that these approaches are all concerned with *upper bounds*. For example, in [2] it is shown that for graphs with highway dimension h, after a preprocessing step, the number of considered nodes during a CH query is $O(h \log n)$, which for polylogarithmic h is polylogarithmic in the network size. While small (i.e., constant or polylogarithmic) highway dimension is often assumed for real-world networks, it is important to note that even a simple $\sqrt{n} \times \sqrt{n}$ grid has highway dimension $h = \Theta(\sqrt{n})$, so the upper bound guaranteed by [2] is $O(\sqrt{n} \log n)$. Similarly, an analysis based on the bounded growth property [7] shows an upper bound of $O(\sqrt{n} \log n)$. In this work are concerned with two specific speed-up techniques, namely contraction hierarchies [10], and hub labels [1] and provide *lower bounds*.

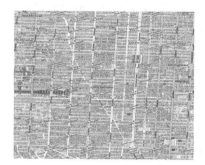 

Fig. 1. Grid-like substructures in real networks (Manhattan on the left, Urbana-Champaign on the right) (by OpenStreetMap).

As grid-like substructures are quite common in real-world road networks, see Fig. 1, one might ask whether better upper bounds for such networks are impossible in general or whether a polylogarithmic upper bound could be shown via more refined proof or CH construction techniques. Our work settles this question for contraction hierarchies as well as hub labels up to a logarithmic factor. We show that for CH, no matter what contraction order is chosen, and for HL, no matter how the hub labels are generated, there are grid networks for which the average number of nodes to be considered during a query is $\Omega(\sqrt{n})$.

The insights of our theoretical lower bound analysis also allow us to devise a scheme to compute *instance-based* lower bounds, that is, for a given concrete road network, we algorithmically compute a lower bound on the average search space size. Note that such an instance-specific lower bound is typically much stronger than an analytical lower bound.

1.1 Related Work

In [2], a graph property called *highway dimension* is proposed to analyze shortest path speed-up schemes. Intuitively, the highway dimension h of a graph is small if there exist sparse local hitting sets for shortest paths of a certain length. For contraction hierarchies and hub labels, a search space size of $O(h \log n)$ was proven (using a NP-hard preprocessing phase; polynomial time preprocessing increases this by a $\log h$ factor). While one might hope that real road networks exhibit a 'small' highway dimension, e.g., constant or polylogarithmic, it is known that $h \in \Omega(\sqrt{n})$ holds for grids. For hub labels, the so-called *skeleton dimension* k of [13] has been instrumented to prove a search space size of $O(k \log n)$. Still, for grids, we have $k \in \Omega(\sqrt{n})$. In [5], CH were analyzed for graphs with treewidth t, and a query time of $O(t \log n)$ was shown. Yet, for grids we again have $t \in \Omega(\sqrt{n})$. Finally, the bounded growth model was introduced in [9], which also led to a search space size of $O(\sqrt{n} \log n)$ for realistic graphs including grids. Specifically for planar graphs, the search space is $O(\sqrt{n})$ by combining the planar separator theorem [14] with nested dissection [5]. Therefore, our lower bound for the presented grid graph will be tight. In [17], White constructs for any given highway dimension h a family of graphs $G_{t,k,q}$ (as introduced in [15]) of highway dimension h, such that hub labelling requires a label size of $\Omega(h \log n)$ and CH a query time of $O((h \log n)^2)$. Unfortunately, the graphs $G_{t,k,q}$ according to the author himself "are not representative of real-world graphs. For instance, the graphs do not have small separators and are not planar". In fact this could be more an indication of the unsuitability of the notion of highway dimension to characterize real-world road networks rather than a weakness of [17]. For transit nodes [3], instance-based lower bounds based on an LP formulation and its dual were derived in [8]. We are not aware of results regarding instance-based lower bound constructions for HL or CH.

Our Contribution and Outline

In this paper we prove a lower bound on the search space (and hence the processing time) of the query phase of contraction hierarchies as well as hub labels. More concretely, we define so-called *lightheaded grids* for which we show that the average search space size is $\Omega(\sqrt{n})$, irrespectively of what contraction order or whatever hub labelling scheme was employed. Based on a $\sqrt{n} \times \sqrt{n}$ grid, our graph is planar and has bounded degree. Our lower bound applies to CH [10] and HL [1] schemes. Based on our insights from the lower bound proof, we also show how to construct *instance-based* lower bounds algorithmically. Our experiments indicate, that current CH constructions yield search space sizes close to optimal.

We first introduce basic properties of contraction hierarchies and hub labels and then present our main theoretical result. After showing how to algorithmically construct instance-based lower bounds, we conclude with some experimental results and open problems.

2   Preliminaries

Contraction Hierarchies. The contraction hierarchies approach [10] computes an overlay graph in which so-called shortcut edges span large sections of shortest paths. This reduces the hop length of shortest paths and therefore allows a variant of Dijkstra's algorithm to answer queries more efficiently.

The preprocessing is based on the so-called *node contraction* operation. Here, a node v as well as its adjacent edges are removed from the graph. In order not to affect shortest path distances between the remaining nodes, shortcut edges are inserted between two neighbors u and w of v, if and only if uvw was a shortest path (which can easily be checked via a Dijkstra run). The cost of the new shortcut edge (u, w) is the sum of costs of (u, v) and (v, w). In the preprocessing phase, all nodes are contracted one-by-one in some order. The rank of the node in this contraction order is also called the *level* of the node.

Having contracted all nodes, the new graph $G^+(V, E^+)$ contains all original edges of G as well as all shortcuts that were inserted in the contraction process. An edge $e = (v, w)$ – original or shortcut – is called upwards, if the level of v is smaller than the level of w, and downwards otherwise. By construction, the following property holds: For every pair of nodes $s, t \in V$, there exists a shortest path in G^+ that consists of a sequence of upward edges followed by a sequence of downward edges. This property allows us to search for the optimal path with a bidirectional Dijkstra only considering upwards edges in the search starting at s, and only downwards edges in the reverse search starting in t. This reduces the search space significantly and allows for answering of shortest path queries within the *milliseconds* range compared to *seconds* on a country-sized road network. We call this algorithm the CH-Dijkstra.

For our lower bound later on, the notion of *direct search space (DSS)* (as defined in [9]) is useful. A node w is in $DSS(v)$, if on the shortest path from v to w all nodes have level at most $level(w)$. Hence, w will be settled with the correct distance $d(v, w)$ in the CH-Dijkstra run. As a remark, the set of nodes considered during a CH-Dijkstra is usually a superset of $DSS(v)$ as also nodes on monotonously increasing (w.r.t. level) but non-shortest paths are considered. Our construction will lower bound the size of $DSS(v)$ which obviously also lower bounds the number of nodes that have to be considered during CH-Dijkstra.

Hub Labels. Hub labelling is a scheme to answer shortest path distance queries which differs fundamentally from graph search based methods. Here the idea is to compute for every $v \in V$ a *label* $L(v)$ such that for given $s, t \in V$ the distance between s and t can be determined by just inspecting the labels $L(s)$ and $L(t)$. All the labels are determined in a preprocessing step (based on the graph G), later on, the graph G can even be discarded. There have been different approaches to compute such labels (even in theory); we will be concerned with hub labels that work well for road networks, following the ideas in [1]. To be more concrete, the labels we are interested in have the following form:

$$L(v) = \{(w, d(v, w)) : w \in H(v)\}$$

Here we call $H(v)$ a set of *hubs* – important nodes – for v. The hubs must be chosen such that for any s, t, the shortest path from s to t intersects $H(s) \cap H(t)$.

If such label sets are computed, the computation of the shortest path distance between s and t boils down to determining the node $w \in H(s) \cap H(t)$ minimizing the summed distance. If the labels $L(.)$ are stored lexicographically sorted, this can be done in a cache-efficient manner in time $O(|L(s)| + |L(t)|)$.

There is a very efficient method to generate hub labels based on CH as proposed in [1], but this is not the only method. Our lower bound applies to any hub labelling scheme. While hub labels are amongst the to-date fastest distance lookup schemes for shortest paths (query times in the *microseconds* range), their main drawback is the quite huge space requirement (up to 100 node-distance pairs have to be stored for each node as label in a country-sized network).

3   Theory: A Lower Bound Construction

In this section of the paper we first provide a simple graph construction, which is essentially a slightly modified $\sqrt{n} \times \sqrt{n}$ grid graph with some desirable properties. Then we provide a lower bound on the direct search space size of *any* contraction order via an amortized analysis. A slight variation of the analysis also yields a lower bound for any hub labelling scheme. For the sake of simplicity, we assume without loss of generality that n is always a square number and a multiple of 4 for our analysis. Furthermore, our construction assumes an *undirected* graph, yet generalization to the directed case is quite straightforward.

3.1   The Lightheaded $\sqrt{n} \times \sqrt{n}$-grid G_{lh}

The basis of our construction is a regular $\sqrt{n} \times \sqrt{n}$ grid with uniform edge costs. We then modify only costs of the *horizontal* edges such that they become 'lighter towards the head', hence the name *lightheaded grid*. More precisely, the horizontal edges in row i ($i = 0, 1, \ldots, \sqrt{n} - 1$, counted from top to bottom) have cost $1 + i\epsilon$ for some small enough $\epsilon < 1$. See Fig. 2, for an example.

Shortest Path Trees in Lightheaded Grids. For small enough choice of ϵ, the following Lemma holds:

Lemma 1. *For $\epsilon < 1/n$, the shortest path between some s and t in a lightheaded grid G_{lh} is unique and always consists of a horizontal and a vertical part, where the horizontal part is on the height of the higher of the nodes s and t.*

Proof. If a shortest path between s and t in the *unweighted* grid has cost d, the modified edge costs add a cost of less than 1 for the same path, hence all shortest paths in G_{lh} have (unweighted) Manhattan distance d. Let $d = d_v + d_h$ where d_v is the vertical, d_h the horizontal distance. Any shortest Manhattan path must be composed of d_h horizontal and d_v vertical segments. In G_{lh}, horizontal edges towards the top have lower cost, hence the shortest path must have all its horizontal edges on the height of the higher of the nodes s and t.   □

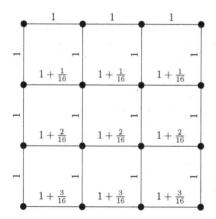

Fig. 2. Lightheaded grid for $n = 16, \epsilon = 1/16$.

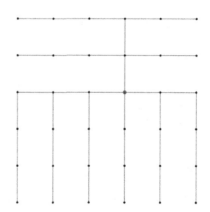

Fig. 3. Shortest path tree from red source. (Color figure online)

See Fig. 3 for an illustration of a shortest path tree in the lightheaded grid. Observe that for all targets in the lower left and the upper right parts, the shortest path has an upper left corner, whereas for all targets in the upper left and lower right part, it has an upper right corner.

3.2 Lower Bounding the Direct Search Space

Let us now assume that the contraction hierarchy has been created with an *arbitrary* contraction order. We will show that no matter what this contraction order is, the average size of the direct search space is $\Omega(\sqrt{n})$.

In our analysis we only consider shortest path trees rooted in the top right quarter of the grid (there are $n/4$ of them). For these shortest path trees, their lower left part always contains a subtree like Fig. 5 of size $\Theta(n)$.

The idea of the analysis is to identify pairs (x, v) such that $v \in DSS(x)$. We will consider each of the shortest path trees rooted at s in the top right quarter as depicted in Fig. 4 and for each identify $\Theta(\sqrt{n})$ such pairs (not necessarily with $x = s$). The main challenge will be to make sure that no double counting of pairs occurs when considering all these shortest path trees.

Let us focus on one shortest path tree rooted at s and the subtree of the lower left part as shown in Fig. 5. By construction, we have $\Theta(\sqrt{n})$ vertical branches in this subtree. Consider one branch, spawning at its corner node c. Furthermore, let s' be the node in the branch which has the same distance to c as c to s. One can think of s being mirrored at c, see Fig. 5. Let w be the highest-level node on the shortest path from s to s'. There are two cases: (a) w lies on the vertical branch including c (this is depicted as w_1 in Fig. 5). (b) w lies on the horizontal part of the shortest path from s to s' excluding c (this is depicted as w_2 in the Figure). In case (a) we generate (s, w_1) since obviously $w_1 \in DSS(s)$. In case

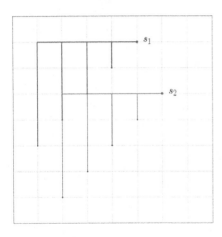

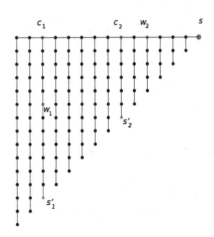

Fig. 5. Subtree in the lower left of a shortest path tree and charging argument.

Fig. 4. Relevant lower left parts of shortest path trees rooted in top right quarter.

(b), we cannot simply generate (s, w_2) since the same pair might be identified when considering other branches in the shortest path tree from s leading to double counting. But we also know that w_2 lies in the direct search space of s_2'. Intuitively, we charge s_2' by generating the pair (s_2', w_2).

So when considering the shortest path tree of a node s we generate exactly one pair (x, v) for each of the $\Theta(\sqrt{n})$ vertical branches to the lower left. Let us first show that we did not double count in this process.

Lemma 2. *No pair (x, v) is generated more than once.*

Proof. Consider a pair (s, v) that was generated according to case (a), i.e., s lies to the upper right or the right of v. Clearly, the same pair cannot be generated according to case (a) again, since the vertical branch in which v resides is only considered once from source s. But it also cannot be generated according to case (b), since these pairs have always s to the lower left of v.

A pair (s, v) generated according to case (b) has s to the lower left of v, hence cannot be generated by case (a) as these pairs have s to the upper right or right of v. As (s, v) was generated according to case (b), it was generated when inspecting the shortest path tree from a source s' which is the node s mirrored at the corner vertex of the shortest path from s to v. But this source s' is uniquely determined, so (s, v) can only be generated when the shortest path tree rooted at s' with the vertical branch containing v was considered. $\square$

Now we are ready to prove the first main result of this paper.

Theorem 1. *The average direct search space of G_{lh} is $\Omega(\sqrt{n})$.*

Proof. In our process we considered $n/4$ shortest path trees, in each of which we identified $\Omega(\sqrt{n})$ pairs (x, v) where $v \in DSS(x)$ and no pair appears twice. Hence we have identified $\Omega(n\sqrt{n})$ such pairs, which on average yields $\Omega(\sqrt{n})$.   □

3.3   Lower Bounding of Hub Label Sizes

Note that the above argument and proof can be modified to also cover label sizes in a hub labelling scheme. Assume hub labels according to an arbitrary hub labelling scheme have been constructed. Then, when considering the shortest path tree rooted at s and the node s' in a vertical branch, we define w to be a node in $H(s) \cap H(s')$. A pair (x, v) corresponds to $v \in H(x)$. Exactly the same arguments as above apply and we obtain the following second main result:

Theorem 2. *The average hub label size of G_{lh} is $\Omega(\sqrt{n})$.*

4   Practice: Instance-Based Lower Bounds

Our theoretical lower bound as just proven only applies to the lightheaded grid G_{lh} as defined before. Yet, even though similar substructures appear in real-world road networks, see Fig. 1, the typical road network is certainly not a lightheaded grid and hence our lower bound proof does not apply.

Still, we can use the insights from the lower bound proof to construct *instance-based* lower bounds algorithmically. Concretely, for a given road network instance, we aim at computing a certificate which *for this instance* proves that the average search space size of a CH query, or the average hub label size cannot be below some lower bound, no matter what CH or HL construction was used.

Note that while for the previous lower bound proof we assumed an undirected graph for sake of a simpler exposition, we will now also include the more general case of a directed graph. To address the bidirectional nature of the CH-Dijkstra here we do now also have to differentiate between forward and backward search space. In the same vein we refer to the forward and backward shortest path tree as $SPT_{\rightarrow}(v)$ and $SPT_{\leftarrow}(v)$, respectively. Both the CH as well as the HL scheme can be easily generalized to directed graphs; in case of HL, compute for each node two labels, an *outlabel* $L^{out}(v)$ storing distances *from* v to hubs and an *inlabel* $L^{in}(v)$ storing distances from hubs *to* v. A query from s to t is then answered by scanning the outlabel $L^{out}(s)$ and the inlabel $L^{in}(t)$. CH also generalizes in a straightforward manner to the directed case, see [10].

4.1   Witness Triples

In our lower bound proof for the lightheaded grid we identified pairs (x, v) such that $v \in DSS(x)$, making use of a concrete (but arbitrary) CH (or HL) for the lightheaded grid to actually identify those pairs (x, v). We cannot do this for a given instance of a road network, since we would have to consider all possible CH/HL constructions. So instead of pairs (x, v) let us now try to identify *witness triples* (x, c, v) where c is again a node on the (directed) shortest path from x to v. The intuition for (x, c, v) is the following: On the shortest path $\pi(x, v) = x \ldots c \ldots v$, some node of the suffix $c \ldots v$ of π must be in the forward search space of x, or some node on the prefix $x \ldots c$ must be in the backward search space of v. This intuition mimics the proof idea in the previous section but also allowing for *directed* graphs and leaving the choice c open.

In the following, we sometimes treat paths just as sets of nodes to simplify presentation. Let us first define the notion of a conflict between two triples.

Definition 1 (conflict between triples). *We say two triples (x, c, v) and (x', c', v') are in conflict if at least one of the following conditions holds true:*

1. $x = x'$ and $\pi(c, v) \cap \pi(c', v') \neq \emptyset$
2. $v = v'$ and $\pi(x, c) \cap \pi(x', c') \neq \emptyset$

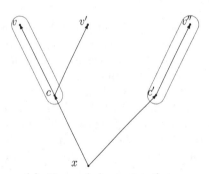 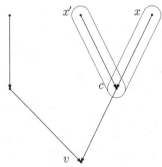

(a) Forward shortest path tree $SPT_\rightarrow(x)$: The witness triple (x, c, v) where the relevant nodes for the conflict check are in the green bubble. There is no conflict with (x, c, v'') and its relevant nodes in the red bubble.

(b) $SPT_\leftarrow(v)$: The green witness triple (x, c, v) of the left Figure shown in the backward tree of v. It is in conflict with the blue triple (x', c, v) because they share the node c.

Fig. 6. Examples for witness triples in shortest path trees.

See Fig. 6a for a non-conflicting and Fig. 6b for a conflicting example. Our goal will be to identify an as large as possible set W of *conflict-free* witness triples. The following Lemma proves that the size of such a set W lower bounds the average search space and label size.

Lemma 3. *If W is a set of conflict-free witness triples, then $|W|$ lower bounds the sum of (backward and forward) search spaces of all nodes in the network.*

Proof. Consider a triple (x, c, v) and the following two cases:

1. it accounts for a node in $\pi(c, v)$ in the forward search space of x. Nodes in the forward search space of x can only be accounted for by triples (x, c', v'). But since $\pi(c, v) \cap \pi(c', v') = \emptyset$ due to W being conflict-free, we have not doubly accounted for it.
2. it accounts for a node in $\pi(x, c)$ in the backward search space of v. Nodes in the backwards search space of v can only be accounted for by triples (x', c', v). But since $\pi(x, c) \cap \pi(x', c') = \emptyset$ due to W being conflict-free, we have not doubly accounted for it. $\qquad\square$

For any witness set W, $|W|/(2n)$ yields a lower bound on the average size of an in- or outlabel in case of the HL scheme, and on the average size of the direct search space from a single node in case of the CH scheme.

So it remains to compute a large conflict-free witness set W. Enumerating all potential triples in $V \times V \times V$ seems somewhat inefficient (having to inspect $\Theta(n^3)$ such triples), and still leaves the problem of identifying a large conflict-free subset amongst them unsolved. Hence, in the following we propose a simple greedy heuristic. Again, let us emphasize that the resulting lower bound holds for *any* CH/HL construction for the given road network.

4.2   Generation of Witness Triples

The high-level idea is as follows: We first compute and store all $2n$ shortest path trees within the network (both forward and reverse). We then enumerate candidate shortest paths in increasing hop length [1] $\ell = 1, 3, 7, 15, \ldots$ (always doubling and adding 1) from the shortest path trees, greedily adding corresponding triples to W in case no conflict is created. As center node c, we always choose the center node of a candidate path (in terms of hops), see, e.g., Fig. 6a.

This specific choice for c as well as the decision to always roughly double the lengths of the considered candidate paths was motivated by the observation in practice that when considering a candidate triple (x, c, v), checking for conflict all triples $(., ., v)$ and $(x, ., .)$ already present in W becomes quite expensive for large W. Intuitively, the latter decision ensures that triples with different hop lengths can never be in conflict at all, the former that not too many triples with the same source or target are generated.

Our actual greedy algorithm is stated in Algorithm 1. Since we made sure that conflicts can only occur between triples with identical hop length, we can restrict the check for conflicts to the candidate set of triples of identical hop length. For an example of our greedy heuristic execution, look at the shortest path trees in Fig. 6: Let us assume we collected the set of candidates for length 3 which contains (x, c, v), (x, c, v'), (x, c', v'') and (x', c, v) beside other triples. We now pick the candidate (x, c, v) and add it to W. Clearly, we remove this from our candidate set but we also have to remove (x, c, v') because it would lead to a conflict in the forward shortest path tree $SPT_{\rightarrow}(x)$ and also remove (x', c, v) because of the conflict in the backward tree $SPT_{\leftarrow}(v)$.

---

[1] We define hop length as the number of nodes here.

Algorithm 1. The algorithm to find a set of non-conflicting witness triples.

```
 1: procedure FINDWITNESSES(G)
 2:     SPTs ← COMPUTESHORTESTPATHTREES(G)
 3:     W ← ∅
 4:     ℓ ← 1
 5:     while ℓ ≤ Diameter(G) do
 6:         WC ← collectCandidatesOfLength(SPTs, ℓ)
 7:         while WC ≠ ∅ do
 8:             (x, c, v) ← WC[0]
 9:             W ← W ∪ (x, c, v)
10:             WC ← WC\(x, c, v)
11:             PRUNECONFLICTINGTRIPLES(WC, (x, c, v))
12:         ℓ ← 2ℓ + 1
13:     return W
```

Storing and computing the shortest path trees require $\Theta(n^2)$ space and $\Omega(n(n \log n + m))$ time when using Dijkstra's algorithm (here, $m = |E|$). Generation and pruning of candidate triples can be bounded by $O(n^3)$, yet in practice the computation of the shortest path trees clearly dominates the running time.

4.3 Experimental Results

We implemented our witness search heuristic in C++ and evaluated it on several graphs. Besides a very small *test graph* for sanity checking, we extracted real-world networks graphs based on Open Street Map (OSM) data. We picked *lower manhattan* to get a grid-like graph. For the street network of the german federal city-state *Bremen*, we created a large version *car* which includes all streets passable by car, as well as the versions *fast car* and *highway* only containing streets allowing at least a speed of 50 km/h and 130 km/h respectively. The code was run on a 24-core Xeon(R) CPU E5-2650v4, 768 GB RAM machine.

To assess the quality of our lower bounds, we constructed a CH where the contraction order was defined by the so called *edge-difference* heuristic, that is, nodes that introduce few additional shortcuts are contracted first. This is one of the most popular contraction order heuristics. From this CH, we calculated the average $|DSS|$ and compared it with the lower bound.

In Table 1 we list our results. Our biggest graph has over a quarter of a million edges and almost $120k$ nodes. As expected, the space consumption is quadratic which makes our current algorithm infeasible for continental-sized networks. For the large bremen graph, our 24-core machine took 32.5 h to complete the lower bound construction via our greedy heuristic and used 354 GB of RAM. Most important is the quotient $\frac{LB}{|DSS|}$ which gives us a hint about the quality of our computed lower bound: For the *highway* version of bremen, we achieve even 83%, that is, the average search spaces in the computed CH are guaranteed to be less than a factor of 1.25 above the optimum. Note however, that this graph is atypical for road networks in the sense that it is far from being fully connected.

Table 1. Experimental results.

	test graph manhattan	lower (fast car)	bremen (car)	bremen (highway)	bremen
# nodes	22	2,828	40,426	119,989	1,781
# edges (org)	52	4,020	64,663	227,567	1,766
# edges (CH)	77	7,752	126,055	400,038	3,340
LB-construction space	5.5 KB	233 MB	40 GB	354 GB	84 MB
LB-construction time	<1 s	36 s	100 m	32.5 h	<1 s
LB	6.18	12.43	19.11	22.75	6.31
$\|DSS\|(avg.)$	10.27	29.85	61.99	78.34	7.58
$\frac{LB}{\|DSS\|}$	0.602	0.417	0.308	0.290	0.832

The value of $\frac{LB}{\|DSS\|}$ decreases for our bigger graphs down to around 30%. The decrease can have several reasons: On one hand, our CH is based on a heuristic and is not necessarily optimal (no optimal algorithm is known), so even if we had brute-forced a maximum sized triple-set we would most certainly not achieve 100%. Indeed, our contribution is to show that the gap between the heuristic and the perfect CH is at most $1 - \frac{LB}{DSS}$. On the other hand, the results strongly indicate that the missing of some long triples which do not have a length of 2^i becomes more relevant in bigger graphs. Note that it could also mean that the edge-difference heuristic performs worse on bigger graphs.

It has been shown that constructing the optimal CH (even though according to a different optimality criterion) is NP-hard, see [15], so we actually conjecture that finding the largest conflict-free set of triples would also turn out to be NP-hard if investigated further.

5   Conclusions and Future Work

In this paper we have proven a strong theoretical lower bound on the number of nodes that have to be considered during a CH search or HL lookup. Our lower bound instance is not too far from road network structures that occur in the real world. Our theoretical results imply that existing CH or HL construction schemes are essentially optimal (up to a logarithmic factor) for such grid-like networks. More on the practical side, we instrumented the insights from our lower bound proof to come up with a construction scheme for *instance-based lower bounds* for CH and HL, if a concrete road network instance is given. For moderately sized networks, we could show that current CH and HL construction schemes indeed yield average search spaces not too far away from the optimum (less than a factor of 4). In future work, we aim at making the respective algorithms constructing the instance-based lower bounds more scalable to allow the lower bound construction even for country- or continent-sized networks. Furthermore, it might be worth investigating for which graph classes our lower bounding technique has the potential to compute tight lower bounds.

References

1. Abraham, I., Delling, D., Goldberg, A.V., Werneck, R.F.: Hierarchical hub labelings for shortest paths. In: Epstein, L., Ferragina, P. (eds.) ESA 2012. LNCS, vol. 7501, pp. 24–35. Springer, Heidelberg (2012). https://doi.org/10.1007/978-3-642-33090-2_4
2. Abraham, I., Fiat, A., Goldberg, A.V., Werneck, R.F.: Highway dimension, shortest paths, and provably efficient algorithms. In: Proceedings of 21st Annual ACM-SIAM Symposium on Discrete Algorithms (SODA), pp. 782–793. SIAM (2010)
3. Bast, H., Funke, S., Matijevic, D., Sanders, P., Schultes, D.: In transit to constant time shortest-path queries in road networks. In: ALENEX. SIAM (2007)
4. Bast, H., Funke, S., Sanders, P., Schultes, D.: Fast routing in road networks with transit nodes. Science **316**(5824), 566–566 (2007)
5. Bauer, R., Columbus, T., Rutter, I., Wagner, D.: Search-space size in contraction hierarchies. In: Fomin, F.V., Freivalds, R., Kwiatkowska, M., Peleg, D. (eds.) ICALP 2013. LNCS, vol. 7965, pp. 93–104. Springer, Heidelberg (2013). https://doi.org/10.1007/978-3-642-39206-1_9
6. Bauer, R., Delling, D.: SHARC: fast and robust unidirectional routing. In: ALENEX, pp. 13–26. SIAM (2008)
7. Blum, J., Funke, S., Storandt, S.: Sublinear search spaces for shortest path planning in grid and road networks. In: Proceedings of the 32nd AAAI Conference on Artificial Intelligence (AAAI), pp. 6119–6126. AAAI Press (2018)
8. Eisner, J., Funke, S.: Transit nodes - lower bounds and refined construction. In: Proceedings of the 14th Workshop on Algorithm Engineering and Experiments (ALENEX), pp. 141–149. SIAM/Omnipress (2012)
9. Funke, S., Storandt, S.: Provable efficiency of contraction hierarchies with randomized preprocessing. In: Elbassioni, K., Makino, K. (eds.) ISAAC 2015. LNCS, vol. 9472, pp. 479–490. Springer, Heidelberg (2015). https://doi.org/10.1007/978-3-662-48971-0_41
10. Geisberger, R., Sanders, P., Schultes, D., Vetter, C.: Exact routing in large road networks using contraction hierarchies. Transp. Sci. **46**(3), 388–404 (2012)
11. Goldberg, A.V., Kaplan, H., Werneck, R.F.: Reach for A^*: efficient point-to-point shortest path algorithms. In: ALENEX, pp. 129–143. SIAM (2006)
12. Gutman, R.J.: Reach-based routing: a new approach to shortest path algorithms optimized for road networks. In: ALENEX, pp. 100–111. SIAM (2004)
13. Kosowski, A., Viennot, L.: Beyond highway dimension: small distance labels using tree skeletons. In: Proceedings of the 28th Annual ACM-SIAM Symposium on Discrete Algorithms (SODA), pp. 1462–1478. SIAM (2017)
14. Lipton, R.J., Tarjan, R.E.: A separator theorem for planar graphs. SIAM J. Appl. Math. **36**(2), 177–189 (1979)
15. Milosavljević, N.: On optimal preprocessing for contraction hierarchies. In: Proceedings of the 5th ACM SIGSPATIAL IWCTS, pp. 33–38. ACM (2012)
16. Sanders, P., Schultes, D.: Engineering highway hierarchies. ACM J. Exp. Algorithmics **17**(1) (2012)
17. White, C.: Lower bounds in the preprocessing and query phases of routing algorithms. In: Bansal, N., Finocchi, I. (eds.) ESA 2015. LNCS, vol. 9294, pp. 1013–1024. Springer, Heidelberg (2015). https://doi.org/10.1007/978-3-662-48350-3_84

Kernelization of Arc Disjoint Cycle Packing in α-Bounded Digraphs

Abhishek Sahu[1(✉)] and Saket Saurabh[1,2(✉)]

[1] The Institute of Mathematical Sciences, HBNI, Chennai, India
{asahu,saket}@imsc.res.in
[2] Department of Informatics, University of Bergen, Bergen, Norway

Abstract. In the Arc Disjoint Cycle Packing problem, we are given a directed graph (digraph) G, a positive integer k, and the task is to decide whether there exist k arc disjoint cycles. The problem is known to be W[1]-hard on general digraphs. In this paper we show that the problem admits a polynomial kernel on α-bounded digraphs. That is, we give a polynomial time algorithm, that given an instance (D, k) of Arc Disjoint Cycle Packing, outputs an equivalent instance (D', k') of Arc Disjoint Cycle Packing, such that $k' \leq k$ and the size of D' is upper bounded by a polynomial function of k. For any integer $\alpha \geq 1$, the class of α-bounded digraphs, denoted by $\mathcal{D}_\alpha$, contains a digraph D such that the maximum size of an independent set in D is at most α. That is, in D, any set of $\alpha + 1$ vertices has an arc with both end-points in the set. For $\alpha = 1$, this corresponds to the well-studied class of tournaments. Our results generalizes the recent result by Bessy et al. [MFCS 2019] about Arc Disjoint Cycle Packing on tournaments.

1 Introduction

In this paper we study the Arc Disjoint Cycle Packing problem in the realm of parameterized complexity and in particular from the view-point of kernelization. A problem instance I of size n with a parameter k is fixed-parameter-tractable (FPT) if it has an algorithm with running time $f(k)n^{\mathcal{O}(1)}$ for some computable function f. A kernelization algorithm is a polynomial time algorithm that takes an instance I of size n along with parameter k as inputs and returns an equivalent instance I' of size $f(k)$ with parameter $g(k)$. If f and g are both polynomial functions, then instance I is said to have a *polynomial kernel*. For convenience, the running time $f(k)n^{\mathcal{O}(1)}$ where f grows super-polynomially with k is denoted by $\mathcal{O}^*(f(k))$. For further details on parameterized complexity, we refer to [7,10,13].

The history of packing cycles in undirected and directed graphs is very old. We first brief about cycle packing in undirected graphs (Cycle Packing). Since

This project has received funding from the European Research Council (ERC) under the European Unions Horizon 2020 research and innovation programme (grant no. 819416), and the Swarnajayanti Fellowship grant DST/SJF/MSA-01/2017-18.

© Springer Nature Switzerland AG 2020
H. Fernau (Ed.): CSR 2020, LNCS 12159, pp. 367–378, 2020.
https://doi.org/10.1007/978-3-030-50026-9_27

the publication of the well known Erdős-Pósa theorem in 1965 [11], CYCLE PACKING problems have been studied with great attention. The fixed-parameter-tractability of CYCLE PACKING follows from the Robertson-Seymour theorem [20], a fact observed by Fellows and Langston in the 1980s [12]. In 1994, Bodlaender showed that CYCLE PACKING can be solved in $\mathcal{O}^*(2^{\mathcal{O}(k^2)})$ time [4]. A feedback vertex set is a set of vertices whose deletion results in a forest. The Erdős-Pósa theorem states that there exists a function $f(r) = \mathcal{O}(r \log r)$ such that for each non-negative integer r, every undirected graph either contains r vertex disjoint cycles or has a feedback vertex set consisting of $f(r)$ vertices [11]. It is well known that the treewidth (**tw**) of a graph is not larger than the size of its feedback vertex set, and a naive dynamic programming scheme solves CYCLE PACKING in time $\mathcal{O}^*(2^{\mathcal{O}(\mathbf{tw} \log \mathbf{tw})})$ [7]. Thus, the existence of an $\mathcal{O}^*(2^{\mathcal{O}(k \log^2 k)})$ time algorithm can be viewed as a direct consequence of the Erdős-Pósa theorem. Only recently, Lokshtanov et al. [18] designed an algorithm for CYCLE PACKING running in time $\mathcal{O}^*(2^{\mathcal{O}(\frac{k \log^2 k}{\log \log k})})$. Finally, let us remark about the kernelization complexity of these problems. While the VERTEX DISJOINT CYCLE PACKING does not admit a polynomial kernel unless $\mathsf{coNP} \subseteq \mathsf{NP/poly}$, the EDGE DISJOINT CYCLE PACKING has a polynomial kernel of size $\mathcal{O}(k \log k)$ [5].

In 1996, Reed et al. [19] showed that there exists a function $f(r)$ such that for each non-negative integer r, every undirected graph either contains r vertex disjoint directed cycles or has a directed feedback vertex set consisting of $f(r)$ vertices. However, the similarity with undirected graphs stops here. The parameterized complexity of packing directed cycles is different than that on undirected graphs. On directed graphs both these problems (ARC DISJOINT CYCLE PACKING and VERTEX DISJOINT CYCLE PACKING) are W[1]-hard [6,9]. On the very well known digraph class of tournaments (orientation of complete graphs), VERTEX DISJOINT CYCLE PACKING is same as VERTEX DISJOINT TRIANGLE PACKING which has a $\mathcal{O}(k^2)$ kernel [1]. Recently ARC DISJOINT CYCLE PACKING was studied on tournaments and bipartite tournaments from a parameterized perspective. The problem was shown to have a linear kernel in tournaments [2,3,16] and a quadratic kernel in bipartite tournaments [15]. Independence number played a crucial role in solving the problem on tournaments. So a natural question one can ask is there a relationship between the independence number and a polynomial size kernel? Can we extend the polynomial kernel on tournaments to the class of α-bounded digraphs that contains tournaments? Formally, for any integer $\alpha \geq 1$, the class of α-bounded digraphs, denoted by $\mathcal{D}_\alpha$, is defined as follows [14].

$$\mathcal{D}_\alpha = \{D \ : \ D \text{ is a digraph and the maximum size of an}$$
$$\text{independent set in } D \text{ is at most } \alpha \ \}.$$

Interested in this question we study the ARC DISJOINT CYCLE PACKING problem on $\mathcal{D}_\alpha$. Notice that an α-bounded digraph is a directed graph where the graph induced on any $\alpha + 1$ vertices has at least one arc. First, in Sect. 3 of our paper we prove a theorem analogous to the Erdős-Pósa theorem to bound the feedback vertex set size. In Sect. 4 we find an approximate feedback vertex set as well as state the notions and results of a *cut-preserving set* [17]. Working with the feedback vertex set, in Sect. 5 we give an algorithm to find the desired polynomial kernel.

2 Preliminaries

For a directed graph G, $V(G)$ and $E(G)$ denote the set of vertices and arcs. uv is a directed arc if there is an arc going from u to v. We also denote such an arc with the ordered pair (u,v). For a set $S \subseteq V(G)$, $G - S$ denotes the graph obtained by deleting S from G and $G[S]$ denotes the subgraph of G induced on S. A path $P = (v_1, \ldots v_i, \ldots v_j, \ldots, v_p)$ is a sequence of distinct vertices where every vertex has an arc to the next vertex in the sequence. We say that P starts at v_1 and ends at v_p. The vertices (or vertex set) of P, denoted by $V(P)$, is the set $\{v_1, \ldots, v_p\}$. The endpoints of P is the set $\{v_1, v_p\}$ and the internal vertices of P is the set $V(P) \setminus \{v_1, v_p\}$. The arcs of P, denoted by $E(P)$ is the arc set $\{(v_1, v_2), \ldots (v_{(p-1)}, v_p)\}$. The *length* of P is defined as $|V(P)|$. We also denote such a path with $(v_1 \ldots \to v_i \ldots \to v_j \ldots \to v_p)$. Any $(v_i \to \ldots \to v_j)$ is a subpath of P. In a path $P = (v_1 \ldots \to v_i \to v_{i+1} \ldots v_{j-1} \to v_j \ldots v_p)$, the subpath S between v_i and v_j is the path $(v_{i+1} \to \ldots \to v_{j-1})$. Path P is contained inside a vertex set V' iff $\{v_1, \ldots v_p\} \subseteq V'$. P is vertex disjoint from V' iff $V(P) \cap V' = \emptyset$. A cycle C is a sequence $(v_1, \ldots, v_c, v_1)$ of vertices such that $(v_1, \ldots, v_c)$ is a path and $v_c v_1$ is an arc. The same cycle is also denoted as $(v_1 \to \ldots v_c \to v_1)$. The vertex set $V(C)$ is $\{v_1, \ldots v_c\}$ and the arc set $E(C)$ is $\{(v_1, v_2), \ldots (v_{(c-1)}, v_c), (v_c, v_1)\}$. Subpaths of a cycle are defined in a similar fashion to those of a path. For a collection $\mathcal{P}$ of paths (or cycles), $V(\mathcal{P})$ denotes the set $\bigcup_{P_i \in \mathcal{P}} V(P_i)$. $E(\mathcal{P}), V(\mathcal{C}), E(\mathcal{C})$ are defined in a similar manner.

In this paper we interchangeably use modulator and feedback vertex set. We use $\times$ for cross product and $\uplus$ for disjoint union. For a set A and an ordered set B, by i elements of A that appear first in B, we mean the first i elements of B that are present in A. For graph theoretic terms and definitions not stated explicitly here, we refer to [8].

3 An Erdős-Pósa Type Theorem for α-Bounded Digraphs

In this section we show that there exists a function $f(r) = \mathcal{O}(2\alpha^2 r^2)$ such that for each non-negative integer r, every digraph $G \in \mathcal{D}_\alpha$ either contains r vertex disjoint cycles or has a directed feedback vertex set consisting of $f(r)$ vertices. We start by showing a density lemma about digraphs in $\mathcal{D}_\alpha$ and then use it to obtain the desired result.

Lemma 1. *Any α-bounded digraph G on n vertices has at least $n^2/2\alpha^2$ arcs.*

Proof. Let $S \subseteq V(G)$ be any set of size $\alpha + 1$. Then by definition all the vertices in S can not be independent in G, i.e., there must be at least one arc between some two vertices in S. Let us call this arc a witness for S (if there are more than one arcs, pick any arbitrary arc as a witness). G has a total of $\binom{n}{\alpha+1}$ many vertex sets of size $\alpha + 1$. And each of them must have a witness arc. Hence there are at least $\binom{n}{\alpha+1}$ witness arcs (not necessarily different arcs). But any arc xy can witness at most $\binom{n-2}{\alpha-1}$ sets of size $\alpha + 1$, since x and y are forced to be present in the set. This implies

$$|E(G)|\binom{n-2}{\alpha-1} \geq \text{ no. of witnesses } \geq \binom{n}{\alpha+1}$$

$$\implies |E(G)| \geq \frac{\binom{n}{\alpha+1}}{\binom{n-2}{\alpha-1}} \geq (n^2/2\alpha^2) \tag{1}$$

Hence G has at least $\frac{n^2}{2\alpha^2}$ many arcs. $\square$

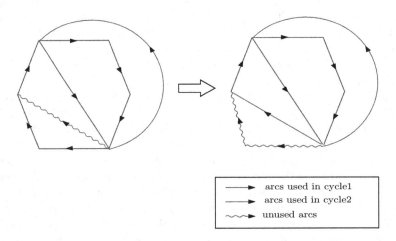

$\longrightarrow$	arcs used in cycle1
$\longrightarrow$	arcs used in cycle2
$\rightsquigarrow$	unused arcs

Fig. 1. Replacement procedure to get a *nice* collection of cycles

Now we are ready to present our main result of this section (Fig. 1).

Theorem 1. *Any α-bounded digraph G that does not have k arc disjoint cycles, has a feedback vertex set (FVS) of size at most $2\alpha^2k^2$. Furthermore, there exists a family of k arc disjoint cycles, where each cycle has length at most $2\alpha^2k$.*

Proof. Suppose the graph G has a maximum of k' many arc disjoint cycles where $k' < k$. Let $\mathcal{C} = \{C_1, \ldots, C_{k'}\}$ be a *nice* collection of k' arc disjoint cycles, i.e., there is no other set of k' arc disjoint cycles, which has less arcs than $\mathcal{C}$. Suppose C_i is the longest cycle in $\mathcal{C}$ with length l. We know that any induced subgraph of an α-bounded digraph is also α-bounded. Hence from Lemma 1, it follows that $G[V(C_i)]$ has at least $\frac{l^2}{2\alpha^2}$ edges. So cycle C_i has $(\frac{l^2}{2\alpha^2} - l)$ many internal chords as the only arcs that are not chords are the l edges used by the cycle.

Let the cycle be $C_i = (v_1 \to v_2 \to v_3 \ldots \to v_l \to v_1)$. We try to find a replacement cycle C^* in the following manner. Let $v_i v_j$ be any internal chord of C_i. If $i < j$, then $C^* = (v_1 \to \ldots \to v_i \to v_j \to v_{j+1}, \ldots \to v_l \to v_1)$, otherwise $C^* = (v_j \to v_{j+1} \ldots \to v_i \to v_j)$. Now clearly C^* has length strictly smaller than the length of C_i. But $\mathcal{C}$ was a nice solution. The only reason we can not replace the cycle C_i in $\mathcal{C}$ with C^* is because there must be some other cycle C_j such that $\{(v_i, v_j)\} \in E(C_j)$. Since the longest cycle in $\mathcal{C}$ has size l, the total number of edges used by all other cycles in $\mathcal{C}$ is at most $l(k'-1)$. But all the internal chords inside cycle C_i must be used by other cycles (otherwise $\mathcal{C}$ is not a nice collection of cycles). This implies

$$\frac{l^2}{2\alpha^2} - l \leq l(k'-1) \tag{2}$$
$$\implies l \leq 2\alpha^2 k' \leq 2\alpha^2 k$$

So there is a feedback vertex set $F' = \bigcup_{i=1}^{k'} V(C_i)$ of size at most $2\alpha^2 k^2$ and in the nice collection $\mathcal{C}$ each cycle has length at most $2\alpha^2 k$. In Sect. 4, in fact we improve the bound on size of FVS to $\alpha^2 k^2$ when the graph does not have k arc disjoint cycles.   □

4   Algorithm to Find an FVS

In this section we make the proof of Theorem 1 algorithmic. We will use this directed feedback vertex set to design our kernel. We first state the algorithm in the box.

Algorithm 1$(G \in D_\alpha, k)$

1. Initialize $F = \emptyset$, $G' = G$, $i = 0$.
2. Run Breadth First Search on each vertex of G'. Find the shortest cycle C_i. If G' is acyclic goto Step 5.
3. If the shortest cycle C_i has length more than $2\alpha^2(k-i)$, return G has k arc disjoint cycles.
4. $F = F \cup V(C_i)$, $G' = G[V(G') \setminus V(C_i)]$, $i = i + 1$.
5. If $i = k$, return G has k arc disjoint cycles.
6. If G' is acyclic, return FVS F, otherwise goto Step 2.

From Theorem 1, any α-bounded digraph H which does not have k arc disjoint cycles, has a cycle of size at most $2\alpha^2 k$. In other words any α-bounded digraph H, where the smallest cycle has length more than $2\alpha^2 k$, has at least k arc disjoint cycles. In the ith iteration of Step 2 of the above algorithm, we have already found i arc disjoint cycles $\{C_1, \ldots C_i\}$. If the shortest cycle in $G[V(G) \setminus \bigcup_{j=1}^{i} V(C_j)]$ has length more than $2\alpha^2(k-i)$, then it has at least $k-i$

arc disjoint cycles. But then the original graph G definitely has k arc disjoint cycles. This proves the correctness of the 3rd Step of the algorithm. In Step 5, if we can get k vertex disjoint cycles then of course G also has k arc disjoint cycles. If the graph G does not have k arc disjoint cycles, then in any ith iteration, the graph G' does not have more than $k-i$ arc disjoint cycles. So it also has a cycle of length at most $2\alpha^2(k-i)$ from Theorem 1. This, together with the fact that the algorithm runs at most k many iterations implies the FVS F that we get at the end of the algorithm has a maximum size of $2\alpha^2((k-1)+(k-2)+\ldots+1) \le \alpha^2 k^2$. Since each step of the algorithm takes poly(n) time and each step is also executed at most poly(n) times, the entire algorithm runs in poly(n) time.

4.1   A Cut-Preserving Set

Definition 1 ([17]). *For any digraph G, a positive integer k and $x, y \in V(G)$, we say that $\mathcal{Z} \subseteq V(G)$ is a k-cut-preserving set for (x, y) in G, if the following properties hold. Let $L = V(G) \setminus \mathcal{Z}$. For any path P from x to y in G, there exist paths $P_1, P_2, \ldots, P_e$ and a set of lists of k path $L_1, \ldots, L_e$ with the following properties:*

- *For every $i \in [e]$, P_i is a subpath of P from s_i to t_i.*
- *The P_is are internally disjoint and contain all vertices in $P \cap L$ as inner vertices.*
- *for every $i \in [e]$, L_i is a set of k vertex disjoint paths from s_i to t_i using only vertices of $\mathcal{Z}$.*
- *Replacing in P each P_i by one of the paths in L_i yields a path of $\mathcal{Z}$ from x to y.*

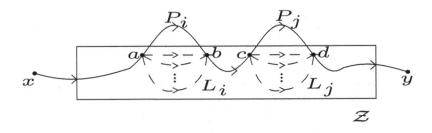

Fig. 2. A cut-preserving set $\mathcal{Z}$ for (x, y)

Figure 2 gives an easy depiction of a cut-preserving set. Lochet et al. [17] have recently shown that in an α-bounded digraph G, from any x to y a k-cut-preserving set of size $f(k, \alpha)$ can be found in polynomial time where $f(k, \alpha) = (22k^5)^{4^\alpha}$.

Theorem 2 ([17]). *Let D be an α-bounded acyclic digraph and $x, y \in V(D)$ such that any (x,y)-vertex- cut in D has size at least $k + 1$. Then one can, in polynomial time, compute a k-cut-preserving (x, y) in D of size at most $(22k^5)^{4^\alpha}$. Moreover in polynomial time one can obtain $k + 1$ vertex disjoint paths from u to v where each path has length at most $2\alpha + 1$.*

5   Algorithm to Compute the Kernel

In this section we gather everything and design our kernel. We start with the description and then prove its correctness and finally give the size bound.

Algorithm 2

1. Initialize TCL (total cycle length)$=2\alpha^2 k^2$, $i = 1$, $Kernel = F$, ℓ (max cycle length)$= 2\alpha^2 k$.
2. Let σ' be an ordered set on $F \times F$ such that $\sigma' = ((u_1, u_1)(u_1, u_2), (u_1, u_3), \ldots (u_{|F|}, u_{|F|}))$. Now fix an ordered set σ of size $k|F|^2$ such that, $\forall j \in [k|F|^2]$, $\sigma(j) = \sigma'(\lceil j/k \rceil)$. In σ each element from σ' is repeated k consecutive times, i.e., $\sigma=((u_1, u_1), (u_1, u_1), \ldots (u_1, u_2), (u_1, u_2), \ldots (u_{|F|}, u_{|F|}), (u_{|F|}, u_{|F|}))$.
3. Initialize $i = 1$.
4. Let $\sigma(i) = (x, y)$. Get a $(\mathcal{K} = TCL + \ell(2\alpha + 1))$-cut-preserving set $\mathcal{Z}$ from x to y of size $f(\mathcal{K}, \alpha)$. $Kernel = Kernel \cup \mathcal{Z}$, $TCL = \mathcal{K}$.
5. if $i=k|F|^2$ (all elements of σ are exhausted) stop, else $i = i + 1$, go to Step 4.
6. Return $Kernel$.

5.1   Running Time and Kernel Size Analysis

Step 1 and Step 2 takes $n^{\mathcal{O}(1)}$ time. Step 4 and 5 of the algorithm are executed $k|F|^2$ many times and in each iteration, we spend at most $n^{\mathcal{O}(1)}$ time to get the cut-preserving set $\mathcal{Z}$ [17]. Hence our algorithm runs in polynomial time.

Next we determine the final size of $Kernel$ set. Let $g(i)$ and $TCL(i)$ denote the size of $Kernel$ and TCL in the ith iteration. We get the following recurrence equations from the above algorithm and solve them to get a bound on the $Kernel$ size:

1. $g(i) = g(i - 1) + f(TCL(i - 1) + \ell(2\alpha + 1)), \alpha)$.
2. $TCL(i) = TCL(i - 1) + \ell(2\alpha + 1)$.
3. $g(0) = \alpha^2 k^2, TCL(0) = 2\alpha^2 k^2$.

Notice that TCL and g are strictly increasing functions. We compute their maximum values below.

$$\begin{aligned} TCL(i) &= 2\alpha^2 k^2 + i(\ell(2\alpha + 1)) \\ &= 2\alpha^2 k^2 + i(2\alpha^2 k(2\alpha + 1)) \end{aligned} \tag{3}$$

$$TCL(k|F|^2) = TCL(\alpha^4 k^5)$$
$$= 2\alpha^2 k^2 + \alpha^4 k^5 (2\alpha^2 k(2\alpha + 1)) \qquad (4)$$
$$\leq 5\alpha^7 k^6.$$

Since g and f both are increasing functions, $g(k|F|^2)$ has the maximum value.

$$g(k|F|^2) \leq g(0) + k|F|^2 . f(TCL(k|F|^2), \alpha)$$
$$\leq \alpha^2 k^2 + \alpha^4 k^5 f(5\alpha^7 k^6, \alpha)$$
$$\leq 2\alpha^4 k^5 f(5\alpha^7 k^6, \alpha) \qquad (5)$$
$$\leq 2\alpha^4 k^5 (110\alpha^{35} k^{30})^{4^\alpha}$$
$$= P_1(k, \alpha)$$

Hence g admits a maximum value ($Kernel$ size) of $P_1(k, \alpha)$. Now if we can show that $G[Kernel]$ has k arc disjoint cycles iff G has k arc disjoint cycles, then the problem indeed admits a kernel of size $P_1(k, \alpha)$.

5.2 Correctness of the Algorithm

In the forward direction, if $G[Kernel]$ has k arc disjoint cycles, then the graph G also has the same k arc disjoint cycles. We will use induction to prove the reverse direction. Suppose G has k arc disjoint cycles. Using the arguments in Theorem 1, we know there is a nice set of k arc disjoint cycles $\mathcal{C} = \{C_i\}_{i=1}^k$, where the total number of arcs in $\mathcal{C}$ is at most $2\alpha^2 k^2$ and any cycle C_i has length $l_i(\leq 2\alpha^2 k)$. Let us define the notions of segment and subsegment for our proof. Segments for $\mathcal{C}$ are the paths from x to y (where $x, y \in F$) and there is no other vertices of F in between. Subsegments are the maximal subpaths of segments that lie outside the $Kernel$. Refer to Fig. 3.

$F = \{x, y, z\}$ $\sigma = \{(x, x), (x, x), (x, x), (x, y), (x, y), (x, y), ..., (z, z)\}$
$k = 3$

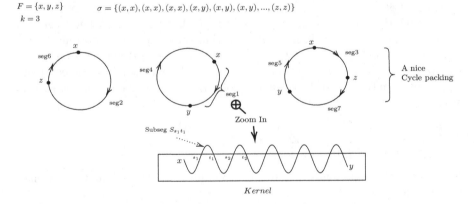

Fig. 3. Segments and subsegments

Let $C_i = (u_{i_1} \to \mathcal{S}_{i_{12}} \to u_{i_2} \to \mathcal{S}_{i_{23}} \to u_{i_3} \dots u_{i_{k_i}} \to \mathcal{S}_{i_{k_i(k_i+1)}} \to u_{i_{(k_i+1)}})$, where u_{i_j} vertices are from the modulator F and $u_{i_1} = u_{i_{(k_i+1)}}$. $\mathcal{S}_{i_{j(j+1)}}$ denotes the segment of cycle C_i from the vertex u_{i_j} to $u_{i_{(j+1)}}$. In each induction step we will replace a segment between two modulator vertices with another segment that is completely contained inside $Kernel$, while maintaining the property that even after the replacement, the cycles in $\mathcal{C}$ are still arc disjoint. Notice each cycle in a nice collection can have at most $|F|$ many segments and hence there are at most $\alpha^2 k^3$ many segments in $\mathcal{C}$. If we are able to replace all the segments (a maximum of $\alpha^2 k^3$ many), then $Kernel$ actually will have k arc disjoint cycles contained in it.

Let $S_i = \bigcup_{j=1}^{k_i} (u_{i_j}, u_{i_{(j+1)}})$ and $S = \biguplus_{i=1}^{k} S_i$. Observe that a pair (x, y) can appear at most k many times in S (If any vertex appears more than once in a cycle, we can get another cycle that uses a strict subset of the arcs used by the original cycle). We use induction below to prove the correctness. Our induction properties will be as follows:

1. In qth step, we are able to replace the segments between the first q pairs of vertices of σ that appear in S, with segments that are completely contained inside $Kernel$.
2. There exists a collection of arc disjoint cycles with the replaced segments whose total length is not more than $TCL(q)$.

Induction Step 1 (First Segment Replacement from $Kernel$)

Let $\mathcal{C}$ be a nice collection of arc disjoint cycles and $(x = u_{i_j}, y = u_{i_{(j+1)}})$ be the pair of vertices in S that appears first in σ. We will replace the segment from x to y with a segment completely contained inside $Kernel$ if it already isn't, while still keeping it arc disjoint from all other segments. If the segment $\mathcal{S}_{i_{j(j+1)}}$ is completely inside the computed $\mathcal{Z}$ in round 1, then it satisfies both the induction properties from Theorem 1. Otherwise for the pair (x, y), in $Kernel$ we have stored enough vertices $(\mathcal{Z})$ to get a $(TCL(0) + \ell(2\alpha + 1))$-cut-preserving set.

From the cut-preserving set properties we have segment $\mathcal{S}_{i_{j(j+1)}} = P = (x = s_0 \to \mathcal{S}_{s_0 s_1} \to s_1 \to \mathcal{S}_{s_1 t_1} \to t_1 \to \mathcal{S}_{t_1 s_2} \dots s_l \to \mathcal{S}_{s_l t_l} \to t_l \to \mathcal{S}_{t_l t_f} \to t_f = y)$, where $V(P) \setminus \mathcal{Z} = \cup_{j=1}^{l} V(\mathcal{S}_{s_j t_j})$ is the set of vertices from P that are not in $Kernel$. Notice that $l \leq k_i \leq \ell$. And $\mathcal{Z}$ is a $(TCL(0) + \ell(2\alpha + 1))$-cut-preserving set. Then from Theorem 2 we can get $(TCL(0) + (\ell(2\alpha + 1))$ many vertex disjoint paths each with length at most $2\alpha + 1$ from s_j to t_j for any j that are completely inside $Kernel$.

First Subsegment Replacement

Hence from s_1 to t_1 there is a path that is vertex disjoint from $V(\mathcal{C})$, since $|V(\mathcal{C})| = 2\alpha^2 k^2 = TCL(0)$. Let this path be L_1 that has length at most $2\alpha + 1$ and is completely inside $\mathcal{Z}$. Now replace the subsegment $\mathcal{S}_{s_1 t_1}$ with L_1 in P to get a new segment from s_0 to t_f, $P' = (s_0 \to \mathcal{S}_{s_0 s_1} \to s_1 \to L_1 \to t_1 \to \mathcal{S}_{t_1 s_2} \dots \to t_f)$ and get a new set of arc disjoint cycles by replacing the segment $\mathcal{S}_{i_{j(j+1)}}$ with P', i.e., $\mathcal{C} = \mathcal{C} \setminus \{C_i\} \cup \{C_i = (u_{i_1} \to S_{i_{12}} \dots u_{i_j} \to P' \to u_{i_{(j+1)}} \dots \to S_{i_{k_i(k_i+1)}} \to u_{i_{(k_i+1)}})\}$. Now the updated $\mathcal{C}$ has at most $2\alpha^2 k^2 + 2\alpha + 1$ many vertices (or arcs) and L_1 is completely contained inside $Kernel$.

Second Subsegment Replacement

Similarly from s_2 to t_2 there are at least $(TCL(0) + \ell(2\alpha + 1))$ vertex disjoint paths in $\mathcal{Z}$, each with length at most $2\alpha + 1$. In the updated $\mathcal{C}$, there are at most $(TCL(0) + 2\alpha + 1)$ vertices, hence there is a path L_2 from s_2 to t_2 that is vertex disjoint from $V(\mathcal{C})$ and is completely inside $\mathcal{Z}$. We replace the subsegment $S_{s_2 t_2}$ with L_2 and get a new path $P' = (s_0 \to S_{s_0 s_1} \to s_1 \to L_1 \to t_1 \to S_{t_1 s_2} \to s_2 \to L_2 \to t_2 \dots \to t_f)$. We also get a new set of arc disjoint cycles $\mathcal{C} = \mathcal{C} \setminus \{C_i\} \cup \{C_i = (u_{i_1} \to S_{i_{12}} \dots u_{i_j} \to P' \to u_{i_{(j+1)}} \dots \to S_{i_{k_i(k_i+1)}} \to u_{i_{(k_i+1)}})\}$. The new $\mathcal{C}$ has at most $2\alpha^2 k^2 + 2(2\alpha + 1)$ many arcs. Moreover L_1 and L_2 are completely contained inside $Kernel$.

All l Subsegments Replacement

But $l \leq \ell$, as every cycle in the beginning had length at most $2\alpha^2 k$. Hence we will be able to apply the above replacement procedure for all $S_{s_i t_i}$, where $i \leq l$. And get a new segment $P' = (s_0 \to S_{s_0 s_1} \to s_1 \to L_1 \to t_1 \dots \to s_l \to L_l \to t_l \to S_{t_l t_f} \to t_f)$ which is completely contained inside $Kernel$. Now in C_i replacing the segment from u_{i_j} to $u_{i_{(j+1)}}$ by P' and updating the $\mathcal{C}$, we get a set of arc disjoint cycles where the segment from u_{i_j} to $u_{i_{(j+1)}}$ is completely contained inside $Kernel$. The updated set of arc disjoint cycles $\mathcal{C}$ uses at most $2\alpha^2 k^2 + \ell(2\alpha + 1)$ many arcs. This proves the correctness for the first step of the induction.

Let the induction properties hold true for all $j < q$. So we are able to successfully replace(or keep) the segments between the first $q - 1$ pairs of vertices in S that appear in σ with segments completely inside $Kernel$, such that the new collection of arc disjoint cycles with the replaced segments uses at most $TCL(q - 1)$ number of arcs.

Induction Step q (qth Segment Replacement from $Kernel$)

Let $(x' = u_{i'_{j'}}, y' = u_{i'_{(j'+1)}})$ be the qth pair of vertices of σ that appear in S. The $(q - 1)$ other pairs from S that appear in σ before (x', y'), their segments have already been replaced in $\mathcal{C}$ in the first $(q - 1)$ steps of induction. Let $\mathcal{Z}$ be a $\mathcal{K}(= TCL(q - 1) + \ell(2\alpha + 1))$-cut-preserving set from x' to y' of size $f(\mathcal{K}, \alpha)$. Let the collection of cycles after the $(q - 1)$ replacements be $\mathcal{C} = \{C_i\}_{i=1}^{k}$. Let $C_{i'} = (u_{i'_1} \to S'_{i'_{12}} \to u_{i'_2} \dots \to u_{i'_{k_{i'}}} \to S'_{i'_{k_{i'}(k_{i'}+1)}} \to u_{i'_{(k_{i'}+1)}})$, where $u_{i'_{j'}}$ and $u_{i'_{(j'+1)}}$ vertices are from the modulator F and $u_{i'_1} = u_{i'_{(k_{i'}+1)}}$. The segment of cycle C_i from the vertex $u_{i'_{j'}}$ to $u_{i'_{(j'+1)}}$ is denoted by $S'_{i'_{j'(j'+1)}}$. For the pair (x', y'), in $Kernel$ we have stored enough vertices $(\mathcal{Z})$ to get a $\mathcal{K}$-cut-preserving set.

If the segment $S'_{i'_{j'(j'+1)}}$ is completely inside $\mathcal{Z}$, we do not need to replace the segment at all and we can move onto the next segments. All the q segments are completely contained in $Kernel$ and the total length of all cycles in $\mathcal{C}$ is at most $TCL(q - 1) \leq TCL(q)$. This satisfies the induction properties for qth step.

If it is not completely contained in $\mathcal{Z}$ then from Theorem 2, we get the segment of the form $S'_{i'_{j'(j'+1)}} = P = (x' = s'_0 \to S'_{s'_0 s'_1} \to s'_1 \to S'_{s'_1 t'_1} \cdots \to S'_{s'_r t'_r} \to t'_r \to S'_{t'_r t'_f} \to t'_f = y')$, where $S'_{s'_i t'_i}$ are the subsegments that are not in $\mathcal{Z}$, i.e., $V(P) \setminus A = \cup_{i=1}^r V(S'_{s'_i t'_i})$. Notice that $r \le k_{i'} \le \ell$. But since $\mathcal{Z}$ is a $\mathcal{K}$-cut-preserving set, we can get $\mathcal{K}$ many vertex disjoint paths each with length at most $2\alpha + 1$ from s'_j to t'_j for all j.

First Subsegment Replacement

Hence from s'_1 to t'_1 there is a path that is vertex disjoint from $V(\mathcal{C})$ since $|V(\mathcal{C})| = TCL(q-1)$. Let this path be L'_1 that has length at most $2\alpha + 1$ and is completely inside $\mathcal{Z}$. Now replace the subsegment $S'_{s'_1 t'_1}$ with L'_1 in P to get a new path from s'_0 to t'_f. Let $P' = (s'_0 \to S'_{s'_0 s'_1} \to s'_1 \to L'_1 \to t'_1 \cdots \to t'_f)$. We get a new set of arc disjoint cycles by replacing the segment $S_{i'_{j'(j'+1)}}$ with P', i.e., $\mathcal{C} = \mathcal{C} \setminus C_i \cup \{C_i = (u_{i'_1} \to S'_{i'_{12}} \ldots u_{i'_{j'}} \to P' \to u_{i'_{j'+1}} \cdots \to S'_{i'_{k_{i'}(k_{i'}+1)}} \to u_{i'_{(k_{i'}+1)}})\}$. The updated $\mathcal{C}$ has at most $(TCL(q-1) + 2\alpha + 1)$ many arcs and L'_1 is completely contained inside $Kernel$.

Second Subsegment Replacement

Similarly from s'_2 to t'_2, there are at least $\mathcal{K}$ vertex disjoint paths in $\mathcal{Z}$ each with length $(2\alpha + 1)$. In the updated $\mathcal{C}$, there are at most $TCL(q-1) + (2\alpha + 1)$ arcs. Hence there is a subsegment(path) L'_2 from s'_2 to t'_2 that is vertex disjoint from $V(\mathcal{C})$ and completely contained inside $\mathcal{Z}$. We replace the subsegment $S'_{s'_2 t'_2}$ with L'_2 and get a new path from s'_0 to t'_f, $P' = (s'_0 \to S'_{s'_0 s'_1} \to s'_1 \to L'_1 \to t'_1 \to S'_{t'_1 s'_2} \to s'_2 \to L'_2 \to t'_2 \ldots \to t'_f)$. We get a new set of arc disjoint cycles by replacing the segment $\mathcal{C} = \mathcal{C} \setminus \{C_i\} \cup \{C_i = (u_{i'_1} \to S'_{i'_{12}} \ldots u_{i'_{j'}} \to P' \to u_{i'_{(j'+1)}} \cdots \to u_{i'_{(k_{i'}+1)}})\}$. The new $\mathcal{C}$ has at most $TCL(q-1) + 2(2\alpha + 1)$ many arcs. Moreover L'_1 and L'_2 are completely contained inside $Kernel$.

All r Subsegments Replacement

But $r \le \ell$, as every cycle in the beginning had length at most $2\alpha^2 k$. Hence we will be able to apply the above replacement procedure for all $S'_{s'_i t'_i}$, where $i \le r$. And get a new segment $P' = (s'_0 \to S'_{s'_0 s'_1} \to s'_1 \to L'_1 \to t'_1 \ldots \to s'_r \to L'_r \to t'_r \to S'_{t'_r t'_f} \to t'_f)$ which is entirely contained inside $Kernel$. Now in $C_{i'}$ replacing the segment from x' to y' by P' and updating $\mathcal{C}$, we get the set of arc disjoint cycles where the segment from $u_{i'_{j'}}$ to $u_{i'_{(j'+1)}}$ is completely contained inside $Kernel$. The updated set of arc disjoint cycles $\mathcal{C}$ uses at most $TCL(q-1) + \ell(2\alpha + 1) = TCL(q)$ many arcs. This proves the correctness of the induction.

Theorem 3. Arc disjoint cycle packing *in α-bounded digraphs, when parameterized by the number of cycles k, admits a kernel of size $P_1(k, \alpha)$ where* $P_1(k, \alpha) = 2\alpha^4 k^5 (110\alpha^{35} k^{30})^{4^\alpha}$.

Acknowledgments. We are grateful to William Lochet for the invaluable suggestions and discussions.

References

1. Abu-Khzam, F.N.: An improved kernelization algorithm for r-set packing. Inf. Process. Lett. **110**(16), 621–624 (2010)
2. Bessy, S., et al.: Packing arc-disjoint cycles in tournaments. In: 44th International Symposium on Mathematical Foundations of Computer Science, MFCS 2019, 26–30 August 2019, Aachen, Germany. LIPIcs, vol. 138, pp. 27:1–27:14 (2019)
3. Bessy, S., Bougeret, M., Thiebaut, J.: (Arc-disjoint) cycle packing in tournament: classical and parameterized complexity. Technical report ArXive CoRR 1802.06669v1 [cs.DM], Cornell University, February 2018
4. Bodlaender, H.L.: On disjoint cycles. Int. J. Found. Comput. Sci. **5**(1), 59–68 (1994)
5. Bodlaender, H.L., Thomassé, S., Yeo, A.: Kernel bounds for disjoint cycles and disjoint paths. Theoret. Comput. Sci. **412**(35), 4570–4578 (2011)
6. Caprara, A., Panconesi, A., Rizzi, R.: Packing cycles in undirected graphs. J. Algorithms **48**(1), 239–256 (2003)
7. Cygan, M., et al.: Parameterized Algorithms. Springer, Cham (2015). https://doi.org/10.1007/978-3-319-21275-3_15
8. Diestel, R.: Graph Theory. Springer, Heidelberg (2006)
9. Dorninger, D.: Hamiltonian circuits determining the order of chromosomes. Discrete Appl. Math. **50**(2), 159–168 (1994)
10. Downey, R.G., Fellows, M.R.: Fundamentals of Parameterized Complexity. TCS. Springer, London (2013). https://doi.org/10.1007/978-1-4471-5559-1
11. Erdős, P., Pósa, L.: On independent circuits contained in a graph. Can. J. Math. **17**, 347–352 (1965)
12. Fellows, M.R., Langston, M.A.: Nonconstructive tools for proving polynomial-time decidability. J. ACM **35**(3), 727–739 (1988)
13. Flum, J., Grohe, M.: Parameterized Complexity Theory. Springer, Heidelberg (2006). https://doi.org/10.1007/3-540-29953-X
14. Fradkin, A.O., Seymour, P.D.: Edge-disjoint paths in digraphs with bounded independence number. J. Comb. Theory Ser. B **110**, 19–46 (2015). https://doi.org/10.1016/j.jctb.2014.07.002
15. Jacob, A.S., Krithika, R.: Packing arc-disjoint cycles in bipartite tournaments. In: Rahman, M.S., Sadakane, K., Sung, W.-K. (eds.) WALCOM 2020. LNCS, vol. 12049, pp. 249–260. Springer, Cham (2020). https://doi.org/10.1007/978-3-030-39881-1_21
16. Krithika, R., Sahu, A., Saurabh, S., Zehavi, M.: The parameterized complexity of packing arc-disjoint cycles in tournaments. Technical report ArXive CoRR 1802.07090, Cornell University (2018). http://arxiv.org/abs/1802.07090
17. Lochet, W., Lokshtanov, D., Misra, P., Saurabh, S., Sharma, R., Zehavi, M.: Fault tolerant subgraphs with applications in kernelization. In: Vidick, T. (ed.) 11th Innovations in Theoretical Computer Science Conference, ITCS 2020, 12–14 January 2020, Seattle, Washington, USA. LIPIcs, vol. 151, pp. 47:1–47:22. Schloss Dagstuhl - Leibniz-Zentrum für Informatik (2020). https://doi.org/10.4230/LIPIcs.ITCS.2020.47
18. Lokshtanov, D., Mouawad, A.E., Saurabh, S., Zehavi, M.: Packing cycles faster than Erdős-Pósa. SIAM J. Discrete Math. **33**(3), 1194–1215 (2019)
19. Reed, B.A., Robertson, N., Seymour, P.D., Thomas, R.: Packing directed circuits. Combinatorica **16**(4), 535–554 (1996)
20. Robertson, N., Seymour, P.D.: Graph minors. XIII. The disjoint paths problem. J. Comb. Theory Ser. B **63**(1), 65–110 (1995)

On Subquadratic Derivational Complexity of Semi-Thue Systems

Alexey Talambutsa[⊠] [iD]

Steklov Mathematical Institute of RAS, 8 Gubkina St., Moscow 119991, Russia
altal@mi-ras.ru

Abstract. We prove that for any rational number $\alpha > 1$ there exists a semi-Thue system with derivational complexity function belonging to the asymptotic class $\Theta(n^\alpha)$. In particular, we answer a question of Y. Kobayashi, whether there exists a semi-Thue system whose derivational complexity function is in the class $\Theta(n^\alpha)$ with $\alpha \in (1, 2)$.

Keywords: Semi-Thue system · Rewriting system · Derivational complexity

1 Introduction

Let $A = \{a_1, \ldots, a_n\}$ be a finite alphabet and denote by A^* the set of all finite words in this alphabet. If W is a word then we denote its length by $\ell(W)$. A semi-Thue system $(A, \mathcal{R})$ is given by the alphabet A and a *set of rules* $\mathcal{R} \subset A^* \times A^*$. For two words $U, V \in A^*$ we say that $U \to V$ is a *rewrite step* in the semi-Thue system $(A, \mathcal{R})$ if for some words $P, Q \in A^*$ and a rule $(X, Y) \in \mathcal{R}$ we have that $U = PXQ$ and $V = PYQ$.

A finite sequence $W_1 \to W_2 \to \ldots \to W_n$ is called a *derivation sequence* if for each $i \in \{1, 2, \ldots, n - 1\}$ there exists a rewrite step $W_i \to W_{i+1}$. In this case we will also use a shorthand notation $W_1 \twoheadrightarrow W_n$. Following the paper [5], the *derivational depth* is defined as a function $\delta : A^* \to \mathbb{N} \cup \{\infty\}$ as follows:

$$\delta(W) = \max\{L \mid W = W_0 \to W_1 \to \ldots \to W_L\}.$$

The *derivational complexity* of the system $(A, \mathcal{R})$ is a function $\Delta : \mathbb{N} \to \mathbb{N} \cup \{+\infty\}$ which is defined as

$$\Delta_{(A, \mathcal{R})}(t) = \max\{\delta(W) \mid \ell(W) = t\}.$$

If $\Delta(t) < \infty$ for all $t \in \mathbb{N}$, then the system is *terminating*.

It can be quite complicated to obtain an estimate from above for the derivational complexity, even for a semi-Thue system that looks simple. For example,

This work was supported by a research grant from Russian Science Foundation, project no. 16-11-10252.

H. Fernau (Ed.): CSR 2020, LNCS 12159, pp. 379–392, 2020.
https://doi.org/10.1007/978-3-030-50026-9_28

H. Zantema suggested a system over a 3-letter alphabet $\{a, b, c\}$ with 3 rules $\{aa \to bc, bb \to ac, cc \to ab\}$, which became notorious for impeding the automated tools in proving its termination. Eventually, it was shown to be terminating in the work [6] of D. Hofbauer and J. Waldmann, but proving a polynomial upper bound (in fact, quadratic) needs an elaborate combinatorial analysis of all possible sequences of derivation (see paper [1] by S. I. Adian).

In general, it was shown by Y. Kobayashi in [7, Theorem 2] that given a finite semi-Thue system, it is undecidable whether its derivational complexity is in the class $O(n^2)$. Another result of Kobayashi [8, Theorem 4.3] shows that for many computable functions $f(n)$ belonging to the class $\Omega(n^2)$, one may construct a semi-Thue system with the derivational complexity belonging to the class $\Theta(f(n))$. It follows, then, that for any algebraic number $\alpha \geq 2$, there exists a semi-Thue system such that its derivational complexity is equivalent to n^α. However, the general method of this paper does not produce functions that are in the asymptotic class $o(n^2)$. For this reason, Kobayashi asks [8, page 12] whether one can find a finite semi-Thue system whose derivational complexity function belongs to the class $\Theta(n^\alpha)$, when $\alpha \in (1, 2)$, in particular for $\alpha = 3/2$.

The main result of this paper is the positive answer to this question. Theorem 2 shows that, given a fraction a/b with $a, b > 0$, one can find an explicit example of a semi-Thue system in an alphabet of $b + 14$ letters with $2b + 14$ rules, whose derivational complexity belongs to the class $\Theta(n^{1+\frac{a}{b}})$.

Quite related, yet much more complicated, are the questions about possible asymptotics for Dehn functions of groups, which in some sense measure the derivational complexity of proving a relation in a group. Notably, A. Yu. Olshanskii proved in [9] that there is an $(1, 2)$-interval gap for the asymptotic classes $\Theta(n^\alpha)$, and there are no other interval gaps, due to N. Brady and M. Bridson (see [2]). M. Sapir, J.-C. Birget and E. Rips [4] constructed a huge class of computable Dehn functions, that are in class $\Omega(n^4)$, and this result was significantly strengthened to the functions of the class $\Omega(n^2)$ by Olshanskii in the recent paper [10].

2   System S_0 with Derivational Complexity in $\Theta(n \log n)$

In this section we present an explicit example of a length-preserving semi-Thue system $S_0 = (A_0, \mathcal{R}_0)$ which has derivational complexity of $\Theta(n \log n)$. This system is an essential part of the main construction from Sect. 3 and can be considered as its baby case. We provide detailed proofs of the lemmas for the system S_0 so that the reader will be prepared to deal with more complicated systems $S_{a,b}$ later.

Our goal is to emulate a "unary logarithm computation" performed on a Turing machine by the semi-Thue system S_0. First, we describe the process that we need and then formally describe the system S_0. We start with a word $W_0 = h1^n = h1 \ldots 1$ where h is a working head. The head starts moving to the right border of this word and replaces each second letter 1 it meets by 0. In the end of this passage, one obtains the word $0101 \ldots h$. Then the head in n steps

moves to the left border of the word and we obtain the word $W_1 = h0101\ldots$.
In such shuttling the head does $2n$ moves and "thins out" the word, namely
it reduced the number of letters 1 nearly twice. Then the head performs the
thinning out again and after $2n$ moves returns with the resulting word $W_2 =$
$h00010001\ldots$. The head continues this shuttling until there are no more letters 1
in the considered word. Since the shuttle does $\Theta(\log n)$ passages then, and each
of them consists of $2n$ moves, altogether it will be $\Theta(n \log n)$ moves, which is
our goal.

The alphabet A_0 of the system S_0 consists of 9 letters. First, consider its
subsets $D_0 = \{0, 1, \bar{0}, \bar{1}\}$ and $H_0 = \{h_0, h_1, h_2, c\}$ and then $A_0 = D_0 \cup H_0 \cup \{w\}$.

We briefly describe the relation of these letters to the process described above.
The letters 0,1 represent the digits, and $\bar{0}, \bar{1}$ are their marked versions. The letters
$h_0, h_1, h_2, c \in H_0$ represent different states of the working head in our process,
and we call them as *heads* of system S_0. The initial state is h_0 and then the state
of the head changes according to the sequence $h_0 \to h_1 \to h_2 \to h_1 \to h_2 \to \cdots$
with each letter 1 met on its way. Once the head h_i reaches the right border, it
turns into the head c which moves to the left. The remaining letter $w \in A_0$ is a
separator for the working zone of the head.

Now we formally describe the set of rules $\mathcal{R}_0$. It consists of 4 parts, which
we describe along explaining what these parts actually mean.

$$h_0 1 \to \bar{0} h_1 \qquad h_1 1 \to \bar{1} h_2 \qquad h_2 1 \to \bar{0} h_1$$
$$h_0 0 \to \bar{0} h_0 \qquad h_1 0 \to \bar{0} h_1 \qquad h_2 0 \to \bar{0} h_2 \tag{1}$$

These rules are used to execute a transformation $h_0 B \to B' h_i$, where B is a
binary word in the alphabet $\{0, 1\}$ and B' is a word in the alphabet $\{\bar{0}, \bar{1}\}$. The
word B' is obtained from the word B by replacing the odd (from the left) digits 1
by the digits 0 and also putting bars over all digits.

The next subset has 2 rules, which describe how the right-moving heads h_1, h_2
change to the cleaning head c in the right end of the working zone:

$$h_1 w \to cw \qquad h_2 w \to cw \tag{2}$$

Note that the head h_0 does not transform into the cleaning head c, since we stop
the transformations if the head has not met a letter 1 on its the way.

After the cleaning head gets created, it starts moving to the left and removes
all bars over digits $\bar{0}, \bar{1}$ on its way. This is done by the following two commands:

$$\bar{0} c \to c0 \qquad \bar{1} c \to c1 \tag{3}$$

Once the cleaning head arrives to the left border of the zone, it is transformed
into initial head h_0, which we describe by the command

$$wc \to wh_0. \tag{4}$$

Theorem 1. *Let $S_0 = (A_0, \mathcal{R}_0)$ be a semi-Thue system with the alphabet A_0
and the set of rules $\mathcal{R}_0$ that is given by the union of sets (1), (2), (3), (4). Then
the derivational complexity function $\Delta_{S_0}(n)$ belongs to the class $\Theta(n \log n)$.*

A remarkable property of the system $S_0 = (A, \mathcal{R}_0)$ is that this system is *length-preserving*, i.e., for each rule $(X, Y) \in \mathcal{R}_0$ one has $\ell(X) = \ell(Y)$. Hence, for any derivation sequence $X = X_1 \to X_2 \to \ldots \to X_k = Y$ also $\ell(X) = \ell(Y)$.

We will be using a particular notion of a *determined derivation*. Usually, for a semi-Thue system there are many different sequences that start from a given word. However, there could be cases where there are no rules applicable to a word or when there is a single rule $(X, Y) \in \mathcal{R}$ and a single position where the change $Z_1 = PXQ \to PYQ = Z_2$ can happen. In this case we say that the *derivation is determined* for the word Z_1. If the word Z_2 and all subsequent words in the possible continuation of the derivation sequence $Z_1 \to Z_2 \to Z_3 \to \ldots$ have this property then we say that the *derivation sequence is determined* for the word Z_1. Obviously, in this case the derivational depth $\delta(Z_1)$ can be easily computed by the applications of rules to the word Z_1 and next results until possible.

Lemma 1. *If $U, V \in D_0^*$ then the derivational depth of the word $Z = Uh_iV$ ($i \in \{0, 1, 2\}$) is subject to the inequality $\delta(Z) \leq \ell(V)$.*

Proof. One can see that initially there is only one rule that can be applied to the word Z, and this is a rule from the set (1). In each of these rules, the letter h_i moves to the right (possibly changing its index), and there is only one way to apply a rule of the system S_0. The resulting word Z' can also be written as $Z' = U'h_{i'}V'$, which is of the same form as in the statement of the Lemma. Hence, the derivation sequence is determined for the word Z. In this sequence the head h_i moves to the right (if possible), reaches the right side of the word and no continuation to the derivation sequence is possible then, as the marker w is absent. All these moves sum up to at most $\ell(V)$ operations, so the lemma is proved. □

In complete analogy (arguing about the rules (3)), one proves the next lemma.

Lemma 2. *If $U, V \in D_0^*$ then $\delta(UcV) \leq \ell(U)$.*

The next words, which we want to consider, are similar to those above, but have a letter w on the left or right side. These words also determine the sequence of derivations, which may though use the rules of the form (2) and (4). Tracing the moves of the letters h_i and c, one can easily obtain the following statement.

Lemma 3. *If $U, V \in D_0^*$ and Z is a word having the form $UxVw$ or the form $wUxV$, where $x \in H_0$ then $\delta(Z) \leq 2\ell(Z)$.*

The next step is to consider a similar word which has w on both ends.

Lemma 4. *If Z is a word having the form $wUxVw$, where $x \in H_0$ and $U, V \in D_0^*$ then $\delta(Z) \leq 2\ell(Z)(\lfloor \log_2(\ell(Z)) \rfloor + 5/2)$.*

Proof. For the word Z of the described type, the derivation sequence is again determined. After at most $2\ell(Z)$ steps either the derivation sequence terminates or it has a word of the form $Z_1 = w\, h_0 T_1(0, 1)\, w$, where $\ell(T_1) = \ell(U) + \ell(V)$. If the sequence terminates, the Lemma is proved, so we consider the other case.

After applying the rules to the word Z_1 at most $2\ell(Z)$ times, either the derivation sequence terminates, or we obtain a word $Z_2 = w\, h_0 T_2(0, 1)\, w$, so that T_2 is a word with at least twice less letters 1 than the word T_1. Then in general, arguing as before, we see that after at most $2\ell(Z)k$ operations to the word Z_1, the derivation sequence either terminates or brings us to the word $w\, h_0 T_{k+1}(0, 1)\, w$ with at least 2^k times less letters 1 than the word T_1. Eventually, after $2\ell(Z)(\lfloor \log_2(\ell(Z)) \rfloor + 1)$ steps (starting from Z_1) there are no letters 1 left, hence after not more than $2\ell(Z)(\lfloor \log_2(\ell(Z)) \rfloor + 5/2)$ total steps the derivation sequence terminates. $\qquad\square$

The same observations as in the previous paragraph provide a lower bound for the derivational complexity of the system S_0. We state this in the following lemma.

Lemma 5. *If $Z = w\, h_0 1^k w$ then $\delta(Z) = k(2(\lceil \log_2 k \rceil + 1) + 1)$.*

As an immediate consequence, we obtain:

Corollary 1. $\Delta_0(n)$ *belongs to the class* $\Omega(n \log n)$

Now we will prove the most important lemma for obtaining the upper bounds.

Lemma 6. *If Z is a word having the form $w\, T\, w$, such that $T \in (A_0 \setminus \{w\})^*$ and the word T has at least 2 letters from the set H_0, then $\delta(Z) \le 2\ell(Z) + 2$.*

Proof. First, we mark all letters from the set H_0 in the word T by x_i, so that the word T is written uniquely as $T = T_1 x_1 T_2 x_2 \ldots x_k T_{k+1}$ with $k \ge 2$ and $T_i \in D_0^*$.

In the contrary to the previous lemmata, the word wTw does not necessarily determine the derivation sequence, because now there are several occurrences of heads c, h_i. Still, in each of the rules, we have only one head letter in the left and right side, so again we may speak of the position of the head and of its image. In the following, for a word W we denote its i-th letter from the left as $W[i]$.

Consider the longest possible derivation sequence

$$Z = Z_1 \to Z_2 \to \ldots \to Z_m. \tag{5}$$

Let $p_i(s)$ be the positions of the image of x_i in the word Z_s. One can see that

$$p_1(s) < p_2(s) < \ldots < p_k(s) \text{ for each } s \in \{1, \ldots, m\}. \tag{6}$$

This inequality is straightforward for $s = 0$ and follows for other s by induction. One just needs to look through all the rules $\mathcal{R}_0$ and check the property.

Now we consider how the position $p_1(s)$ is changing along the course of the sequence (5), i.e., with the increase of s. There are two cases.

Case 1. $x_1 = h_i$, where $i \in \{0, 1, 2\}$. In this case $p_1(s)$ can not decrease, and it can increase only up to $p_2(1) - 1$. Indeed, otherwise consider the smallest s for which $p_1(s) = p_2(1)$. Since from (6) we have $p_2(s) > p_1(s)$, then by induction on s one can show $Z_{s-1}[p_2(1)] \in \{\bar{0}, \bar{1}\}$. It follows that the move of x_1 to the position $p_2(1)$ is impossible.

Case 2. $x_1 = c$. Here $p_1(s)$ can decrease (it takes exactly $p_1(s) - 1$ step), then change to h_0 due to the rule $wc \rightarrow wh_0$ and then continue as in the Case 1.

Summing up, we see that altogether all applications of rules to x_1 make at most $p_1(s) + p_2(s) - 1$ transformations. We remark that we can rearrange the sequence (5) so that all moves of the head x_1 happen in its beginning.

Analyzing in a manner similar as in the two cases above, we can see that one can apply to the head x_k at most $(\ell(T) - p_k(1)) + (\ell(T) - p_{k-1}(1) - 2)$ transformations. The moves of x_k are happening either to the left (finishing at the position $p_{k-1}(1) + 1$); or to the right (up to the position $\ell(T) - 1$), then applying the rule (2) and then to the left, stopping at the position $p_{k-1}(1) + 1$.

All the heads x_i (where $1 < i < k$) never reach positions 2 and $\ell(V) - 1$ due to the inequalities (6), so they are never subject to the rules (2) and (4). Hence, each head x_i can move one direction, stopping at the position of $p_{i-1}(1)$ or $p_{i+1}(1)$ (depending on the direction of its moving). Then, the head x_i and its images can be moving for at most $p_{i+1}(1) - p_{i-1}(1)$ steps.

Summing up the described amounts of possible derivations for $x_1, x_2, \ldots, x_k$ we obtain the following cumulative upper bound:

$$p_1(1) + p_2(1) + \sum_{i=2}^{k-1}(p_{i+1}(1) - p_{i-1}(1)) + (2\ell(T) + 2 - p_k(1) - p_{k-1}(1)).$$

Splitting the sum in the center for odd and even i we see that most summands cancel except for the first and the last, so the whole bound can be rewritten as

$$p_1(1) + p_2(1) + (p_m(1) - p_1(1)) + (p_{m'}(1) - p_2(1)) + (2\ell(T) + 2 - p_k(1) - p_{k-1}(1)),$$

where m and m' are the biggest odd and even number not exceeding k. Then the sum above is actually equal to $2\ell(T) + 2$, and the inequality of the lemma follows.   □

Now we are ready to make the final step in proving Theorem 1.

Lemma 7. *If* $Z \in A_0^*$ *then* $\delta(Z) \leq 2(\ell(Z) + 2)(\log \ell(Z) + 6)$.

Proof. Write out Z with all the letters w marked, then it has a form

$$Z = T_1 w\, T_2\, w \ldots w\, T_k \quad \text{for some } k \geq 1,$$

and $T_1, \ldots, T_k$ are some words in the alphabet $A_0 \backslash \{w\} = \{0, 1, \bar{0}, \bar{1}, c, h_0, h_1, h_2\}$.

Consider any derivation sequence $Z = Z_1 \rightarrow Z_2 \rightarrow \ldots \rightarrow Z_m$.

Observing the set of rules, we see that the letters w never disappear and never appear, if they were not present; also we see that these letters keep their positions. Hence, in each word Z_i all the letters w have the same positions as in the word Z. Since there are no rules of length larger than 2, we can deduce that all derivations of the word Z in the sequence above can be seen independently in k words $T_1 w, wT_2 w, \ldots, wT_k$ and therefore

$$\delta(Z) = \delta(T_1 w) + \delta(wT_2 w) + \ldots + \delta(wT_{k-1} w) + \delta(wT_k).$$

From this, using obvious inequalities $\delta(wT_1) \leq \delta(wT_1w)$ and $\delta(T_kw) \leq \delta(wT_kw)$, we obtain that

$$\delta(Z) \leqslant \sum_{i=1}^{k} \delta(wT_iw). \tag{7}$$

For each word T_i we have that $\delta(wT_iw) \leq 2(\ell(T_i) + 2)(\lceil \log(\ell(T_i) + 2) \rceil + 3)$ due to Lemma 4 and Lemma 6 (depending on the number of the head letters). Summing over all i we obtain an upper bound for the sum of derivational depths

$$\sum_{i=1}^{k} \delta(wT_iw) \leq \sum_{i=1}^{k} 2(\ell(T_i) + 2)(\lceil \log(\ell(T_i) + 2) \rceil + 3) \leq$$
$$2\sum_{i=1}^{k} (\ell(T_i) + 2)(\log(\ell(T_i) + 2)) + 8\sum_{i=1}^{k}(\ell(T_i) + 2). \tag{8}$$

First, note that $\sum_{i=1}^{k+1}(\ell(T_i)+2) = \ell(Z)+2$. Since the function $f(n) = n \log n$ is convex and $f(0) = 0$, then f is super-additive for two positive arguments, i.e., $f(a_1 + a_2) \geq f(a_1) + f(a_2)$, which implies that $f(\sum_{i=1}^{k} a_i) \geq \sum_{i=1}^{k} f(a_i)$ for $a_i \geq 0$. Hence, the left part in the last line of (8) can be estimated as

$$2\sum_{i=1}^{k+1} (\ell(T_i) + 2)(\log(\ell(T_i) + 2)) \leq 2(\ell(Z) + 2)(\log(\ell(Z) + 2)),$$

whilst the right part $8\sum_{i=1}^{k}(\ell(T_i) + 2)$ is just equal to $8(\ell(Z) + 2)$.

Putting these two things together we obtain that

$$\sum_{i=1}^{k+1} \delta(wT_iw) \leq 2(\ell(Z) + 2)(\log(\ell(Z) + 6)).$$

Then, using the inequality (7), we deduce that the lemma holds true.   □

An obvious consequence of this lemma is

Corollary 2. $\Delta_0(n)$ *belongs to the asymptotic class* $\Omega(n \log n)$.

Finally, we sum up that Corollaries 1 and 2 give us Theorem 1.

3   A System with Derivational Complexity in $\Theta(n^{1+\frac{a}{b}})$

The semi-Thue system $S_{a,b}$ which has the demanded derivational complexity from the class $\Theta(n^{1+\frac{a}{b}})$ is obtained by an elaboration of the system S_0, presented in the previous section. The system $S_{a,b}$ emulates a certain process, which we explain alongside its formal description. Throughout all section, we suppose that a and b are some fixed natural numbers greater than or equal to 1.

The alphabet A_b of the system $S_{a,b}$ is a union $A_b = A_0 \cup B_b$, where A_0 is the alphabet of the system S_0 and B_b is an alphabet of $5 + b$ letters. To define B_b first we consider two subsets: $H_b = \{d, p_0, p_1, \ldots, p_b\}$ and $D_q = \{q, \bar{q}\}$, and then set $B_b := D_q \cup H_b \cup \{v\}$.

The process which we emulate (with a derivation sequence of longest length) deals with *split words* of the form $wXvYw$, where X is a word in the alphabet $S_0 \setminus \{w\}$ and Y is a word in the alphabet $B_b \setminus \{v\}$. The subword wXv in essence works as the system S_0 and serves as the "timer" which emits signals for the second part of the machine. The second part is emulated with a subword vYw that receives the signals from the first part "through the splitter" v, counts them and after each b signals performs an elongation of an auxiliary word 2^a times.

The meaning of the letters from the alphabet B_b is following: v is a splitter, q and $\bar{q}$ serve as the letters of the auxiliary word; and the letters of H_b serve as working heads, where p_0 moves to the right and elongates the auxiliary word, d moves to the left and erases the marks, whilst the heads $p_1, p_2, \ldots, p_b$ cannot move and they are used to count signals from the timer. We call the letters from the set $H_0 \cup H_b$ as *heads* of the system $S_{a,b}$.

To describe the due set of rules $\mathcal{R}_{a,b}$ we modify the set $\mathcal{R}_0$ as follows:

First we substitute the set of rules (2) by a new set consisting of $2b$ rules:

$$
\begin{aligned}
h_i v p_j &\to c v p_{j+1} \quad (i = 1, 2; \ j = 1, 2, \ldots, b-1), \\
h_i v p_b &\to c v p_0 \quad (i = 1, 2).
\end{aligned} \tag{9}
$$

Here, instead of just bouncing at the left border w (as done in system S_0), the head h_i transmits a signal through the "splitter" v, which performs the counting $1 \to 2 \to \ldots \to b \to 0$ of the machine state.

Second, we introduce the rule which is elongating the words:

$$
p_0 q \to (\bar{q})^{2^a} p_0, \tag{10}
$$

The idea of this rule is similar to the rules of (1). The head p_0 moves to the right and transforms all letters q, which it passes, by blocks of marked letters $(\bar{q})^{2^a}$.

Once the head p_0 reaches the letter w on the right side of the word, it "bounces", meaning that it transforms to the head d, which moves to the left and removes the bars over the letters q. We realize this part by the two rules:

$$
p_0 w \to dw, \qquad \bar{q} d \to dq. \tag{11}
$$

Once the cleaning head d reaches the splitter v, it turns into the head p_1.

$$
vd \to vp_1 \tag{12}
$$

Actually, this rule can be seen as a supplemental command to the series (9).

The main result of the paper is then given by the following statement.

Theorem 2. *Let $S_{a,b}$ be a semi-Thue system with the alphabet A_b and the set of rules $\mathcal{R}_{a,b}$ that is the union of rules (1), (3), (4), (9), (10), (11). Then the derivational complexity function $\Delta_{a,b}(n)$ of this system is in the class $\Theta(n^{1+\frac{a}{b}})$.*

First we show, that the function $\Delta_{a,b}(n)$ can be estimated from below with the desired asymptotic behaviour. This we do by proving the following statement.

Lemma 8. *If $k \geq 1$ then for the word $U_k = wh_0 1^k vp_0 q^k w$ of length $2k + 5$ the derivational depth $\delta(U_k)$ is at least $k^{1+\frac{a}{b}}$.*

Proof. Similarly to Lemma 5, the exact computation of the derivation length can be done by a direct application of the rules. However, here the derivation sequence is not necessarily determined for the word U_k. Indeed, the word U_k possesses two options: to apply the rule (1) in the left part $wh_0 1^k v$ and to apply the rule (10) in the right part $vp_0 q^k w$. However, the transformations of the left and the right part can be studied independently until both heads arrive to the splitter v, and we obtain a word which has a form

$$wY h_i vp_1 q^L w, \text{ where } Y \in \{\bar{0}, \bar{1}\}^*. \tag{13}$$

For the left head it takes k operations and for the right head it takes $(k+2+k2^a)$ operations. After this arrival, one can only apply the rule $h_i vp_1 \rightarrow cvp_2$ and then the sequence is determined for some time again: the left head should shuttle to the left border and back $(b - 2)$ times until we again face the ambiguity of rule application in the word of the form $wXcvp_0 q^L w$, where $X \in \{\bar{0}, \bar{1}\}^*$.

After this, again the steps of the left head and the right head can be counted independently until again we face a word of the form (13), and then the derivation continues uniquely for some time as before.

Observing the moves of the left head the right head independently, we can count them as follows. The left head (similarly to the shuttling of S_0) can make $\lfloor \log_2 k \rfloor + 1$ travels from the left to the splitter and one less travel back. Altogether it sums up to $k(2\lfloor \log_2 k \rfloor + 1)$ operations, where the applications of type (9) are also counted. Hence, the right head can make

$$R = R(k) = 1 + \lfloor (\lfloor \log_2 k \rfloor + 1)/b \rfloor$$

travels to the right border and back. This can be counted with geometric series since after each travel the auxiliary word q^L is elongated 2^a times. Hence, all travels of p_0 to the right border (i.e., operations of type (10)) take

$$k(1 + 2^a + \ldots + (2^a)^R) = k(2^{a(R+1)} - 1)/(2^a - 1)).$$

steps, and the travels back (operations $\bar{q}d \rightarrow dq$) take 2^a times more. To obtain the exact number we would also need to add precisely R transform operations $p_0 w \rightarrow dw$, but in this lemma we care only about some lower bound. Therefore, we can roughly estimate from below the number of steps that the right header can make as $k2^{aR}$ and also use an obvious estimation $R(k) \geq \lceil (\lceil \log_2 k \rceil + 1)/b \rceil \geq (\log_2 k)/b$ to conclude that $\delta(U_k) \geq k2^{(a \log_2 k)/b} = k^{1+\frac{a}{b}}$, as required. $\qquad\square$

As in Sect. 2, we also need to show upper bounds, namely that for any word W one can obtain the derivation sequence of length at most $O(\ell(W)^{1+\frac{a}{b}})$.

Similarly to Lemma 4, we start our analysis by considering split words which have exactly one head l in the left and one head r in the right part.

Lemma 9. *Let words $X_1, X_2 \in D_0^*$ and $Y_1, Y_2 \in D_q^*$ and let letters $l \in H_0$ and $r \in H_b$. Then there exists $C > 0$ such that the word*

$$Z = wX_1lX_2vY_1rY_2w \tag{14}$$

has derivational depth $\delta(Z) \le C\ell(Z)^{1+\frac{a}{b}}$.

Proof. The two heads l and r give two positions in the word Z where the rules of $S_{a,b}$ can be (possibly) applied. After any application of the rules, the result has the form (14), so this form is preserved in the derivation sequence. Therefore, as in Lemma 8 we can count the rules applied in the left and in the right part.

Since the left head after the first return to the left border can perform at most $S := 1 + \lceil \log_2(\ell(X_1) + \ell(X_2)) \rceil$ runs to the right and one less back, the total number of its steps is at most $N := (\ell(X_1) + \ell(X_2))(3 + 2\lceil \log_2(\ell(X_1) + \ell(X_2)) \rceil)$.

The right head can shuttle to the right border and back and then the length of the word in the right part increases at most 2^a times, hence it becomes at most equal to $Q := 1 + 2^a(\ell(Y_1) + \ell(Y_2))$. After the first shuttling, each next shuttle of the right head needs b "signals" from the left head, hence it will be able to make at most $\lfloor S/b \rfloor$ such shuttles.

Counting the geometric series as in Lemma 8, we obtain that the right head performs at most $T := (Q + 2^aQ)(1 + 2^a + \ldots + 2^{a\lfloor S/b \rfloor})$ travelling steps and $1 + \lfloor S/b \rfloor$ transform steps of type $p_0w \to dw$. Obviously, $T = \frac{2^a+1}{2^a-1}Q(2^{a(\lfloor S/b \rfloor + 1)})$.

The number N, which counts the transform steps and the moves of the left head, can be estimated by $(\ell(X_1) + \ell(X_2))(4 + \lceil \log_2(\ell(X_1) + \ell(X_2)) \rceil)$, which is $O((\ell(X_1) + \ell(X_2)) \log((\ell(X_1) + \ell(X_2))))$, hence also $O((\ell(X_1) + \ell(X_2))^{1+\frac{a}{b}})$.

We are left to estimate T. Obviously, the product $Q(2^a + 1)/(2^a - 1)$ is just $O(\ell(Y_1) + \ell(Y_2))$. The remaining factor $2^{a(\lfloor S/b \rfloor + 1)}$ is at most $2^{(a/b)S + a}$, and since $S < (2 + \log_2(\ell(Y_1) + \ell(Y_2)))$, we obtain that this factor can be estimated as $O((\ell(X_1) + \ell(X_2))^{a/b})$. Then $T = O((\ell(Y_1) + \ell(Y_2))(\ell(X_1) + \ell(X_2))^{a/b}) = O(\ell(Z)\ell(Z^{a/b})) = O(Z^{1+\frac{a}{b}})$.

Putting together the estimates of N and T obtained in two paragraphs above, we get the inequality for $\delta(Z)$ as claimed in the lemma. $\qquad \square$

The next lemma is the main tool to estimate the length of derivation sequences.

Lemma 10. *If the word Z is a word having form wXw, where $X \in (A_b \setminus \{w\})^*$ then $\delta(Z) < C_1\ell(Z)^{1+\frac{a}{b}}$ for some $C_1 > 0$ that is independent of the word Z.*

Proof. Consider all occurrences of letters v in the word Z. There are three cases.

Case I. There are no letters v in the word Z. As usual, we call letters from the alphabet $H_0 \cup H_a$ heads. If there are no heads in X, then no rule can be applied to the word Z, and the lemma trivially follows.

If there is a single head x, then the derivation sequence is determined, and we consider two subcases.

Case I.1.a. $x \in H_b$. The head can only bounce on the right border, hence it bounces at most once. Then the longest derivation sequence has at most $(1 + 2^a)\ell(X)$ steps, where the factor $1 + 2^a$ comes from a possible elongation of the word, if the head x initially moves to the right, bounces and moves back.

Case I.1.b. $x \in H_0$. The head is not changing the length of the word, and it can travel freely along the word X only if all letters of X (except for the head) are from D_0, and in this case $\delta(Z) = O(\ell(Z) \log_2(\ell(Z)))$. If there are other letters on the tape, they should be digits from D_q as the rest is excluded; then the head stops after at most $2\ell(Z)$ steps, once it meets a letter from D_q on its way.

Now, we turn to the case when there are at least 2 heads in the word x.

Case I.2. Let $x_1, x_2, \ldots, x_k$ be all heads in the word Z, and $k \geq 2$. Then $X = w Z_1 x_1 Z_2 x_2 \ldots Z_k x_k Z_{k+1}$ for some words $Z_i \in (D_0 \cup D_q)^*$, where $i = 1, 2, \ldots, k$.

We will use an argument similar to the one of Lemma 6. One can observe that for $1 < i < k$ each head x_i can move inside the part Z_i or Z_{i+1}, depending on its initial direction, and cannot change the direction as x_i never reaches w or v. It follows that x_i can make at most $\max(Z_i, Z_{i+1}) \leq \ell(Z_i) + \ell(Z_{i+1})$ steps.

The leftmost head x_1 can bounce in the left border, but it happens at most once, and only in the case $x_1 = d$. In this particular case x_1 makes at most $2\ell(Z_1) + \ell(Z_2)$ steps, and in other cases at most $\ell(Z_1) + \ell(Z_2)$ steps are possible.

The rightmost head x_k can bounce from the right border also at most once, and only in the case $x_k = h_0$. In this case x_k makes at most $(1 + 2^a)\ell(Z_k) + \ell(Z_{k-1})$ steps, and in all other cases not more than $\ell(Z_k) + \ell(Z_{k-1})$ steps are possible.

Summing up the number of possible steps of all heads x_i for $i = 1, 2, \ldots, k$, we obtain a total bound for the number of steps

$$\ell(Z_1) + \sum_{i=1}^{k-1}(\ell(Z_i) + \ell(Z_{i+1})) + (1 + 2^a)\ell(Z_k) \leq (1 + 2^a)\sum_{i=1}^{k}\ell(Z_i) \leq (1 + 2^a)\ell(Z),$$

which finishes the consideration of the Case I.

Now, we consider the case when there is a single letter v in X.

Case II. The word Z can be written as $wXvYw$, where $X, Y \in A_b \setminus \{v, w\}$. Similarly to Lemma 8 and Lemma 9, we separately count the number of steps made on the left of the splitter v and on the right of it. Let λ be the number of heads in the left part wXv and μ be the number of heads in the right part vYw.

If $\lambda = 0$ or $\mu = 0$, no rule of the series (9) can ever be used, so only one left bounce in the left part and one right bounce in the right part is possible. In both situations one considers the part with heads and count steps similarly to Case I.2, obtaining at most $2\ell(X)$ steps, when the heads are in the left part and $(1 + 2^a)\ell(Y)$ when the heads are in the right part.

If $\lambda \geq 2$, then by a similar argument to the Case I.2, we see that at most one right bounce in wXv is possible. Indeed, the leftmost and rightmost heads can bounce at most once, and the remaining heads never change their direction and work in a restricted space. It follows that in the right part at most one left bounce in vYw, as it happens only with a right bounce in the left part. Also similarly to the Case I.2. the number of steps in the left part can be bound linearly by $2\ell(X)$, and in the right part the number of steps is at most $(1 + 2^a)\ell(Y)$.

If $\mu \geq 2$, then similarly to the above, at most one right bounce is possible for the left part, henceforth for the right part we have at most one bounce, so we obtain the same bounds $(1 + 2^a)\ell(Y)$ for the number of steps in the right part and $2\ell(X)$ for the number of steps in the left.

The remaining case is $\lambda = 1$ and $\mu = 1$. The only case, when the rule (9) can be used more than once in the derivation is very much restricted. This happens when $X = X_1 l X_2$ with $l \in H_0$ and $X_1, X_2 \in D_0^*$ and simultaneously $Y = Y_1 r Y_2$ with $r \in H_a$ and $Y_1, Y_2 \in D_q^*$. Then we can use Lemma 9 to obtain the inequality we need. In all other cases either left or right head will get blocked during the first shuttling, so that at most one bounce at the splitter is possible, and in this case we obtain a linear bound $2\ell(X) + (1 + 2^a)\ell(Y)$ for the total number of steps.

Finally, we study the case, when the word Z has two or more letters v.

Case III. $Z = wXvY_1v\ldots vY_kw$, $k \geq 2$ and $X, Y_1, \ldots, Y_k \in (A_b \setminus \{v, w\})^*$.

In this case at each splitter v we cannot have more than one use of the rule (9). Indeed, in each part vY_iv either we have more than one head, and then this follows from the argument of Case I.2. If the head is sole, it either belongs to H_0 or to H_b. In the first case it can only bounce with v on the right side, and in the second case it bounces with v only on the left side.

For the reason described above, in each part vY_iv, where $1 \leq i < k$, we can estimate the number of steps by $2\ell(Y_i)$. The part wXv also gives at most $2\ell(X)$ steps, and the part vY_kw gives at most $(1 + 2^a)\ell(Y_k)$. In total, the number of steps in the derivation can be estimated by $(1 + 2^a)\ell(Z)$.   □

Now, we generalize the previous lemma to the case of all possible words in $S_{a,b}$.

Lemma 11. *For any word $Z \in A_b^*$ one has inequality $\delta(Z) < C_1\ell(Z)^{1+\frac{a}{b}}$ for some $C_1 > 0$ that is independent of the word Z.*

Proof. Consider all occurrences of letters w in the word Z and write it out as $Z = A_1wA_2w\ldots wA_k$, where $k \geq 1$ and $A_i \in A_b^* \setminus \{w\}$. Note that the words A_i can be empty. In any derivation sequence $Z = Z_1 \to Z_2 \to \ldots \to Z_m$ the rules are applied independently in the parts $A_1w, wA_2w, \ldots, wA_{k-1}w, wA_k$, hence $\delta(wZw) = \delta(A_1w) + \delta(wA_k) + \sum_{i=2}^{k-1} \delta(wA_iw)$. Obviously, $\delta(A_1w) \leq \delta(wA_1w)$ and also $\delta(A_kw) \leq \delta(wA_kw)$, therefore we have $\delta(wZw) \leq \sum_{i=1}^{k} \delta(wA_iw)$, and applying Lemma 10 to each word wA_iw we obtain

$$\delta(Z) \leq \sum_{i=1}^{k} C_1\ell(A_i)^{1+\frac{a}{b}}. \tag{15}$$

Eventually, super-additivity of the convex function $f(x) = x^{1+a/b}$ shows that the sum in (15) is at most equal to $C_1(\sum_{i=1}^{k} \ell(A_i))^{1+\frac{a}{b}} \leq \ell(Z)^{1+\frac{a}{b}}$, as required.   □

Now we can see that Theorem 2 follows from Lemma 11 and Lemma 8, with a note that the words of odd length having long derivational depth can be obtained by adding any final letter to the words of even length considered in Lemma 8.

4   Further Questions

In connection to the results of this paper, it seems interesting to put forward the following two problems:

Problem 1. *Do there exist finite length-preserving semi-Thue systems, that have derivational complexity functions $\Delta_S(n)$ in the class $o(n \log n)$, but not in the class $\Theta(n)$?*[1]

Problem 2. *Does there exist a finitely presented monoid Π, whose Dehn function $f_\Pi(n)$ belongs to the class $o(n^2)$, but is not in the class $\Theta(n)$?*

Acknowledgements. The author would like to thank Jean-Camille Birget (Rutgers University at Camden) and Daria Smirnova (Université de Genève) for stimulating discussions. The author is also indebted to several anonymous referees whose comments significantly improved the quality of this paper.

References

1. Adian, S.I.: On a method for proving exact bounds on derivational complexity in Thue systems. Math. Notes **92**(1), 3–15 (2012)
2. Brady, N., Bridson, M.R.: There is only one gap in the isoperimetric spectrum. Geom. Funct. Anal. GAFA **10**(5), 1053–1070 (2000)
3. Birget, J.-C.: Time-complexity of the word problem for semigroups and the Higman embedding theorem Internat. J. Algebra Comput. **8**(2), 235–294 (1998)
4. Sapir, M., Birget, J.-C., Rips, E.: Isoperimetric and isodiametric functions of groups. Ann. Math. **156**(2), 345–466 (2002)
5. Hofbauer, D., Lautemann, C.: Termination proofs and the length of derivations. In: Dershowitz, N. (ed.) RTA 1989. LNCS, vol. 355, pp. 167–177. Springer, Heidelberg (1989). https://doi.org/10.1007/3-540-51081-8_107
6. Hofbauer, D., Waldmann, J.: Termination of $aa \rightarrow bc$, $bb \rightarrow ac$, $cc \rightarrow ab$. Inform. Process. Lett. **98**, 156–158 (2006)
7. Kobayashi, Y.: Undecidability of the complexity of rewriting systems, Algebraic system, Logic, Language and Computer Science, Kyoto University Research Information Repository 2008, pp. 47–51 (2016). https://repository.kulib.kyoto-u.ac.jp/dspace/handle/2433/231547
8. Kobayashi, Y.: The derivational complexity of string rewriting systems. Theoret. Comput. Sci. **438**, 1–12 (2012)

---

[1] After the preliminary version of this paper became partly available to the public, several examples of systems, *which do not* preserve length, have been constructed. An anonymous referee suggested a construction of a system having complexity $\Theta(n \log^* n)$ and recently Y. Kobayashi announced an example of a system with derivational complexity function from the class $\Theta(n \log \log n)$.

9. Olshanskii, A.Y.: Hyperbolicity of groups with subquadratic isoperimetric inequality. Int. J. Algebra Comput. **1**(3), 281–289 (1991)

10. Yu Olshanskii, A.: Polynomially-bounded Dehn functions of groups. J. Comb. Algebra **2**(4), 311–433 (2018)

The Untold Story of SBP

Ilya Volkovich$^{(\boxtimes)}$ ⓘ

CSE Division, University of Michigan, Ann Arbor, MI, USA
ilyavol@umich.edu

Abstract. In the seminal work of [4], Babai has introduced *Arthur-Merlin Protocols* and in particular the complexity classes MA and AM as randomized extensions of the class NP. While it is easy to see that NP ⊆ MA ⊆ AM, it has been a long standing open question whether these classes are actually different. In [5], Böhler et al. introduced the probabilistic class of SBP and showed that MA ⊆ SBP ⊆ AM. Indeed, this is the only known natural complexity class that lies between MA and AM. In this work we study the relations between these classes further, partially answering some open questions posed in [5].

Keywords: Arthur-Merlin Protocols · Randomized complexity theory · NP problems with bounded number of solutions · SZK

1 Introduction

For more than three decades, the question of whether the classes MA and AM are different had remained open. While it was shown that under widely-believed derandomization assumptions [14,15] MA = AM and, moreover, both to collapse to NP, there has been only a mild advancement on this front. In particular, Arvind et al. [3] showed that AM = MA if NP ⊆ P/poly. Yet, the same premises imply a collapse of the Polynomial Hierarchy (see, e.g., [13]) and hence are not *believed* to be true.

In [5], Böhler et al. introduced the class of SBP, which stands for *small bounded-error probability*. They also showed that SBP lies between MA and AM. To the best of our knowledge, SBP is the only natural class with this property. However, the only known conditional collapse results of either AM to SBP or SBP to MA are actually those that collapse AM all the way to MA.

SZK (Statistical Zero Knowledge) is the class of decision problems for which a "yes" answer can be verified by a statistical zero-knowledge proof protocol. Rather than providing a formal definition, the class can be captured by its complete (promise) problem known as *Statistical Difference* [19]: given two polynomial-size circuits, C_0 and C_1 on n variables, decide if the statistical distance between the induced probability distributions is either at most $1/3$ or at least $2/3$. This problem is denoted as $\mathrm{SD}^{(2/3,1/3)}$. Similarly to SBP, SZK ⊆ AM (see, e.g., [7]).

© Springer Nature Switzerland AG 2020
H. Fernau (Ed.): CSR 2020, LNCS 12159, pp. 393–405, 2020.
https://doi.org/10.1007/978-3-030-50026-9_29

A different line of work [1,16,20] has been involved with the study of the computational power of NP machines with bounded number of accepting paths (or witnesses). In [20], Valiant introduced the complexity class UP that consists of NP problems with unique solutions. For instance, UNIQUE-SAT stands for a version of the SAT problem in which the given Boolean formula φ is either unsatisfiable or contains **exactly** one satisfying assignment. Another natural example of such class is FewP, introduced by Allender in [1], which consists of NP problems with polynomially-many solutions. More generally, we consider the class SOLUTIONS$[f(n)]$ that consists of NP problems with at most $f(n)$ solutions, for inputs of size of n. For a formal definition, see Definition 10.

1.1   Our Results

Our first main result links the aforementioned lines of work:

Theorem 1. *Suppose there exists $\varepsilon > 0$ such that* 3-SAT $\in$ SOLUTIONS$[2^{n^{1-\varepsilon}}]$. *Then* AM $=$ SBP.

In other words, if there exists an NP machine that decides 3-SAT with somewhat less than trivial number of accepting paths, then AM collapses to SBP. In particular, the result holds if 3-SAT $\in$ UP or even if 3-SAT $\in$ FewP. To put the result in the correct context, note that even a subexponential number of accepting paths is not known to imply a (deterministic) subexponential-time algorithm. In fact, such an implication is not even known for the case of a unique path (i.e., SAT $\in$ UP).

We now would like to elaborate on the premises. For a 3-CNF φ of size s, the NP machine is required to have at most $2^{s^{1-\varepsilon}}$ accepting paths for some $\varepsilon > 0$. This requirement is trivially met when $s = n^{1+\delta}$, for $\delta > 0$. Indeed, the main challenge is to satisfy the requirement for formulas of linear and slightly superlinear sizes (i.e., when $s = \mathcal{O}(n)$ or $s = n \cdot \text{polylog}(n)$). Furthermore, we observe that the requirement is met for formulas of size n iff it is met for formulas of size $n \cdot \text{polylog}(n)$. This in turn allows us to define the size of a formula as the number of clauses as opposed to the encoding size, as these two notions are within poly-logarithmic factor from each other. For more details see Sect. 2.3

In terms of oracle separation, in [21], Vereshchagin has shown that while MA $\subseteq$ PP for every oracle, there exists an oracle that separates AM from PP. In [5], Böhler et al. have extended this inclusion to MA $\subseteq$ SBP $\subseteq$ PP $\cap$ AM. As a corollary, they have concluded that the same oracle also separates SBP from PP. Furthermore, it is an easy exercise to see that AM is closed under union and intersection. Yet, whether SBP is closed under intersection remains an open question. In [11], Göös et al. have shown an oracle relative to which SBP is not closed under intersection. In conclusion, the collapse of AM to SBP should evade numerous relativization barriers. In [18], Rubinstein have shown an oracle relative to which SAT is not in UP and FewP. Since the proof of Theorem 1 is relativizable, we obtain a further oracle separation as a corollary:

Corollary 1. *There exists an oracle relative to which for any $\varepsilon > 0$, SAT $\notin$* SOLUTIONS$[2^{n^{1-\varepsilon}}]$.

This result partially answers an open question posed in [5], whether one could extend the oracle separations to collapse consequences.

Relations Between SBP and SZK. Our next result studies the relation between SBP and SZK. To that end, we consider the *general Statistical Difference* problem: for functions $\alpha(n) > \beta(n)$, SD$^{(\alpha(n),\beta(n))}$ is the (promise) problem of deciding whether the statistical distance is either at most $\beta(n)$ or at least $\alpha(n)$ (for a formal definition, see Definition 8). Our next main result exhibits a non-trivial problem in the intersection of the promised versions of SZK and SBP.

Theorem 2. $\overline{\text{SD}}^{\left(1-\frac{1}{2^{n+3}}\,,\,\frac{1}{2^{n+3}}\right)} \in$ PromiseSBP.

First of all, it is to be noted that since PromiseSZK is closed under complement (see, e.g., [17]), $\overline{\text{SD}}^{\left(1-\frac{1}{2^{n+3}}\,,\,\frac{1}{2^{n+3}}\right)} \in$ PromiseSZK. Furthermore, the problem represents a somewhat more general version of $\overline{\text{SD}}^{(1\,,\,0)}$, which is complete for the class of the so-called problems with "V-bit" perfect zero knowledge protocols [12]. While $\overline{\text{SD}}^{(1\,,\,0)} \subseteq$ PromiseNP and hence is clearly in PromiseSBP, to the best of our knowledge, $\overline{\text{SD}}^{\left(1-\frac{1}{2^{n+3}}\,,\,\frac{1}{2^{n+3}}\right)}$ is not known to lie in any subclass of PromiseSBP (not even PromiseMA). In that sense, the proposed problem constitutes the first known non-trivial problem in PromiseSZK $\cap$ PromiseSBP. Indeed, this result partially answers another open question posed in [5], whether there is a natural problem in SBP that is not contained in MA. It is to be noted that Watson [22] has shown another natural problem complete for PromiseSBP.

Relation to Polarization. The *polarization lemma*, introduced by Sahai and Vadhan in [19], is an efficient transformation that takes as input a pair of Boolean circuits (C_0, C_1) and an integer k and coverts them into a new pair of circuits (D_0, D_1) such that:

$$\Delta(C_1, C_2) \geq 2/3 \implies \Delta(D_1, D_2) \geq 1 - 2^k$$
$$\Delta(C_1, C_2) \leq 1/3 \implies \Delta(D_1, D_2) \leq 2^k$$

We would like to highlight one important aspect of this transformation: if the input circuits C_1 and C_2 are defined on n variables, i.e., the distributions are samplable using n random bits, the resulting circuits D_1 and D_2 are defined on poly$(k) \cdot n$ variables, thus requiring more random bits. Similar phenomenon occurs when one tries to naively amplify the success probability of a BPP algorithm by a simple repetition. Indeed, if a given BPP algorithm achieves an error probability of $1/3$ using r random bits, one could drive down the error probability to 2^{-t} using $\mathcal{O}(t) \cdot r$ random bits. More efficient amplification procedures (see, e.g., [8,23]) allow us to achieve a similar probability bound using only $\mathcal{O}(t) + r$ random bits. This raises a natural question: could we obtain a "randomness-efficient" polarization procedure? Our Theorem 2 suggests that in a very efficient regime of parameters, the existence of such a procedure implies that SZK $\subseteq$ SBP.

Corollary 2 (Informal). *If there exists a randomness-efficient polarization,* *then* SZK ⊆ SBP.

Nonetheless, we believe that this result should be regarded as evidence that a "randomness-efficient" polarization may not be possible. Since while polarization is an inherently relativizable procedure, there exists an oracle that separates SZK from SBP (and, in fact, from PP. See [6]).

1.2 Ideas and Techniques

We show the collapse of AM to SBP by identifying complete sets of AM and SBP[1]. In [9], Goldwasser & Sipser considered the problem of determining whether a set S is of size at least m or at most $m/2$, where membership of x in S can be efficiently determined given a witness w. They show an AM protocol for the problem. Our first observation is that this problem is, in fact, hard for the class AM. In conclusion, we obtain a natural AM-complete problem, WSSE (see Definition 5 for more details). Next, we observe that the class SBP corresponds to a simpler version of the problem, SSE. As before, we would like to determine whether a set S is of size at least m or at most $m/2$. Yet, in this version of the problem, the membership of x can be efficiently determined given (just) x (see Definition 6 for more details).

In what follows, we show a polynomial-time reduction from WSSE to SSE. The natural approach would be to regard the set S as a set of tuples (x, w) such that w is a witness for membership of x in S. By definition, each $x \in S$ has at least one witness w associated with it. Yet, the actual number of such witnesses could be arbitrary. To illustrate this, consider the following two sets: S_1 contains only one element x_1 with $K \gg 2$ witnesses of membership; S_2 contains two elements e_1, e_2 with 1 witnesses of membership each. Suppose $m = 2$. Viewing S_1 and S_2 as above introduces order inversion between S_1 and S_2 as we will obtain sets with K and 2 elements, respectively. One approach to overcome this issue could be to actually count the number of witnesses. However this task turns out to be a #P-hard problem. We take a slightly different approach.

Rather than counting witnesses, we would like to ensure that each element x has only a "small" number of witnesses. For the sake of intuition, let us assume that SAT $\in$ UP. Fix x and consider the set W_x of witnesses associated with x. By definition, if $x \notin S$ then $W_x = \emptyset$; otherwise, if $x \in S$ then $|W_x| \geq 1$. Moreover, observe that the membership of w in W_x can be efficiently determined given (just) w. Will now run the unique-solution NP machine A on the predicate (formula) that corresponds to W_x. Observe that in the former case (i.e., $W_x = \emptyset$) A has zero accepting paths and in the latter case A has exactly one accepting path. In other words, every $x \in S$ will have a unique witness w of its membership in S. To handle the more general case of SAT $\notin$ SOLUTIONS$[2^{n^{1-\epsilon}}]$ we "preprocess" the circuit by applying sequential repetition thus increasing the gap between number of witnesses in the "yes" and the "no" cases. See Lemma 4 for the formal proof.

---

[1] Technically, we are looking at the promise versions of AM and SBP.

In order to related SZK and SBP we study the relation between Statistical and the Collision Distances. See Lemma 2 for the formal proof.

1.3 Organization

We start by some basic definitions and notation in Sect. 2. In Sect. 3 we prove our main results. In fact, we prove somewhat more general and technical statements. Finally, we discuss some open questions in Sect. 4.

2 Preliminaries

For a unary relation $R(x)$, we define $\#_x R \overset{\Delta}{=} |\{x \mid x \in R\}|$. For a binary relation $R(x, w)$, we define $\#_x \exists_w R \overset{\Delta}{=} |\{x \mid \exists w \text{ s.t. } (x, w) \in R\}|$. For $k \in \mathbb{N}$, we define $R^{\otimes k}$, the *tensor power of R*, as

$$R^{\otimes k} \overset{\Delta}{=} R(x_1, w_1) \wedge R(x_2, w_2) \wedge \ldots \wedge R(x_k, w_k)$$

where $x_1, \ldots, x_k$ and $w_1, \ldots w_k$ are k disjoint copies of x and w, respectively.

Observation 1. *Let $\bar{x} = (x_1, \ldots, x_k)$ and $\bar{w} = (w_1, \ldots, w_k)$. Then $\#_{\bar{x}} \exists_{\bar{w}} R^{\otimes k} = (\#_x \exists_w R)^k$.*

We will require the following technical lemma.

Lemma 1. *For any $s > 1$ and $0 < \varepsilon < 1$ it holds:*

1. $s^{-\varepsilon} \leq (s^{-1} - 1)\varepsilon + 1$.
2. $s^{\frac{1}{\varepsilon}} \geq 1 + \ln s \cdot \frac{1}{\varepsilon}$.

2.1 Probability Distributions and Circuits

Let X and Y be two random variables taking values in some finite domain $\mathcal{U}$.
    We define the *support* of a random variable X as

$$\text{Supp}(X) \overset{\Delta}{=} \{a \in \mathcal{U} \mid \Pr[X = a] > 0\}.$$

Definition 1 (Distances Between Distributions). *The Statistical Distance between X and Y is defined as*

$$\Delta(X, Y) = \max_{\mathcal{U}' \subseteq \mathcal{U}} \Pr[X \in \mathcal{U}'] - \Pr[Y \in \mathcal{U}'].$$

The equality is attained for $\mathcal{U}' = \mathcal{U}_X \overset{\Delta}{=} \{a \in \mathcal{U} \mid \Pr[X = a] \geq \Pr[Y = a]\}$.
    The Collision Distance between X and Y is defined as $\text{Col}(X, Y) \overset{\Delta}{=} \Pr[X = Y]$.

We prove two properties relating the Statistical and the Collision Distances.

Lemma 2. *Let $k = |\text{Supp}(X) \cup \text{Supp}(Y)|$. Then $\frac{1}{k} \leq \Delta(X,Y) + \text{Col}(X,Y) \leq 1$.*

Proof. Let $\mathcal{U}' = \text{Supp}(X) \cup \text{Supp}(Y)$, $\mathcal{U}_X = \{a \in \mathcal{U} \mid \Pr[X=a] \geq \Pr[Y=a]\}$ and $\mathcal{U}_Y = \mathcal{U}' \setminus \mathcal{U}_X$. For the first inequality:

$$
\begin{aligned}
\text{Col}(X,Y) &= \sum_{a \in \mathcal{U}'} \Pr[X=a] \Pr[Y=a] \\
&= \sum_{a \in \mathcal{U}'} (\Pr[X=a])^2 - \sum_{a \in \mathcal{U}'} \Pr[X=a](\Pr[X=a] - \Pr[Y=a]) \\
&\geq \frac{\left(\sum_{a \in \mathcal{U}'} \Pr[X=a] \right)^2}{|\mathcal{U}'|} - \sum_{a \in \mathcal{U}_X} \Pr[X=a](\Pr[X=a] - \Pr[Y=a]) \\
&\geq \frac{1}{|\mathcal{U}'|} - \sum_{a \in \mathcal{U}_X} (\Pr[X=a] - \Pr[Y=a]) = \frac{1}{|\mathcal{U}'|} - \Delta(X,Y).
\end{aligned}
$$

We now move to the second inequality.

$$
\begin{aligned}
&\text{Col}(X,Y) + \Delta(X,Y) - 1 \\
&= \sum_{a \in \mathcal{U}'} \Pr[X=a] \Pr[Y=a] + \sum_{a \in \mathcal{U}_X} (\Pr[X=a] - \Pr[Y=a]) - 1 \\
&\quad \Pr[X \in \mathcal{U}_X] \Pr[Y \in \mathcal{U}_X] + \Pr[X \in \mathcal{U}_Y] \Pr[Y \in \mathcal{U}_Y] + \\
&\quad \Pr[X \in \mathcal{U}_X] - \Pr[Y \in \mathcal{U}_X] - 1 \\
&= (\Pr[X \in \mathcal{U}_X] - 1)(\Pr[Y \in \mathcal{U}_X] + 1) + (1 - \Pr[X \in \mathcal{U}_X])(1 - \Pr[Y \in \mathcal{U}_X]) \\
&= 2(\Pr[X \in \mathcal{U}_X] - 1) \Pr[Y \in \mathcal{U}_X] \leq 0.
\end{aligned}
$$

Observe that $\Pr[X \in \mathcal{U}_Y] = 1 - \Pr[X \in \mathcal{U}_X]$ and $\Pr[Y \in \mathcal{U}_Y] = 1 - \Pr[Y \in \mathcal{U}_X]$. □

We complement our result by observing that for a pair of variables X and Y with disjoint supports it holds that: $\Delta(X,Y) = 1$ and $\text{Col}(X,Y) = 0$, and hence $\Delta(X,Y) + \text{Col}(X,Y) = 1$. In addition, for any $n \geq 1$ and $\varepsilon \geq 0$, consider a random variable X over $\{0,1\}^n$ defined as follows: For $\bar{a} \in \{0,1\}^n$:

$$
\Pr[X = \bar{a}] = \frac{1+\varepsilon}{2^n}, \text{ if } a_n = 0 \text{ and } \Pr[X = \bar{a}] = \frac{1-\varepsilon}{2^n}, \text{ otherwise.}
$$

Observe that $k = 2^n$, $\Delta(X, \bar{1} - X) = \varepsilon$ and $\text{Col}(X, \bar{1} - X) = \frac{1-\varepsilon^2}{k}$.

A Boolean circuit $C : \{0,1\}^n \to \{0,1\}^m$ induces a probability distribution on $\{0,1\}^m$ by evaluating C on a uniformly chosen input in $\{0,1\}^n$. For two Boolean circuits, C_1 and C_2, we will use the notations $\Delta(C_1, C_2)$ and $\text{Col}(C_1, C_2)$ to denote the corresponding distances between the induced distributions.

2.2   Complexity Classes and Promise Problems

We will be mostly concerned with the two following complexity classes. We refer the reader to [2] for the definitions of other standard complexity classes.

Definition 2 ([4]). *A language L is in* AM *if there exists a polynomial-time computable predicate $A(x, r, w)$ such that:*

$$x \in L \implies \Pr_r[\exists w : A(x, r, w) = 1] \geq 2/3$$
$$x \notin L \implies \Pr_r[\exists w : A(x, r, w) = 1] \leq 1/3.$$

Definition 3 ([5]). *A language L is in* SBP *if there exists $\varepsilon > 0, k \in \mathbb{N}$ and a polynomial-time computable predicate $B(x, r)$ such that:*

$$x \in L \implies \Pr_r[B(x, r)) = 1] \geq (1 + \varepsilon) \cdot \frac{1}{2^{n^k}}$$
$$x \notin L \implies \Pr_r[B(x, r)) = 1] \leq (1 - \varepsilon) \cdot \frac{1}{2^{n^k}}.$$

where $n = |x|$.

For technical reasons we will need to consider promise problems. A *promise problem* is a relaxation of a language. Formally:

Definition 4 (Promise Problems). $\Pi = (\Pi_{YES}, \Pi_{NO})$ *is a promise problem if $\Pi_{YES} \cap \Pi_{NO} = \emptyset$.*

In [9], Goldwasser & Sipser consider the problem of determining whether a set S is of size at least m or at most $m/2$, where membership of x can be efficiently determined given x and witness w. Formally, we define the following promise problem.

Definition 5 (Witnessed Set-Size Estimation).
   WSSE $\overset{\Delta}{=}$ (WSSE$_{YES}$, WSSE$_{NO}$), *where*

$$\text{WSSE}_{YES} = \{(C, m) \mid \#_x \exists_w C \geq m\}, \quad \text{WSSE}_{NO} = \{(C, m) \mid \#_x \exists_w C \leq m/2\}.$$

Here $C(x, w)$ is a Boolean circuit and m is an integer given in binary representation.

In the same paper, an AM protocol for the problem was given. In other words, it was shown that WSSE $\in$ PromiseAM. We begin by observing that WSSE is also hard for the class AM. Recall Definition 2. Let $L \in$ AM and suppose $r \in \{0, 1\}^{\ell}$. Furthermore, set $A_x(r, w) \overset{\Delta}{=} A(x, r, w)$ and $m = 2^{\ell+1}/3$. We observe that:

$$x \in L \implies \#_r \exists_w A_x \geq m$$
$$x \notin L \implies \#_r \exists_w A_x \leq m/2.$$

Corollary 3. WSSE *is* PromiseAM*-complete.*

In this paper, we also study a simpler version of the problem. As before, we would like to determine whether a set S is of size at least m or at most $m/2$. Yet, in this version of the problem, the membership of x can be efficiently determined given (just) x. Formally, we define the following promise problem.

Definition 6 (Set-Size Estimation). $\text{SSE} \triangleq (\text{SSE}_{YES}, \text{SSE}_{NO})$, *where*
$\text{SSE}_{YES} = \{(C, m) \mid \#_x C \geq m\}$, $\text{SSE}_{NO} = \{(C, m) \mid \#_x C \leq m/2\}$.

Here $C(x)$ is a Boolean circuit and m is an integer given in binary represen-tation.

Lemma 3 (Implicit in [5]). SSE *is* PromiseSBP-*complete.*

Indeed, SSE and WSSE capture the complexity classes SBP and AM, respec-tively. Indeed, AM corresponds to the class of all languages that reduce to WSSE. Likewise, SBP is the class of all languages that reduce to SSE. We now define the class SZK in a similar fashion.

Definition 7 (Statistical Difference, see [19]). *Let $\alpha(n) : \mathbb{N} \to \mathbb{N}$ and $\beta(n) : \mathbb{N} \to \mathbb{N}$ be computable functions, such that $\alpha(n) > \beta(n)$.*
Then $\text{SD}^{(\alpha(n)\,,\,\beta(n))} \triangleq (\text{SD}_{YES}^{(\alpha(n)\,,\,\beta(n))}, \text{SD}_{NO}^{(\alpha(n)\,,\,\beta(n))})$, *where*

$$\text{SD}_{YES}^{(\alpha(n)\,,\,\beta(n))} = \{(C_1, C_2) \mid \Delta(C_1, C_2) \geq \alpha(n)\},$$
$$\text{SD}_{NO}^{(\alpha(n)\,,\,\beta(n))} = \{(C_1, C_2) \mid \Delta(C_1, C_2) \leq \beta(n)\}.$$

Here, C_1 and C_2 are Boolean circuits $C_1, C_2 : \{0,1\}^n \to \{0,1\}^m$ of size $\text{poly}(n)$.

Definition 8 (Statistical Zero Knowledge). SZK *is defined as class of all languages that reduce to* $\text{SD}^{(2/3\,,\,1/3)}$.

We remark that originally SZK was defined in by Goldwasser et al. in [10] as the class of decision problems for which a "yes" answer can be verified by a statistical zero-knowledge proof protocol. The alternate characterization via the complete problem was given in [19].

In order the explore the relation between SZK and SBP further, we define a sparse version of the Statistical Difference problem.

Definition 9 (Sparse Statistical Difference). *For a computable function $t(n) : \mathbb{N} \to \mathbb{N}$, $t(n)$-$\text{SSD}^{(\alpha(n)\,,\,\beta(n))}$ is a specialization of SD to the case where the support size of distributions induced by C_1 and C_2 is bounded by $t(n)$. Formally: $|\text{Supp}(C_1)|\,,|\text{Supp}(C_2)| \leq t(n)$.*

2.3 SOLUTIONS[$f(n)$]

In this section we formally define the class SOLUTIONS[$f(n)$] and discuss some of its properties. Indeed, SOLUTIONS[$f(n)$] constitutes a subclass of NP with a bounded number of solutions.

Definition 10. *Let $f : \mathbb{N} \to \mathbb{N}$ be a computable function. We say that a lan-guage L is in the class SOLUTIONS[$f(n)$], if there exists a polynomial-time com-putable predicate $A(x, y)$ such that:*

$$x \in L \implies 1 \leq \#_y A_x \leq f(|x|)$$
$$x \notin L \implies \#_y A_x = 0.$$

where $A_x(y) = A(x, y)$.

In other words, SOLUTIONS$[f(n)]$ is special case of NP where for each $x \in L$ there are at most $f(n)$ witnesses. Observe that: UP = SOLUTIONS$[1] \subseteq$ FewP = SOLUTIONS$[\text{poly}(n)] \subseteq$ NP.

Remark: We note the definition would not change if we relaxed the requirement "to have of at most $f(n)$ solutions" to hold only for sufficiently large values of n. Next, we would like to point out a property of the SOLUTIONS$[f(n)]$ in a subexponential regime of parameters.

Observation 2. *Suppose there exists $\varepsilon > 0$ such that $L \in 2^{n^{1-\varepsilon}}$. Then there exist $\varepsilon' > 0$ and an NP machine that decides instances of size $n^{1+\varepsilon}$ of L with at most $2^{n^{1-\varepsilon'}}$ solutions.*

Proof. For instances of size $n^{1+\varepsilon}$, there are at most $2^{(n^{1+\varepsilon})^{1-\varepsilon}} = 2^{n^{1-\varepsilon^2}}$ solutions. $\square$

We conclude this section by presenting two facts about transforming Turing machines into Boolean circuits and Boolean circuits into Boolean formulas.

Fact 3. *There exists a polynomial-time algorithm that, given a Turing machine M that computes a Boolean predicate $A(z)$ in time $t(|z|)$ and input length n, outputs a Boolean circuit C of size $\text{poly}(t(n))$ on n inputs such that $C(z) = A(z)$ for every $z \in \{0,1\}^n$.*

Fact 4. *There exists a polynomial-time algorithm that, given a Boolean circuit C of size s, transforms it into a 3-CNF formula φ of size $\mathcal{O}(s)$ such that φ is satisfiable iff C is satisfiable.*

Combined with Observation 2, we obtain that wlog we can use various notions of size (i.e., number of gates in the circuit, number of clauses in a formula, bit-size complexity, etc.) interchangeably as they are with poly-log factor from each other and $s \cdot \text{polylog}(s) = o(s^{1+\varepsilon})$ for any $\varepsilon > 0$.

Corollary 4. *There exists $\varepsilon > 0$ such that CKT-SAT $\in$ SOLUTIONS$[2^{n^{1-\varepsilon}}]$ iff there exists $\varepsilon' > 0$ such that 3-SAT $\in$ SOLUTIONS$[2^{n^{1-\varepsilon'}}]$.*

3   Proofs of the Main Results

In this section we prove our main results Theorems 1 and 2. In fact, we prove somewhat more general and technical results.

Lemma 4. *Suppose there exists $\varepsilon > 0$ such that CKT-SAT $\in$ SOLUTIONS $[2^{n^{1-\varepsilon}}]$. Then PromiseAM = PromiseSBP.*

Proof. We show that PromiseAM $\subseteq$ PromiseSBP by showing that WSSE $\leq_p$ SSE. In particular, let $C(x,w)$ be a circuit of size s. We map an instance $(C(x,w), m)$ of WSSE to $\left(\hat{C}(\bar{x}, y), m^k \right)$, where $\hat{C}(\bar{x}, y)$ is a circuit of size $\text{poly}(sk)$, for $k = s^{\frac{1}{\varepsilon}}$.

Let $A(C', y)$ be a polynomial-time computable predicate that given a Boolean circuit $C'(z)$ of size s satisfies:

$$\#_z C' \geq 1 \implies 1 \leq \#_y A_{C'} \leq 2^{s^{1-\varepsilon}}$$
$$\#_z C' = 0 \implies \#_y A_{C'} = 0.$$

where $A_{C'}(y) = A(C', y)$. Consider the following Boolean predicate $\hat{A}(\bar{x}, y)$, where $\bar{x} = (x_1, \ldots, x_k)$:

---
1 $C'_{\bar{x}}(\bar{w}) \leftarrow C^{\otimes k}(\bar{x}, \bar{w})$; /* Taking k-th tensor power of the circuit
   $C(x, w)$ and plugging in the value of $\bar{x}$. Here
   $\bar{w} = (w_1, \ldots, w_k)$.   */
2 return $A(C', y)$

Let $\hat{C}(\bar{x}, y)$ denote the circuit that results from converting $\hat{A}(\bar{x}, y)$ into a Boolean circuit (applying Fact 3). The claim about the runtime is clear. We now analyze the reduction.

- Suppose that $\#_x \exists_w C \geq m$. By Observation 1: $\#_{\bar{x}} \exists_{\bar{w}} C^{\otimes k} \geq m^k$. In other words, there exist at least m^k different inputs $\bar{x}$ for which $\#_{\bar{w}} C'_{\bar{x}} \geq 1$. By the properties of A, for each such $\bar{x}$ there exists y such that $\hat{C}(\bar{x}, y) = 1$. Therefore, $\#_{(\bar{x}, y)} \hat{C} \geq m^k$.

- Suppose that $\#_x \exists_w C \leq m/2$. By Observation 1: $\#_{\bar{x}} \exists_{\bar{w}} C^{\otimes k} \leq (m/2)^k$. In other words, there exist at most $(m/2)^k$ different inputs $\bar{x}$ for which $\#_{\bar{w}} C'_{\bar{x}} \geq 1$. Since $C'_{\bar{x}}$ is a circuit of size at most sk, by the properties of A, for each such $\bar{x}$ there exist at most $2^{(sk)^{1-\varepsilon}}$ witnesses y such that $\hat{C}(\bar{x}, y) = 1$. Therefore, $\#_{(\bar{x}, y)} \hat{C} \leq (m/2)^k \cdot 2^{(sk)^{1-\varepsilon}} \leq m^k/2$. To justify the last inequality, assume wlog that $s > 4$ and recall Lemma 1:

$$(sk)^{1-\varepsilon} - k = s^{(1+\frac{1}{\varepsilon})(1-\varepsilon)} - s^{\frac{1}{\varepsilon}} = s^{\frac{1}{\varepsilon} - \varepsilon} - s^{\frac{1}{\varepsilon}} \leq s^{\frac{1}{\varepsilon}}[(s^{-1} - 1)\varepsilon + 1 - 1]$$
$$= s^{\frac{1}{\varepsilon}}(s^{-1} - 1)\varepsilon \leq (1 + \ln s \cdot \frac{1}{\varepsilon})(s^{-1} - 1)\varepsilon \leq (\varepsilon + \ln s)(s^{-1} - 1) < -1$$

$\square$

Theorem 1 follows from the lemma combined with Corollary 4.

Lemma 5. *Let $\beta(n)$ and $t(n)$ be such that $\beta(n) \cdot t(n) \leq 1/6$. Then*

$$\overline{t(n)\text{-SSD}}^{(1-\beta(n), \beta(n))} \in \text{PromiseSBP}.$$

Proof. Given two circuits $C_1, C_2 : \{0, 1\}^n \to \{0, 1\}^m$, the algorithm will try to find a collision. Namely, pick $x, x' \in \{0, 1\}^n$ uniformly at randomly and accept iff $C_1(x) = C_2(x')$. Observe that the success probability of the algorithm is exactly $\text{Col}(C_1, C_2)$. We now analyze this probability.

- Suppose $\Delta(C_1, C_2) \leq \beta(n)$. Observe that $|\text{Supp}(C_1) \cup \text{Supp}(C_2)| \leq 2t(n)$. Therefore, by Lemma 2, $\text{Col}(C_1, C_2) \geq \frac{1}{2t(n)} - \beta(n) \geq 3\beta(n) - \beta(n) = 2\beta(n)$.

- Suppose $\Delta(C_1, C_2) \geq 1 - \beta(n)$. By Lemma 2, $\mathrm{Col}(C_1, C_2) \leq \beta(n)$.

$\square$

Theorem 2 follows by observing that for circuits defined over n bits we have: $|\mathrm{Supp}(C_1)|, |\mathrm{Supp}(C_2)| \leq 2^n$, and instantiating the lemma to $t(n) = 2^n$ and $\beta(n) = \frac{1}{2^{n+3}}$.

4 Discussion and Open Question

The major widely-believed derandomization assumption of [14] that implies the collapse of AM and MA to NP is that some language in $\mathrm{NE} \cap \mathrm{coNE}$ requires SAT-oracle circuits of size $2^{\Omega(n)}$. Later in [15], the assumption of SAT-oracle circuits was relaxed to nondeterministic circuits. Can one prove that the premises of Theorem 1 are implies by a weaker assumption? For example, the assumption of [14,15] that some language in E requires SAT-oracle circuits of size $2^{\Omega(n)}$ implies a deterministic version of the argument of [9]. Could one utilize this connection?

Another natural question is to identify a corresponding MA-complete problem in the flavor of WSSE for AM and SSE for SBP. Could the presented collapse, then, be extended to MA? Conversely, could one show that any collapse either AM, SBP or MA to a subclass implies the premises of Theorem 1? Perhaps under an even stronger assumption that $\mathrm{NP} \subseteq \mathrm{P/poly}$?

Finally, we note that setting $\beta(n) = 0$ in the statement of Lemma 5, will recover the class $\overline{\mathrm{SD}}^{(1,\,0)}$. Could we identify a natural problem that reduces to $\overline{t(n)\text{-SSD}}^{(1-\beta(n)\,,\,\beta(n))}$ that does not reduce to $\overline{\mathrm{SD}}^{(1,\,0)}$ (with $\beta(n) \cdot t(n) \leq 1/6$)? Such a problem will attest the non-triviality of the SSD problem.

Acknowledgment. The author would like to extend his gratitude to Thomas Watson and Ryan Williams for useful conversations. Finally, the author would like to thank Henning Fernau and the anonymous referees for their detailed comments and suggestions.

References

1. Allender, E.W.: The complexity of sparse sets in P. In: Selman, A.L. (ed.) Structure in Complexity Theory. LNCS, vol. 223, pp. 1–11. Springer, Heidelberg (1986). https://doi.org/10.1007/3-540-16486-3_85
2. Arora, S., Barak, B.: Computational Complexity: A Modern Approach. Cambridge University Press, Cambridge (2009)
3. Arvind, V., Köbler, J., Schöning, U., Schuler, R.: If NP haspolynomial-size circuits, then MA = AM. Theor. Comput. Sci. **137**(2), 279–282 (1995). https://doi.org/10.1016/0304-3975(95)91133-B
4. Babai, L.: Trading group theory for randomness. In: Proceedings of the 17th Annual ACM Symposium on Theory of Computing (STOC), pp. 421–429 (1985). https://doi.org/10.1145/22145.22192

5. Böhler, E., Glaßer, C., Meister, D.: Error-bounded probabilistic computations between MA and AM. J. Comput. Syst. Sci. **72**(6), 1043–1076 (2006). https://doi.org/10.1016/j.jcss.2006.05.001

6. Bouland, A., Chen, L., Holden, D., Thaler, J., Vasudevan, P.: On the power of statistical zero knowledge. In: 58th IEEE Annual Symposium on Foundations of Computer Science, FOCS 2017, Berkeley, CA, USA, 15–17 October 2017, pp. 708–719 (2017). https://doi.org/10.1109/FOCS.2017.71

7. Fortnow, L.: The complexity of perfect zero-knowledge. Adv. Comput. Res. **5**, 327–343 (1989)

8. Goldreich, O.: A sample of samplers: a computational perspective on sampling. In: Goldreich, O. (ed.) Studies in Complexity and Cryptography. Miscellanea on the Interplay between Randomness and Computation. LNCS, vol. 6650, pp. 302–332. Springer, Heidelberg (2011). https://doi.org/10.1007/978-3-642-22670-0_24

9. Goldwasser, S., Sipser, M.: Private coins versus public coins in interactive proof systems. In: Proceedings of the 18th Annual ACM Symposium on Theory of Computing (STOC), pp. 59–68 (1986). https://doi.org/10.1145/12130.12137

10. Goldwasser, S., Micali, S., Rackoff, C.: The knowledge complexity of interactive proof systems. SIAM J. Comput. **18**(1), 186–208 (1989). https://doi.org/10.1137/0218012

11. Göös, M., Lovett, S., Meka, R., Watson, T., Zuckerman, D.: Rectangles are non-negative juntas. SIAM J. Comput. **45**(5), 1835–1869 (2016). https://doi.org/10.1137/15M103145X

12. Kapron, B.M., Malka, L., Srinivasan, V.: A framework for non-interactive instance-dependent commitment schemes (NIC). Theor. Comput. Sci. **593**, 1–15 (2015). https://doi.org/10.1016/j.tcs.2015.05.031

13. Karp, R.M., Lipton, R.J.: Some connections between nonuniform and uniform complexity classes. In: Proceedings of the 12th Annual ACM Symposium on Theory of Computing, Los Angeles, California, USA, 28–30 April 1980, pp. 302–309 (1980). https://doi.org/10.1145/800141.804678

14. Klivans, A., van Melkebeek, D.: Graph nonisomorphism has subexponential sizeproofs unless the polynomial-time hierarchy collapses. SIAM J. Comput. **31**(5), 1501–1526 (2002). https://doi.org/10.1137/S0097539700389652

15. Miltersen, P.B., Vinodchandran, N.V.: Derandomizing Arthur-Merlin games using hitting sets. Comput. Complex. **14**(3), 256–279 (2005). https://doi.org/10.1007/s00037-005-0197-7

16. Moran, S.: On the accepting density hierarchy in NP. SIAM J. Comput. **11**(2), 344–349 (1982). https://doi.org/10.1137/0211026

17. Okamoto, T.: On relationships between statistical zero-knowledge proofs. J. Comput. Syst. Sci. **60**(1), 47–108 (2000). https://doi.org/10.1006/jcss.1999.1664

18. Rubinstein, R.: Structural Complexity Classes of Sparse Sets: Intractability, Data Compression, and Printability. Ph.D. thesis, Northeastern University, Department of computer Science (1988)

19. Sahai, A., Vadhan, S.P.: A complete problem for statistical zero knowledge. J. ACM **50**(2), 196–249 (2003)

20. Valiant, L.G.: Relative complexity of checking and evaluating. Inf. Process. Lett. **5**(1), 20–23 (1976). https://doi.org/10.1016/0020-0190(76)90097-1

21. Vereshchagin, N.K.: On the power of PP. In: Proceedings of the Seventh Annual Structure in Complexity Theory Conference, pp. 138–143 (1992). https://doi.org/10.1109/SCT.1992.215389
22. Watson, T.: The complexity of estimating min-entropy. Comput. Complex. **25**(1), 153–175 (2014). https://doi.org/10.1007/s00037-014-0091-2
23. Zuckerman, D.: Simulating BPP using a general weak random source. Algorithmica **16**(4/5), 367–391 (1996). https://doi.org/10.1007/BF01940870

Weighted Rooted Trees: Fat or Tall?

Yaokun Wu[ORCID] and Yinfeng Zhu[✉][ORCID]

School of Mathematical Sciences and MOE-LSC, Shanghai Jiao Tong University,
Shanghai 200240, China
{ykwu,fengzi}@sjtu.edu.cn

Abstract. Let V be a countable set, let T be a rooted tree on the vertex set V, and let $\mathcal{M} = (V, 2^V, \mu)$ be a finite signed measure space. How can we describe the "shape" of the weighted rooted tree $(T, \mathcal{M})$? Is there a natural criterion for calling it "fat" or "tall"? We provide a series of such criteria and show that every "heavy" weighted rooted tree is either fat or tall, as we wish. This leads us to seek hypergraphs such that regardless of how we assign a finite signed measure on their vertex sets, the resulting weighted hypergraphs have either a "heavy" large matching or a "heavy" vertex subset that induces a subhypergraph with small matching number. Here we also must develop an appropriate definition of what it means for a set to be heavy in a signed measure space.

Keywords: Dilworth's Theorem · Down-set · Path

1   Background

Roughly speaking, Heisenberg's uncertainty principle for position and momentum says that, for a good function on the real line, either its variance is large or the variance of its Fourier transform is large [15, Theorem 4.1]. This kind of weak duality or orthogonality [7] also happens in combinatorics. The most famous example may be the Erdös-Szekeres subsequence theorem [6], which says that each sequence of $rs+1$ real terms contains an increasing subsequence of $r+1$ terms or a decreasing subsequence of $s + 1$ terms or both. Though the Erdös-Szekeres subsequence theorem has a short self-contained proof, it also easily follows from either Dilworth's Theorem [3] or Mirsky's Theorem [11]. The two equalities results, Dilworth's Theorem and Mirsky's Theorem, are generalized from posets to digraphs as two inequalities results, the Gallai–Milgram theorem [9] and the Gallai-Roy Theorem [8,12]. As a consequence of the Gallai–Milgram theorem or the Gallai-Roy Theorem, we know that $\lambda(D)\alpha(D) \geq |\mathrm{V}(D)|$ for any finite digraph D, where we use $\mathrm{V}(D)$, $\lambda(D)$ and $\alpha(D)$ for the vertex set of D, the length of a longest path in D and the maximum size of an independent set in D. This means that, we either have a 'long' path or a 'large' independent set in D, provided we define 'long' and 'large' in an appropriate way. Note that $\lambda(D)$ measures the length of a path by giving each vertex of the path a weight 1 while

Supported by NSFC (11671258, 11971305) and STCSM (17690740800).

H. Fernau (Ed.): CSR 2020, LNCS 12159, pp. 406–418, 2020.
https://doi.org/10.1007/978-3-030-50026-9_30

$\alpha(D)$ measures the size of a stable set by giving each vertex in the stable set a weight 1. Are there similar orthogonality results in which we assign weight to a set by something other than the usual cardinality function? For Dilworth's Theorem, a weighted version is recently discovered by Hujdurović, Husić, Milanič, Rizzi and Tomescu [10, Theorem 3.3]. Recall that a poset is just a transitive acyclic digraph and a rooted tree poset is a poset in which there is a special root vertex and a unique saturated path from the root vertex to any other vertex. Among the very few such weighted results which we are aware of, another one is a result of Song, Ward and York [14, Theorem 1.2], which improves a result of Bonamy, Bousquet and Thomassé [2, Lemma 3]: for any weighted rooted tree poset, either there is a long path or there are two heavy vertex subsets A and B for which there are no arcs between A and B. In some sense, this conforms to the above-mentioned result about the orthogonality of independent sets and paths: we are now considering vertex subsets which are pairwise independent of each other, instead of merely considering a set of pairwise nonadjacent vertices. We will present a vast generalization of the result of Song et al. (see Corollary 1) and discuss the possibility of going further along this line. Note that our proof strategy is totally different with the approach of Song et al.

2   Fat or Tall?

A graph G consists of a pair $(V(G), E(G))$ where $E(G) \subseteq \binom{V(G)}{2}$. We call $V(G)$ and $E(G)$ the vertex set and edge set of G, respectively. A graph G is countable if $V(G)$ is a countable set and is finite if $V(G)$ is finite. Recall that each countable set is either finite or denumerable (infinite). Let V be a countable set and let $\mathcal{M} = (V, 2^V, \mu)$ be a finite signed measure space, that is, $\sum_{v \in V} |\mu(v)| < \infty$ and $\mu(A) = \sum_{a \in A} \mu(a)$ holds for all $A \subseteq V$. If G is a graph with $V(G) = V$, the graph G together with the finite signed measure space $\mathcal{M}$ gives us a **weighted graph**, in which we think of μ as the **weighting function**. If $r \in V = V(G)$, we call the triple $\mathbb{G} = (G, r, \mu)$ a weighted rooted graph, where r is the root of $\mathbb{G}$ and μ, as known to be the weighting function of (G, μ), is also referred to as the **weighting function** of $\mathbb{G}$. In the case that G is a tree, (G, μ) is called a weighted tree and (G, r, μ) is called a **weighted rooted tree**.

For any integer k, we use $[k]$ for the set of positive integers between 1 and k. For any two vertices u and v of a tree T, the set of vertices on the unique path connecting u and v in T is denoted by $T[u, v]$. A **down-set** of a rooted tree (T, r) is a subset A of $V(T)$ such that, for each $a \in A$, the set A contains all components of $T - a$ to which r does not belong. A **chain** in a rooted tree (T, r) is a subset A of $V(T)$ such that $T[u, r]$ and $T[v, r]$ are comparable elements in the Boolean lattice $2^{V(T)}$ for every $u, v \in A$. A down-set of a rooted tree is a union of geodesic rays to the ends of the tree; A chain in a rooted tree is a subset of one geodesic ray.

Given a weighted rooted tree, when would you call it a "tall" tree and when would you call it a "fat" tree? If you think of being tall and being fat as two interesting properties, when do you expect to see an interesting weighted rooted tree? We suggest the following definitions so that you will not encounter any boring tree.

Definition 1. *Let* $\mathbb{T} = (T, r, \mu)$ *be a weighted rooted tree and let* $d_1, \dots, d_k, c$ *be* $k + 1$ *reals. We call* $\mathbb{T}$ $(d_1, \dots, d_k)$*-fat provided we can find* k *disjoint down-sets* $D_1, \dots, D_k$ *of* (T, r) *such that* $\mu(D_i) \geq d_i$ *for* $i \in [k]$. *We call* $\mathbb{T}$ c*-tall provided we can find a chain* C *in* (T, r) *such that* $\mu(C) \geq c$.

The next result says that as long as your weighted rooted tree is heavy enough, it is inevitable for it to be either fat or tall. It illustrates the spirit of Ramsey theory: You always find interesting structures when you are entering a large but otherwise arbitrary space. Note that Theorem 1 just recalls [14, Theorem 1.2] when μ is a probability measure, $k = 2$ and $d_1 = d_2 = c = \frac{1}{3}$.

Theorem 1. *Let* $\mathbb{T} = (T, r, \mu)$ *be a weighted rooted tree and let* k *be a positive integer. If* $d_1, \dots, d_k, c$ *are* $k + 1$ *positive reals such that*

$$\mu(\mathrm{V}(T)) \geq \sum_{i=1}^{k} d_i + (k - 1)c, \tag{1}$$

then $\mathbb{T}$ *is either* $(d_1, \dots, d_k)$*-fat or* c*-tall or both.*

Let k be a positive integer and let $d_1, \dots, d_k, c$ be $k + 1$ positive reals. If $d_1, \dots, d_k$ take at least two different values, we are not aware of any general way of constructing a weighted rooted tree $\mathbb{T} = (T, r, \mu)$ with

$$\mu(\mathrm{V}(T)) < \sum_{i=1}^{k} d_i + (k - 1)c,$$

which is neither $(d_1, \dots, d_k)$-fat nor c-tall. However, if we assume that $d_1, \dots, d_k$ take a constant value, we can give one such construction below, demonstrating the tightness of (1) in Theorem 1.

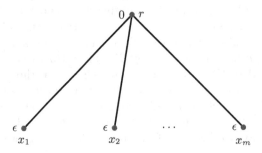

Fig. 1. A weighted rooted tree with weights indicated on the left of each node.

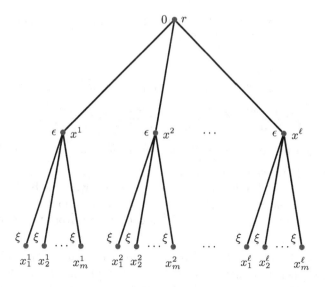

Fig. 2. A weighted rooted tree.

Example 1. Let k be a positive integer and let $d_1, \ldots, d_k, c$ and δ be $k+2$ positive reals. Let $K = \sum_{i=1}^k d_i$ and assume that $K + (k-1)c \geq \delta$.

CASE 1. $k = 1$.

Take a positive integer m such that $mc > K - \delta$ and let $\epsilon = \frac{K-\delta}{m}$. The weighted rooted tree $\mathbb{T} = (T, r, \mu)$ as shown in Fig. 1 satisfies

$$\mu(V(T)) = m\epsilon = K - \delta = K + (k-1)c - \delta$$

and is neither $(d_1, \ldots, d_k)$-fat nor c-tall.

CASE 2. $k = \ell + 1 \geq 2$.

Pick a positive integer m such that $(m+1)\delta > 2K$. Let $\epsilon = c - \frac{\delta}{2\ell}$ and $\xi = \frac{2K-\delta}{2m\ell}$. We now consider the weighted rooted tree $\mathbb{T} = (T, r, \mu)$ as displayed in Fig. 2. Note that

$$\mu(V(T)) = \ell(m\xi + \epsilon) = (K - \frac{\delta}{2}) + (\ell c - \frac{\delta}{2}) = K + (k-1)c - \delta.$$

Firstly, every chain C in (T, r) satisfies $\mu(C) \leq \epsilon + \xi = c - \frac{\delta}{2\ell} + \frac{2K-\delta}{2m\ell} < c$, showing that $\mathbb{T}$ is not c-tall.

Secondly, we assume that $D_1, \ldots D_k$ are k disjoint down-sets of (T, r). Under the additional condition that d_i takes a constant value d for each $i \in [k]$, let us show that $d_j = d > \mu(D_j)$ holds for at least one $j \in [k]$, which says that $\mathbb{T}$ is not $(d_1, \ldots, d_k)$-fat.

For $i \in [\ell]$, we denote the set $\{x_j^i : j \in [m]\}$ by L_i. Note that, for all $i \in [\ell]$, a down-set of (T, r) containing x^i must also contain L_i. By the pigeonhole principle, without loss of generality, we assume that $\bigcup_{j\in[t]} D_j \subseteq \bigcup_{i\in[t-1]} L_i$ holds for some integer t satisfying $2 \leq t \leq k$. This gives

$$\frac{dt}{m(t-1)} \geq \frac{d(\ell+1)}{m\ell} = \frac{dk}{m\ell} > \frac{2dk - \delta}{2m\ell}$$

$$= \frac{2K - \delta}{2m\ell} = \xi = \frac{\mu(\bigcup_{i\in[t-1]} L_i)}{m(t-1)} \geq \frac{\sum_{j\in[t]} \mu(D_j)}{m(t-1)},$$

yielding that, for at least one $j \in [t]$, it happens $d_j = d > \mu(D_j)$, as desired. $\square$

In Theorem 1, we can surely allow any one of $d_1, \ldots, d_k, c$ to be zero. This does not make any real difference, as the empty set, which is both a chain and a down-set, has measure zero. But Theorem 1 may not hold if we allow any of $d_1, \ldots, d_k, c$ to take a negative value. This can be seen from the following easy example.

Example 2. Let $V = \{r\}$, let T be the unique tree on V and let μ be the measure on V such that $\mu(r) = 0$. Let $c = d_1 = 1$ and $d_2 = -2$. Note that $\mu(V(T)) = 0 \geq 0 = (d_1 + d_2) + (2 - 1)c$. For (T, r, μ), surely the μ-measure of each chain is less than $c = 1$ and the μ-measure of each down-set is less than $d_1 = 1$.

Question 1. A function μ on 2^V is **submodular** provided $\mu(A \cup B) + \mu(A \cap B) \leq \mu(A) + \mu(B)$. In Theorem 1, what will happen if μ is not a measure but only a submodular function?

A **hypergraph** H is a pair $(V(H), E(H))$, where $V(H)$ is the vertex set of H and $E(H) \subseteq 2^{V(H)}$ is known as the edge set of H. To emphasize that we are considering a hypergraph, we often call each edge of the hypergraph H an **hyperedge** of H. For each positive integer k, a k-**matching** of H is a set of k disjoint hyeredges of H, while a k-**antimatching** of H is a subset C of $V(H)$ which is disjoint from at least one member of any $(k+1)$-matching of H. Note that a k-antimatching is just a set which cannot be a transversal of any $(k+1)$-matching.

Let V be a countable set and let $(V, 2^V, \mu)$ be a finite signed measure space. Let H be a hypergraph with $V(H) = V$ and $E(H) \subseteq 2^V$. For real numbers $d_1, \ldots, d_k$, we say that (H, μ) is $(d_1, \ldots, d_k)$-**fat** provided we can find a k-matching of H, say $\{e_1, \ldots, e_k\}$, such that $\mu(e_i) \geq d_i$ for $i \in [k]$. For any real number c and positive integer t, we say that (H, μ) is (c, t)-**tall** provided we can find a t-antimatching W of H such that $\mu(W) \geq c$. These concepts allow us to formulate the next conjecture, which coincides with Theorem 1 when $t = 1$.

Conjecture 1. Let $\mathbb{T} = (T, r, \mu)$ be a weighted rooted tree and let k, t be two positive integers. Let H be the hypergraph with $V(H) = V(T)$ and with the set of all down-sets of (T, r) as $E(H)$. If $d_1, \ldots, d_k, c$ are $k + 1$ positive reals such that

$$\mu(V(T)) \geq \sum_{i=1}^{k} d_i + \lceil \frac{k-1}{t} \rceil c,$$

then (H,μ) is either $(d_1,\ldots,d_k)$-fat or (c,t)-tall or both.

Let (X,μ) be a Borel measure space, namely X is a topological space and μ a Borel measure on the Borel sets of X. For the rooted tree case, we are indeed considering the topological space on its vertex set with all down-sets as open sets. In general, you can consider an Alexandroff space, which is essentially the same as a poset antimatroid[1], and a corresponding Borel measure space. Can we go further to talk about the hypergraph consisting of all open sets of X and conclude under certain assumption that it is fat or tall or both as in Conjecture 1?

Let P be a poset of countably many elements. For any $x \in P$, we write $x \downarrow_P$ for the set of elements which are less than or equal to x in P and we write $x \uparrow_P$ for the set of elements which are greater than or equal to x in P. A **down-set** of P is a subset A of P such that $x \downarrow_P \subseteq A$ for all $x \in P$, and an **up-set** of P is the complement set of a down-set of P. When the poset is finite, sending a down-set to its set of maximal elements yields a one-to-one correspondence between down-sets and antichains of the poset. If $(P, 2^P, \mu)$ is a finite signed measure space, we call (P, μ) a **weighted poset**. Each rooted tree (T, r) naturally gives rise to a poset $(V(T), \prec)$, called its **ancestral poset**, in which $x \prec y$ if and only if $y \in T[x, r] \setminus \{x\}$. One natural question is to ask to what extent Theorem 1 can be extended to general weighted posets. A **filter** in a poset $(Q, \prec)$ is a nonempty subset F such that

- if $x \in F$ and $x \prec y$, then $y \in F$;
- if $x, y \in F$, then there exists $z \in F$ with $z \prec x$ and $z \prec y$.

Note that each filter in the ancestral poset of a rooted tree has to be a path. The next example tells us that we cannot always expect to see either a heavy antimatching/filter or a heavy matching in a general weighted poset.

Example 3. Take two positive integers k and n. Let $V_1 = [k] \times [n]$ and $V_0 = [n]^{[k]}$. You can think of V_0 as the set of all vertices (atoms) of the n-ary k-dimensional cube and think of V_1 as the set of all facets (coatoms) of the n-ary k-dimensional cube. Let P be the poset on $V_0 \cup V_1$ in which $x > y$ if and only if $x = (\ell, h) \in V_1$ and $y \in V_0$ is a function satisfying $y(\ell) = h$. Let H be the hypergraph consisting of all downsets of P. Choose any nonnegative real δ. We define a signed measure μ on P such that

$$\mu(x) = \begin{cases} \frac{1-n^k\delta}{nk} & \text{if } x \in V_1; \\ \delta & \text{if } x \in V_0. \end{cases}$$

A subset of $V_0 \cup V_1$ is a 1-antimatching of H if and only if it is a filter in P if and only if it is of the form $x \uparrow_P$ for some $x \in P$. But for any $x \in P$, it holds $\mu(x \uparrow_P) \leq \delta + \frac{1-n^k\delta}{n} = \frac{1}{n} + \delta(1 - n^{k-1}) \leq \frac{1}{n}$.

Let $\{Q_1, Q_2\}$ be a 2-matching in H. If one of them is contained in V_0, then $\min(\mu(Q_1), \mu(Q_2)) \leq \mu(V_0) = n^k\delta$. If both Q_1 and Q_2 are not contained in V_0,

---

[1] More precisely, an Alexandroff space is the set of down-sets of a preorder.

then there exists $\ell \in [k]$ such that $Q_1 \cup Q_2 \subseteq V_0 \cup (\{\ell\} \times [n])$. This means that $\mu(Q_1 \cup Q_2) \leq n\frac{1-n^k\delta}{nk} + n^k\delta = \frac{1}{k} + \frac{(k-1)n^k\delta}{k}$ and so $\min(\mu(Q_1), \mu(Q_2)) \leq \frac{1}{2k} + \frac{(k-1)n^k\delta}{2k}$. To summarize, when δ is small enough, say $\delta = 0$, we have $\min(\mu(Q_1), \mu(Q_2)) \leq \frac{1}{k}$. □

One reason that we like to study trees is that they are really visible so that we may easily say many simple facts on them and then there are many directions to go for possible generalizations. For a rooted tree and a measure on its vertex set, we can add a new root vertex and join it to the old root vertex and then naturally produce a measure on the edge set of the new graph from the existing measure on the old tree. This operation allows us view the claim in Theorem 1 as a statement on an undirected branching greedoid [1, 13]. We think that we should be quite close to a proof of the following conjecture. Besides branching greedoid addressed in Conjecture 2, one may even consider possible generalizations to multiply-rooted graphs [4].

Conjecture 2. Let $\mathcal{F}$ be an undirected branching greedoid on a countable ground set E. Let H be the hypergraph on E whose edge set is $\{E - X : X \in \mathcal{F}\}$. Let μ be a measure on the power set of E and let k be a positive integer. If $d_1, \ldots, d_k, c$ are $k + 1$ positive reals such that

$$\mu(E) \geq \sum_{i=1}^{k} d_i + (k-1)c,$$

then (H, μ) is either $(d_1, \ldots, d_k)$-fat or c-tall or both.

We mention that Theorem 1 is self-strengthening. The next two easy corollaries of Theorem 1 both have it as a special case.

Corollary 1. *Let (P, μ) be a weighted poset and $r \in P$. Assume that, for each $y \in r \downarrow_P$, the number of saturated chains from r to y, denoted by n_y, is a finite number. For any $k + 1$ nonnegative reals $c, d_1, \ldots, d_k$ satisfying $(k-1)c + \sum_{i=1}^{k} d_k \leq \mu(r \downarrow_P)$, either there exists a saturated chain C of $r \downarrow_P$ starting from its maximum element r such that $\sum_{u \in C} \frac{\mu(u)}{n_u} \geq c$, or there exist pairwise disjoint down-sets $D_1, \ldots, D_k$ of $r \downarrow_P$ such that $\sum_{u \in D_i} \frac{\mu(u)}{n_u} \geq d_i$ for all $i \in [k]$.*

Corollary 2. *Let V be a countable set and let $(V, 2^V, \mu)$ be a finite signed measure space. Let W be a subset of V and let T be a tree on V. Let $d_1, \ldots, d_k, c$ be $k + 1$ positive reals such that Eq. (1) holds. Then there are either k disjoint subsets $D_1, \ldots, D_k$ such that $\mu(D_i) \geq d_i$ and $T - D_i$ is a tree containing W for all $i \in [k]$, or there is a vertex u such that $\mu(C) \geq c$ where C is the the convex hull of $\{u\} \cup W$ in T.*

Erdős and Hajnal [5] conjectured that for every graph H, there exists a constant c_H such that every graph G on n vertices which does not contain an induced copy of H has a clique or a stable set of size n^{c_H}. Since clique and stable set of a graph correspond to chain and antichain in a poset, this conjecture is also in the spirit of Dilworth's Theorem which we discuss in Sect. 1. The work of Song et al. [14, Theorem 1.2] is to verify a conjecture posed by Bonamy et al. in their study of the Erdös-Hajnal Conjecture [2]. We finally present a result, Theorem 2, as an application of Theorem 1. Note that the proof can be done by following the proof of [2, Theorem 6] with our Theorem 1 playing the role of [2, Lemma 3] there.

Let G be a graph. For any $X \subseteq V(G)$, the **neighborhood** of X in G, denoted by $N_G(X)$, is the set of vertices from $V(G) \setminus X$ which are adjacent to at least one element of X in G, and the **closed neighborhood** of X in G, denoted by $\overline{N}_G(X)$, is defined to be $N_G(X) \cup X$.

Theorem 2. *Let k be a positive integer and let (G, μ) be a connected countable weighted graph. If $d_1, \ldots, d_k, c$ are $k + 1$ positive integers such that*

$$\mu(V(G)) \geq \sum_{i=1}^{k} d_i + (k - 1)c,$$

then either there exists a subset A of $V(G)$ such that $G[A]$ is a path and $\mu(\overline{N}_G(A)) \geq c$, or there are k disjoint subsets $X_1, \ldots, X_k$ of $V(G)$ such that $\mu(X_i) \geq d_i$ for all $i \in [k]$ and that there are no edges between X_i and X_j for all $\{i, j\} \in \binom{[k]}{2}$.

3 Up and Down in a Rooted Tree

The purpose of this section is to prove Theorem 1.

For each poset P and each subset D of P, we write $D \uparrow_P$ for the minimum up-set of P which contains D and we write $D \downarrow_P$ for the minimum down-set of P which contains D. Let $\mathcal{T} = (T, r)$ be a rooted tree. We will naturally regard $\mathcal{T}$ as a poset in which $x > y$ if and only if $x \in T[y, r] \setminus \{y\}$. For any $x \in V(T)$, let $S_{\mathcal{T}}(x)$ be the set of neighbors y of x in T such that $x \in T[y, r]$, which we call the **shadow** of x in $\mathcal{T}$. Surely, it holds $x \downarrow_{\mathcal{T}} \supseteq S_{\mathcal{T}}(x)$ for all $x \in V(T)$.

Definition 2. *Let (P, μ) be a weighted poset. For any two nonnegative real numbers α and β, we say that a down-set D of P is an (α, β) down-set of (P, μ) provided $\mu(D) \geq \beta$ and $\mu(D \uparrow_P) \leq \alpha + \beta$.*

An (α, β) down-set D is like a good watermelon, where D really stands for the pulp of the watermelon and $D \uparrow_P$ represents its closure, namely the pulp together with the peel.

Let us explore the condition under which we can find an (α, β) down-set in a weighted rooted tree. We first do this for finite trees in Lemma 1. Then we strengthen Lemma 1 to Lemma 3, which makes the same statement for countable trees.

Lemma 1. *Let $\mathcal{T} = (T, r)$ be a finite rooted tree and let μ be a weighting function on T. Let α and β be two nonnegative reals such that $\mu(V(T)) \geq \alpha + \beta$ and $\mu(x \uparrow_T) \leq \alpha$ for all $x \in V(T)$. Then the weighted rooted tree $\mathbb{T} = (\mathcal{T}, \mu)$ has an (α, β) down-set.*

Proof. We intend to find a down-set D of $\mathcal{T}$ such that $\mu(D) \geq \beta$ and $\mu(D \uparrow_T) \leq \alpha + \beta$. We will demonstrate its existence by induction on $|V(T)|$.

If $|V(T)| = 1$, then $\beta = 0$ and we can set $D = \{r\}$.

Assume now $|V(T)| > 1$ and that the result holds when $|V(T)|$ is smaller. List the elements in $S_T(r)$ as $x_1, \ldots, x_s$. Let $V_i := x_i \downarrow_T$ for $i \in [s]$ and put $\epsilon := \alpha - \mu(r) \geq 0$. There are three cases to consider.

CASE 1. $\beta \leq \mu(V_1) \leq \beta + \epsilon$.

Take $D = V_1$, which is a down-set of $\mathcal{T}$. Then $\mu(D) = \mu(V_1) \geq \beta$ and $\mu(D \uparrow_T) = \mu(V_1) + \mu(r) \leq (\beta + \epsilon) + (\alpha - \epsilon) = \alpha + \beta$.

CASE 2. $\beta + \epsilon < \mu(V_1)$.

Define a signed measure space $(V_1, 2^{V_1}, \mu')$ by requiring

$$\mu'(A) = \begin{cases} \mu(A) + \mu(r) & \text{if } x_1 \in A \subseteq V_1, \\ \mu(A) & \text{if } A \subseteq V_1 \setminus \{x_1\}. \end{cases}$$

Let T' be the subtree of T induced by V_1. Note that

$$\mu'(V(T')) = \mu'(V_1) = \mu(V_1) + \mu(r) > (\beta + \epsilon) + (\alpha - \epsilon) = \alpha + \beta. \tag{2}$$

By induction hypothesis for (T', x_1, μ'), we have a down-set D of (T', x_1) such that

$$\mu'(D) \geq \beta \quad \text{and} \quad \mu'(D \uparrow_{T', x_1}) \leq \alpha + \beta. \tag{3}$$

Comparing (3) with (2) yields $D \uparrow_{T', x_1} \subsetneq V_1 = x_1 \downarrow_{T, x}$ and so $x_1 \notin D$ follows. We now see that $D = D \downarrow_{T, r}$ satisfies $\mu(D) = \mu'(D) \geq \beta$ and $\mu(D \uparrow_{T, r}) = \mu'(D \uparrow_{T', x_1}) \leq \alpha + \beta$.

CASE 3. $\mu(V_1) < \beta$.

Let T' be the tree obtained from T by deleting V_1 and let μ' be the restriction of μ on $2^{V(T) \setminus V_1}$. Let $\alpha' = \alpha$ and $\beta' = \beta - \mu(V_1) > 0$. Note that $\mu'(V(T')) = \mu(V(T)) - \mu(V_1) \geq \alpha' + \beta'$. Applying induction assumption on (T', r, μ'), we can find a down-set D' of (T', r) such that $\mu(D') \geq \beta' = \beta - \mu(V_1)$ and $\mu(D' \uparrow_{T', r}) \leq \alpha' + \beta' = \alpha + \beta - \mu(V_1)$. We thus see that $D = D' \cup V_1$ is a down-set of $\mathcal{T}$, as required. □

Lemma 2. *Let $\mathcal{T} = (T, r)$ be a countable rooted tree and let μ be a weighting function on T. Take $x \in V(T)$ and any positive real ϵ. Then $S_T(x)$ can be partitioned into two sets A and B such that $|B| < \infty$ and $|\mu(A \downarrow_T)| < \epsilon$.*

Proof. If $S_T(x)$ is a finite set, we can choose $A = \emptyset$ and $B = S_T(x)$. Otherwise, we can enumerate the elements of $S_T(x)$ as $x_1, x_2, \ldots$. Note that $\sum_{i=1}^{\infty} |\mu(x_i \downarrow_T)| < \infty$ and so there exists a positive integer N such that $\sum_{i=N}^{\infty} |\mu(x_i \downarrow_T)| < \epsilon$. Now, let $A = \{x_i : i \geq N\}$ and $B = \{x_i : i \in [N-1]\}$. □

Lemma 3. *Let $\mathcal{T} = (T, r)$ be a countable rooted tree. Let α and β be two non-negative real numbers. Consider a weighted rooted tree $\mathbb{T} = (\mathcal{T}, \mu)$ satisfying $\mu(V(T)) \geq \alpha + \beta$ while $\mu(x \uparrow_{\mathcal{T}}) \leq \alpha$ for all $x \in V(T)$. Then $\mathbb{T}$ has an (α, β) down-set.*

Proof. If $\mu(V(T)) = \alpha + \beta$, clearly $D = V(T)$ itself is an (α, β) down-set of $\mathbb{T}$.

In the sequel, we turn to the case that $\mu(V(T)) > \alpha + \beta$. Take any ϵ such that $0 < \epsilon \leq \mu(V(T)) - \alpha - \beta$. We claim that we can find an $(\alpha + \epsilon, \beta)$ down-set D_ϵ of $\mathbb{T}$. If this really holds, a compactness argument tells us that there exists an (α, β) downset D of $\mathbb{T}$, as wanted.

For any nonnegative integer i, let L_i be the set of vertices of T which are of distance i from r in T. By assumption,

$$\sum_{x \in V(T)} |\mu(x)| < \infty.$$

Consequently, there exists a nonnegative integer N such that

$$\sum_{i=N}^{\infty} \sum_{x \in L_i} |\mu(x)| < \epsilon. \tag{4}$$

By Lemma 2, we can associate to each $x \in V(T)$ a partition of $S_{\mathcal{T}}(x)$ into two sets A_x and B_x such that $|B_x|$ is finite and $|\mu(A_x \downarrow_{\mathcal{T}})| < \epsilon$. In view of (4), we will require that $B_x = \emptyset$ for $x \in L_{N-1}$. For all $x \in V(T)$, we denote the set $A_x \downarrow_{\mathcal{T}}$ by Λ_x. Let $\mathcal{A}$ be the set system

$$\{\Lambda_x : x \in \bigcup_{i=0}^{N-1} L_i\}.$$

Observe that $\mathcal{A}$ forms a hierarchy, namely $A \cap A' \in \{\emptyset, A, A'\}$ for all $A, A' \in \mathcal{A}$. Let $\overline{\mathcal{A}}$ be the set of maximal elements of $\mathcal{A}$, namely

$$\overline{\mathcal{A}} = \{A \in \mathcal{A} : \nexists A' \in \mathcal{A}, A' \supsetneq A\}$$
$$= \{A \in \mathcal{A} : A' \cap A \in \{A', \emptyset\}, \forall A' \in \mathcal{A}\}.$$

It is clear that the elements of $\overline{\mathcal{A}}$ are pairwise disjoint and we write Σ for the union of them. Let

$$W = \left(\bigcup_{i=0}^{N-1} L_i\right) \setminus \Sigma$$

and let T^* be the subtree of T induced by W. For each element $\Lambda_x \in \overline{\mathcal{A}}$, we add a new vertex λ_x and connect it to $x \in W$ and thus obtain from (T^*, r) a new rooted tree (T°, r).

For $i = 0, \ldots, N-1$, let $W_i = W \cap L_i$. Of course, $W_0 = \{r\}$ is a finite set. For each integer ℓ satisfying $0 \leq \ell \leq N - 2$, we have $W_{\ell+1} = \bigcup_{x \in W_\ell} B_x$ and so the finiteness of $W_{\ell+1}$ is guaranteed by the finiteness of W_ℓ. We can thus conclude now that W is a finite set, and henceforth (T°, r) is a finite rooted tree.

We let μ° be the weighting of T° such that

$$\begin{cases} \mu^\circ(x) = \mu(x) & \text{if } x \in W, \\ \mu^\circ(\Lambda_x) = \mu(\Lambda_x) & \text{if } \Lambda_x \in \overline{\mathcal{A}}. \end{cases}$$

We can see that $\mu^\circ(V(T^\circ)) = \mu(V(T)) \geq \alpha + \beta + \epsilon$ and that $\mu^\circ(x \uparrow_{T^\circ}) \leq \alpha + \epsilon$ for all $x \in V(T^\circ)$. Applying Lemma 1 yields the claim that we can find an $(\alpha + \epsilon, \beta)$ down-set D° of (T°, μ°). Finally, letting $\mathcal{D}$ be the subset of D° consisting of all elements of the form λ_x, the required $(\alpha + \epsilon, \beta)$ down-set D_ϵ of $\mathbb{T}$ can be chosen to be

$$D_\epsilon = (D^\circ \setminus \mathcal{D}) \cup \left(\bigcup_{\lambda_x \in \mathcal{D}} \Lambda_x \right).$$

This is the end of the proof. $\qquad\qquad\qquad\qquad\qquad\qquad\qquad\qquad\qquad$ $\square$

Proof (of Theorem 1). We prove the statement by induction on k. Let $\mathcal{T} = (T, r)$.

When $k = 1$, we can simply choose D_1 to be the down-set $V(T)$. Since $\mu(D_1) \geq d_1$, we see that (T, r, μ) is d_1-fat, and so the base case holds true.

Assume now $k > 1$ and the result holds for smaller k. Let us suppose that (T, r, μ) is not c-tall and try to find k disjoint down-sets $D_1, \ldots, D_k$ of $\mathcal{T}$ such that $\mu(D_i) \geq d_i$ for all $i \in [k]$.

Taking $\alpha = c$ and $\beta = d_k$, it follows from Lemma 3 that there exists a down-set D of $\mathcal{T}$ such that $\mu(D) \geq \beta = d_k$ and $\mu(D \uparrow_T) \leq \alpha + \beta = c + d_k$. Consider the finite measure space $(V(T), 2^{V(T)}, \mu')$ where

$$\mu'(x) := \begin{cases} 0 & \text{if } x \in D \uparrow_T; \\ \mu(x) & \text{if } x \in V(T) \setminus (D \uparrow_T). \end{cases}$$

Note that $(k - 2)c + \sum_{i=1}^{k-1} d_i \leq \mu'(V(T))$. By induction hypothesis, there exist $k - 1$ down-sets of $\mathcal{T}$, say $D'_1, \ldots, D'_{k-1}$, such that $\mu'(D'_i) \geq d_i$ holds for all $i \in [k - 1]$. For $i \in [k]$, define

$$D_i := \begin{cases} D'_i \setminus (D \uparrow_T) & \text{if } i \in [k - 1]; \\ D & \text{if } i = k. \end{cases}$$

Clearly, $D_1, \ldots, D_k$ are pairwise disjoint sets with $\mu(D_i) \geq d_i$ for all $i \in [k]$. To verify that $D_1, \ldots, D_k$ are what we need, it is sufficient to show that D_i is a down-set of $\mathcal{T}$ for every $i \in [k-1]$. Pick $i \in [k-1]$ and $x \in D_i \downarrow_T \subseteq D'_i$. Then there exists $x' \in D_i$ such that $x' \in T[x, r]$. Since $D_i = D'_i \setminus D \uparrow_T$, we have $x' \notin D \uparrow_T$ and therefore $x \notin D \uparrow_T$. Consequently, we arrive at $x \in D'_i \setminus (D \uparrow_T) = D_i$, finishing the proof. $\qquad\qquad\qquad\qquad\qquad$ $\square$

We have been playing up and down in a tree to deduce our main result. It is interesting to see for which structure more general than trees we can play up and down analogously. We conclude the paper by giving a couterexample for the "poset version" of Lemma 1.

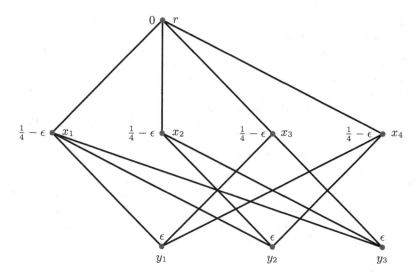

Fig. 3. Hasse diagram of a poset in which x covers y when x is depicted higher than y and xy is drawn as an edge. See Example 4.

Example 4. Let ϵ be a positive real such that $\epsilon < \frac{1}{12}$, let $\alpha = \frac{3}{4}$ and let β be a number inside the open interval $(\epsilon, \frac{1}{4} - 2\epsilon)$. For the weighted poset depicted in Fig. 3, $\mu(P) = 1 - \epsilon > \alpha + \beta$ and $\mu(x \uparrow_T) \leq \frac{3}{4} - 2\epsilon \leq \alpha$ for all $x \in V(T)$. But there is no (α, β) down-set in P. $\square$

References

1. Björner, A., Ziegler, G.M.: Introduction to greedoids. In: Matroid Applications, Encyclopedia of Mathematics and its Applications, vol. 40, pp. 284–357. Cambridge University Press, Cambridge (1992). https://doi.org/10.1017/CBO9780511662041.009
2. Bonamy, M., Bousquet, N., Thomassé, S.: The Erdős-Hajnal conjecture for long holes and antiholes. SIAM J. Discrete Math. **30**(2), 1159–1164 (2016). https://doi.org/10.1137/140981745
3. Dilworth, R.P.: A decomposition theorem for partially ordered sets. Ann. Math. **51**(1), 161–166 (1950). https://doi.org/10.2307/1969503
4. Eaton, L., Tedford, S.J.: A branching greedoid for multiply-rooted graphs and digraphs. Discrete Math. **310**(17), 2380–2388 (2010). https://doi.org/10.1016/j.disc.2010.05.007
5. Erdős, P., Hajnal, A.: Ramsey-type theorems. Discrete Appl. Math. **25**(1–2), 37–52 (1989). https://doi.org/10.1016/0166-218X(89)90045-0
6. Erdős, P., Szekeres, G.: A combinatorial problem in geometry. Compos. Math. **2**, 463–470 (1935). http://eudml.org/doc/88611
7. Felsner, S.: Orthogonal structures in directed graphs. J. Combin. Theory Ser. B **57**(2), 309–321 (1993). https://doi.org/10.1006/jctb.1993.1023
8. Gallai, T.: On directed paths and circuits. In: Theory of Graphs (Proc. Colloq., Tihany, 1966), pp. 115–118. Academic Press, New York (1968)

9. Gallai, T., Milgram, A.N.: Verallgemeinerung eines graphentheoretischen Satzes von Rédei. Acta Sci. Math. (Szeged) **21**, 181–186 (1960)
10. Hujdurović, A., Husić, E., Milanič, M., Rizzi, R., Tomescu, A.I.: Perfect phylogenies via branchings in acyclic digraphs and a generalization of Dilworth's theorem. ACM Trans. Algorithms **14**(2), Art. 20, 26 (2018). https://doi.org/10.1145/3182178
11. Mirsky, L.: A dual of Dilworth's decomposition theorem. Amer. Math. Mon. **78**, 876–877 (1971). https://doi.org/10.2307/2316481
12. Roy, B.: Nombre chromatique et plus longs chemins d'un graphe. Rev. Française Informat. Recherche Opérationnelle **1**(5), 129–132 (1967)
13. Schmidt, W.: A characterization of undirected branching greedoids. J. Combin. Theory Ser. B **45**(2), 160–184 (1988). https://doi.org/10.1016/0095-8956(88)90067-6
14. Song, Z.X., Ward, T., York, A.: A note on weighted rooted trees. Discrete Math. **338**(12), 2492–2494 (2015). https://doi.org/10.1016/j.disc.2015.06.014
15. Stein, E.M., Shakarchi, R.: Fourier Analysis: An Introduction, Princeton Lectures in Analysis, vol. 1. Princeton University Press, Princeton (2003)

Groupoid Action and Rearrangement Problem of Bicolor Arrays by Prefix Reversals

Akihiro Yamamura[1(✉)], Riki Kase[1], and Tatiana B. Jajcayová[2]

[1] Faculty of Engineering Science, Akita University, Akita, Japan
yamamura@ie.akita-u.ac.jp, m8018304@s.akita-u.ac.jp
[2] Faculty of Mathematics, Physics and Informatics, Comenius University,
Bratislava, Slovak Republic
jajcayova@fmph.uniba.sk

Abstract. We consider a rearrangement problem of two-dimensional bicolor arrays by prefix reversals as a generalization of the burnt pancake problem. An equivalence relation on the set of bicolor arrays is induced by prefix reversals, and the rearrangement problem is to characterize the equivalence classes. While previously studied the rearrangement problem for unicolor arrays made use of the classical group theoretic tools, the present problem is quite different. For bicolor arrays a rearrangement can be described by partial injections, and thus we characterize the equivalence classes in terms of a groupoid action. We also outline an algorithm for rearrangement by prefix reversals and estimate a minimum number of rearrangements needed to rearrange bicolor arrays by prefix reversals.

Keywords: Groupoids · Groupoid actions · Rearrangement problem · Burnt pancake problem · Prefix reversals

1 Introduction

The problem of rearranging by prefix reversals, known as the *pancake sorting problem*, is to sort randomly piled pancakes of different sizes. A prefix reversal is an operation of reversing elements in a sublist including the first element. The minimum number of prefix reversals to sort a given list of integers into an ascending order was asked by Gates and Papadimitriou and an upper bound $\frac{5}{3}n$ for sorting a list of length n was given in [6]. They conjectured that $\frac{19}{16}n$ is required, however, it is disproved in [7]. A better bound $\frac{18}{11}n$ was given in [4]. On the other hand, the pancake sorting problem is shown to be NP-hard in [3]. A similar problem of sorting a permutation by reversals is studied by many authors (e.g., [1,2]). Another variant called the *burnt pancake problem* has been also studied by many authors (e.g., [5,8]). Instead of just integers, signed integers are filled in a list in the case of the burnt pancake sorting problem and signs of the corresponding integers are inverted when a prefix reversal is performed. A two-dimensional variant of the pancake sorting problem is introduced and studied in [11]. Unlike in the pancake sorting problem, in the two-dimensional variant it is not always possible to rearrange one array into another. A necessary and sufficient condition for two given arrays to

© Springer Nature Switzerland AG 2020
H. Fernau (Ed.): CSR 2020, LNCS 12159, pp. 419–431, 2020.
https://doi.org/10.1007/978-3-030-50026-9_31

be rearranged one to the other is given in terms of symmetric groups and alternating groups.

In this paper we introduce a two-dimensional variant of the burnt pancake problem. This is a natural and interesting generalization of the original problem. Given two $n \times m$ arrays filled with letters colored by either blue or red, we examine whether given two arrays can be rearranged by performing a sequence of prefix reversals. The concept of groupoid action is introduced and applied to characterize bicolor arrays which can be rearranged to each other. It establishes a connection with well-studied notions from algebra. We argue that groupoids are the proper structure to work with here.

1.1 Rearrangement of Arrays

An $n \times m$ array A consists of $n \times m$ cells and the entry in the (i, j) position is denoted by a_{ij}. We employ the standard matrix representation (a_{ij}) to denote the array A (Fig. 1), and denote $A = \frac{A_1}{A_2}$ if A consists of an upper block A_1 and a lower block A_2, or $A = A_3|A_4$ if A consists of a left block A_3 and a right block A_4. For an $n \times m$ array $A = (a_{ij})$, the *reversal* of A is the $n \times m$ array (b_{ij}) such that $b_{ij} = a_{n-i+1, m-j+1}$ for every (i, j) and is denoted by $Rev(A)$ (Fig. 2). A transformation $\frac{A_1}{A_2} \Rightarrow \frac{Rev(A_1)}{A_2}$ or $A_3|A_4 \Rightarrow Rev(A_3)|A_4$ is called a *prefix reversal*.

Suppose that n and m are positive integers and X is a set of cardinality nm. Let A be an $n \times m$ array in which a letter from X is placed on each (i, j) position so that no letter appears more than once. Suppose that σ belongs to the symmetric group S_X on X. Sometimes we loosen our notation a bit and instead of S_X we write S_{nm}, where $nm = |X|$. Similarly, the alternating group A_X on X is sometimes denoted as A_{nm}, where $nm = |X|$. See [10] for the theory of groups. The $n \times m$ array obtained from A by performing σ on each letter on A is called a *rearrangement* of A by σ, and denoted by $\sigma(A)$ (Fig. 3). The *rearrangement problem for two-dimensional arrays* is to ask whether a given $n \times m$ array $\sigma(A)$, where σ belongs to the symmetric group S_X, can be rearranged to the array A by performing sequence of prefix reversals. If so, we also ask what is the minimum number of prefix reversals needed to rearrange. Recall that an orbit of an array A under the action of the symmetric group S_X is $S_X(A)$.

a_{11}	a_{12}	$\cdots$	a_{1m-1}	a_{1m}
a_{21}	a_{22}	$\cdots$	a_{2m-1}	a_{2m}
$\vdots$	$\vdots$	$\vdots$	$\vdots$	$\vdots$
a_{n1}	a_{n2}	$\cdots$	a_{nm-1}	a_{nm}

Fig. 1. $n \times m$ array A

a_{nm}	a_{nm-1}	$\cdots$	a_{n2}	a_{n1}
$\vdots$	$\vdots$	$\vdots$	$\vdots$	$\vdots$
a_{2m}	a_{2m-1}	$\cdots$	a_{22}	a_{21}
a_{1m}	a_{1m-1}	$\cdots$	a_{12}	a_{11}

Fig. 2. Rev(A)

$\sigma(a_{11})$	$\sigma(a_{12})$	$\cdots$	$\sigma(a_{1m-1})$	$\sigma(a_{1m})$
$\sigma(a_{21})$	$\sigma(a_{22})$	$\cdots$	$\sigma(a_{2m-1})$	$\sigma(a_{2m})$
$\vdots$	$\vdots$	$\vdots$	$\vdots$	$\vdots$
$\sigma(a_{n1})$	$\sigma(a_{n2})$	$\cdots$	$\sigma(a_{nm-1})$	$\sigma(a_{nm})$

Fig. 3. $\sigma(A)$

Theorem 1 (*[11]*). *Let n and m be positive integers. For given $n \times m$ arrays A and B, A can be transformed to B by a sequence of prefix reversals if and only if $B \in G(A)$ under action of a group G specified in the table below according to the parameter n and m.*

$n(\bmod 4)\backslash m(\bmod 4)$	0	1	2	3
0	A_{nm}	S_{nm}	S_{nm}	S_{nm}
1		S_{nm}	S_{nm}	S_{nm}
2		S_{nm}	S_{nm}	S_{nm}
3		S_{nm}	S_{nm}	S_{nm}

2 Rearrangement of Bicolor Arrays

2.1 Bicolor Arrays

Suppose n and m are positive integers. Let X be a set of cardinality nm. We call X an *alphabet of unicolor letters*. Let X_B and X_R be disjoint sets such that $|X_B| = |X_R| = |X| = nm$. We call X_B and X_R an *alphabet of blue letters* and an *alphabet of red letters*, respectively. Suppose $\phi : X_B \to X_R$, $\mathbf{b} : X \to X_B$, and $\mathbf{r} : X \to X_R$ are bijections, respectively, making the diagram below commutative.

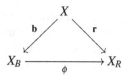

If x belongs to X, that is, x is a unicolor letter, then we regard $\mathbf{b}(x)$ as a blue letter and $\mathbf{r}(x)$ as a red letter. We define a mapping π of $X_B \cup X_R$ onto X by $\pi(z) = x$ if $z = \mathbf{b}(x) \in X_B$ and $\pi(z) = y$ if $z = \mathbf{r}(y) \in X_R$, and it is called the *color deleting operator*. Suppose A is an array with the alphabet $X_B \cup X_R$. Then we denote the set of all the letters appearing on A by $Char(A)$ and an array A is called *bicolor* if $\pi(Char(A)) = X$.

The mapping ϕ and its inverse mapping ϕ^{-1} play a role in changing colors of letters. We extend ϕ to a bijection ψ of $X_B \cup X_R$ onto $X_B \cup X_R$ by $\psi(x) = \phi(x)$ if $x \in X_B$ and $\psi(y) = \phi^{-1}(y)$ if $y \in X_R$ and call ψ the *color changing operator*. Note that ψ inverts color but does not change letters, that is, the image under π is unaltered. We extend ψ on the set of bicolor arrays; for a bicolor array $A = (a_{ij})$, $\psi(A)$ is defined to be the array $A = (\psi(a_{ij}))$, that is, $\psi(A)$ is obtained from A by inverting color of every letter appearing in A. We should note that if $\pi(Char(A)) = X$ then $\pi(Char(\psi(A))) = X$ and so $\psi(A)$ is also a bicolor array.

2.2 Rearrangement Problem by Prefix Reversals

Suppose A is an $n \times m$ bicolor array and $A = \frac{A_1}{A_2} = A_3|A_4$. The transformation $\frac{A_1}{A_2} \Rightarrow \frac{\psi(Rev(A_1))}{A_2}$ or $A_3|A_4 \Rightarrow \psi(Rev(A_3))|A_4$ is called a *prefix reversal*. Note that the resulting array is bicolor as well.

Example: A bicolor array $\begin{bmatrix} \mathbf{b}(a) & \mathbf{b}(b) & \mathbf{b}(c) \\ \mathbf{b}(d) & \mathbf{b}(e) & \mathbf{b}(f) \end{bmatrix}$ can be rearranged from $\begin{bmatrix} \mathbf{r}(c) & \mathbf{r}(f) & \mathbf{r}(e) \\ \mathbf{r}(b) & \mathbf{b}(d) & \mathbf{b}(a) \end{bmatrix}$:

$$\begin{bmatrix} \mathbf{r}(c) & \mathbf{r}(f) & \mathbf{r}(e) \\ \mathbf{r}(b) & \mathbf{b}(d) & \mathbf{b}(a) \end{bmatrix} \Rightarrow \begin{bmatrix} \mathbf{b}(e) & \mathbf{b}(f) & \mathbf{b}(c) \\ \mathbf{r}(b) & \mathbf{b}(d) & \mathbf{b}(a) \end{bmatrix} \Rightarrow \begin{bmatrix} \mathbf{r}(d) & \mathbf{b}(b) & \mathbf{b}(c) \\ \mathbf{r}(f) & \mathbf{r}(e) & \mathbf{b}(a) \end{bmatrix} \Rightarrow \begin{bmatrix} \mathbf{r}(a) & \mathbf{b}(e) & \mathbf{b}(f) \\ \mathbf{r}(c) & \mathbf{r}(b) & \mathbf{b}(d) \end{bmatrix}$$

$$\Rightarrow \begin{bmatrix} \mathbf{r}(f)\ \mathbf{r}(e)\ \mathbf{b}(a) \\ \mathbf{r}(c)\ \mathbf{r}(b)\ \mathbf{b}(d) \end{bmatrix} \Rightarrow \begin{bmatrix} \mathbf{r}(d)\ \mathbf{b}(b)\ \mathbf{b}(c) \\ \mathbf{r}(a)\ \mathbf{b}(e)\ \mathbf{b}(f) \end{bmatrix} \Rightarrow \begin{bmatrix} \mathbf{b}(a)\ \mathbf{b}(b)\ \mathbf{b}(c) \\ \mathbf{b}(d)\ \mathbf{b}(e)\ \mathbf{b}(f) \end{bmatrix}$$

Let $\mathbf{A}$ be the set of $n \times m$ bicolor arrays with the alphabet $X_B \cup X_R$, where X is an alphabet of cardinality nm. Suppose $A, B \in \mathbf{A}$. We define $A \sim B$ if there exists a sequence of prefix reversals that transforms A to B. Clearly, $\sim$ is an equivalence relation on $\mathbf{A}$ and it gives a partition of $\mathbf{A}$. For each $A \in \mathbf{A}$, we denote the equivalence class containing A by $[A]$.

Rearrangement problem is to decide for two given $n \times m$ bicolor arrays whether one can be rearranged into the other by applying a sequence of prefix reversals. Unlike in the burnt pancake sorting problem, which can be regarded as a rearrangement problem of $1 \times m$ bicolor arrays, it is not always possible to rearrange and so there may be more than one $\sim$ equivalence class. Our objective is to characterize all the equivalence classes and to estimate the minimum number of prefix reversals for rearrangement.

2.3   Groupoids and Groupoid Actions

A *groupoid* is a small category in which every morphism is invertible (see [9]). In other words, it is an algebraic system G with partial multiplication satisfying the associative law in which each element has an inverse: (1) a multiplication of G is defined only partially, however, if a, b, c are in G and ab and bc are defined then $(ab)c$ and $a(bc)$ are also defined and the associative law $(ab)c = a(bc)$ holds. (2) there exists an inverse a^{-1} for each element $a \in G$, such that aa^{-1} and $a^{-1}a$ satisfy properties that $baa^{-1} = b$ if ba is defined and $a^{-1}ac = c$ if ac is defined.

Let X be a non-empty set. A partial injection σ of X is an injection such that $\mathrm{Dom}(\sigma) \subset X$ and $\mathrm{Ran}(\sigma) \subset X$. Let $D(k)$ be the set of partial injections of X in which the domain and range have the cardinality k, where $1 \le k \le |X|$. Then $D(k)$ has a groupoid structure if we define a partial multiplication $\alpha\beta$ for $\alpha, \beta \in D(k)$ by composition $\alpha \circ \beta$ provided that $Dom(\alpha) = Ran(\beta)$, otherwise $\alpha\beta$ is undefined. We show that $D(k)$ has a groupoid structure. Evidently the associative law is satisfied because composition of mappings does so. Suppose α belongs to $D(k)$ and is expressed as $\begin{pmatrix} a_1\ a_2\ \cdots\ a_{k-1}\ a_k \\ b_1\ b_2\ \cdots\ b_{k-1}\ b_k \end{pmatrix}$ in Cauchy's two line notation. Then the inverse mapping α^{-1} is expressed as $\begin{pmatrix} b_1\ b_2\ \cdots\ b_{k-1}\ b_k \\ a_1\ a_2\ \cdots\ a_{k-1}\ a_k \end{pmatrix}$ and also belongs to $D(k)$. Note that $\alpha^{-1}\alpha$ is the identity mapping of $\{a_1, a_2, \ldots, a_{k-1}, a_k\}$ and so $\alpha^{-1}\alpha\beta = \beta$ for any β if $\alpha\beta$ is defined, that is, $Ran(\beta) = \{a_1, a_2, \ldots, a_{k-1}, a_k\}$. Similarly $\beta\alpha\alpha^{-1} = \beta$ for any β if $\beta\alpha$ is defined. It follows that $D(k)$ forms a groupoid. We remark that $D(k)$ is a $\mathscr{D}$-class of the symmetric inverse semigroup on X (see [9]).

Let $\mathbf{X}$ be a nonempty set and G a groupoid. A *groupoid action* of G on $\mathbf{X}$ is a partial mapping $f : G \times \mathbf{X} \to \mathbf{X}$ satisfying the conditions: (1) If $e \in G$ is an idempotent, that is, $e^2 = e$ and $f(e, x)$ are defined for e and $x \in \mathbf{X}$, then $f(e, x) = x$ and (2) If $f(g, x)$ is defined and hg is defined for $h, g \in G$ and $x \in \mathbf{X}$, then $f(hg, x)$ and $f(h, f(g, x))$ are also defined and $f(hg, x) = f(h, f(g, x))$ holds. A groupoid action f is partial in the sense that $f(g, x)$ is not necessarily defined for every pair $(g, x) \in G \times \mathbf{X}$.

2.4 Groupoid Action on Bicolor Arrays

Suppose $D(nm)$ is the groupoid of partial injections of $X_B \cup X_R$ in which both the domain and range have the cardinality nm. Suppose $A \in \mathbf{A}$ and $\sigma \in D(nm)$. Recall that $\pi(Char(A)) = X$ since A is a bicolor array. If $Dom(\sigma) = Char(A)$, then we define $\sigma(A)$ to be the array obtained from A by replacing every entry a in A by $\sigma(a)$, otherwise $\sigma(A)$ is undefined. In other words, if $A = (a_{ij})$ then $\sigma(A) = (\sigma(a_{ij}))$ provided that $Dom(\sigma) = Char(A)$.

Note that if $\sigma(A)$ is defined then $Dom(\sigma) = Char(A)$ and so $\pi(Char(\sigma(A))) = \pi(Ran(\sigma))$. Since A is a bicolor array, we have $\pi(Char(A)) = X$. Thus, $\pi(Dom(\sigma)) = X$. If σ also satisfies $\pi(Dom(\sigma)) = \pi(Ran(\sigma))$ then $\pi(Ran(\sigma)) = \pi(Dom(\sigma)) = X$. In such a case, we have $\pi(Char(\sigma(A))) = \pi(Ran(\sigma)) = X$ and so $\sigma(A)$ is also a bicolor array, that is, $\sigma(A) \in \mathbf{A}$.

We define G_1 to be $\{\sigma \in D(nm) \mid \pi(Dom(\sigma)) = \pi(Ran(\sigma)) = X\}$. It is easy to check that G_1 is a proper subgroupoid of $D(nm)$. A groupoid action f of G_1 on $\mathbf{A}$ is naturally defined; for $\sigma \in G_1$ and $A \in \mathbf{A}$, $f(\sigma,A) = \sigma(A)$ if $Dom(\sigma) = Char(A)$, and otherwise undefined. Note that if $f(\sigma,A)$ is defined then $f(\sigma,A) \in \mathbf{A}$. It is easy to check that f satisfies (1) for $\sigma \in G_1$ and $A \in \mathbf{A}$, if σ is an idempotent and $f(\sigma,A)$ is defined then $f(\sigma,A) = A$, and (2) if $\sigma\rho$ is defined and $f(\rho,A)$ is defined for $\sigma,\rho \in G_1$ and $A \in \mathbf{A}$ then $f(\sigma\rho,A)$ and $f(\sigma,f(\rho,A))$ are also defined and we have $f(\sigma\rho,A) = f(\sigma,f(\rho,A))$. Hence, f is a groupoid action of G_1 on $\mathbf{A}$. An *orbit* of a bicolor array A under the action of G_1 is defined to be the set $\{\sigma(A) \mid \sigma \in G_1, Dom(\sigma) = Char(A)\}$ and is denoted by $G_1(A)$.

3 Equivalence Classes of 2×2 Bicolor Arrays

We characterize $\sim$ equivalence classes of 2×2 bicolor arrays, where X is a set of cardinality 4.

Theorem 2. *Suppose $\{w,x,y,z\}$ is any subset of $X_B \cup X_R$ satisfying $\pi(\{w,x,y,z\}) = X$.*

(1) The equivalence class $[A]$ of an array $A = \begin{bmatrix} w & x \\ y & z \end{bmatrix}$ consists of the following 12 arrays and their $(\frac{\pi}{2}, \pi, \frac{3\pi}{2})$ rotations without inverting colors:

(i) $\begin{bmatrix} w & x \\ y & z \end{bmatrix}, \begin{bmatrix} w & x \\ \psi(z) & \psi(y) \end{bmatrix}, \begin{bmatrix} w & y \\ x & z \end{bmatrix}, \begin{bmatrix} w & y \\ \psi(z) & \psi(x) \end{bmatrix}, \begin{bmatrix} w & \psi(z) \\ x & \psi(y) \end{bmatrix}, \begin{bmatrix} w & \psi(z) \\ y & \psi(x) \end{bmatrix},$

(ii) $\begin{bmatrix} \psi(w) & \psi(x) \\ \psi(y) & \psi(z) \end{bmatrix}, \begin{bmatrix} \psi(w) & \psi(x) \\ z & y \end{bmatrix}, \begin{bmatrix} \psi(w) & \psi(y) \\ \psi(x) & \psi(z) \end{bmatrix}, \begin{bmatrix} \psi(w) & \psi(y) \\ z & x \end{bmatrix}, \begin{bmatrix} \psi(w) & z \\ \psi(x) & y \end{bmatrix}, \begin{bmatrix} \psi(w) & z \\ \psi(y) & x \end{bmatrix}.$

(2) There are exactly eight $\sim$ equivalence classes whose representatives are given by

$$\begin{bmatrix} w & x \\ y & z \end{bmatrix}, \begin{bmatrix} \psi(w) & x \\ y & z \end{bmatrix}, \begin{bmatrix} w & \psi(x) \\ y & z \end{bmatrix}, \begin{bmatrix} w & x \\ \psi(y) & z \end{bmatrix},$$
$$\begin{bmatrix} w & x \\ y & \psi(z) \end{bmatrix}, \begin{bmatrix} w & x \\ \psi(y) & \psi(z) \end{bmatrix}, \begin{bmatrix} w & \psi(x) \\ y & \psi(z) \end{bmatrix}, \begin{bmatrix} w & \psi(x) \\ \psi(y) & z \end{bmatrix}.$$

Proof. (1) Clearly, 6 arrays in (i) belong to $[A]$. We can rotate an array $\frac{\pi}{2}$ degree without inverting colors;
$$\begin{bmatrix} w & x \\ y & z \end{bmatrix} \Rightarrow \begin{bmatrix} \psi(x) & \psi(w) \\ y & z \end{bmatrix} \Rightarrow \begin{bmatrix} \psi(y) & \psi(w) \\ x & z \end{bmatrix} \Rightarrow \begin{bmatrix} \psi(z) & \psi(x) \\ w & y \end{bmatrix} \Rightarrow \begin{bmatrix} x & z \\ w & y \end{bmatrix}$$
for each of six arrays. Therefore, the arrays obtained from an array in $[A]$ by $(\frac{\pi}{2}, \pi, \frac{3\pi}{2})$ rotations without inverting colors also belong to $[A]$. We can also invert color of each letter in an array by using $\frac{\pi}{2}$ degree rotations twice and prefix reversal;
$$\begin{bmatrix} w & x \\ y & z \end{bmatrix} \Rightarrow \begin{bmatrix} z & y \\ x & w \end{bmatrix} \Rightarrow \begin{bmatrix} \psi(w) & \psi(x) \\ \psi(y) & \psi(z) \end{bmatrix}.$$
Therefore, the arrays in (ii) also belong to $[A]$ since these are obtained from the arrays in (i) by inverting color of all letters. We shall show that no other arrays belong to $[A]$.

Since every prefix reversal preserves a parity of the number of letters in X_B, A can be rearranged only from an array with the same parity of the number of letters in X_B as has A. Therefore, we exclude the arrays in which odd number of letters in $\{w, x, y, z\}$ are inverted. We also note that every prefix reversal preserves color pattern of the array
$$\begin{bmatrix} p & \psi(q) \\ \psi(r) & s \end{bmatrix} \text{ and } \begin{bmatrix} \psi(p) & q \\ r & \psi(s) \end{bmatrix}$$
where $\{p, q, r, s\} = \{w, x, y, z\}$, that is the patterns with the letters on one diagonal of original color and the letters on the other diagonal of inverted color. It follows that A cannot be rearranged from any such array and thus $[A]$ does not contain this type of arrays.

Now, without loss of generality, we only have to check arrays with w in $(1, 1)$ position, and letters on the first row of the original color and the letters on the second row inverted. The only arrays to consider are
$$\begin{bmatrix} w & x \\ \psi(y) & \psi(z) \end{bmatrix}, \begin{bmatrix} w & y \\ \psi(x) & \psi(z) \end{bmatrix}, \begin{bmatrix} w & z \\ \psi(x) & \psi(y) \end{bmatrix},$$
$$\begin{bmatrix} w & z \\ \psi(y) & \psi(x) \end{bmatrix}, \text{ since } \begin{bmatrix} w & x \\ \psi(z) & \psi(y) \end{bmatrix} \text{ and } \begin{bmatrix} w & y \\ \psi(z) & \psi(x) \end{bmatrix}$$
belong to $[A]$. We show that the array $\begin{bmatrix} w & x \\ \psi(y) & \psi(z) \end{bmatrix}$ belongs to $[A]$. (All other cases are done similarly.)

This would mean that there exists a series of prefix reversals to obtain A from our array $\begin{bmatrix} w & x \\ \psi(y) & \psi(z) \end{bmatrix}$. Such operation would invert colors of the letters in the second row of an array without changing arrangement of letters. But this, since $\begin{bmatrix} w & \psi(x) \\ y & \psi(z) \end{bmatrix}$ is from $[A]$ would force $\begin{bmatrix} w & \psi(x) \\ \psi(y) & z \end{bmatrix}$, to belong to $[A]$, which is contradiction.

(2) It is not difficult to see that no two arrays in the list are $\sim$ equivalent. For example, $\begin{bmatrix} w & x \\ \psi(y) & \psi(z) \end{bmatrix}$ and $\begin{bmatrix} w & \psi(x) \\ y & \psi(z) \end{bmatrix}$ are not $\sim$ equivalent; if they were $\sim$ equivalent, then it would be possible to invert colors of letters in the $(2, 1)$ and $(1, 2)$ entries of any array by the same operation of prefix reversals, and then the array $\begin{bmatrix} w & \psi(x) \\ \psi(y) & z \end{bmatrix}$ would belong to $[A]$, which is a contradiction. □

We note that there are $4! \times 2^4 (= 384)$ bicolor arrays. Each equivalence class contains 48 arrays. We can construct a graph whose vertices are 2×2 bicolor arrays and edges correspond to a transform by a prefix reversal. The diameter of this graph gives the minimum number of prefix reversals to rearrange. Each connected component corresponds to a $\sim$ equivalence class and forms a regular graph of degree three with 48 vertices and the diameter is 7. All connected components are isomorphic.

4   Equivalence Classes of $n \times m$ Bicolor Arrays ($n \geq 3$ or $m \geq 3$)

We shall show $\sim$ equivalence classes of $n \times m$ bicolor arrays ($n \geq 3$ or $m \geq 3$) coincide with the orbits of action under some subgroupoids of G_1. Let us extend the color deleting operation π on G_1. Suppose $\sigma \in G_1$. Recall that σ is a partial injection of $X_B \cup X_R$ such that $\pi(Dom(\sigma)) = \pi(Ran(\sigma)) = X$. We define $\pi(\sigma)$ to be the permutation of X obtained from σ by operating π to both the domain and range of σ, that is, if $\sigma = \begin{pmatrix} a_1 \, a_2 \, \cdots \, a_{k-1} \, a_k \\ b_1 \, b_2 \, \cdots \, b_{k-1} \, b_k \end{pmatrix}$, then $\pi(\sigma)$ is the permutation $\begin{pmatrix} \pi(a_1) \, \pi(a_2) \, \cdots \, \pi(a_{k-1}) \, \pi(a_k) \\ \pi(b_1) \, \pi(b_2) \, \cdots \, \pi(b_{k-1}) \, \pi(b_k) \end{pmatrix}$ of X. Since $\pi(Dom(\sigma)) = \pi(Ran(\sigma)) = X$, $\pi(\sigma)$ is a permutation on X.

We define G_2 and G_3 to be $\{\sigma \in G_1 \mid |Dom(\sigma) \cap X_B| \equiv |Ran(\sigma) \cap X_B| \pmod 2\}$ and $\{\sigma \in G_2 \mid \pi(\sigma) \in A_X\}$, where A_X stands for the alternating group on X, respectively. It is easy to see that G_2 and G_3 are subgroupoids of G_1 and $[A] \subset G_1(A)$ for any $n \times m$ bicolor array A.

Lemma 1. *If both n and m are even, then the parity of the number of letters of X_B on an $n \times m$ bicolor array is unaltered by operating any prefix reversal.*

Proof. Let A, B be an $n \times m$ bicolor array. We show that if B is rearranged from A by a prefix reversal, then parity of the number of letters in X_B is unaltered. Suppose B is rearranged from A by a horizontal prefix reversal; $A = \frac{A_1}{A_2} \Rightarrow B = \frac{\psi(Rev(A_1))}{A_2}$, where A_1 is an $n_1 \times m$ upper block ($1 \leq n_1 \leq n$) of A. Suppose the number of letters in X_B on A_1 and A_2 are t_1 and t_2, respectively. The number of letters in X_B on $\psi(Rev(A_1))$ is equal to $n_1 m - t_1$ that is the number of letters in X_R on A_1. Thus, the number of letters in X_B on B is $n_1 m - t_1 + t_2$. Since m is even, the difference $(t_1 + t_2) - (n_1 m - t_1 + t_2) = n_1 m - 2t_1$ is even and so, parity of letters on X_B is unaltered by a prefix reversal. Similarly, we can show that a vertical prefix reversal does not change the parity of the number of letters in X_B.

$\square$

Therefore, if both n and m are even we have $[A] \subset G_2(A)$ for any bicolor array A. Further, if $n \equiv m \equiv 0 \pmod 4$ and $\sigma \in G_2$ then $\pi(\sigma)(\pi(A))$ can be rearranged to A provided that $\pi(\sigma)$ belongs to the alternating group A_X by Theorem 1. It follows that $[A] \subset G_3(A)$ in this case. In the next theorem we shall prove the opposite inclusion holds in all the three cases.

Theorem 3. *Let n and m be integers such that $n \geq 2$, $m \geq 2$ and at least one of them is larger than 2. For any $n \times m$ bicolor array A, we have $[A] = G_i(A)$, where G_i is a groupoid specified in the table below according to the parameter n and m.*

$n(\bmod 4)\backslash m(\bmod 4)$	0	1	2	3
0	G_3	G_1	G_2	G_1
1	G_1	G_1	G_1	G_1
2	G_2	G_1	G_2	G_1
3	G_1	G_1	G_1	G_1

Proof. We divide the proof in five cases: (1) $(n,m) = (2,2k+1)$, (2) $(n,m) = (2,2k)$, (3) $n,m \geq 3$ and either n or m is odd, (4) $n,m \geq 3$, both n and m are even and either $n \not\equiv 0 \pmod 4$ or $m \not\equiv 0 \pmod 4$, and (5) $n \equiv m \equiv 0 \pmod 4$. We shall show that $[A] = G_1(A)$ for (1) and (3), $[A] = G_2(A)$ for (2) and (4), and $[A] = G_3(A)$ for (5). Note that $[A] \subset G_i(A)$ always holds in each case, and so, we have to prove the opposite inclusion.

In the case (1) and (3), we suppose A is a bicolor array and $\sigma \in G_1$ such that $\sigma(A)$ is defined. We shall show $\sigma(A) \in [A]$. By Theorem 1, $\pi(\sigma)(\pi(A))$ can be rearranged from $\pi(A)$ by a certain sequence Γ_1 of prefix reversals. Note that $\pi(\sigma) \in S_{nm}$. Let B be the bicolor array rearranged from A by operating the same sequence Γ_1 of prefix reversals. Then we have $\pi(B) = \pi(\sigma)(\pi(A))$. On the other hand, we have $\pi(\sigma)(\pi(A)) = \pi(\sigma(A))$. Therefore, $\pi(B) = \pi(\sigma(A))$. Hence, the arrangements of letters on B and $\sigma(A)$ are same if we ignore colors. See the diagram below.

$$
\begin{array}{ccccc}
A & \xrightarrow{\ \Gamma_1\ } & B & \xrightarrow{\ \Gamma_2\ } & \sigma(A) \\
\downarrow{\scriptstyle \pi} & & \downarrow{\scriptstyle \pi} & & \downarrow{\scriptstyle \pi} \\
\pi(A) & \xrightarrow{\ \Gamma_1\ } & \pi(\sigma)(\pi(A)) & =\!=\!= & \pi(\sigma(A))
\end{array}
$$

We shall prove there exists a sequence Γ_2 of prefix reversals that rearranges B to $\sigma(A)$, and then it follows $\sigma(A) \in [A]$. We have only to show that any inversion of color of any letter on an array can be realized by prefix reversals without changing anything else.

In the case (2) and (4), we suppose A is a bicolor array and $\sigma \in G_2$ such that $\sigma(A)$ is defined. A similar argument shows there exists an array B rearranged from A by prefix reversals and $\pi(B) = \pi(\sigma(A))$. Since a prefix reversal does not change the parity of the number of letters in X_B and $\sigma \in G_2$, the parity of the number of letters in X_B is same for B and $\sigma(A)$. Under such an assumption we have only to show that inversion of colors of a couple of any two letters on an array can be simultaneously realized by prefix reversals without changing anything else.

In the case (5), we suppose A is a bicolor array and $\sigma \in G_3$ such that $\sigma(A)$ is defined. Since $\sigma \in G_3$, we have $\pi(\sigma) \in A_X$. Therefore, $\pi(\sigma)(\pi(A))$ can be obtained from $\pi(A)$ by a certain sequence of prefix reversals by Theorem 1. Then, as the case (4), we have only to show that inversions of colors of a couple of any two letters on an array can be simultaneously realized by prefix reversals without changing anything else.

(1) $2 \times (2k+1)$Arrays: We consider a $2 \times (2k+1)$ bicolor array $\begin{bmatrix} A & x & B \\ C & y & D \end{bmatrix}$, where A, B, C, D are words on $X_B \cup X_R$ of length k, and x and y are letters in $X_B \cup X_R$. We shall show that we can invert the color of x without affecting any other letters by prefix reversals.

$$\begin{bmatrix} A & x & B \\ C & y & D \end{bmatrix} \Rightarrow \begin{bmatrix} \psi(Rev(C)) & x & B \\ \psi(Rev(A)) & y & D \end{bmatrix} \Rightarrow \begin{bmatrix} \psi(Rev(B)) & \psi(x) & C \\ \psi(Rev(A)) & y & D \end{bmatrix} \Rightarrow \begin{bmatrix} A & \psi(x) & C \\ B & y & D \end{bmatrix}$$

$$\Rightarrow \begin{bmatrix} A & \psi(x) & C \\ \psi(Rev(D)) & \psi(y) & \psi(Rev(B)) \end{bmatrix} \Rightarrow \begin{bmatrix} A & \psi(x) & B \\ \psi(Rev(D)) & \psi(y) & \psi(Rev(C)) \end{bmatrix} \Rightarrow \begin{bmatrix} A & \psi(x) & B \\ C & y & D \end{bmatrix},$$

where $\psi(W)$ represents the word obtained from a word W by inverting the color of each letter appearing in W. The color of x is inverted to $\psi(x)$ and nothing else is changed. We can invert color of a letter in any position by transposing the letter and the letter at the $(1, k+1)$ position and carry out the operation above, and then do the same prefix reversals in the opposite direction. Note that it is possible to transpose a letter in any position in an array and in the $(1, k+1)$ position by prefix reversals. Consequently we can invert color of a letter in any position. Therefore, $[A] = G_1(A)$.

(2) $2 \times 2k$ Arrays ($k > 2$): Suppose $\sigma \in G_2$. We have noted that $[A] \subset G_2(A)$ for a $2 \times 2k$ array A. Conversely, we show that $\sigma(A)$ can be rearranged from A by prefix reversals for any $\sigma \in G_2$, that is, $\sigma(A) \in [A]$ provided that $\sigma(A)$ is defined. By the argument above, we have an array B that can be rearranged from A by prefix reversals and $\pi(B) = \pi(\sigma(A))$. Then the arrangement of letters in B is same as that of $\sigma(A)$ if we ignore colors. Therefore, we have only to show that we can invert color of the letters on B to obtain $\sigma(A)$ by prefix reversals. On the other hand, $\sigma \in G_2$ and so the parity of the numbers of letters in X_B in B and $\sigma(A)$ are same since a prefix reversal does not change the parity by Lemma 1. There are even number of differences of colors between B and $\sigma(A)$. It is sufficient to show that we can simultaneously invert colors of any pair of two letters in any array B.

(2-1) The $(1, 1)$ and $(1, 2k)$ letters on B can be transposed with inverting colors by prefix reversals (w is in the $(1, 1)$ position and x is in the $(2, 2k)$ position):

$$\begin{bmatrix} w & A & x \\ y & B & z \end{bmatrix} \Rightarrow \begin{bmatrix} w & A & \psi(z) \\ y & B & \psi(x) \end{bmatrix} \Rightarrow \begin{bmatrix} w & A & \psi(z) \\ x & \psi(Rev(B)) & \psi(y) \end{bmatrix}$$

$$\Rightarrow \begin{bmatrix} \psi(x) & A & \psi(z) \\ \psi(w) & \psi(Rev(B)) & \psi(y) \end{bmatrix} \Rightarrow \begin{bmatrix} \psi(x) & A & \psi(z) \\ y & B & w \end{bmatrix} \Rightarrow \begin{bmatrix} \psi(x) & A & \psi(w) \\ y & B & z \end{bmatrix}$$

First, we show that a pair of two letters can be transposed with inverting colors by prefix reversals. There are four cases to be considered according to the positions of the letters. Note that there exist at least 4 columns in the array.

(2-2) t and u lie on the first row:

$$\begin{bmatrix} s & A & t & B & u & C & v \\ w & D & x & E & y & F & z \end{bmatrix} \Rightarrow \begin{bmatrix} \psi(x) & \psi(Rev(D)) & \psi(w) & B & \psi(z) & \psi(Rev(F)) & \psi(y) \\ \psi(t) & \psi(Rev(A)) & \psi(s) & E & \psi(v) & \psi(Rev(C)) & \psi(u) \end{bmatrix}$$

$$\Rightarrow \begin{bmatrix} t & \psi(Rev(D)) & \psi(w) & B & \psi(z) & \psi(Rev(F)) & u \\ x & \psi(Rev(A)) & \psi(s) & E & \psi(v) & \psi(Rev(C)) & y \end{bmatrix}$$

$$\Rightarrow \begin{bmatrix} \psi(u) & \psi(Rev(D)) & \psi(w) & B & \psi(z) & \psi(Rev(F)) & \psi(t) \\ x & \psi(Rev(A)) & \psi(s) & E & \psi(v) & \psi(Rev(C)) & y \end{bmatrix}$$

$$\Rightarrow \begin{bmatrix} \psi(x) & \psi(Rev(D)) & \psi(w) & B & \psi(z) & \psi(Rev(F)) & \psi(y) \\ u & \psi(Rev(A)) & \psi(s) & E & \psi(v) & \psi(Rev(C)) & t \end{bmatrix} \Rightarrow \begin{bmatrix} s & A & \psi(u) & B & \psi(t) & C & v \\ w & D & x & E & y & F & z \end{bmatrix}$$

We use (2-1) in the second transformation.

(2-3) x and y lie on the second row:

$$\begin{bmatrix} s\ A\ t\ B\ u\ C\ v \\ w\ D\ x\ E\ y\ F\ z \end{bmatrix} \Rightarrow \begin{bmatrix} \psi(z)\ \psi(Rev(F))\ \psi(y)\ \psi(Rev(E))\ \psi(x)\ \psi(Rev(D))\ \psi(w) \\ \psi(v)\ \psi(Rev(C))\ \psi(u)\ \psi(Rev(B))\ \psi(t)\ \psi(Rev(A))\ \psi(s) \end{bmatrix}$$

$$\Rightarrow \begin{bmatrix} \psi(z)\ \psi(Rev(F))\quad x\quad \psi(Rev(E))\quad y\quad \psi(Rev(D))\ \psi(w) \\ \psi(v)\ \psi(Rev(C))\ \psi(u)\ \psi(Rev(B))\ \psi(t)\ \psi(Rev(A))\ \psi(s) \end{bmatrix}$$

$$\Rightarrow \begin{bmatrix} s\ A\quad t\quad B\quad u\quad C\ v \\ w\ D\ \psi(y)\ E\ \psi(x)\ F\ z \end{bmatrix}$$

We use (2-2) in the second transformation.

(2-4) v and y lie on a same column:

$$\begin{bmatrix} u\ A\ v\ B\ w \\ x\ C\ y\ D\ z \end{bmatrix} \Rightarrow \begin{bmatrix} \psi(y)\ \psi(Rev(C))\ \psi(x)\ B\ w \\ \psi(v)\ \psi(Rev(A))\ \psi(u)\ D\ z \end{bmatrix} \Rightarrow \begin{bmatrix} v\ \psi(Rev(C))\ \psi(x)\ B\ w \\ y\ \psi(Rev(A))\ \psi(u)\ D\ z \end{bmatrix}$$

$$\Rightarrow \begin{bmatrix} u\ A\ \psi(y)\ B\ w \\ x\ C\ \psi(v)\ D\ z \end{bmatrix}$$

(2-5) u and z lie on a different row and a different column:

Since $k > 1$, there are at least 4 columns in the array and so we can choose an extra column except for the columns on which u and z lie. Suppose that w lies on the same row as u and on the same column as z and that v lies on the same row as u and w but v lies on a different column from u and z. We use (2-1) and (2-4).

$$\begin{bmatrix} A\ u\ B\ v\ C\ w\ D \\ E\ x\ F\ y\ G\ z\ H \end{bmatrix} \Rightarrow \begin{bmatrix} A\ \psi(v)\ B\ \psi(u)\ C\ w\ D \\ E\quad x\quad F\quad y\quad G\ z\ H \end{bmatrix} \Rightarrow \begin{bmatrix} A\ \psi(v)\ B\ \psi(w)\ C\ u\ D \\ E\quad x\quad F\quad y\quad G\ z\ H \end{bmatrix}$$

$$\Rightarrow \begin{bmatrix} A\ \psi(v)\ B\ \psi(w)\ C\ \psi(z)\ D \\ E\quad x\quad F\quad y\quad G\ \psi(u)\ H \end{bmatrix} \Rightarrow \begin{bmatrix} A\ \psi(v)\ B\ z\ C\quad w\quad D \\ E\quad x\quad F\ y\ G\ \psi(u)\ H \end{bmatrix}$$

$$\Rightarrow \begin{bmatrix} A\ \psi(z)\ B\ v\ C\quad w\quad D \\ E\quad x\quad F\ y\ G\ \psi(u)\ H \end{bmatrix}$$

Second, we show that the colors of any two letters can be inverted by prefix reversals without affecting anything else. Since $k > 1$, there are at least 4 columns in the array and so we can choose an extra column. Suppose that w lies on a column different from that of u or y. The color of u and y can be reversed using (2-2) and (2-5) as follows.

$$\begin{bmatrix} A\ u\ B\ v\ C\ w\ D \\ E\ x\ F\ y\ G\ z\ H \end{bmatrix} \Rightarrow \begin{bmatrix} A\ \psi(w)\ B\ v\ C\ \psi(u)\ D \\ E\quad x\quad F\ y\ G\quad z\quad H \end{bmatrix} \Rightarrow \begin{bmatrix} A\ \psi(w)\ B\ v\ C\ \psi(y)\ D \\ E\quad x\quad F\ u\ G\quad z\quad H \end{bmatrix}$$

$$\Rightarrow \begin{bmatrix} A\ y\ B\ v\ C\ w\ D \\ E\ x\ F\ u\ G\ z\ H \end{bmatrix} \Rightarrow \begin{bmatrix} A\ \psi(u)\ B\quad v\quad C\ w\ D \\ E\quad x\quad F\ \psi(y)\ G\ z\ H \end{bmatrix}$$

Consequently, we can invert simultaneously colors of a pair of two letters in any positions. It follows that $[A] = G_2(A)$. We remark that the operation (2-5) is impossible for a 2×2 array since it does not have enough columns to perform the operation, and so the case of 2×2 arrays is discussed separately in Sect. 3.

(3) $n, m \geq 3$ **and Either** n **or** m **is Odd:** We classify operations realized by operating prefix reversals in order to invert color of letters. We assume the condition "either n or m is odd" only in (3-6) but the condition is not assumed for the cases (3-1) through (3-5). The proof is based on the technique in [11] and we omit the details due to lack of space.

(3-1) Twin transpositions along diagonal with fixing colors: It is easy to check that the twin transposition along diagonal fixes colors of the corresponding 4 letters on the same two rows and two columns.

$$
\begin{bmatrix}
\cdots & \cdots & \cdots & \cdots \\
\cdots & w & \cdots & x & \cdots \\
\cdots & \cdots & \cdots & \cdots \\
\cdots & y & \cdots & z & \cdots \\
\cdots & \cdots & \cdots & \cdots
\end{bmatrix}
\rightarrow
\begin{bmatrix}
\cdots & \cdots & \cdots & \cdots \\
\cdots & z & \cdots & y & \cdots \\
\cdots & \cdots & \cdots & \cdots \\
\cdots & x & \cdots & w & \cdots \\
\cdots & \cdots & \cdots & \cdots
\end{bmatrix}
$$

(3-2) Twin transpositions along row and column with inverting colors: It is easy to check that the twin transposition along diagonal changes colors of the corresponding 4 letters.

$$
\begin{bmatrix}
\cdots & \cdots & \cdots & \cdots \\
\cdots & w & \cdots & x & \cdots \\
\cdots & \cdots & \cdots & \cdots \\
\cdots & y & \cdots & z & \cdots \\
\cdots & \cdots & \cdots & \cdots
\end{bmatrix}
\rightarrow
\begin{bmatrix}
\cdots & \cdots & \cdots & \cdots \\
\cdots & \psi(y) & \cdots & \psi(z) & \cdots \\
\cdots & \cdots & \cdots & \cdots \\
\cdots & \psi(w) & \cdots & \psi(x) & \cdots \\
\cdots & \cdots & \cdots & \cdots
\end{bmatrix}
$$

(3-3) Twin transpositions along row and column with fixing colors: Contrary to the twin transposition along diagonal, a twin transpositions along row or column changes colors of corresponding letters as is. Likewise, the twin transposition along row fixing colors can be realized.

$$
\begin{bmatrix}
\cdots & \cdots & \cdots & \cdots \\
\cdots & w & \cdots & x & \cdots \\
\cdots & \cdots & \cdots & \cdots \\
\cdots & y & \cdots & z & \cdots \\
\cdots & \cdots & \cdots & \cdots
\end{bmatrix}
\rightarrow
\begin{bmatrix}
\cdots & \cdots & \cdots & \cdots \\
\cdots & y & \cdots & z & \cdots \\
\cdots & \cdots & \cdots & \cdots \\
\cdots & w & \cdots & x & \cdots \\
\cdots & \cdots & \cdots & \cdots
\end{bmatrix}
$$

(3-4) Inverting colors of two letters on the same row: We can invert colors of two letters on the same row and nothing else is changed. Likewise, we can change colors of two letters on the same column.

$$
\begin{bmatrix}
\cdots & \cdots & \cdots & \cdots \\
\cdots & w & \cdots & x & \cdots \\
\cdots & \cdots & \cdots & \cdots \\
\cdots & y & \cdots & z & \cdots \\
\cdots & \cdots & \cdots & \cdots
\end{bmatrix}
\rightarrow
\begin{bmatrix}
\cdots & \cdots & \cdots & \cdots \\
\cdots & \psi(w) & \cdots & \psi(x) & \cdots \\
\cdots & \cdots & \cdots & \cdots \\
\cdots & y & \cdots & z & \cdots \\
\cdots & \cdots & \cdots & \cdots
\end{bmatrix}
$$

(3-5) Inverting colors of two letters in arbitrary locations: We can invert colors of two letters in arbitrary positions and nothing else is changed.

$$
\begin{bmatrix} \cdots \cdots \cdots \cdots \cdots \\ \cdots\ w\ \cdots\ x\ \cdots \\ \cdots \cdots \cdots \cdots \cdots \cdots \\ \cdots\ y\ \cdots\ z\ \cdots \\ \cdots \cdots \cdots \cdots \cdots \cdots \end{bmatrix}
\rightarrow
\begin{bmatrix} \cdots\ \ \cdots\ \ \cdots\ \ \cdots\ \ \cdots \\ \cdots\ \psi(w)\ \cdots\ x\ \cdots \\ \cdots\ \ \cdots\ \ \cdots\ \ \cdots\ \ \cdots \\ \cdots\ y\ \ \cdots\ \psi(z)\ \cdots \\ \cdots\ \ \cdots\ \ \cdots\ \ \cdots\ \ \cdots \end{bmatrix}
$$

(3-6) Inverting a color of a letter in the center of a row: Suppose m is odd and $m = 2k+1$. We can invert color of the letter in the center of a row and nothing else is changed.

$$
\begin{bmatrix} x_1\ x_2\ \cdots\ x_k\ x_{k+1}\ x_{k+2}\ \cdots\ x_{2k}\ x_{2k+1} \\ \cdots \cdots \cdots \cdots\ \ \cdots\ \ \cdots \cdots \cdots\ \ \cdots \\ \cdots \cdots \cdots \cdots\ \ \cdots\ \ \cdots \cdots \cdots\ \ \cdots \end{bmatrix}
$$

$$
\rightarrow
\begin{bmatrix} x_1\ x_2\ \cdots\ x_k\ \psi(x_{k+1})\ x_{k+2}\ \cdots\ x_{2k}\ x_{2k+1} \\ \cdots \cdots \cdots \cdots\ \ \cdots\ \ \cdots \cdots \cdots\ \ \cdots \\ \cdots \cdots \cdots \cdots\ \ \cdots\ \ \cdots \cdots \cdots\ \ \cdots \end{bmatrix}
$$

We can invert color of any letter in an array using (3-6). It follows that $[A] = G_1(A)$.

(4) $n, m \geq 3$, **Both n and m Are Even, and either** $n \not\equiv 0 \pmod 4$ **or** $m \not\equiv 0 \pmod 4$: We can invert colors of any couple of two letters on an array using (3-5). It follows that $[A] = G_2(A)$.

(5) $n \equiv m \equiv 0 \pmod 4$: We can invert colors of any couple of two letters on an array using (3-5). It follows that $[A] = G_3(A)$. $\square$

5   Estimation of Minimum Number of Prefix Reversals

A theoretical estimate of an upper bound of the minimum number of prefix reversals for $n \times m$ unicolor arrays is obtained in [11], see Table 1. Here we give an estimate of an upper bound for the minimum number of prefix reversals to rearrange an $n \times m$ bicolor array to another array from a $\sim$ equivalence class. Our estimate is based on the algorithm for rearrangement given in the proof of Theorem 3. We make use of the unicolor estimate from Table 1, since as it follows from Theorem 1 a bicolor array A is rearranged to a certain array B such that $\pi(\sigma(A))$ coincides with $\pi(B)$. Then we have to rearrange B to $\sigma(A)$. Recall that B and $\sigma(A)$ differ only by color, that is, $\pi(B)$ coincides with $\pi(\sigma(A))$. Hence, we need to invert colors of at most nm letters to rearrange B

Table 1. Upper bound of the minimum number of prefix reversals for unicolor arrays

			$m > 2$		
			$m \equiv 0 \pmod 4$	$m \equiv 2 \pmod 4$	$m \equiv 1,3 \pmod 4$
$n \geq 2$	$n = 2$		$34m - 17$		
	$n > 2$	$n \equiv 0 \pmod 4$	$280(nm - 1)$		
		$n \equiv 2 \pmod 4$	$(38n + 37)(nm - 1)$		
		$n \equiv 1,3 \pmod 4$	$(13n + 100)(nm - 1)$		

Table 2. Upper bound of the minimum number of prefix reversals for bicolor arrays

			$m > 2$		
			$m \equiv 0 \pmod 4$	$m \equiv 2 \pmod 4$	$m \equiv 1,3 \pmod 4$
$n \geq 2$	$n = 2$		$330m - 17$		$70m - 17$
	$n > 2$	$n \equiv 0 \pmod 4$	$608nm - 280$		
		$n \equiv 2 \pmod 4$	$(38n + 37)(nm - 1) + 328nm$		
		$n \equiv 1,3 \pmod 4$	$(13n + 100)(nm - 1) + 328nm + \dfrac{n-1}{2}(66(n-1) + 1397)$		

into $\sigma(A)$. Combining these two estimates gives us a theoretical upper bound for the minimum number of prefix reversals needed. We summarize our results in Table 2.

These are very basic estimates that are not tight, which is evident also from computer simulations. It is our plan for our future study to reduce these upper bounds of number of prefix reversals. An improvement can be obtained by more detailed analysis of particular steps of our algorithm, or by switching to analysis of the graph of transformation of bicolor arrays using best-first search algorithms.

References

1. Berman, P., Karpinski, M.: On some tighter inapproximability results (extended abstract). In: Wiedermann, J., van Emde Boas, P., Nielsen, M. (eds.) ICALP 1999. LNCS, vol. 1644, pp. 200–209. Springer, Heidelberg (1999). https://doi.org/10.1007/3-540-48523-6_17
2. Berman, P., Hannenhalli, S., Karpinski, M.: 1.375-approximation algorithm for sorting by reversals. In: Möhring, R., Raman, R. (eds.) ESA 2002. LNCS, vol. 2461, pp. 200–210. Springer, Heidelberg (2002). https://doi.org/10.1007/3-540-45749-6_21
3. Bulteau, L., Fertin, G., Rusu, I.: Pancake flipping is hard. In: Rovan, B., Sassone, V., Widmayer, P. (eds.) MFCS 2012. LNCS, vol. 7464, pp. 247–258. Springer, Heidelberg (2012). https://doi.org/10.1007/978-3-642-32589-2_24
4. Chitturi, B., et al.: An $(18/11)n$ upper bound for rearranging by prefix reversals. Theoret. Comput. Sci. **410**(36), 3372–3390 (2009)
5. Cohen, D.S., Blum, M.: On the problem of rearranging burnt pancakes. Discrete Appl. Math. **61**(2), 105–120 (1995)
6. Gates, W., Papadimitriou, C.: Bounds for sorting by prefix reversal. Discrete Math. **79**, 47–57 (1979)
7. Heydari, M.H., Sudborough, I.H.: On the diameter of the pancake network. J. Algorithms **25**(1), 67–94 (1997)
8. Kaplan, H., Shamir, R., Tarjan, R.E.: Faster and simpler algorithm for sorting signed permutations by reversals. In: ACM-SIAM SODA 1997, pp. 178–187 (1997)
9. Lawson, M.V.: Inverse Semigroups: The Theory of Partial Symmetries. World Scientific, Singapore (1998)
10. Rotman, J.J.: An Introduction to the Theory of Groups, 4th edn. Springer, New York (1994). https://doi.org/10.1007/978-1-4612-4176-8
11. Yamamura, A.: Rearranging two dimensional arrays by prefix reversals. In: Bojańczyk, M., Lasota, S., Potapov, I. (eds.) RP 2015. LNCS, vol. 9328, pp. 153–165. Springer, Cham (2015). https://doi.org/10.1007/978-3-319-24537-9_14

Author Index

Printed in the United States
By Bookmasters